Gérard Belmont
Roland Grappin
Fabrice Mottez
Filippo Pantellini
Guy Pelletier

Collisionless Plasmas in Astrophysics

Related Titles

Irwin, J.

Astrophysics – Decoding the Cosmos

2007
Print ISBN: 978-0-470-01306-9

Rybicki, G.B., Lightman, A.P.

Radiative Processes in Astrophysics

1979
Print ISBN: 978-0-471-82759-7

Foukal, P.V.

Solar Astrophysics
2nd edn

2004
Print ISBN: 978-3-527-40374-5

Woods, L.C.

Physics of Plasmas

2004
Print ISBN: 978-3-527-40461-2

Stacey, W.M.

Fusion Plasma Physics

2005
Print ISBN: 978-3-527-40586-2

d'Agostino, R., Favia, P., Kawai, Y., Ikegami, H., Sato, N., Arefi-Khonsari, F. (eds.)

Advanced Plasma Technology

2008
Print ISBN: 978-3-527-40591-6

Hippler, R., Kersten, H., Schmidt, M., Schoenbach, K.H. (eds.)

Low Temperature Plasmas
Fundamentals, Technologies and Techniques

2008
Print ISBN: 978-3-527-40673-9

Stock, R. (ed.)

Encyclopedia of Applied High Energy and Particle Physics

2009
Print ISBN: 978-3-527-40691-3

Smirnov, B.M.

Cluster Processes in Gases and Plasmas

2010
Print ISBN: 978-3-527-40943-3

Stacey, W.M.

Fusion
An Introduction to the Physics and Technology of Magnetic Confinement Fusion, 2nd edn

2010
Print ISBN: 978-3-527-40967-9

Gérard Belmont
Roland Grappin
Fabrice Mottez
Filippo Pantellini
Guy Pelletier

Collisionless Plasmas in Astrophysics

WILEY-VCH
Verlag GmbH & Co. KGaA

The Authors

Dr. Gérard Belmont
CNRS
Lab de Physique des Plasmas/
Ecole Polytechnique

Dr. Roland Grappin
CNAP
Lab de Physique des Plasmas/
Ecole Polytechnique

Dr. Fabrice Mottez
CNRS
LUTH/
Observatoire de Paris-Meudon

Dr. Filippo Pantellini
CNAP
LESIA/
Observatoire de Paris-Meudon

Dr. Guy Pelletier
Université Joseph Fourier
LAOG/
Observatoire de Grenoble

Library of Congress Card No.:
applied for

British Library Cataloguing-in-Publication Data:
A catalogue record for this book is available from the British Library.

Bibliographic information published by the Deutsche Nationalbibliothek
The Deutsche Nationalbibliothek lists this publication in the Deutsche Nationalbibliografie; detailed bibliographic data are available on the Internet at http://dnb.d-nb.de.

Print ISBN 978-3-527-41074-3
ePDF ISBN 978-3-527-65625-7
ePub ISBN 978-3-527-65624-0
mobi ISBN 978-3-527-65623-3
oBook ISBN 978-3-527-65622-6

Cover Design Adam-Design, Weinheim
Typesetting le-tex publishing services GmbH, Leipzig
Printing and Binding Markono Print Media Pte Ltd, Singapore

Printed on acid-free paper

Contents

About the Authors

Gerard Belmont works as a "Directeur de Recherches" at the French CNRS for twenty years. He is a specialist of collisionless media, and their description through kinetic and fluid theories.

Roland Grappin is Astronomer at the Paris Observatory since 1979. His scientific activity covers turbulence in fluids and plasmas, dynamics of the solar wind, corona and transition region.

Fabrice Mottez is a scientist at the Paris Observatory. He has devoted his career to collisionless space plasmas, the terrestrial and Jovian magnetospheres, fundamental plasma physics, and numerical simulation.

Filippo Pantellini is a scientist at the Paris Observatory. His main research fields cover the theoretical and numerical investigation of collisionless and weakly collisional space plasmas, with a particular interest for the solar wind and the solar corona.

Guy Pelletier is a professor at the University Joseph Fourier in Grenoble. He founded the theoretical group of the Laboratory for Astrophysics. He accessed to all the levels of professorship and got the status of Emeritus Professor in 2009.

1
Introduction

1.1
Goals of the Book

A plasma is an assembly of charged particles, making its behavior inseparable from that of the electromagnetic field. When a plasma includes neutrals and when the collisions are numerous enough between charged and neutral particles, it causes the plasma to behave more like a neutral gas. This book will focus on the fully ionized plasmas, so emphasizing more the specific plasma properties.

Plasma evolution is governed by a loop: the charged particles move under the effect of the electromagnetic fields, and the particles, by their density and their velocities, create collective electromagnetic fields. This is true for any kind of plasma, collisional or not, fully ionized or not, and whatever the plasma and field parameters.

This "plasma loop" is sketched in Figure 1.1. One can observe on this sketch that two subloops can exist.

1. There is an electromagnetic loop, which can exist even in the absence of particles. In this case, the fields $\boldsymbol{E}$ and $\boldsymbol{B}$ are related to each other only by the vacuum Maxwell equations. The local source of the magnetic field is then just the displacement current $\varepsilon_0 \partial_t \boldsymbol{E}$ since there is no electric current due to particle motions. The signature of this electromagnetic loop is the existence of the electromagnetic waves in vacuum.
2. There is a collisional loop, which can exist in the absence of a collective field, and even with neutral particles (although the notion of collision is then different). The collisions between particles also allows information to propagate. The signature of this collisional loop is the existence of pressure (/sound) waves in the medium.

The general loop of Figure 1.1 can exist even with negligible collisions and with negligible displacement current. It is clear from this that in these conditions, any plasma evolution, for instance, any plasma wave, must always involve both kinds of evolution: particle and fields. In neutral gas like air, we are familiar with an almost complete separation between electromagnetic waves (light, radio, and so on), only involving $\boldsymbol{E}$ and $\boldsymbol{B}$, and sound waves, only involving the gas properties like mass

Collisionless Plasmas in Astrophysics, First Edition. Gérard Belmont, Roland Grappin, Fabrice Mottez, Filippo Pantellini, and Guy Pelletier.

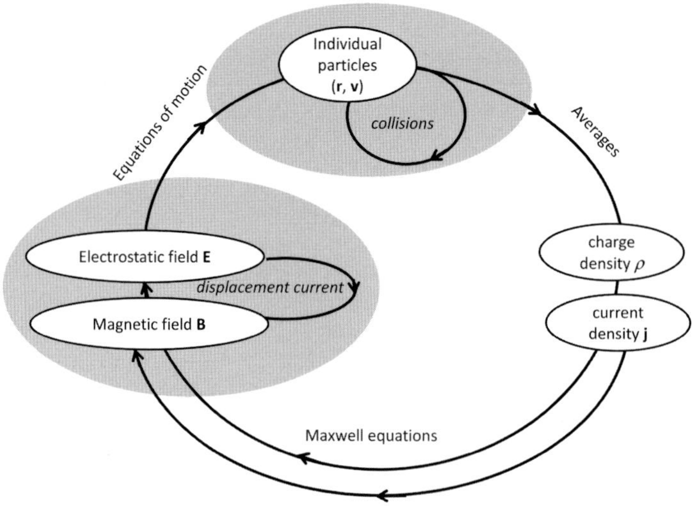

Figure 1.1 The plasma loop. The electromagnetic fields indicated in the sketch are the "collective" ones, that is, where the collision fields have been subtracted. The shaded areas concern, respectively, the electromagnetic subloop and the collisional subloop, which can exist in the absence of charged particles, but which are then not coupled to each other.

density ρ, fluid velocity $\boldsymbol{u}$, and pressure P. This separation is of course prohibited in a plasma.

The sketch of Figure 1.1 makes use of the notion of a "collective field". This notion will be defined in detail, but how can it be understood first from an intuitive point of view? In a small volume, the difference between the electron and ion densities makes a collective charge density which is a source for the electrostatic field, and the difference between their mean velocities makes a current which is a source for the magnetic field and the induced electric field. This loop is the intrinsic plasma loop. The displacement current, if not negligible, is never essential: it is just an additional complication to the fundamental phenomenon. Similarly, the presence of collisions is not essential to plasma phenomena, even if they bring specific properties to the plasma, which can allow for simplified modeling.

The collisions, when present, insure a continuous velocity redistribution between particles. It is the reason why they can allow simplified statistical descriptions: they make the thermodynamical functions such as entropy meaningful. In the absence of collisions, on the contrary, all these notions must be used with care. This book will particularly emphasize the collisionless limits of plasmas, in order to focus on the most intrinsic properties of the plasmas and understand what the descriptions are that remain valid without collisions and those which are specific to the collisional hypothesis.

The question of the collisionless limit is particularly crucial when considering the so-called fluid models. These models, such as MHD (magnetohydrodynamics), allow describing the plasma with a small number of macroscopic parameters, typically density, fluid velocity and pressure. Such a description is of course a huge

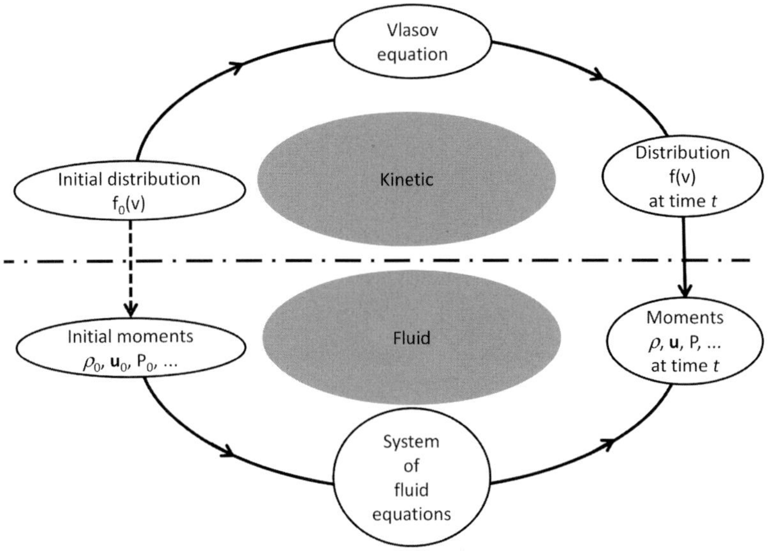

Figure 1.2 Principles of the fluid and kinetic methods in the case of an initial value problem. Note that, if the moments ρ, $\boldsymbol{u}$, P, can always be calculated from the distribution function $f(\nu)$, the opposite is not feasible without strong hypotheses.

reduction if compared to the description of all individual particles, but it is still an extremely big reduction with respect to the kinetic one, which describes the particle populations by their distribution function $f(\boldsymbol{v})$, that is, the density of probability of each velocity in small volumes. This reduction is, however, necessary, for computation time reasons, for any complex problem, in particular large scale and 3D. The validity of the fluid models is well established in the collisional case, but not in a general manner in the collisionless one. It is, therefore, important to understand what is universal in these models and what has to be questioned. We will show that all the weaknesses of these models lie in the so-called closure equation and emphasize the consequences of different choices for this equation.

Figure 1.2 shows the two main methods for modeling the behavior of a particle population. Both methods assume that one knows a valid kinetic equation, that is, a differential equation which describes the variations of the distribution function $f(\boldsymbol{v})$ with time t and space $\boldsymbol{r}$. In a collisionless plasma, this equation is the "Vlasov equation." In a collisional plasma, several equations such as the Boltzmann equation, can be used depending what approximate modeling has been adopted to describe the collisions.

Supposing, for instance, that we have to solve an initial value problem, the principles of the two methods are as follows:

1. Kinetic. To use this method, one is supposed to know the distribution function $f(t = 0)$ in the initial condition. The kinetic equation then allows one to determine $f(t)$ at any later time. Finally, as one is generally interested in the macroscopic parameters such as $\rho(t)$, $\boldsymbol{u}(t)$ and $P(t)$, the resulting distribution

function has to be integrated over velocities to determine them (they will be shown to be moments of $f(t)$).
2. Fluid. Starting for the initial macroscopic parameters such as $\rho(t=0)$, $\boldsymbol{u}(t=0)$ and $P(t=0)$, one solves a differential system relating the variations of the moments to each other. The result is then directly the values of the moments at time t: $\rho(t)$, $\boldsymbol{u}(t)$ and $P(t)$.

For comparing the two methods, first one has to know the relationship between the "fluid moment" system and the original kinetic equation. We will show that all the equations of this system, except one, can be derived directly from the kinetic equation by integration. These moment equations are, therefore, as exact as the initial kinetic equation and do not introduce any further approximation with respect to it. Nevertheless, we will see that the systematic integration actually provides an infinity of moment equations (continuity equation, transport of momentum, pressure, and so on), but that each of them relate the moment of order n to the moment of order $n+1$ (for instance, the pressure temporal variations to the heat flux spatial ones). For this reason, for solving a closed system with a finite number of equations, one is obliged to add a "closure equation", which is not obtained by integration. This is where all the approximation lies.

Another difference between the two methods has to be outlined, however: the fluid method supposes that the moments are known in the initial condition, while the kinetic one demands that the full distribution function is known. This makes a big difference. If only the initial moments are known, the later evolution is a priori not unique since a finite number of moments does not determine a unique distribution function. We will see that some evolutions are much more probable than others, but it is clear from this remark that, whatever the closure equation, the fluid method selects a particular class of distribution function perturbations. We will show in Chapter 5 that this point is crucial to understand why waves in a collisionless plasma are always damped with ordinary initial conditions (Landau damping [13]).

This book intends to be a basic textbook of plasma physics, and it, therefore, covers most classical topics of the domain, such as turbulence (weak/strong), magnetic reconnection, linear waves, instabilities, and nonlinear effects. In each domain, it starts from zero and tries to lead in a self sufficient manner to a view in accordance with the 2013 state of the art. Its main specificity is, however, to pay particular attention, in each domain, to the collisionless limit and the consequences of the different modelings, fluid or kinetic in this case. Many kinetic results in the collisionless limit may appear counterintuitive. For instance, it may appear surprising that the nondissipative Vlasov equation always leads to a damping of the waves; it is surprising as well to find a heat flux in the low solar corona, in a sense opposite to the temperature gradient. The main reason for all these surprises, is that our intuition, for many fundamental physical notions such as irreversibility, has been built in the more usual strongly collisional limit. This makes separating the universal concepts from those that are linked to this limit difficult. These basic notions, for this reason, are specially developed in the book, beginning with the nontrivial notion of collision and of mean free path in a plasma.

The book is designed for an audience of students and researchers. Those who discover the domain should find the essential basic notions. Those who already know them should find the necessary perspective to approach some profound questions concerning the collisionless limit. The book should also help understanding the necessary compromises to be made for modeling plasmas in different circumstances, the global fluid modeling being often necessary to complement the kinetic one, the latter being easily handleable only at small scales and for simple geometries such as 1D.

Most of the examples of the book for illustrating the theoretical concepts are taken in space physics (planetary magnetospheres) and in solar wind. Some others examples concern more remote astrophysical objects (see, for instance, Chapter 8). This choice of "natural plasmas" has been done for insuring homogeneity of the book and respecting the specialties of the authors. However, these examples must be understood only as illustrations. The concepts that are so illustrated are universal and of course not limited to them. Researchers working on laboratory plasmas, in particular, on magnetic confinement for nuclear fusion, are expected to find their interest as well in the presentation.

1.2 Plasmas in Astrophysics

1.2.1 Plasmas Are Ubiquitous

Most of the baryonic matter in the universe resides in the stars, whose hot interiors are made of plasma. Apart from the coolest ones, most star atmospheres are made of plasmas, as are their coronas. The outer parts of stellar coronas are made of tenuous plasmas generally in expansion, called stellar winds. Some of the gas clouds in galaxies can be ionized by neighboring stars. This is the case in the HII regions, forming vast clouds of hot and tenuous hydrogen rich plasmas. On a larger scale, in clusters of galaxies, the development of X-ray astronomy has revealed huge clouds of hot plasma filling the space between the galaxies.

If most of the planetary materials are made of neutral atoms and molecules, their nucleus is composed of a very dense nucleus of degenerate plasma partly supported by the Fermi pressure of free electrons. On the opposite side, the outskirts of the planetary atmospheres are an ionosphere, and possibly a magnetosphere, made of dilute plasmas in interaction with the wind of their star.

The physics of the fully ionized collisionless plasmas is the key element to understanding the corona and wind of stars (including the Sun), the magnetosphere of the planets, and a large variety of shock waves present in various astrophysical contexts.

1.2.2 The Magnetosphere of Stars

The lower layers of a star atmosphere are made of collisional plasma, and the ambient magnetic field has a complex structure involving many scales. It is often represented as a global simple magnetic field superimposed with a multiplicity of open or closed magnetic flux tubes. Some groups of magnetic flux tubes can be isolated, and constitute relatively coherent systems dominated by the plasma pressure forces and the magnetic field. Both the plasma and the magnetic fields evolve; they constitute a dynamical system.

It is not possible to measure directly the magnetic field in the star magnetosphere (it is only possible on the photosphere). Therefore, analytical and numerical models play an important role in their study. The models are generally based on the theory of dissipative plasmas, and the dissipation is attributed to collisions.

As the distance to the star increases, the density is reduced, while the temperature tends to increase (above the chromosphere) and then remains at a high level (typically 10^6 K). Therefore, farther from the star, the magnetosphere is less and less collisional. At the altitude where the magnetic flux tubes are open, the plasma flow velocity is high and supersonic; it is called a *stellar wind*. The *solar wind* is a collisionless plasma. With spacecrafts, in situ measurements of the magnetic field of waves, chemical composition and particle distribution functions have been performed down to sun distances of 0.3 au. As far as it has been measured, the solar wind was always supersonic and faster than MHD waves (see Section 5.1) and noncollisional.

From a theoretical point of view, the boundary conditions that define a star corona are a hot and collisional plasma with a generally complex magnetic field at its base, and a fast expanding plasma wind expanding into the interstellar medium on the other side. The star rotation must be taken into account for the consideration of the overall structure of the magnetosphere.

1.2.3 Shock Waves

As soon as a stellar wind meets another kind of medium, there is an interaction that is preceded by a shock, as long as the difference of velocities between the wind and the object exceeds the speed of sound and/or MHD waves. In collisionless plasmas, shock waves do not have visual signatures, but they can be radio emitters. Therefore, some of them can be studied remotely. In the solar wind, shock waves happen when a stream of fast wind reaches a slower one. They also develop upstream of planets or comets, provided that they are surrounded by an atmosphere. The shocks upstream of a solid obstacle are called *bow shocks*. The bow shock of the Earth is at a distance of about 10 earth radii, in a purely collisionless plasma. Bow shocks upstream of nonmagnetized planets such as Mars or Venus are, along the Sun–planet direction, very close to or inside the ionosphere that is a collisional plasma.

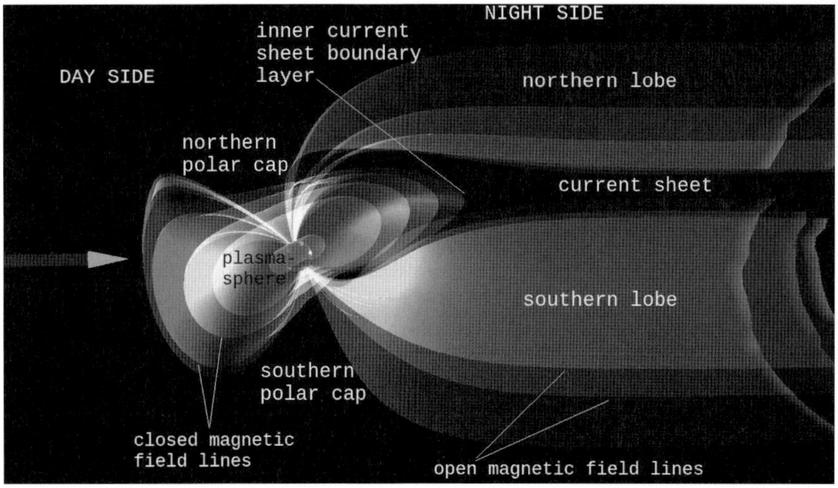

Figure 1.3 Shells of magnetic field lines (starting at the same magnetic latitude from the Earth's surface) showing various regions of the Earth's magnetosphere. The solar wind comes from the left-hand side. Image: courtesy of Bruno Katra, computed with the model [1].

Far from the Sun, the interface between the solar wind and the interstellar plasma is expected to include two shock waves. One has already been crossed by the Voyager 1 and 2 spacecrafts. This interface is called the *heliopause*.

The phenomenology of collisionless shocks is associated with particle acceleration, radio wave emissions and turbulence. They also exist on the borders of fast plasma flows associated with the remnants of supernovae. These shocks are potential sources of galactic cosmic rays (see Chapter 8).

1.2.4
Planetary Magnetospheres

From a theoretical point of view, planetary magnetospheres are the interface between a rotating spherical and magnetized body with a conducting surface (usually an ionosphere) and a stellar wind. The Earth and all the giant planets (Jupiter, Saturn, Uranus, Neptune) have a magnetic field; they are all surrounded by a magnetosphere. The largest magnetosphere in the solar system is that of Jupiter. It itself contains the smaller magnetosphere of its magnetized satellite Ganymede.

The most explored magnetosphere is, of course, that of the Earth, represented in Figure 1.3. As with other magnetospheres, it is first preceded by a *bow shock*, mentioned in Section 1.2.3. Behind the bow shock is a region of fast and turbulent plasma called the *magnetosheath*. In the magnetosheath, the majority of the magnetic field lines are convected in the same direction as the magnetic field (see Section 1.3.1.3 for the explanation of field lines motion), and they are not connected to the Earth. Then, a sharp transition is met: the *magnetopause*. Behind the magnetopause, all the magnetic field lines are connected to the Earth, at least on one end.

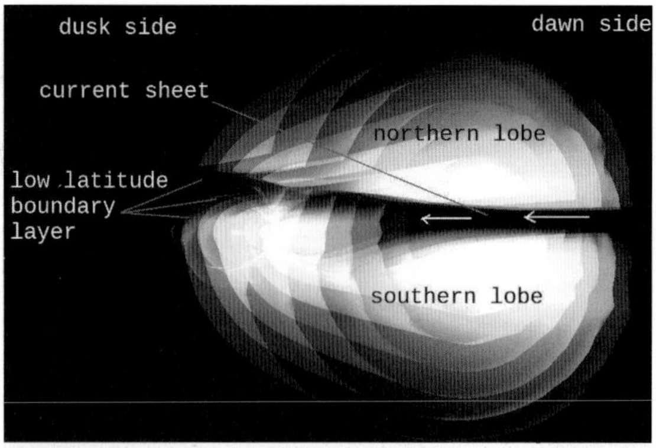

Figure 1.4 Shells of magnetic field lines (starting at the same magnetic latitude from the Earth's surface). A cut is made in the plane perpendicular to the solar wind direction. The white arrows represent the current density direction in the current sheet. On the dusk side, the low latitude boundary layer, which bounds the current sheet, is shown. There is a similar boundary on the dawn side. Image: courtesy of Bruno Katra, computed with the model [1].

The region enclosed by the magnetopause is properly called the magnetosphere. The magnetopause has a few singularities, where magnetic field lines connected in these regions to the Earth can easily (from a topological point of view) be connected to solar wind field lines. The most well known are the *polar caps,* but there are also the flanks of the magnetopause at low latitude, also called the *low latitude boundary layer* (Figure 1.4). The magnetosphere has an asymmetric profile. On the dayside, its extent is of the order of 10 earth radii. On the nightside, the magnetosphere is very elongated, forming the *magnetotail.* Two vast regions, where the magnetic field is almost aligned with the Earth–Sun direction, on the northern and southern sides have a very low density, and are called the *lobes.* The lobes are among the least dense regions of the solar system (about 0.1 particle/cm^3). Between the two lobes is a region of inversion of the direction of the magnetic field; it is the *neutral sheet* (see Figure 1.4). Because an electric current oriented in the east–west direction supports this magnetic field inversion, it is also called the *current sheet.* But this region is also much denser than the lobes, and it is called the *plasma sheet.* The various names of this region are a token of its importance in regards to the physics of the magnetosphere.

Closer to the Earth, and at low latitudes (below the polar cap) there is a denser region of plasma that corotates with the Earth, called the *plasmasphere.* The inner boundary of the plasma sheet is close to the nightside of the plasmasphere boundary, at a distance of about 6–10 earth radii. At higher latitudes, in a region where the magnetic field lines are still connected to the magnetotail, there is an occasional plasma acceleration that causes polar auroras on the ionosphere. This area (from the ionosphere up to a few earth radii of altitude along the field lines) is called the *auroral region.* At even higher altitudes, the plasma is connected to the solar wind via open field lines, in the *polar cap* and the *cusp* regions.

At the distance of the giant planets the solar wind is weaker (especially because of its density varying as d^{-2} where d is the Sun–planet distance). It, therefore, exerts a weaker pressure than on the Earth. Planetary rotation is another source of energy. For Jupiter, with a 10 h rotation and a strong magnetic field, the effect of the rotation dominates those of the solar wind. The plasma in corotation extends quite far from the planet, and the particles inertia in the rotating motion favors the settlement of an extended *ring current* region.

A ring current also exists around the Earth and is associated with the magnetospheric compression by the solar wind; therefore, its origin is of a different nature than fast rotating planets.

1.3 Upstream of Plasma Physics: Electromagnetic Fields and Waves

1.3.1 Electromagnetic Fields

The Maxwell equations describe the time and space variations of the electromagnetic field due to its sources: the charge density ϱ and the current density $\boldsymbol{j}$. In classical physics and in special relativity, the electromagnetic field can be split into two different fields, the electric field $\boldsymbol{E}$ and the magnetic (or induction) field $\boldsymbol{B}$. The four Maxwell equations relating their variations are respectively called the Maxwell–Gauss, Maxwell–Ampère, Maxwell–Faraday, and divergence-free equations, and they are:

$$\text{Gauss} \qquad \nabla \cdot (\varepsilon_0 \boldsymbol{E}) = \varrho \tag{1.1}$$

$$\text{Ampère} \qquad \nabla \times (\boldsymbol{B}/\mu_0) = \boldsymbol{j} + \partial_t(\varepsilon_0 \boldsymbol{E}) \tag{1.2}$$

$$\text{Faraday} \qquad \nabla \times \boldsymbol{E} = -\partial_t \boldsymbol{B} \tag{1.3}$$

$$\text{div-free} \qquad \nabla \cdot \boldsymbol{B} = 0\,. \tag{1.4}$$

The constants ε_0 et μ_0 are called, respectively, the “dielectric permittivity” and “magnetic permeability” of the vacuum, and they appear in the Gauss and Ampère equations, which explicitly relate the fields to the ϱ and $\boldsymbol{j}$ sources. They are linked by the relation: $\varepsilon_0 \mu_0 c^2 = 1$, which makes the constant c (speed of light) enter the system. In the Maxwell–Ampère equation, the term $\partial_t(\varepsilon_0 \boldsymbol{E})$ is called the displacement current. The pure “Ampère equation”, applicable in magnetostatic fields, does not include this term. Nevertheless, in short, we use here the name “Ampère equation” even in the nonstationary case.

In vacuum, the source terms ϱ and $\boldsymbol{j}$ are zero and the electromagnetic field is made of harmonic functions of space for each field, superposed with a linear superposition of electromagnetic waves propagating with the speed c. When sources are present (in particular in plasmas), the charge density changes the electric field, adding an "electrostatic" component; the electric current modifies the magnetic field and also the electric field via an "induced" component (whenever the magnetic field varies in time). The two kinds of sources are always related by the equation of charge conservation:

$$\nabla \cdot \boldsymbol{j} + \partial_t \varrho = 0 \,. \tag{1.5}$$

This equation can of course be derived from the equations of motion of the source charges, but also from the above Maxwell equations (divergence of Eq. (1.2) and temporal derivative of Eq. (1.4)), which outlines the necessary consistency between the electromagnetic fields and its sources.

From the Maxwell equations, an equation can be derived for the electromagnetic energy:

$$\partial_t E_{\mathrm{em}} + \nabla \cdot \boldsymbol{S} = -\boldsymbol{j} \cdot \boldsymbol{E} \,. \tag{1.6}$$

It relates the temporal variations of the electromagnetic energy

$$E_{\mathrm{em}} = \varepsilon_0 E^2/2 + B^2/2\mu_0 \tag{1.7}$$

to the divergence of the Poynting flux vector (energy arriving through the boundaries of a volume):

$$\boldsymbol{S} = \boldsymbol{E} \times \boldsymbol{B}/\mu_0 \tag{1.8}$$

and to the term $-\boldsymbol{j} \cdot \boldsymbol{E}$, which represents the energy exchanges in volume between the electromagnetic field and the matter (for example the plasma).

1.3.1.1 The Scalar and Vector Potentials

The last two Maxwell equations (Faraday and divergence-free) are independent of the sources. They can usefully be integrated once, the former with respect to time, the latter with respect to space. This allows replacing the original fields $\boldsymbol{E}$ and $\boldsymbol{B}$ by two other functions: the scalar and vector potentials, Φ and $\boldsymbol{A}$, defined as:

$$\boldsymbol{B} = \nabla \times \boldsymbol{A} \tag{1.9}$$

$$\boldsymbol{E} = -\nabla \Phi - \partial_t \boldsymbol{A} \,. \tag{1.10}$$

With this formalism, the two last Maxwell equations are automatically satisfied and the first two (Gauss and Ampère) become:

$$\nabla^2 \Phi + \partial_t (\nabla \cdot \boldsymbol{A}) = -\varrho/\varepsilon_0 \tag{1.11}$$

$$\nabla^2 \boldsymbol{A} - \partial_t^2 \boldsymbol{A}/c^2 = -\mu_0 \boldsymbol{j} + \nabla \left(\nabla \cdot \boldsymbol{A} + \partial_t \Phi / c^2\right) \,. \tag{1.12}$$

The potentials are prime integrals of the original fields; therefore, they are not unique. Each particular choice is characterized by a "gauge". The two most famous ones are called the Coulomb and Lorentz gauges. The Coulomb gauge is the simplest. It is defined by:

$$\nabla \cdot \boldsymbol{A} = 0 \,. \tag{1.13}$$

With this choice, Gauss and Ampère equations simplify into:

$$\nabla^2 \Phi = -\varrho/\varepsilon_0 \tag{1.14}$$

$$\nabla^2 \boldsymbol{A} - \partial_t^2 \boldsymbol{A}/c^2 = -\mu_0 \boldsymbol{j} + \nabla \left(\partial_t \Phi/c^2\right) \,. \tag{1.15}$$

In this gauge, the scalar potential is simply a solution of a Poisson equation (no propagation involved for Φ).

The Coulomb gauge is invariant in the nonrelativistic case (it keeps the same form in any inertial frame change), but not in relativity. The Lorentz gauge has better properties in this respect. It is defined by:

$$\nabla \cdot \boldsymbol{A} + \partial_t \Phi/c^2 = 0 \,. \tag{1.16}$$

It is slightly less simple, but it has the great advantage of dissociating Φ and $\boldsymbol{A}$ in their relations with the source terms. Indeed, Gauss and Ampère equations become in this case:

$$\nabla^2 \Phi - \partial_t^2 \Phi/c^2 = -\varrho/\varepsilon_0 \tag{1.17}$$

$$\nabla^2 \boldsymbol{A} - \partial_t^2 \boldsymbol{A}/c^2 = -\mu_0 \boldsymbol{j} \,. \tag{1.18}$$

In vacuum, the potentials defined in the Lorentz gauge just propagate at speed c. Moreover, this gauge is indeed invariant by any inertial frame change.

1.3.1.2 Changes of Reference Frame

The Maxwell equations are invariant in any change of reference frame, but the fields are not. In the nonrelativistic case, going from a frame R to a frame R' moving at a velocity $\boldsymbol{V}$ relative to R, the fields change as:

$$\boldsymbol{E}' = \boldsymbol{E} + \boldsymbol{V} \times \boldsymbol{B} \tag{1.19}$$

$$\boldsymbol{B}' = \boldsymbol{B} \,. \tag{1.20}$$

In special relativity, one has to distinguish between the directions longitudinal and transverse relative to the velocity $\boldsymbol{V}$ of the frame change (respectively subscripts l and t):

$$\begin{aligned} E_\mathrm{l}' &= E_\mathrm{l} \\ \boldsymbol{E}_\mathrm{t}' &= \gamma(\boldsymbol{E}_\mathrm{t} + \boldsymbol{V} \times \boldsymbol{B}) \end{aligned} \tag{1.21}$$

$$\begin{aligned} B_\mathrm{l}' &= B_\mathrm{l} \\ \boldsymbol{B}_\mathrm{t}' &= \gamma\left(\boldsymbol{B}_\mathrm{t} - \boldsymbol{V} \times \boldsymbol{E}/c^2\right) \,. \end{aligned} \tag{1.22}$$

In these relations, the constant c (speed of light) appears explicitly, and also through the relativistic Lorentz factor $\gamma = 1/\sqrt{1 - V^2/c^2}$. It can be noted that the nonrelativistic case ($V \ll c$) corresponds to taking $\gamma = 1$ and neglecting $\boldsymbol{V} \times \boldsymbol{E}/c^2$ relatively to $\boldsymbol{B}_t$.

Like the original fields, their first integrals, the potentials Φ and $\boldsymbol{A}$ are changed by a frame change. In the nonrelativistic case, they become:

$$\Phi' = \Phi - A_l V \tag{1.23}$$

$$\boldsymbol{A}' = \boldsymbol{A} \,. \tag{1.24}$$

In special relativity:

$$\Phi' = \gamma(\Phi - A_l V) \tag{1.25}$$

$$A'_l = \gamma\left(A_l - \Phi\, V/c^2\right) \tag{1.26}$$

$$\boldsymbol{A}'_t = \boldsymbol{A}_t \,. \tag{1.27}$$

Finally, in the same referential change, the source terms become, in the nonrelativistic case:

$$\varrho' = \varrho - j_l V/c^2 \tag{1.28}$$

$$\boldsymbol{j}' = \boldsymbol{j} - \varrho \boldsymbol{V} \tag{1.29}$$

and in special relativity:

$$\varrho' = \gamma\left(\varrho - j_l V/c^2\right) \tag{1.30}$$

$$j'_l = \gamma(j_l - \varrho V) \tag{1.31}$$

$$\boldsymbol{j}'_t = \boldsymbol{j}_t \,. \tag{1.32}$$

An important remark has to be made concerning Eqs. (1.28) and (1.30). As usual, the nonrelativistic case derives from the relativistic one by taking $\gamma = 1$. But it must be emphasized that the term $j_l V/c^2$ exists in both relativistic and nonrelativistic cases: this term cannot be neglected with respect to ϱ in general. Neither the current nor the charge density remain invariant in an inertial referential change. The change in charge density at zero order in V/c is actually consistent with the change in electric field: its electrostatic part appears due to the appearance of the electric charge.

These changes of reference frame are exact when V is a time invariant uniform velocity. When $V = V(t)$ is variable, one has to associate, at every time t, a tangent change of reference frame associated with the instantaneous value of $V(t)$. In that case, the local equations can be still used. This is the case in Eqs. (1.19)–(1.22). (Because V is a parameter of the Lorentz transform, it is considered local.) When the derivatives (charge and current densities) or integrals (potentials) are considered, the corresponding derivatives and integrals of V can introduce terms that do not appear in the above formulas. This is illustrated in Section 1.3.1.4 where the case of a rotating plasma is considered.

1.3.1.3 Notion of "Magnetic Velocity"

The Maxwell equations involve the constant parameter c, which has the dimension of speed. But there is actually another speed which derives directly from the fields themselves:

$$\boldsymbol{v}_\mathrm{m} = \boldsymbol{E} \times \boldsymbol{B}/B^2 \,. \tag{1.33}$$

This velocity can be called the "magnetic velocity" since, locally, the electric field is zero in the frame moving at $\boldsymbol{v}_\mathrm{m}$; this means that the electromagnetic field is purely magnetic in this frame. This property gives to the velocity $\boldsymbol{v}_\mathrm{m}$ a major importance in plasma physics. In particular, in a quasi-homogeneous field, it is known that the particles rotate with a negligible drift velocity in this frame. This means that they follow, on average, the magnetic motion so defined. This is the origin of the so-called ideal Ohm's law used in MHD (see Chapter 3).

From the purely electromagnetic point of view, the magnetic velocity is also important for allowing, in certain circumstances, to define a "magnetic field line motion". The magnetic field is indeed often represented by its field lines. They are by definition tangent to $\boldsymbol{B}$. In rectangular coordinates they are the solutions of

$$\frac{dx}{B_x} = \frac{dy}{B_y} = \frac{dz}{B_z} \,. \tag{1.34}$$

They can always be defined at any time and any point (except at the null points, that is points where $B = 0$, if any). The concept of "magnetic field line motion" is meaningful in all cases when the lines are equipotential, that is when the component $E_\parallel$ parallel to the magnetic field is zero all along of them. In these conditions (and even in conditions slightly more general, see next section), one can prove that the field lines "move at velocity $\boldsymbol{v}_\mathrm{m}$". This means that if all points of a given line are moved at velocity $\boldsymbol{v}_\mathrm{m}$, they still are all on the same field line at any time later. In the limits of validity of the condition $E_\parallel = 0$, it is, therefore, a usual – and quite useful – concept to consider that the line at different times is the same, which just moves. In this way, one gives an identity to the field lines, which can be viewed as kinds of "rubbers" moving and deforming. This concept is particularly important when studying low frequency fluctuations in a plasma ("MHD range"). This property of "freezing" of the field lines in the velocity field $\boldsymbol{v}_\mathrm{m}$ is purely electromagnetic since no plasma parameter is involved, neither in the velocity definition nor in the demonstration (Maxwell equations are sufficient). Nevertheless, the condition of validity $E_\parallel = 0$ is actually imposed – or not – by the plasma. We will see in Chapter 3 that this condition is actually verified in a plasma at sufficiently large scales.

It can also be noted that the velocity $\boldsymbol{v}_\mathrm{m}$ is collinear and close (within a factor of 2) to the velocity of propagation $\boldsymbol{V}_\mathrm{em}$ of the electromagnetic energy E_em. The relation between $\boldsymbol{v}_\mathrm{m}$ and $\boldsymbol{V}_\mathrm{em}$ is evident if the Poynting flux, which is defined as $\boldsymbol{S} = \boldsymbol{E} \times \boldsymbol{B}/\mu_0$ is written as $\boldsymbol{S} = \boldsymbol{V}_\mathrm{em} E_\mathrm{em}$, with $E_\mathrm{em} = \varepsilon_0 E^2/2 + B^2/2\mu_0$.

Proof that the field lines move at velocity $\boldsymbol{v}_\mathrm{m}$ when $E_\parallel = 0$. Let $\delta \boldsymbol{l}$ be a vector that connects two points P_1 and P_2 located on the same field line and separated by an infinitesimal distance. Because $\delta \boldsymbol{l}$ is parallel to the magnetic field, $\boldsymbol{C} = \delta \boldsymbol{l} \times \boldsymbol{B} = 0$. If this vector can be shown to be invariant in motion at velocity $\boldsymbol{v}_{\mathrm{m}}$, this will prove that $\delta \boldsymbol{l}$ is always parallel to $\boldsymbol{B}$, and, therefore, that P_1 and P_2 will remain on the same field line. Noting $D_t = \partial_t + \boldsymbol{v}_{\mathrm{m}} \cdot \boldsymbol{\nabla}$, one has:

$$D_t(\boldsymbol{C}) = D_t(\delta \boldsymbol{l} \times \boldsymbol{B}) = D_t(\delta \boldsymbol{l}) \times \boldsymbol{B} + \delta \boldsymbol{l} \times D_t \boldsymbol{B} \, .$$

The first term can be expressed thanks to the definition of D_t:

$$D_t(\delta \boldsymbol{l}) = \boldsymbol{v}_{\mathrm{m}}(\boldsymbol{l} + \delta \boldsymbol{l}) - \boldsymbol{v}_{\mathrm{m}}(\boldsymbol{l}) = \delta \boldsymbol{l} \cdot \boldsymbol{\nabla} \boldsymbol{v}_{\mathrm{m}} \, .$$

The second term can be expressed thanks to Faraday's equation:

$$D_t \boldsymbol{B} = -\boldsymbol{\nabla} \times \boldsymbol{E} + \boldsymbol{v}_{\mathrm{m}} \cdot \boldsymbol{\nabla} \boldsymbol{B} \, .$$

Using the definition of $\boldsymbol{v}_{\mathrm{m}}$, one can express the electric field $\boldsymbol{E}$ as:

$$\boldsymbol{E} = \boldsymbol{E}_{\parallel} - \boldsymbol{v}_{\mathrm{m}} \times \boldsymbol{B} \, .$$

Replacing $\boldsymbol{E}$ by this expression and developing the curl of the cross product as explained in Appendix A.1, a little algebra provides:

$$\delta \boldsymbol{l} \times D_t \boldsymbol{B} = -\delta \boldsymbol{l} \times (\boldsymbol{\nabla} \times \boldsymbol{E}_{\parallel}) - B \delta l \boldsymbol{b} \cdot \boldsymbol{\nabla}(\boldsymbol{v}_{\mathrm{m}}) \times \boldsymbol{b} \tag{1.35}$$

$$D_t(\delta \boldsymbol{l}) \times \boldsymbol{B} = B \delta l \boldsymbol{b} \cdot \boldsymbol{\nabla}(\boldsymbol{v}_{\mathrm{m}}) \times \boldsymbol{b} \, . \tag{1.36}$$

The vector $\boldsymbol{b}$ is the unit vector of the field line. The sum of the two equations finally provides the variation of C we were looking for:

$$D_t(\boldsymbol{C}) = -\delta \boldsymbol{l} \times (\boldsymbol{\nabla} \times \boldsymbol{E}_{\parallel}) \, . \tag{1.37}$$

We can, therefore, conclude that $E_{\parallel} = 0$ is a sufficient condition to get the freezing-in property of the field lines in the $\boldsymbol{v}_{\mathrm{m}}$ velocity field: if P_1 and P_2 move with the magnetic field velocity $\boldsymbol{v}_{\mathrm{m}}$, they remain on the same field line. The condition $\delta \boldsymbol{l} \times (\boldsymbol{\nabla} \times \boldsymbol{E}_{\parallel}) = 0$, more general and slightly less restrictive, is rarely used because, if the condition $E_{\parallel} = 0$ is often satisfied at large scale in a plasma because of the electron motion, there is no such physical justification for the more general condition.

1.3.1.4 **Space Plasmas in Corotation with Their Planet/Star**

The magnetized bodies in rotation, such as planets or stars, are ubiquitous in the universe. Their magnetic field generally comes from an internal "dynamo" source, but it can also be remnant fields in some occasions. These bodies are generally embedded in plasmas of external origin and one is justified to ask whether these plasmas will remain insensitive to the body rotation or if they will be drawn into this rotation. As shown in Chapter 3, the plasma always follows, at large scale, the "magnetic motion" $\boldsymbol{v}_{\mathrm{m}}$ of the field lines. Near the body surface, the magnetic field is generally rigidly anchored to it; consequently, the plasma can be considered in corotation with the magnetized body. Let us first see the consequences of the corotation of a plasma with a magnetized body.

The notion of corotation. Let $\boldsymbol{\Omega}$ be the body rotation velocity. The corotating plasma has a velocity $\boldsymbol{V} = \boldsymbol{\Omega} \times \boldsymbol{r}$. Because the magnetic field lines (close to the body) follow the same motion, $\boldsymbol{V}$ is also the motion of the magnetic field lines defined in Eq. (1.33). This sets the existence of the so-called corotation electric field,

$$\boldsymbol{E} = -\boldsymbol{V} \times \boldsymbol{B} = -(\boldsymbol{\Omega} \times \boldsymbol{r}) \times \boldsymbol{B} = \boldsymbol{B} \cdot \boldsymbol{r}\boldsymbol{\Omega} - \boldsymbol{B} \cdot \boldsymbol{\Omega}\boldsymbol{r} \,. \tag{1.38}$$

This is the electric field that an observer would see in the inertial frame of reference where the body velocity is null.

It is important to mention at this point that if corotation is generally assumed at very close distance to the body, it is not granted at a larger distance. Even when the plasma moves with the magnetic field line velocity $\boldsymbol{v}_{\mathrm{m}}$, this velocity can be different from the corotation velocity $\Omega \times \boldsymbol{r}$, provided that the plasma has an appropriate retroaction on the shape of the magnetic field lines. It is shown in Section 1.3.3.3 that even in vacuum, the shape of the magnetic field lines depend on the rotation rate Ω of the body; therefore, it is easy to understand that this happens too with a plasma.

Terrestrial magnetosphere and ionosphere. Most of the Earth's ionosphere is in corotation. This means that the plasma in the ionosphere is exposed to the same alternation of nights and days as the Earth's surface. When the ionosphere is exposed to sunlight, the UV increase the ionization rate, while it is zero at night. In the range of altitudes above 400 km, the recombination rate of the ions is low in comparison to the duration of the night, and the plasma density of this ionospheric layer remains roughly constant. But in the range of 60 350 km, the recombination rate is higher, and the ionospheric plasma content at these altitudes varies periodically with the same period as the Earth's rotation with a minimum in the morning hours.

Above the ionosphere, in the range of latitudes $\sim \pm 60°$, the plasma is trapped along closed magnetic field lines. The plasma filling this region has escaped from the ionosphere. It is cold ($T \sim 1$ eV) in comparison to the $T \sim 10^2 - 10^4$ eV plasma found in the solar wind and other regions of the magnetosphere. It is in corotation with Earth. As with the ionosphere (also in corotation), it is asymmetric relative to local time. Its extension is typically 7 R_{E} on the evening side after having been refilled with ionospheric plasma, and 4 R_{E} on the morning side (after spending a night above a less dense and colder ionosphere). The plasma there is denser than anywhere else in the magnetosphere. This zone of corotating plasma is the *plasmasphere*. On the nightside of the Earth, the plasmasphere ends where the magnetotail begins. Compared to the magnetotail, the plasmasphere is a rather quiet region. The auroras observed in the midnight sector are magnetically connected to the magnetotail, therefore, at magnetic latitudes above those of the plasmasphere.

Other magnetospheres. Other magnetospheres contain a plasma in corotation. Actually, most of the magnetosphere of Jupiter is in corotation. More precisely, it is subcorotating. This means that the main component of the plasma velocity is $\boldsymbol{\Omega}'(r) \times \boldsymbol{r}$, with $\Omega'(r)$ close to but smaller than the angular velocity Ω of the planet. The plasma corotating with Jupiter contains the orbit of the closest Galilean satellite, Io, situated at a distance of six Jovian radii from Jupiter's surface. The observations tend to show that Jupiter's main auroral oval corresponds to the interaction of the (sub) corotating region with the noncorotating plasma.

From Eq. (1.1) and the Maxwell–Gauss equation, a charge density can be associated with the corotation electric field,

$$\varrho = -2\varepsilon_0 \boldsymbol{B} \cdot \boldsymbol{\Omega} - \varepsilon_0 (\boldsymbol{r} \times \boldsymbol{\Omega}) \cdot \nabla \times \boldsymbol{B} \,. \tag{1.39}$$

(We notice that the first term would not appear with a direct application of Eq. (1.28). This is because the charge density equation is not local, and the space derivative of the velocity $\boldsymbol{V}$, supposed null in Eq. (1.28), has been taken into account.) Considering the Maxwell–Ampère equation, and the fact that the partial time derivative of $\boldsymbol{E}$ is orthogonal to $\boldsymbol{r} \times \boldsymbol{\Omega}$, for a magnetic field that is not associated with an electric current (for instance a dipole field, see Section 1.3.3.1), we find the Goldreich–Julian density

$$\varrho = -2\varepsilon_0 \boldsymbol{B} \cdot \boldsymbol{\Omega} \,. \tag{1.40}$$

For Jupiter, this corresponds to a particle density $\varrho/e \sim 10^{-28}\,\mathrm{cm}^{-3}$ that is totally negligible. That is not the case with pulsars: they have a fast rotation rate (with a period of 1 s or less) and a strong magnetic field (typically 10^8 T) and the Goldreich–Julian charge density can correspond to an excess of 10^{16} electrons or positrons/m^3. The pulsars illustrate the fact that a plasma in corotation around a highly magnetized fast rotating body is nonneutral.

There is an absolute limit to the size of a corotation region, called the light cylinder radius $R_{\mathrm{LC}} = c/\Omega$. This is the distance at which the corotation velocity would be the speed of light. For all the objects in the solar system, R_{LC} is much larger than their magnetosphere. In the case of pulsars, the light cylinder is well inside the magnetosphere. In most models, it defines broadly the frontier between the inner magnetosphere, with a mixture of corotation and poloidal motion, and the wind, where the plasma motion is mostly radial.

1.3.2
Transverse and Longitudinal Electromagnetic Field

Usefully, the electric and magnetic fields can be considered, as any vector fields, as the sum of an irrotational (or longitudinal) component, and of a solenoidal (or transverse) component:

$$\boldsymbol{E} = \boldsymbol{E}_{\mathrm{l}} + \boldsymbol{E}_{\mathrm{t}} \quad \text{and} \quad \boldsymbol{B} = \boldsymbol{B}_{\mathrm{l}} + \boldsymbol{B}_{\mathrm{t}} \,. \tag{1.41}$$

These components are defined by:

$$\nabla \cdot \boldsymbol{E}_\mathrm{t} = 0, \nabla \cdot \boldsymbol{B}_\mathrm{t} = 0 \tag{1.42}$$

$$\nabla \times \boldsymbol{E}_\mathrm{l} = 0, \nabla \times \boldsymbol{B}_\mathrm{l} = 0 \,. \tag{1.43}$$

The names "solenoidal" and "irrotational" are the most general ones. They are here often replaced by "transverse" and "longitudinal" by reference to the simple case of plane variations. These names are then defined with respect to the gradient (for example, the wave vector for a plane wave). It must not be confused with their use in the above section where "transverse" and "longitudinal" are defined with respect to the relative velocity between two different reference frames.

The Faraday equation involves only the transverse fields; the divergence-free and Gauss equations involve only the longitudinal ones. The Ampère equation can be applied separately to the two components. The Darwin approximation (see Section 1.3.10) is based on this decomposition of the electric field:

$$\nabla \times \boldsymbol{B} = \mu_0 \boldsymbol{J}_\mathrm{t} + \frac{1}{c^2} \partial_t \boldsymbol{E}_\mathrm{t} \tag{1.44}$$

$$\mu_0 \boldsymbol{J}_\mathrm{l} + \frac{1}{c^2} \partial_t \boldsymbol{E}_\mathrm{l} = 0 \,. \tag{1.45}$$

Many wave properties can be analyzed in terms of this longitudinal and transverse decomposition.

1.3.3 Electromagnetic Fields in Vacuum

Plasma physics is a coupled system involving electromagnetic fields and charged particles. To better understand how the charged particles rule the electromagnetic field, it is worth recalling what are the properties of this field in the absence of particles, that is, in vacuum. This corresponds to the absence of second members of Maxwell equations: $\varrho = 0$ and $\boldsymbol{J} = 0$.

1.3.3.1 Static Fields

Because of the absence of charge and current density, one has: $\nabla \cdot \boldsymbol{E} = 0$ and $\nabla \times \boldsymbol{B} = 0$. For static fields, one has also: $\nabla \cdot \boldsymbol{B} = 0$ and $\nabla \times \boldsymbol{E} = 0$, which show that electric and magnetic fields have the same properties. Then, both fields can be treated as gradients of scalar fields $\boldsymbol{E} = -\nabla \Phi$ and $\boldsymbol{B} = \nabla \Psi$.

For astrophysical spherical bodies (planets, stars), it is often useful to approximate at large distance the magnetic field as a dipole field. In spherical coordinates, $\boldsymbol{B}$ derives from $\Psi = M\cos(\theta)/r^2$. Then $B_r = -2M\cos(\theta)r^{-3}$, $B_\theta = -M\sin(\theta)r^{-3}$, $B_\phi = 0$, and $B = Mr^{-3}[1 + 3\sin^2(\theta)]^{1/2}$. A magnetic field line obeys the equation $r = R_\mathrm{B}\sin^2(\theta)/\sin^2(\theta_0)$ where R_B is the body radius and θ_0 determines a particular field line. The radius of curvature of the magnetic field lines is:

$$R_c = \frac{r_0}{3}\frac{\cos\lambda(1+3\sin^2\lambda)^{3/2}}{1+\sin^2\lambda} = \frac{a}{3\cos^2\Lambda}\frac{\cos\lambda(1+3\sin^2\lambda)^{3/2}}{1+\sin^2\lambda}\,. \qquad (1.46)$$

At the equator ($\lambda = 0$), the curvature radius is a third of the distance to the dipole center.

1.3.3.2 Waves

Time dependent electromagnetic fields are the solutions of vectorial d'Alembert equations, directly deriving from Ampère and by the Faraday's laws.

$$\nabla^2 \boldsymbol{E} = \frac{1}{c^2}\partial^2_{tt}\boldsymbol{E}, \quad \text{and} \quad \nabla^2 \boldsymbol{B} = \frac{1}{c^2}\partial^2_{tt}\boldsymbol{B}\,. \qquad (1.47)$$

Looking for sinusoidal plane waves, these equations lead to $\omega^2/(k^2c^2) = 1$, showing that these perturbations propagate at the speed of light in any direction. Going back to the original Ampère and Faraday's laws, it is easy checking that their amplitudes also have the following properties: $\boldsymbol{E}_0 = c\boldsymbol{B}_0$, $\boldsymbol{k}\cdot\boldsymbol{E}_0 = 0$, $\boldsymbol{k}\cdot\boldsymbol{B}_0 = 0$ and $\boldsymbol{E}_0\cdot\boldsymbol{B}_0 = 0$. As all plane waves can be described as the sum of sinusoidal functions (linear equations), all plane waves have these same properties. For each particular sinusoidal solution, the amplitudes and the phases of $\boldsymbol{E}$ or $\boldsymbol{B}$ determine the polarization.

1.3.3.3 Electromagnetic Wave of a Rotating Neutron Star, Pulsar Families

Magnetized planets and stars are generally considered as conducting bodies in rotation. We have seen in Section 1.3.1.4 what happens when they are surrounded by a corotating plasma. We now examine the situation when there is no plasma in their environment, apart from a conducting plasma rotating with the body on its surface. In this simpler case, it is possible to compute the magnetic field far from the surface of the body.

The magnetic field near the surface is approximated as an inclined dipole making an angle I with the rotation axis. It turns with a frequency Ω that is equal, or closely related, to the spin frequency of the body. On the surface, the material is a conductor that behaves like a corotating plasma (it can be a thin ionosphere for planets, or a metal crust for pulsars). Therefore, there is a corotation electric field. Because the body is not surrounded by a plasma, the electromagnetic field outside is a solution of Eq. (1.47). The time derivative in Eq. (1.47) is caused by the rotation with a frequency Ω; it is:

$$-\left(\frac{\Omega}{c}\right)^2 \boldsymbol{B} = \nabla^2\boldsymbol{B}\,. \qquad (1.48)$$

The characteristic length $R_L = c/\Omega$ is called the *light cylinder* radius. It is the distance, projected onto the equatorial plane, at which the corotation speed would equal the speed of light.

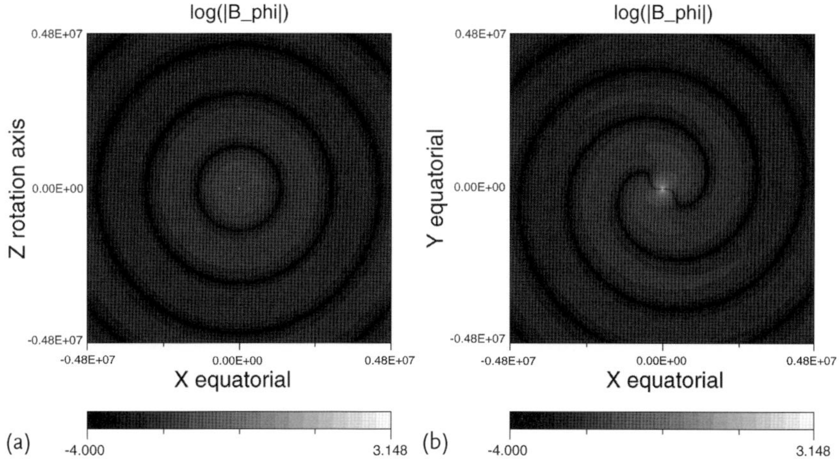

Figure 1.5 The azimuthal component of the magnetic field associated with a rotating conducting sphere with a dipole. The grey level scale represents $\log \| B_\varphi \|$. (The minimal value -4 is set arbitrarily when B_φ becomes smaller.) The tilt angle between the magnetic and rotation axes is 50°, the star's radius is 10^4 m, and its rotation period is 10 ms. We can notice spiral shaped regions separated by a zone of low B_φ value. On each side of these low magnetic field regions, the sign of B_φ changes. The thickness of these spiral shaped regions is $2\pi R_{LC}$. This computation is based on the full solution, and not only the asymptotic solution. (a) $\log \| B_\varphi \|$ on a meridional plane (relatively to the rotation axis). The vertical axis z is the rotation axis. (b) $\log \| B_\varphi \|$ on the equatorial plane.

The complete calculation of the solution is rather tedious but can be completed fully analytically making use of spherical harmonics. At large distances $r \gg R_L$, the result can be put under the simplified form:

$$
\begin{aligned}
B_r &= B_0 \frac{\Omega R^3}{c r^2} \sin I \sin\theta \sin\Phi \\
B_\theta &= B_0 \frac{\Omega^2 R^3}{2c^2 r} \sin I \cos\theta \cos\Phi \\
B_\varphi &= -B_0 \frac{\Omega^2 R^3}{2c^2 r} \sin I \sin\Phi \\
E_r &= 0 \\
E_\theta &= -B_0 \frac{\mu_0 \Omega^2 R^3}{2cr} \sin I \sin\Phi \\
E_\varphi &= -B_0 \frac{\mu_0 \Omega^2 R^3}{2cr} \sin I \cos\theta \cos\Phi\ .
\end{aligned}
\tag{1.49}
$$

This solution, displayed in Figure 1.5, is clearly not a plane wave. We have noted R the radius where the boundary solutions are defined (close to the star's surface). The electromagnetic field at these boundary conditions has been supposed to be a magnetic dipole B_s, inclined by an angle I over the rotation axis, associated with the corotation electric field $\boldsymbol{E} = -(\boldsymbol{\Omega} \times \boldsymbol{r}) \times \boldsymbol{B}_s$. The continuity of the vertical magnetic field component, and of the horizontal electric field components are the constraints put on the electromagnetic field in vacuum and on the body surface: the vertical magnetic field is the same as for the dipole, the horizontal electric field is the same as for the corotation electric field. The phase Φ of the star rotation is defined as:

$$\Phi(r, \phi) = \phi + k(r - R)/R_{\mathrm{L}} - \Omega t \,. \tag{1.50}$$

From the above results, it can be checked that $\boldsymbol{E}$ and $\boldsymbol{B}$ vectors are perpendicular and that the wave radiates energy away at the rate

$$d_t E = \frac{2\pi\mu_0}{3c^3} \Omega^4 R^6 B_0^2 \sin^2 I \,. \tag{1.51}$$

It is an important result because it shows that the rate of energy radiated, which is an observable, depends very sharply on the rotation velocity. It can be used in the physics of magnetized fast rotating bodies such as white dwarfs (with $\Omega \sim 10^{-1}\,\mathrm{s}^{-1}$, for which the model was initially derived) and pulsars, with $\Omega \sim 10^1$–$10^4\,\mathrm{s}^{-1}$. It allows estimating the surface magnetic field B_0 of these objects. Thanks to pulsar timing over long lapses of time (with radio telescopes), it is possible to measure both Ω and $\dot{\Omega}$. For standard pulsars $\Omega \sim 6\,\mathrm{s}^{-1}$ and $\dot{\Omega} \sim -10^{-14}\,\Omega^2$. The energy loss $(1/2) M_{\mathrm{I}} \Omega \dot{\Omega}$ (where M_{I} is the momentum of inertia) associated with this slowing down is mostly due to the electromagnetic radiation defined in Eq. (1.51). Considering a radius $R \sim 10\,\mathrm{km}$, a star mass $M \sim 1$–$3\,M_{\mathrm{Sun}}$, one finds for standard pulsars a value $B_0 \sim 10^8\,\mathrm{T}$. This places the pulsars among the most magnetized bodies in the universe. A new class of pulsars with a typical rotation period of 10 s and a large time derivative of the angular velocity are found to reach $B_0 \sim 10^{11}\,\mathrm{T}$. They are called *magnetars*. Oppositely, pulsars with fast rotations rates ($\Omega \sim 10^3\,\mathrm{s}^{-1}$) have a lower $\dot{\Omega}$; they are found to be less magnetized, $B_0 \sim 10^5\,\mathrm{T}$. Several pulsar families can be so defined, depending upon the different correlations between rotation rates and surface magnetic field.

Of course, in reality, pulsars are not surrounded by vacuum. It was shown shortly after the discovery of the first pulsars that the electric force parallel to the magnetic field exceeds the gravitational forces by a factor 10^6. Therefore, pulsars are expected to host a magnetosphere filled with a plasma. Nevertheless, Eq. (1.51) is still used in first approximation to estimate the surface magnetic field. In some particular cases, the magnetic field has been estimated through other means, which confirmed the orders of magnitudes given here.

1.3.3.4 The Plasma as a Dielectric/Diamagnetic/Conducting Medium

Dielectric media The Maxwell–Gauss equation $\nabla \cdot (\varepsilon_0 \boldsymbol{E}) = \varrho$ describes how the spatial variations of the electric field are determined by the charge density in the medium. The electric field can actually be determined when this Maxwell equation is coupled with the equations of the medium which model how the charge density is determined by the electric field.

The charge density of the medium is in general a reaction to the electric field. In standard dielectrics such as silica, this reaction is just a "polarization" of the medium, which means that each atom, initially neutral, is transformed in a small electric dipole. This results in creating a density of electric dipole moment $\boldsymbol{P}$ called "dielectric polarization density". It can be shown that the corresponding charge density is the divergence of this vector field:

$$\varrho_\mathrm{p} = -\nabla \cdot \boldsymbol{P} \,. \tag{1.52}$$

If the polarization varies in time, it corresponds also to a "polarization current" $\boldsymbol{j}_\mathrm{p}$ given by:

$$\boldsymbol{j}_\mathrm{p} = \partial_t \boldsymbol{P} \,. \tag{1.53}$$

If the medium involves other types of charges, for instance free charges, which cannot be easily linked to the notion of electric dipole, this extra charge can be noted ϱ_ex:

$$\varrho = \varrho_\mathrm{p} + \varrho_\mathrm{ex} \,. \tag{1.54}$$

The further step is to assume that the polarization vector $\boldsymbol{P}$ depends on its cause, the electric field $\boldsymbol{E}$, through a simple relation:

$$\boldsymbol{P} = \boldsymbol{\chi}_\mathrm{e} \cdot \boldsymbol{E} \,. \tag{1.55}$$

It is worth noting that this relation is just algebraic and not differential, meaning that the medium polarization is supposed to depend on $\boldsymbol{E}$ linearly (if $\boldsymbol{\chi}_\mathrm{e}$ is constant) or nonlinearly otherwise, but not on the variations of $\boldsymbol{E}$.

When such a relation is satisfied, the medium is said to be "dielectric" and the coefficient $\boldsymbol{\chi}_\mathrm{e}$ is called the dielectric "susceptibility". In standard materials such as silica, the dielectric susceptibility can indeed be defined and measured; it is a known characteristic of the medium. It can, therefore, be used directly to model the field inside the material when it is put in a given environment, for instance, to study the role of a dielectric sheet inside a capacitor. When the polarization charge is independent of the electric field direction, the dielectric susceptibility is scalar: $\boldsymbol{\chi}_\mathrm{e} = \chi_\mathrm{e} \boldsymbol{I}$ ($\boldsymbol{I}$ being the identity tensor). Otherwise (anisotropic media), it is a full tensor.

Whenever such a dielectric susceptibility can be defined, the Maxwell–Gauss can be rewritten in the following form:

$$\nabla \cdot (\boldsymbol{\varepsilon} \cdot \boldsymbol{E}) = \varrho_\mathrm{ex} \,. \tag{1.56}$$

The tensor $\boldsymbol{\varepsilon} = \varepsilon_0(\boldsymbol{I} + \boldsymbol{\chi}_\mathrm{e})$ is called the dielectric "permittivity", and the vector $\boldsymbol{D} = \boldsymbol{\varepsilon} \cdot \boldsymbol{E}$ is named "electric induction". In many instances, the extra charge ϱ_ex is zero.

Diamagnetic media Similar arguments can be used with respect to the magnetic field and its sources, that is concerning the Maxwell–Ampère equation:

$$\nabla \times \boldsymbol{H} = \boldsymbol{j} + \partial_t(\varepsilon_0 \boldsymbol{E}) \,. \tag{1.57}$$

Under this form, one can notice that the notation $\boldsymbol{B}/\mu_0$, used in Eq. (1.2) and everywhere else in this book, has been replaced by the notation $\boldsymbol{H}$. Both notations are of course strictly equivalent as long as the equation is kept under this initial form, with the source term $\boldsymbol{j}$ on the RHS. Nevertheless, in the context of magnetic media, we will see that the notation $\boldsymbol{B}$ is classically reserved to another use, including a part of the current density, as the electric induction $\boldsymbol{D}$ includes a part of the charge density. In this context, $\boldsymbol{H}$ is simply called the "magnetic field", while $\boldsymbol{B}$ is called "magnetic induction".

In standard magnetic media such as ferrites, a part of the current density can be attributed to a "magnetization" of the medium, which can be put under the form:

$$\boldsymbol{j}_{\mathrm{m}} = \nabla \times \boldsymbol{M} . \tag{1.58}$$

The introduction of the magnetization density vector $\boldsymbol{M}$ is quite similar to the introduction of the polarization vector $\boldsymbol{P}$ for the charge density. It corresponds to a density of magnetic dipole moment, exactly as the first one corresponds to a density of electric dipole moment. It can also be viewed as an ensemble of small electric loops (sometimes called "Amperian currents"). In a plasma, it could be related to the circular motion of the particle around the magnetic field.

The magnetization current, which corresponds to spatial variations of the magnetic dipole density, has to be added to the polarization current presented above, which corresponds to time variation of the polarization vector. Other kinds of currents can be involved, in particular due to free charges when the medium is conducting. The corresponding extra current density can be noted $\boldsymbol{j}_{\mathrm{ex}}$. When both kinds of dipoles are present, the current density is:

$$\boldsymbol{j} = \boldsymbol{j}_{\mathrm{m}} + \boldsymbol{j}_{\mathrm{p}} + \boldsymbol{j}_{\mathrm{ex}} . \tag{1.59}$$

It should be noted that the part $\boldsymbol{j}_{\mathrm{m}}$ (magnetization current), contrary to the two other parts, does not correspond to any charge density since it has no divergence. The next step consists once again in supposing a simple relation between $\boldsymbol{M}$ and the magnetic field $\boldsymbol{H}$.

$$\boldsymbol{M} = \boldsymbol{\chi}_{\mathrm{m}} \cdot \boldsymbol{H} . \tag{1.60}$$

The tensor $\boldsymbol{\chi}_{\mathrm{m}}$ is called the magnetic susceptibility. In isotropic media, it is a scalar. Its sign can be positive or negative, which shows that the reaction of the medium can increase (paramagnetism) or decrease (diamagnetism) the magnetic field.

This allows rewriting the Maxwell–Ampère equation under the form:

$$\nabla \times \boldsymbol{H} = \boldsymbol{j}_{\mathrm{ex}} + \partial_t(\boldsymbol{D}) . \tag{1.61}$$

Or equivalently:

$$\nabla \times (\boldsymbol{\mu}^{-1} \cdot \boldsymbol{B}) = \boldsymbol{j}_{\mathrm{ex}} + \partial_t(\boldsymbol{\varepsilon} \cdot \boldsymbol{E}) . \tag{1.62}$$

This second form is equivalent at the condition of taking the new definition of the vector $\boldsymbol{B}$ (magnetic induction):

$$\boldsymbol{B}/\mu_0 = \boldsymbol{H} + \boldsymbol{M} = (\boldsymbol{I} + \boldsymbol{\chi}_{\mathrm{m}}) \cdot \boldsymbol{H} . \tag{1.63}$$

The relation between $\boldsymbol{B}$ and $\boldsymbol{H}$ can, therefore, be written as $\boldsymbol{B} = \boldsymbol{\mu} \cdot \boldsymbol{H}$. The tensor $\boldsymbol{\mu} = \mu_0(\boldsymbol{I} + \boldsymbol{\chi}_{\mathrm{m}})$ is called the magnetic permeability. In standard materials such as ferrites, the magnetic permeability is a known characteristic of the medium. It allows, for instance, to determine the consequences of introducing a ferrite barrel inside a solenoid.

Conducting media In materials such as metals, the conduction current density can be related to the electric field via a relation of the form:

$$\boldsymbol{j} = \boldsymbol{\sigma} \cdot \boldsymbol{E} \, . \tag{1.64}$$

The tensor $\boldsymbol{\sigma}$ is called the conductivity tensor. It can be scalar or not depending whether the medium is isotropic or not. In an isotropic metal, the conductivity is just the inverse of the resistivity: $\boldsymbol{\sigma} = \boldsymbol{I}/\eta$.

It is to be noticed that, due to the charge continuity equation, the Maxwell–Gauss equation can be put also under the form of a relation between $\boldsymbol{j}$ and $\boldsymbol{E}$. There is therefore a relation between the dielectric susceptibility $\boldsymbol{\chi}_{\mathrm{e}}$ and the conductivity $\boldsymbol{\sigma}$. In the simplest case of no extra charge and a monochromatic wave ($\partial_t = -i\omega$), this relation takes the quite simple form: $\epsilon_0 \boldsymbol{\chi}_{\mathrm{e}} = -\boldsymbol{\sigma}/i\omega$.

1.3.3.5 Use of These Concepts in Plasmas

The equations relating ϱ, $\boldsymbol{j}$, $\boldsymbol{E}$, and $\boldsymbol{B}$ in a plasma are demonstrated in this book. They depend on the model used, which itself can depend on the time and space scale range. However, whatever the model, these relations generally do not have the above forms.

Let us show it by a simple example. The plasma models provide a set of differential equations. If one succeeds, for instance, in eliminating all variables but ϱ and $\boldsymbol{E}$ and their derivatives, one would actually find a relation between these variables. But this relation would be a *differential* equation, with derivatives of both variables at different orders, with respect to time and to space. This is not reducible to Eqs. (1.54) and (1.55).

After the previous example, this should prevent use of the notions of dielectric/diamagnetic/conducting medium for a plasma. No charge and no current can be included in full generality on the LHS of the Maxwell equations, and all of them must remain on the RHS as ϱ_{ex} and $\boldsymbol{j}_{\mathrm{ex}}$ source terms. Nevertheless, all these concepts can be used anyway, and often in a very efficient manner, in a particular domain: the calculation of the linear plasma waves. When calculating monochromatic waves, with one ω and one k, all the derivatives can be expressed as functions of these two parameters in an algebraic form and all the linear perturbations can always be expressed as functions of a single one. This allows calculating a relevant dielectric permittivity as well as a magnetic permeability or an electric conductivity as functions of ω and k.

A special mention must be made for the electric conductivity. In collisional and nonmagnetized plasmas, the plasma relation between $\boldsymbol{j}$ and $\boldsymbol{E}$ has actually the form (1.64). This is actually the only exception: as soon as a magnetic field is added, for instance, the simplest "Ohm's law", for a resistive medium, becomes:

$$E = -u \times B + \eta j \ . \tag{1.65}$$

This cannot be written under the form $j = \sigma \cdot E$ without solving the whole system and expressing the variables u and B as functions of E. And the form is then algebraic only for monochromatic waves in the linear limit. It can be noted also that, in this case, the magnetic field introduces an anisotropy to the system, which implies that the effective conductivity is not scalar.

1.3.4
Plane Waves in a Plasma

In collisionless media, as in any material or in a vacuum, waves can propagate, which means that any initial perturbation varying in space will evolve in time and space in a specific manner. We summarize here the most general methods allowing the calculation of the linear waves, their dispersion and their polarization. Some properties are independent of the medium and can be directly derived from the Maxwell equations: we will summarize only these ones in the present section. A more complete description of the waves demands to specify the plasma parameters and the scale range: some of the most important ones can be found in Chapter 5.

The method is based on three main steps:

1. Linearize the physical equation system around an equilibrium state. This does not change the Maxwell equations themselves (which give the electromagnetic fields variations as functions of the sources ϱ and j) since they are linear. But it changes most of the plasma equations (which give the variations of ϱ and j as functions of the fields), because they are generally not linear.
2. Transform this differential system into an algebraic one (usually via Fourier transform)
3. Solve the algebraic system (eigenmodes and eigenvectors).

The first step relies on the fact that one deals with small amplitude fluctuations, which justifies a perturbative approach where only the first order is retained. The equilibrium state is most generally supposed to be a plain homogeneous and stationary state, where all fields and plasma parameters are constant.

The second step relies on the linearity of the above system: thanks to it, any solution is just a linear superposition of a basis of particular solutions. It is, therefore, sufficient to calculate these different particular solutions, each of which is characterized by a small number of free parameters. These parameters are typically the pulsation ω and the wave vector k when choosing monochromatic plane waves as particular solutions, that is, using Fourier transform (which is the most common method when the zero order state is homogeneous and stationary). The algebraic system obtained in this way is homogeneous (that is, without the RHS) since one studies the natural oscillations of the system, without forcing.

The third step then consists of solving the system as a function of ω and k. Usually all the variables can be eliminated except one vector field, that we can call a "reference field". The equation relating the three components of that field with ω and $\boldsymbol{k}$ as parameters is a system of linear equations. Since the system is homogeneous, its determinant must be zero to get nontrivial solutions. This provides a relation between ω and $\boldsymbol{k}$, which is the dispersion equation. Each solution $(\omega, \boldsymbol{k})$ is called an eigenmode of the plasma and the corresponding eigenvector gives its *polarization*, that is, generally speaking, the perturbations of all variables (including the different components of the vectorial variables) as functions of only one.

In practice, the choice of the reference field requires attention. For high-frequency waves, when the displacement current is dominant on the particle current ($E/B \approx v_{\text{phase}}/c \approx 1$), the more convenient reference field is the electric field. It allows doing the calculation with any densities and recovering the electromagnetic waves in vacuum as a limiting case of an evanescent plasma.

On the contrary, in the context of low frequency waves, that is, when the displacement current can be neglected in front of the particle one ($E/B \approx v_{\text{phase}}/c \ll 1$), the choice of $\boldsymbol{E}$ as a reference field is no more convenient. Using a field that is negligible in some of the equations as a reference field carries the risk of an ill posed problem. It is the case in particular for MHD waves. The most usual choice of reference field is then the plasma velocity field $\boldsymbol{u}$. The choice of the magnetic field as a reference variable could seem also possible but it is not so pertinent because it does not allow to compute the electrostatic waves (which have no magnetic component).

It is worth noting that (i) even in the case of a stationary zero order state, the Fourier transform vs. time, at step 2, can sometimes be usefully replaced by a Laplace transform, which allows introducing the initial condition and investigating more clearly the questions of causality (see Chapter 5); (ii) With nonplane waves, step 2 cannot be systematically accomplished, and one remains with a set of both algebraic and differential equations.

1.3.5
Electromagnetic Components of Plane Plasma Waves

In a plasma, collisionless or not, all waves involve fluctuations in both the electromagnetic fields and in the plasma variables. The fields are expressed as functions of the plasma parameters by the Maxwell equations via the source terms ϱ and $\boldsymbol{j}$. Conversely, the plasma parameters depend on the fields via the particle trajectories: this provides a second part of the equation system, which we will call here the *plasma equations*. The plasma equations are generally complex and highly dependent on both the plasma parameters and the scale range of the phenomena under study. It is studied in the different chapters of the book. On the contrary, the Maxwell equations are quite general and some general wave properties can be deduced from them.

When applying the two first steps of the general program above to the Maxwell equations, one obtains:

$$\boldsymbol{k} \cdot \boldsymbol{E}_1 = -i\varrho_1/\varepsilon_0 \tag{1.66}$$

$$\boldsymbol{k} \times \boldsymbol{B}_1 + \omega \boldsymbol{E}_1/c^2 = -i\mu_0 \boldsymbol{j}_1 \tag{1.67}$$

$$\boldsymbol{k} \times \boldsymbol{E}_1 - \omega \boldsymbol{B}_1 = 0 \tag{1.68}$$

$$\boldsymbol{k} \cdot \boldsymbol{B}_1 = 0\,. \tag{1.69}$$

The subscript 1 labels the first order fluctuations, and the notation $\boldsymbol{E}_1$ must be understood as a short one to represent the complex amplitude of the Fourier transform, which should be written $\tilde{\boldsymbol{E}}_{1\omega,k}$ to be complete, and which is a function of ω and $\boldsymbol{k}$. All functions of the Fourier basis are supposed to vary with time and space as $e^{i(\boldsymbol{k}\cdot\boldsymbol{r}-\omega)t}$. This means that, for deriving the above equations from the original differential ones, the derivatives with respect to time have been replaced by multiplications by $-i\omega$ and the nablas by $i\boldsymbol{k}$.

Some general properties can readily be drawn by distinguishing the longitudinal and transverse components of the vectors with respect to $\boldsymbol{k}$ (subscripts l and t). The above system can then be rewritten as:

$$B_{1\mathrm{l}} = 0 \tag{1.70}$$

$$k\,E_{1\mathrm{l}} = -i\varrho_1/\varepsilon_0 \tag{1.71}$$

$$\omega\,E_{1\mathrm{l}}/c^2 = -i\mu_0\,j_{1\mathrm{l}} \tag{1.72}$$

$$\boldsymbol{k} \times \boldsymbol{B}_{1\mathrm{t}} - \omega\,\boldsymbol{E}_{1\mathrm{t}}/c^2 = -i\mu_0\,\boldsymbol{j}_{1\mathrm{t}} \tag{1.73}$$

$$\boldsymbol{k} \times \boldsymbol{E}_{1\mathrm{t}} - \omega\,\boldsymbol{B}_{1\mathrm{t}} = 0\,. \tag{1.74}$$

Equation (1.70) is independent of all the others. It is just the direct consequence of the divergence-free equation: waves have no magnetic field component of their perturbation along $\boldsymbol{k}$. Equations (1.71) and (1.72) relate the longitudinal component of the electric field $E_{1\mathrm{l}}$ to ϱ_1 and $j_{1\mathrm{l}}$. These two equations are equivalent if the charge continuity equation is taken into account. Equations (1.72) and (1.74) relate the transverse components of both fields to the transverse component of the electric current. The last two sets of two equations are independent of each other only if the plasma equations make $j_{1\mathrm{l}}$ independent of $\boldsymbol{E}_{1\mathrm{t}}$ and $\boldsymbol{j}_{1\mathrm{t}}$ independent of $E_{1\mathrm{l}}$. In terms of the conductivity tensor defined above ($\boldsymbol{j} = \boldsymbol{\sigma} \cdot \boldsymbol{E}$), this occurs when $\boldsymbol{\sigma}$ is a scalar. In these conditions, the electrostatic and electromagnetic waves are independent and can be superposed linearly: the electrostatic waves can be calculated by ignoring the magnetic field and the electromagnetic ones by ignoring the electrostatic component.

1.3.6 Some General Properties of Plane Wave Polarization and Dispersion

Keeping the current source term unknown, the magnetic field perturbation $\boldsymbol{B}_1$ can easily be eliminated between Eqs. (1.67) and (1.68). This elimination provides:

$$\left[\left(\omega^2 - k^2c^2\right)\mathbf{1} + \boldsymbol{k}\boldsymbol{k}c^2\right] \cdot \boldsymbol{E}_1 = -i\frac{\omega}{\varepsilon_0}\boldsymbol{j}_1\,. \tag{1.75}$$

Taking $\boldsymbol{j}_1 = 0$, one readily recovers the vacuum solutions: two transverse waves with $\omega^2 - k^2c^2 = 0$ (the longitudinal solution $\omega = 0$ cannot exist in vacuum because of Eq. (1.71).

Beyond the vacuum case, very few results can be considered as really general. To obtain the dispersion relations in plasmas, the complete system, Maxwell plus plasma equations, has to be known. A way to get a step further without specifying the plasma equations is to use the fact that, whatever they are, they can always be noted in linear theory: $\boldsymbol{j}_1 = \boldsymbol{\sigma} \cdot \boldsymbol{E}_1$, using the $\boldsymbol{\sigma}$ "conductivity tensor" presented just above. This tensor will have of course to be calculated in each specific situation to obtain concretely the properties of the waves in a given situation.

Rewriting Eq. (1.75) with this notation, one obtains:

$$\boldsymbol{P}_E \cdot \boldsymbol{E}_1 = 0 \quad \text{with} \quad \boldsymbol{P}_E = \left[\left(\omega^2 - k^2c^2\right) \mathbf{1} + \boldsymbol{k}\boldsymbol{k}c^2 + i\frac{\omega}{\varepsilon_0}\boldsymbol{\sigma} \right] . \tag{1.76}$$

The dispersion relation, therefore, results, in general, from the nullity of the determinant $D(\omega, \boldsymbol{k})$ of a propagation matrix $\boldsymbol{P}_E$ which depends on the plasma equations through the tensor $\boldsymbol{\sigma}$. As already mentioned in Section 1.3.3.4, the conductivity is related to the dielectric susceptibility and permittivity and $\boldsymbol{\sigma}$ can be replaced by its expression in any of these functions.

Let us consider a magnetic field locally aligned in the z direction; the equation $D(\omega, \boldsymbol{k}) = 0$ can be expanded in the following way, as a function of the relative permittivity $\boldsymbol{\epsilon} = \boldsymbol{\varepsilon}/\varepsilon_0$.

$$\begin{vmatrix} \omega^2\epsilon_{xx} - k_\parallel^2 c^2 & \omega^2\epsilon_{xy} & \omega^2\epsilon_{xz} + k_\perp k_\parallel c^2 \\ \omega^2\epsilon_{yx} & \omega^2\epsilon_{yy} - k^2c^2 & \omega^2\epsilon_{yz} \\ \omega^2\epsilon_{zx} + k_\perp k_\parallel c^2 & \omega^2\epsilon_{zy} & \omega^2\epsilon_{zz} - k_\perp^2 c^2 \end{vmatrix} = 0 \, , \tag{1.77}$$

where the wave vector $\boldsymbol{k}$ vector is included in the x, z plan, and $k_\parallel$ and $k_\perp$ are respectively its parallel and perpendicular projections relatively to the magnetic field direction: $\boldsymbol{k} = (k_\perp, 0, k_\parallel)$. One can see on Eq. (1.77) how all the wave properties can be derived from the permittivity tensor. This tensor will be expressed and analyzed more explicitly in Chapter 5, when the role of the particle distribution is specified.

Even in simple cases, with simple plasma equations, a great diversity of waves can appear. They are generally characterized by their frequency ω and their wave vector $\boldsymbol{k}$. The equation relating these two quantities is called the *dispersion relation*. The ratio $v_\phi = \omega/k$ is the *phase velocity*. The ratio $N = kc/\omega = c/v_\phi$ is the *refraction index*. For certain values of the parameters, $N \gg 1$, that is $v_\phi/c \mapsto 0$. This is called a *resonance*. When $N \ll 1$ that is $v_\phi \mapsto \infty$, this is a *cut-off*. Examples will be given hereafter.

1.3.7
Electrostatic Waves

It is worth noting that plasma waves can involve, contrary to the vacuum waves, a longitudinal component E_{1l}. This component can even, for some wave modes such as the famous "Langmuir wave", be predominant. As soon as the transverse component of the electric field is negligible, Eq. (1.74) shows that its magnetic counterpart is also very small. It means that these modes are mainly caused by the space charge density ρ_1, while the electric current associated with the corresponding charge displacement can be neglected. This is why these waves are called "electrostatic waves", although their frequency is not zero. In this case, the linear dispersion equation is simply

$$\epsilon_{\parallel} = \epsilon_0 - \frac{\sigma_{\parallel}}{i\omega} = 0\,. \tag{1.78}$$

These modes depend only on the longitudinal component of the dielectric tensor. The above dispersion relation can be derived directly from Eq. (1.76), or from the consideration of the dominant longitudinal component in Eq. (1.71).

1.3.8
Wave Packets and Group Velocity

With a single plane wave (of real frequency and wave vector), there is no propagation of any form of information, because the wave is present everywhere with the same intensity. In real life, waves do not fill the entire space. Instead of a monochromatic plane wave, let us rather consider packets of plane waves, belonging to a common branch of solutions of the dispersion equation $D(\omega, \boldsymbol{k}) = 0$. They are supposed to be linearly superposed, with their wave vectors spread in a finite interval. When this interval is small, one speaks of quasi-monochromatic waves. The amplitude of wave packets are represented, for any field A in the form

$$A(\boldsymbol{r}, t) = \int \tilde{A}(\boldsymbol{k}) e^{i(\boldsymbol{k}\cdot\boldsymbol{r}-\omega t)} d\boldsymbol{k}\,. \tag{1.79}$$

The phase of each wave is defined as $\phi = \boldsymbol{k}\cdot\boldsymbol{r} - \omega t$. The wave packet has a maximum at a point $\boldsymbol{r}$ where all contributions are in phase, that is, where these phases reach a common value ϕ_0 which is independent of $\boldsymbol{k}$. For a quasi-monochromatic wave packet, where a Taylor expansion can be used, the phase is:

$$\phi = \phi_0 + d\boldsymbol{k}\cdot\nabla_k(\boldsymbol{k}\cdot\boldsymbol{r} - \omega t) = \phi_0 + d\boldsymbol{k}\cdot(\boldsymbol{r} - \nabla_k(\omega)t)\,. \tag{1.80}$$

This shows the point $\boldsymbol{r}$ where the wave packet has its maximum amplitude given by $\boldsymbol{r} = \nabla_k(\omega)t$. This point moves at a velocity called the group velocity and defined by:

$$\boldsymbol{v}_g = \nabla_k\omega\,. \tag{1.81}$$

In a 1D problem, it can be simply written: $v_g = d\omega/dk$.

1.3.9
Propagation of Plane Waves in a Weakly Inhomogeneous Medium

The propagation of waves in an inhomogeneous medium is a complex topic that involves many possible phenomena: wavelength and frequency modifications, partial transmissions, absorption and reflection, loss of planarity, mode conversion (a wave pass from one branch of the dispersion equation solutions to another), and so on. Quite often, in an inhomogeneous medium, the hypothesis of planarity does not hold. Nevertheless, an asymptotic theory called the BGK theory after Bernstein, Green, and Kruskal, describes the propagation of planar waves when the gradient scales in the plasma are much larger than the wavelength.[1] The BGK theory provides a set of equations providing the wavelength, the frequency, the phase velocity, and the group velocity evolutions. It is also possible to do ray tracing.

In a stratified medium, where the propagation index N depends on z only, a ray with an initial vertical angle ϕ_0 in the x, z plane evolves along the path described as

$$x = \pm \int_0^z \frac{\sin \phi_0 d\zeta}{\sqrt{N^2(\zeta) - \sin^2 \phi_0}} . \tag{1.82}$$

1.3.9.1 An Example of Wave Propagation with a Cut-off Frequency

Anticipating Chapter 5, we can investigate the propagation of a wave with a propagation index given by $N^2(z) = 1 - \omega_p^2/\omega^2 = 1 - \alpha n$, where $n = n(z)$ is the plasma density, which is a function of the altitude. This relation is indeed the propagation of an electromagnetic wave in a plasma (ordinary radio wave). We consider the case of the lower ionosphere where the plasma density $n(z)$ increases with altitude. For simplification, we have a linear dependence. Then $N^2(z) = 1 - az$, where a^{-1} is the altitude at which the wave frequency equals the cut-off frequency of its branch of the dispersion relation. Equation (1.82) becomes $x = \pm(c\sqrt{b - z} - c\sqrt{b})$, where $b = (1 - \sin^2 \phi_0)/a$ and $c = \sin \phi_0/\sqrt{a}$. The rays have a parabolic shape, with a reflection below the altitude where $N(z)$ becomes null. More generally, in the vicinity of an altitude where $N^2(z)$ cancels, the expansion $N^2(z) \sim 1 - az$ is valid, and the above solution is quite general in the zone near the altitude of the cut-off. This fact is illustrated in Figure 1.6a that provides a numerical example of Eq. (1.82). We can conclude that a cut-off at a certain altitude corresponds to a reflection of the waves below the altitude where their frequency is the cut-off frequency.

1.3.9.2 Examples of Wave Propagation with a Resonance Frequency

We represent a resonance at the altitude z_0 by a function of the shape $N^2 = a(z - z_0)^{-l}$ where a is an arbitrary constant. This relation does not correspond to a particular plasma dispersion function; it is simply chosen to illustrate the effect of a resonance. For z close to z_0, the function to integrate is equivalent to $\sin \phi_0 (z_0 - \zeta)^{l/2} a^{-1/2}$, and its primitive is proportional to $(z_0 - \zeta)^{(l+2)/2}$. Therefore, in this vicinity, $z \sim z_0 - cx^{l/(l+2)}$. The ray approaches the altitude z_0 for a finite value of x, and has a vertical slope. Beyond this point, there is no propagation; therefore, the wave is absorbed at the resonance altitude. This is illustrated on the right-hand side of Figure 1.6 by a numerical solution of Eq. (1.82) for the case

1) This theory could be developed in the frame of the multiscale expansion introduced in Appendix A and used in Section 1.4.

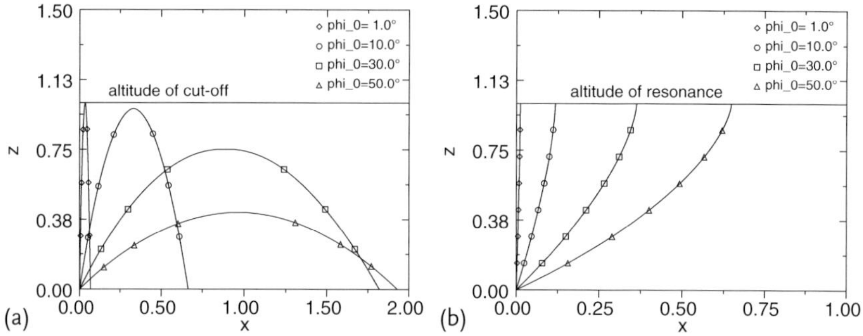

Figure 1.6 Ray tracing in a stratified medium for various initial angles. (a) The dispersion index $N(z) = \sqrt{1-z}$ has a cut-off at the altitude $z = 1$ marked by the horizontal line. We can see that the waves are reflected below this altitude. (b) The dispersion index $N(z) = 1/\sqrt{z-1}$ has a resonance at the altitude $z = 1$ marked by the horizontal line.

$l = 1$. We can see that the waves do not propagate beyond the altitude where the resonance is reached.

1.3.10
Useful Approximations of the Maxwell Equations in Plasma Physics

1.3.10.1 The Approximation of a Null Displacement Current Used in Classical MHD

On many occasions, we have defined a conductivity σ tensor. For dimensional analysis, one can reduce it to a scalar field σ and $J = \sigma E$. This is analogous to the introduction of an impedance Z defined by $ZI = U$ where I is the current circulating in a portion of plasma submitted to a potential drop U. Then, for dimensional analysis, $ZI = ZJS = U = EL$ where L is the length of the plasma portion and $S = L^2$ the area of surface perpendicular to it. Then, $J = E/(ZL)$, and we can write the Ampère equation

$$\nabla \times \mathbf{B} = \frac{Z_0}{Z}\frac{\mathbf{E}}{Lc} + \frac{1}{c^2}\partial_t \mathbf{E} \,, \tag{1.83}$$

where $Z_0 = c\mu_0 = 376.7\ldots\ \Omega$ is the impedance of free space. In terms of plane waves analysis, it is more simply written

$$\mathbf{k} \times \mathbf{B} = k\left(\frac{Z_0}{Z} + \frac{1}{c}\frac{\omega}{k}\right)\frac{\mathbf{E}}{c} \,. \tag{1.84}$$

For the comparison of orders of magnitude we make abstraction of the geometrical operators (see Section 3.3), and

$$B \equiv \left(\frac{Z_0}{Z} + \frac{v}{c}\right)\frac{E}{c} \,. \tag{1.85}$$

where $v/c = \omega/(kc)$ represents the typical velocity of the phenomenons under consideration (the phase velocity in the case of plane waves). Similarly, the Faraday equation is

$$\frac{E}{c} \equiv B\frac{v}{c} \tag{1.86}$$

and the Gauss equations is

$$kE \equiv \frac{\rho}{\epsilon_0} \,. \tag{1.87}$$

Concerning the Maxwell equations, the approximation made in MHD is the following: large spatial scales, low velocities relative to the speed of light, and a high plasma conductivity. The magnetic field amplitude is generally finite. In terms of the above equations, a small parameter ϵ is introduced and allows us to write

$$k = k_0\epsilon \quad \text{and} \quad \frac{v}{c} = \beta\epsilon \quad \text{and} \quad \frac{Z_0}{Z} = \frac{\alpha}{\epsilon} \,, \tag{1.88}$$

where α, β, k_0 are finite, as well as B. Then from Eq. (1.85),

$$B \equiv \left(\frac{\alpha}{\epsilon} + \beta\epsilon\right)\frac{E}{c} \,. \tag{1.89}$$

As a result, the $\mu_0 \boldsymbol{J}$ term in the Ampère equation associated with $\alpha\epsilon$ dominates the displacement current, which is associated with $\beta\epsilon$. Therefore, in that context, the displacement current is neglected, and the Ampère equation reduces to

$$\nabla \times \boldsymbol{B} = \mu_0 \boldsymbol{J} \,. \tag{1.90}$$

Moreover, as the magnetic field is finite, $E/(cB) \equiv \delta\epsilon$ where δ has a finite value; therefore, E is of order one in ϵ. Of course, this does not imply that the electric field is null. But in the MHD equation, when one part of the electromagnetic field has to be eliminated from a set of equations, it is better to keep the magnetic field, because it is of a larger order of magnitude than the (normalized by c) electric field E/c (see Section 5.1).

Is the charge conservation Eq. (1.5) relevant to this MHD approximation? The charge conservation is derived from the divergence of the Ampère equation, combined with the Gauss equation. In the present case, the divergence of the Ampère equation is reduced to $\nabla \cdot \boldsymbol{J} = 0$. (This term is of order ϵ, while those associated with the displacement current is of order ϵ^3.) Therefore, the charge continuity equation is not relevant. Because there is no charge conservation, and because the use of the current density is unavoidable, it is not recommended to consider the charge density. Consistently, the Gauss equation, in the form of Eq. (1.87) shows that the charge density time derivative is of order ϵ^2, and should not be retained. Nevertheless, the Maxwell equations are completed, in MHD by a Ohm's law that generally provides a value for the electric field. Taking its divergence does not necessarily lead to a null charge density, but the charge density should be of order 2 or more.

1.3.10.2 The Approximation of a Null Displacement Current Used with a Multicomponent Plasma

The MHD approximation is made for large spatial scales. When one is also interested in smaller spatial scales, the above set of approximations can be replaced by the following:

$$\frac{Z_0}{Z} = \frac{\alpha}{\epsilon} \quad \text{and} \quad \frac{v}{c} = \beta\epsilon \,, \tag{1.91}$$

where α, β are finite, as well as B and k. Then Eq. (1.89) is still valid. Therefore, the displacement current is still negligible in the Ampère equation, and Eq. (1.90) still applies. Then, when B is finite, E/c is still of order ϵ. But the Gauss equation implies now that ρ is of the same order as E (when it was one order less in the MHD approximation). Then, one has to take ρ explicitly into account, as well as the Gauss equation, because they both involve fields of order ϵ. Because of Eq. (1.90), the charge continuity equation implies that $\rho = 0$. Therefore, in this approximation, Eq. (1.90) is explicitly completed by

$$\rho = 0 \,. \tag{1.92}$$

This is quasi-neutrality. This equation is especially important, because it allows us to set a relation between the densities of the different species. For instance, in the case of a plasma with a single ion species, the electron and the ion densities are related through the relation $n_\text{i} = n_\text{e}$. This is not the case in MHD where the plasma is considered as a single fluid.

The two approximations presented in this section are not compatible with electromagnetic waves in vacuum (because of the large conductivity assumption) and not with the electrostatic ones where ρ is finite and explicitly taken into account.

1.3.10.3 The Electrostatic Approximation

The electrostatic approximation supposes a large plasma impedance and small velocities. Then

$$\frac{Z_0}{Z} = \alpha\epsilon \quad \text{and} \quad \frac{v}{c} = \beta\epsilon \,, \tag{1.93}$$

where α, β, k_0 are finite, as well as E. The approximation on Z_0/Z consists of neglecting the current density. We can see that B is of order ϵ. To order zero, the Ampère equation is reduced to $\nabla \times \boldsymbol{B} = 0$. As $\nabla \cdot \boldsymbol{B} = 0$, the solution is a uniform magnetic field, that cannot vary as a function of time.[2)] The Maxwell equations reduce to

$$\nabla \times \boldsymbol{E} = 0 \quad \text{and} \quad \nabla \cdot \boldsymbol{E} = \rho/\epsilon_0 \,; \quad \boldsymbol{B} = \boldsymbol{B}_0 \,, \tag{1.94}$$

2) The Faraday equation shows that for a uniform magnetic field that depends on time, $\boldsymbol{\nabla} \times \boldsymbol{E}$ is uniform, and, therefore, the electric field diverges at infinity. Indeed, let us consider a sphere V of radius R. For finite values of $\partial_t \boldsymbol{B}$, the integral $\int_V \boldsymbol{\nabla} \times \boldsymbol{E} dV = \int_S \boldsymbol{n} \times \boldsymbol{E} dS$ varies as R^3, the sphere surface varies as R^2; consequently, E diverges as R, involving an infinite amount of electric energy. In order to avoid this divergence, $\partial_t \boldsymbol{B} = 0$ is necessary.

where $\boldsymbol{B}_0$ is constant and uniform. The electrostatic approximation can also be used locally. In that case, there exist solutions where the curl and the divergence of $\boldsymbol{B} = 0$ are null and $\boldsymbol{B}$ is not uniform. For instance, this is the case of a magnetic dipole, when the center of the dipole is out of the domain of validity of the approximation. In that case, for a bounded system,

$$\nabla \times \boldsymbol{E} = -\partial_t \boldsymbol{B} \quad \text{and} \quad \nabla \cdot \boldsymbol{E} = \rho/\epsilon_0 \; ; \quad \nabla \times \boldsymbol{B} = 0 \quad \text{and} \quad \nabla \cdot \boldsymbol{B} = 0 \, . \tag{1.95}$$

The electrostatic approximation does not allow for the propagation of any electromagnetic wave, only purely electric waves can be considered. The continuity of the electric charge cannot be deduced any more from these reduced equations. Nevertheless, when used, this equation shows that $\nabla \cdot \boldsymbol{J}$ can be finite, but it has no consequence because the current density does not appear in the above equations.

The electrostatic approximation has various domains of application. It is widely used in the theory of auroral acceleration. The BGK theory is generally developed in this frame, as well as the theory of electrostatic double layers (see Chapter 6). It is also possible to establish a theory of electrostatic waves.

The propagation of electromagnetic waves into vacuum is not possible, and this is good news for those making numerical simulations who are interested only in low frequency waves: they won't have to bother with the computer time and memory requirements associated with a proper numerical computation of light waves.

1.3.10.4 The Darwin Approximation

The Darwin approximation, mainly used in numerical simulation, is designed to eliminate the light waves that do not interfere with the plasma, because they raise strong constraints on the time step: $c\Delta t < \Delta x$. But it is also designed to retain the physics of the waves that interact with the plasma as much as possible. In particular, it must retain the electrostatic waves and the continuity of the charge density.

The Darwin approximation proceeds in two steps. First, the electric field is cut in a longitudinal (or solenoidal) component $\boldsymbol{E}_\text{l}$, and a transverse component $\boldsymbol{E}_\text{t}$, defined by

$$\boldsymbol{E} = \boldsymbol{E}_\text{t} + \boldsymbol{E}_\text{l} \quad \text{and} \quad \nabla \cdot \boldsymbol{E}_\text{t} = 0 \quad \text{and} \quad \nabla \times \boldsymbol{E}_\text{l} = 0 \, . \tag{1.96}$$

Then, one can notice that the Faraday equation involves only the transverse electric field, and the Gauss equation involves only the longitudinal field. The Ampère equation is applied separately to the two components:

$$\nabla \times \boldsymbol{B} = \mu_0 \boldsymbol{J}_\text{t} + \frac{1}{c^2} \partial_t \boldsymbol{E}_\text{t} \tag{1.97}$$

$$\mu_0 \boldsymbol{J}_\text{l} + \frac{1}{c^2} \partial_t \boldsymbol{E}_\text{l} = 0 \, . \tag{1.98}$$

The second stage in the Darwin approximation consists of neglecting the displacement current only in Eq. (1.97). Thus, it is no longer possible to have light waves.

The divergence of the second equation and the Gauss equation provides the equation of continuity of the charge density. In summary:

$$E = E_t + E_l \,, \tag{1.99}$$

$$\nabla \times E_l = 0 \quad \text{and} \quad \nabla \cdot E_l = \rho/\epsilon_0 \tag{1.100}$$

$$\nabla \cdot E_t = 0 \quad \text{and} \quad \nabla \times E_t = -\partial_t B \,, \tag{1.101}$$

$$\nabla \times B = \mu_0 J_t \,. \tag{1.102}$$

In which order can one solve these equations? The longitudinal electric field can be deduced directly from the charge density. From the current density, one can know only the curl of the magnetic field. It is better to combine the Eqs. (1.101), (1.102), and the null divergence of the (transverse) magnetic field. There remains only a Laplacian equation,

$$\nabla^2 E_t = \mu_0 \partial_t J_t \,. \tag{1.103}$$

But usually, one cannot separate a priori the longitudinal and the transverse components. Therefore, an Ampère equation where this separation is not explicitly done is used:

$$\nabla^2 E_t = \mu_0 \partial_t J - \frac{1}{c^2} \partial_t^2 E_l \,. \tag{1.104}$$

Then, as E_l obtained as the solution of the Gauss equation is known, the only unknown is on the left-hand side of the equation, and it can be solved by inversion of the Laplacian.

This approximation indeed allows the charge conservation and eliminates light waves. Many electrostatic and electromagnetic waves interacting with the plasma are possible, but not all of them. To be convinced of this limitation, we can derive the dispersion equation of linear waves in the Fourier space. The solution is

$$k \times (k \times E) + \frac{\omega^2}{c^2} \frac{\epsilon}{\epsilon_0} \cdot E = 0 \,. \tag{1.105}$$

with

$$\epsilon = \epsilon_0 \left(\frac{k k}{k^2} + i \frac{\sigma}{\omega \epsilon_0} \right) \,. \tag{1.106}$$

Without the Darwin approximation, we would have

$$\epsilon = \epsilon_0 \left(I + i \frac{\sigma}{\omega \epsilon_0} \right) \,. \tag{1.107}$$

The two systems give the same solutions for electrostatic waves (as k and E are parallel). In vacuum ($\sigma = 0$), we can see that the transverse waves (with $k \perp E$) have necessarily a null amplitude. This is consistent with the purpose of this approximation. But for long length waves (therefore, interacting with the plasma) with orthogonal k and E, the solution can be greatly affected by this approximation.

1.4 Upstream of Plasma Physics: The Motion of Charged Particles

The motion of the particles in a fluid is determined by integration of the motion equation in the collective electromagnetic and gravitational fields over a time period that is inversely proportional to the collision frequency. The global motion of the fluid is derived from the particle motion. In a collisional medium, these equations have to be solved over the short time between two collisions, and each collision gives a set of random new initial conditions. When the collision rate is high, this can be treated by a statistical approach. When there is no collision, the particles follow a deterministic trajectory during the whole duration of the phenomenon under study. The initial distribution of positions and motion can be described statistically, but the determination of its evolution is deterministic and requires that we can compute trajectories over long periods of time.

In a collisionless plasma, the full dynamics of the particle is taken into account when the Lorentz force equations of motion are solved,

$$d_t \boldsymbol{r} = \boldsymbol{v} \quad \text{and} \quad d_t \boldsymbol{v} = \frac{q}{m}[\boldsymbol{E}(r) + \boldsymbol{v} \times \boldsymbol{B}(r)] \, . \tag{1.108}$$

It is possible to add to the electric field the addition of other potential forces, for instance, gravitational. But practically, the typical velocities in an astrophysical plasma around a planet or a star exceed by far the gravitational escape velocity. This is why the gravitation forces are often neglected.

1.4.1 The Motion of the Guiding Center

Nevertheless, a full description of the particles' trajectories is not necessary. Quite often, as for a particle into a wave, or in a slowly varying magnetic field, the particle has a rapid and periodic motion, added to a slow variation of the characteristics of the periodic motion, and a drift motion. We are interested by the general characteristics of the motion, without knowing the phase of its fast component. The slow component of the velocity is called the *guiding center* motion.

In many textbooks, the different components of the guiding center velocity are introduced separately as consequences of particular features of the electromagnetic field. Unfortunately, in this way, we are never sure whether we include all the terms, nor whether one was counted twice. This is why we present here a global theory of the guiding center.

A global, theory of the guiding center was published in 1963 in a reference textbook [2]. The so-called rigorous demonstration in that book requires previous knowledge of asymptotic expansion methods (this chapter is then preceded by a not-so-rigorous demonstration for physicists), and the book is no longer available. This is why it is worth presenting here an extensive derivation of the guiding center theory. We base our development on the multitimescales method. No previous knowledge of the theory of asymptotic developments is required. The general method is described in Appendix A.2 and it is applied in Appendix A.2.3 to the guiding center motion.

1.4.1.1 Principles of the Guiding Center Motion

The multiscale expansion method is based on a small parameter ϵ that relates the space and time derivatives of the electromagnetic field. If the electromagnetic field is uniform and constant, $\epsilon = 0$, and the velocity is the sum of a constant drift $\boldsymbol{E} \times \boldsymbol{B}/B^2$ and a rapid circular motion of frequency $\omega_c = qB/m$ (the gyrofrequency) in the plane that is perpendicular to $\boldsymbol{B}$. Therefore, the high frequency of the system (noted ω_0 in Appendix A) is ω_c. Then, ϵ is considered as a small parameter provided that

$$\rho_L|(\nabla B)_\perp| = \frac{v_\perp}{\omega_c}|(\nabla B)_\perp| \ll B \,, \tag{1.109}$$

that is, formally equivalent to

$$m v_\perp |(\nabla B)_\perp| \ll q B^2 \,. \tag{1.110}$$

The characteristic fast time scale is $T_c = 2\pi/\omega_c$, during which the particle must see only a weak variation of the electromagnetic field

$$v_z \left(\frac{2\pi}{\omega_c}\right) |(\nabla B)_\parallel| \ll B \quad \text{and} \quad \left(\frac{2\pi}{\omega_c}\right) |\partial_t B| \ll B \,. \tag{1.111}$$

The parallel electric field $\boldsymbol{E}_\parallel$ is generally not the cause of a fast periodic motion. Therefore, it must induce a weak acceleration in comparison to those induced by the magnetic field

$$(q/m) E_\parallel \ll |\boldsymbol{v} \times \boldsymbol{B}| \sim \rho_L \omega_c^2 \,. \tag{1.112}$$

The velocity equation (Lorentz force) is

$$d_t \boldsymbol{v} = \boldsymbol{e}(\boldsymbol{r}, t_1, \ldots) + \boldsymbol{v} \times \boldsymbol{b}(\boldsymbol{r}, t_1, \ldots) \,, \tag{1.113}$$

where we note $\boldsymbol{e} = q\boldsymbol{E}/m$ and $\boldsymbol{b} = q\boldsymbol{B}/m$. We accept a contribution of order 0 of the magnetic field and the perpendicular electric field.

A development (detailed in the Appendix A) is made of the electromagnetic field

$$\boldsymbol{e}(\boldsymbol{r}, t_1, \ldots) = \boldsymbol{e}_\perp(\boldsymbol{R}, t_1, \ldots) + \epsilon(\boldsymbol{r}_0 \cdot \nabla)\boldsymbol{e}_\perp + \epsilon \boldsymbol{e}_{\parallel 1} + \ldots \tag{1.114}$$

$$\boldsymbol{b}(\boldsymbol{r}, t_1, \ldots) = \boldsymbol{b}(\boldsymbol{R}, t_1, \ldots) + (\boldsymbol{r}_0 \cdot \nabla)\boldsymbol{b} + \ldots \tag{1.115}$$

where the gradients are computed at the position $\boldsymbol{R}$ of the guiding center, and not of the particle.

The velocity equation is then written at orders ϵ^0 and ϵ^1 in the context of the multiscale expansion. This procedure is detailed in Appendix A.2.3. To lowest order, the slow component of the solution is the well known cross field drift velocity, also found in ideal MHD theory,

$$U_0 = \frac{E \times B}{B^2} + U_{\|0} b , \tag{1.116}$$

with $U_{\|0}$ still undetermined at this stage. The fast part at order zero describes a linear oscillator, whose solution is

$$\begin{aligned} u_{0x} &= +u_{\perp 0} \cos(\omega_c t + \psi) \\ u_{0y} &= -u_{\perp 0} \sin(\omega_c t + \psi) \\ u_{0z} &= 0 \,. \end{aligned} \tag{1.117}$$

The slow part of Eq. (A25) is the basis of the derivation of the slow motion at first order of the guiding center velocity.

1.4.1.2 The Perpendicular Velocity of the Guiding Center

It is found that, in terms of ordinary time and space variables,

$$\begin{aligned} U_\perp &= U_0 + \epsilon U_1 \\ &= \frac{E \times B}{B^2} + \frac{m b}{q B} \times \left[d_t \frac{E \times B}{B^2} + U_\| d_t b + \frac{\mu}{m} \nabla B \right] . \end{aligned} \tag{1.118}$$

The first term is the cross field drift $E \times B / B^2$, occurring as soon as the electric and the magnetic fields have components orthogonal to each other. All the contributions to the slow perpendicular drift are the product of $q b / m B$ and a vector that is homogeneous to an acceleration, the effect of a force. We can develop the derivatives appearing in the force terms in order to prove a few effects

$$\begin{aligned} U_\perp &= \frac{m b}{q B} \times \left[\frac{q E}{m} + d_t \frac{E \times B}{B^2} + U_\| d_t b + \frac{\mu}{m} \nabla B \right] \\ &= \frac{m b}{q B} \times \left\{ \frac{q E}{m} + d_t E_\perp \times \frac{B}{B^2} + E_\perp \times d_t \left(\frac{B}{B^2} \right) \right\} \\ &+ \frac{m b}{q B} \times \left\{ U_{\|0} \left[\partial_t b + U_{0\|} b \cdot \nabla b + U_\perp 0 \cdot \nabla b \right] + \frac{\mu}{m} \nabla B \right\} . \end{aligned} \tag{1.119}$$

The force $d_t E_\perp \times B / B^2$ is associated with the polarization drift velocity, that in virtue of the double cross product takes the simple form

$$\frac{m}{q B} b \times \left(d_t E_\perp \times \frac{B}{B^2} \right) = \frac{m}{q B^2} d_t E_\perp \,. \tag{1.120}$$

The polarization drifts, because its proportionality to the mass is mainly affected by ions and is often neglected for electrons, and as a result, it is a cause of electric current density in a plasma.

The development of $U_{\|0} d_t b$ in Eq. (1.119) is associated to the velocities

$$\frac{m}{q B} U_{\|0} b \times \partial_t b + \frac{m}{q B} U_{\|0}^2 b \times (b \cdot \nabla b) + \frac{m}{q B} U_{\|0} b \times (U_{0\perp} \cdot \nabla b) \,. \tag{1.121}$$

The second of these terms is called the curvature drift because it appears when the magnetic field lines have a finite curvature. It is generally the most important term of this group. All these terms are charge and mass dependent, they are potentially the cause of an electric current density.

The last term comes form the mirror force $-\mu \nabla B/m$ connected to the gradient of the modulus of the magnetic field. When the magnetic field increases, this force is repulsive, which is why it is called a magnetic mirror force, as if photons are reflected by a mirror. This term gives an indication about the tendency of plasmas to go preferentially in the regions of a weak magnetic field.

Apart from the curvature drift, the polarization drift, and the mirror force, we can see other terms. They appear in [2], but they are seldom mentioned in many other textbooks, whose approach of the guiding center motion, if pedagogic, is less scrupulous.

1.4.1.3 The Parallel Velocity of the Guiding Center

The equation of the parallel motion is

$$d_t U_{\|0}(t) = \frac{q}{m} E_{\|} - \frac{\mu}{m} \nabla_{\|} B - \frac{\boldsymbol{E} \times \boldsymbol{B}}{B^2} \cdot \frac{\partial \boldsymbol{b}}{\partial t}. \tag{1.122}$$

The first term is the parallel electric field. It is not specific to the guiding center theory. The term derived from the perpendicular magnetic gradient (the gradient of B in the direction perpendicular to $\boldsymbol{B}$) in the equation of the perpendicular velocity

$$\frac{\mu}{qB} \boldsymbol{b} \times \nabla B \tag{1.123}$$

reappears in the equation of the parallel velocity Eq. (1.122).

The last term, connected to the partial time derivative of $\boldsymbol{b}$, is seen in [2] but is not visible in many other textbooks. The Eqs. (1.122) and (1.118) provide a fairly classical representation of the guiding center motion.

1.4.1.4 The Guiding Center Equation That Does Not Separate the Perpendicular and Parallel Components

From Eqs. (1.116) and (A61), it is possible to derive an equation of the global guiding center slow velocity $\boldsymbol{U} = \boldsymbol{U}_0 + \epsilon \boldsymbol{U}_1$, without separating the parallel and the perpendicular component, at order 1.

$$d_t \boldsymbol{U} = \frac{q}{m} [\boldsymbol{E} + \boldsymbol{U} \times \boldsymbol{B}] - \frac{\mu}{m} \nabla B, \tag{1.124}$$

where the electromagnetic field $\boldsymbol{E}$, $\boldsymbol{B}$ is defined at the position $\boldsymbol{R}$ of the guiding center, and not at the instantaneous position $\boldsymbol{r}$ of the particle velocity. This equation determines the evolution of the slow component of the velocity. Equation (1.124) is seldom used in analytical calculations, but it can be very useful in numerical computations [3]. It is interesting when solved in a way that allows us to get rid of the fast varying part (coming from the $\boldsymbol{U} \times \boldsymbol{B}$ term). This can be done with an implicit numerical algorithm that combines a low-pass filtering with the computation of $\boldsymbol{U} \times \boldsymbol{B}$.

1.4.2 Adiabatic Invariants

1.4.2.1 First Adiabatic Invariant and Magnetic Trapping

In the slowly varying electromagnetic field of a magnetized medium, the shortest time scale concerning the particle motion is the gyrofrequency ω_c. An adiabatic invariant, defined in Appendix A.2, can be associated with the gyromotion of the particle. As only the motion defined by $\boldsymbol{p}_\perp$ is periodic, only this one must be retained in the definition of the first adiabatic invariant. Let $T = 2\pi/\omega_c = 2\pi m/qB$ be the period of the fast motion,

$$\mathcal{I} = \int_0^T p\,dq = \int_0^T m v_\perp dr = \int_0^T m v_\perp d v_\perp dt = \frac{2m\pi}{q} \frac{m v_\perp^2}{2B} . \qquad (1.125)$$

Practically, the retained definition of the first invariant is the magnetic moment $\mu = m v_\perp^2/2B$ (see also Eq. (A60)).

If the particle evolves in a purely magnetic field, the total kinetic energy $v_\parallel^2 + v_\perp^2$ as well as μ is conserved. Then the pitch angle α, related to the above quantities through the equation $\sin\alpha = v_\parallel / v e_\perp$, depends directly on the amplitude of the local magnetic field:

$$\frac{\sin^2\alpha}{B} = \text{constant} = \frac{\sin^2\alpha_0}{B_0} . \qquad (1.126)$$

Here, B_0 and α_0 are reference values. They often correspond to the region where the magnetic field value B_0 is minimal, thus defining the minimal value α_0 of the pitch angle. In a magnetosphere, the value B_0 is generally found at the magnetic equator, and α_0 is the equatorial pitch angle.

When a particle moves along a magnetic field line towards a region of increasing magnetic field, the pitch angle tends to increase too, until it reaches the value $\pi/2$. Then the particle can not go any further, otherwise $\sin\alpha$ would be larger than one. Equation (1.122) clearly states that at this stage, the particle bounces back into the region of lower magnetic field. Then, the first adiabatic invariant tells us that the tendency of the particles is to be trapped in the region of a low amplitude magnetic field, unless their initial pitch angle is small enough. This process is called the *mirror effect*. Particles bouncing back because of the increase of the magnetic field are said to be *mirrored*. The others are *untrapped*, or *passing* particles.

The particles with $\sin\alpha_0 < (B_0/B_{\max})^{1/2}$ are not reflected. In the vicinity of a planet, $B_{\max}$ corresponds to the topside ionosphere. Below, the plasma becomes collisional, and the particles are captured through binary interactions in the ionosphere. The capture process, when the particle has enough energy can be a source of visible light, UV, and X-rays. The particles at the origin of the polar auroras, on the Earth and on other magnetized planets, are untrapped particles. They are said to be in the *loss cone* of the velocity distribution function.

For mirrored particles, whose position along the magnetic field line is defined by their curvilinear abscissa s, the bouncing period τ_b between the two mirror points m_1 and m_2 is

$$\tau_b = \int_{m_1}^{m_2} \frac{ds}{v_{\|}(s)} = \frac{2}{v} \int_{m_1}^{m_2} \frac{ds}{\sqrt{1 - B(s)/B_{max}}} . \tag{1.127}$$

The regions of closed field lines contain a trapped plasma. The fact that they contain particles of high energy (a few MeV) was one of the first discoveries about the magnetosphere, made in 1958 with the space probe Explorer I and was very advertised. These regions of trapped plasma are called the *Van Allen belts,* or *radiation belts.* The plasma of low energies (below a few keV) is also trapped and can remain for weeks. The trapped plasma region has a torus shape, centered, in the case of Earth, at a distance of about 5 R_E (where $R_E = 6400$ km is the Earth's radius). This region is inside the plasmasphere mentioned in Section 1.3.1.4. At close distance to the magnetized planets, and not too close to the poles, we have seen that the magnetic field lines are closed on both sides, and that $\boldsymbol{B}$ can be approximated as a dipole (see Section 1.3.3.1). For a dipole, the field lines of equation $r = R_B \cos^2(\lambda)/\cos^2(\lambda_0)$ reach the body surface (radius R_B) at the latitude $\lambda_p = \arccos L^{-1/2}$ where $L = r_0/R_B$ is the *McIllwain parameter,* r_0 being the equatorial distance of the magnetic field line. In the dipole field approximation as well as other models, L is used to characterize a field line. For a dipole field, the pitch angle can be expressed as a function of the magnetic latitude λ

$$\sin^2 \alpha(\lambda) = \frac{\cos^6 \lambda}{\sqrt{1 + 3\sin^2 \lambda}} \sin^2 \alpha_0 . \tag{1.128}$$

Knowing that at the mirror points of latitude λ_m, $\sin^2 \lambda_m = 1$ this equation gives the possibility of computing the latitude of the mirror points. The latitude of the mirror points is independent of L and, therefore, it is independent of the magnetic field line. The boundary l of the loss cone is defined by $\lambda_m > \lambda_e$, that can be expressed as $\sin^2 \alpha_{0l} = (4L^6 - 3L^5)(-1/2)$. The bounce period is

$$\tau_b = \frac{4r_0}{v} \int_0^{\lambda_m} \frac{\cos\lambda(1 + 3\sin^2\lambda)^{1/2}}{\cos\alpha(\lambda)} d\lambda \sim \frac{4r_0}{v} . \tag{1.129}$$

The typical bounce period is $\sim$ 1 s for 1 keV electrons and $\sim$ 1 min for ions trapped around the Earth. The motion of these particles comprises the cross fields drift that coincide with the corotation velocity (provided that the electric field is caused only by corotation). The sum of their gradient and curvature drifts is

$$v_{GC} = \frac{mv^2}{2qBR_c}(1 + \cos^2 \alpha) \tag{1.130}$$

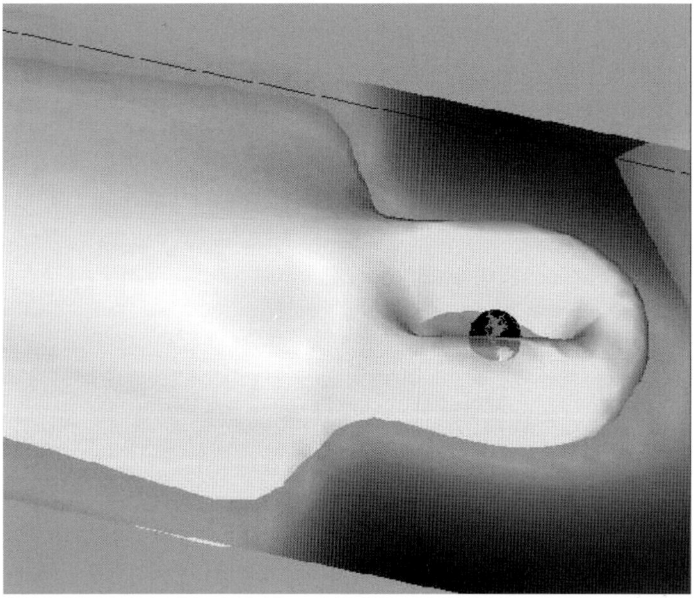

Figure 1.7 Isocontour of the current density magnitude $|j| \propto |\nabla \times B|$ in the inner magnetosphere. The magnetic field is computed with the model [1]. On the left side of the figure is the current sheet layer, extended in the east–west direction. The Earth is surrounded by the ring current clearly visible on the figure. This current is associated with a particle drift in the azimuthal direction. (Courtesy: P. Robert, LPP/CNRS)

where the variables B, R_c, α depend on the latitude λ and R_c is the curvature radius of the magnetic field line $R_c = \|d^2 \mathbf{r}/ds^2\| = (r_0/3)\cos\lambda(1 + 3\sin^2\lambda)(3/2)/(2 - \cos^2\lambda)$. Then, averaged over a bounce period, this corresponds to an azimuthal drift

$$\langle\dot{\phi}\rangle = \frac{4}{\tau_b}\int_0^{\lambda_b} \frac{d\lambda}{\cos\alpha}\frac{ds}{d\lambda} \sim (3/2)\frac{mv^2 r_0}{q B_b R_b^3} \tag{1.131}$$

where B_b is the surface magnetic field on the equator of the body, and R_b the body radius. Since average drift is charge dependent, it is at the origin of an azimuthal electric current that turns completely around the body in the regions of closed magnetic field lines (that is, at low altitude and low latitudes). This current is called the *ring current*, visible in Figure 1.7.

1.4.2.2 **Second Adiabatic Invariant**

When a particle is trapped between two mirror points, its guiding center bounces between the mirror points with the period τ_b. This is the highest frequency involved in the motion of the guiding center (not of the particle). Therefore, an adiabatic invariant can be associated with this periodic motion, as far as the variation of the magnetic field is slow relative to the period τ_b. Following the same procedure as for the first adiabatic invariant, we select the component of the impulsion that has the periodic motion. This time, this is the parallel component. The associated generalized coordinate is the curvilinear abscissa along the magnetic field line. Therefore, the second adiabatic invariant, noted J, is

$$J = \int_{m_1}^{m_2} p_{\parallel} ds \tag{1.132}$$

where m_1 and m_2 are the abscissas of the mirror points. As we have seen, the typical period for a 1 keV particle in the Earth's radiation belt is about 1 sec. Even during sudden events, such as the magnetic field reconfigurations associated with substorms, the typical time of evolution of $\boldsymbol{B}$ is on the order of minutes. Therefore, J can be considered an electron adiabatic invariant. For the ions, the bouncing period is on the order of a minute, and J is not necessary an adiabatic invariant for them. But it can still be an adiabatic invariant for ions of much higher energy (a few MeV for the high-energy component of the radiation belts).

1.4.2.3 Third Adiabatic Invariant and the Magnetic Storms

The azimuthal drift of the guiding centers given explicitly in Eq. (1.131) also induces a periodic motion, which can be associated with a third adiabatic invariant. It can be shown that this third invariant is the magnetic flux enclosed by the guiding center during its azimuthal motion,

$$\Phi = \int A \cdot dl \,. \tag{1.133}$$

In the inner magnetosphere of the Earth, for the particles in the range 1–100 keV, the period associated with the azimuthal motion is typically hours, while the variations of the magnetospheric magnetic field are generally faster. The flux Φ is an adiabatic invariant only for the high-energy particles (a few MeV) of the radiation belts.

1.4.3 The Motion of a Particle in a Wave

In most theories of wave collisionless plasma interaction, the motion of charged particles in the wave electromagnetic field must be evaluated. A complete theory of particles in a wave would involve the simultaneous appearance of various concepts. Here, in simplified contexts, we show a few of the most important effects connected with the motion of a particle in a wave.

We start with particles propagating much slower than the wave phase velocity, dealing with the effect of the modulation of the wave amplitude. This effect, as the guiding center theory, is developed in the mathematical context of the multi-timescale asymptotic expansion.

Then the notion of wave trapping is presented. It appears with quasi-monochromatic waves, even in unmagnetized plasmas, and is important for particles with a velocity of the same order of magnitude as the phase velocity.

1.4.3.1 The Nonresonant Particle in a Modulated Sinusoidal Electromagnetic Field

In this section, the concept of a guiding center is left out. The multiscale expansion method is used in a different context, more related to microphysics: the behavior of a particle into the electromagnetic field of a wave. Here, the wave is supposed to have a phase velocity larger than the particle velocity, but its envelope is not uniform. The wave frequency, considered the high frequency, is ω, and we consider that the magnetic field has no uniform component. The electric field is

$$\boldsymbol{E}(\boldsymbol{r}, t) = \boldsymbol{E}_0(\boldsymbol{r}, t)\cos(\omega t - \boldsymbol{k}\cdot\boldsymbol{r}) \tag{1.134}$$

$E_0(\boldsymbol{r}, t)$ varies slowly according to space and time, and the wave envelope propagates slowly. In terms of multiscale expansion, it is straightforwardly written

$$\boldsymbol{E}(\boldsymbol{r}, t) = \boldsymbol{E}_0(\boldsymbol{r}_1, t_1, \ldots)\cos(\omega t_0 - \boldsymbol{k}_1\cdot\boldsymbol{r}_1)\,. \tag{1.135}$$

An example of such a waveform, taken in the solar wind, is shown in Figure 1.8. The Maxwell–Ampère equation can be solved, giving a magnetic field

$$\boldsymbol{B}(\boldsymbol{r}, t) = \boldsymbol{B}_{c,0}(\boldsymbol{r}_1, t_1, \ldots)\cos(\omega t_0 - \boldsymbol{k}_1\cdot\boldsymbol{r}_1) + \boldsymbol{B}_{s,0}(\boldsymbol{r}_1, t_1, \ldots)\sin(\omega t_0 - \boldsymbol{k}_1\cdot\boldsymbol{r}_1) \tag{1.136}$$

where

$$\boldsymbol{B}_{c,0} = \epsilon\frac{\boldsymbol{k}_1\times\boldsymbol{E}_0}{\omega} \quad \text{and} \quad \boldsymbol{B}_{s,0} = -\epsilon\frac{\nabla_1\times\boldsymbol{E}_0}{\omega}\,. \tag{1.137}$$

Then, the equation of motion involving the Lorentz force is written up to the first order. The zero order equation is solved explicitly,

$$\boldsymbol{u}_0 = \frac{q}{m\omega}\boldsymbol{E}_0\sin\phi \quad \text{and} \quad \boldsymbol{r}_0 = -\frac{q}{m\omega^2}\boldsymbol{E}_0\cos\phi\,. \tag{1.138}$$

We have introduced here the notation $\phi = \omega t_0 - \boldsymbol{k}_1\cdot\boldsymbol{r}_1$. The first order equation is averaged, and one finds

$$\partial_{t_1}\boldsymbol{U}_0 = -\frac{q^2}{2m^2\omega}[\boldsymbol{E}_0\cdot\nabla_1\boldsymbol{E}_0 + (\nabla_1\times\boldsymbol{E}_0)\times\boldsymbol{E}_0] = -\frac{q^2}{2m^2\omega}\nabla_1 E_0^2\,. \tag{1.139}$$

Going back to the ordinary variables, the following equation gives the equation of slow motion,

$$d_t\boldsymbol{U} = -\frac{q^2}{2m^2\omega}\nabla E_0^2\,. \tag{1.140}$$

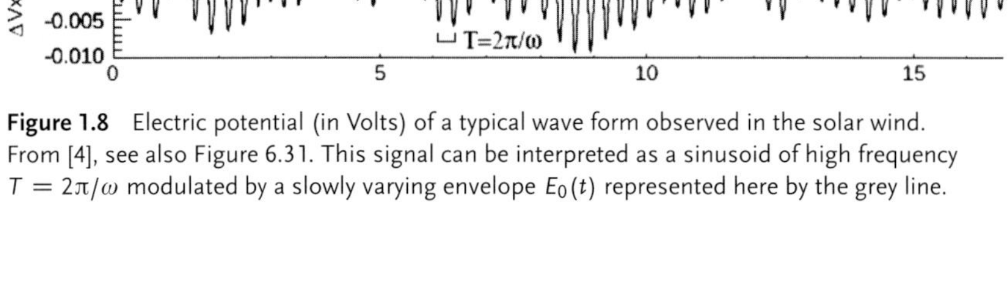

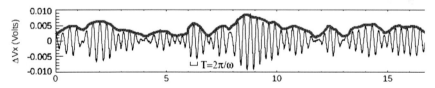

Figure 1.8 Electric potential (in Volts) of a typical wave form observed in the solar wind. From [4], see also Figure 6.31. This signal can be interpreted as a sinusoid of high frequency $T = 2\pi/\omega$ modulated by a slowly varying envelope $E_0(t)$ represented here by the grey line.

It says that the particle is submitted to a force acting on the long run that derives from the modulation of the wave envelope, that pushes the particles in the regions of lower wave energy. This is called the *ponderomotive force*.

1.4.3.2 **The Particle in a Sinusoidal Electrostatic Wave Field**

We consider the simple case of an electrostatic wave, and we compute the motion in the reference frame of the wave (the one that moves at the wave phase velocity). The equation of motion is

$$d_\tau^2 x(\tau) = \frac{q E_0}{m} \sin(k x(\tau)) . \tag{1.141}$$

where E_0 is the wave electric field amplitude, and k is the wave vector. For small wave amplitude E_0, the amplitude of the motion is small too. It is a harmonic oscillation, called bounce motion, of frequency

$$\omega_\mathrm{B} = \left(\frac{q k E_0}{m} \right)^{1/2} . \tag{1.142}$$

In the following development, we do not suppose small oscillation, but the harmonic bounce frequency ω_B is used to parametrize the system.

The equation of motion has a first integral

$$\frac{m}{2} (d_\tau x(\tau))^2 + \frac{q E_0}{k} \cos[k x(\tau)] = W . \tag{1.143}$$

This is the particle energy. This integral is used to characterize the particle trajectories. Two cases must be considered separately, whether $e E_0 < k W$ or $e E_0 > k W$. In the first case, the particle can keep a nonnull velocity whatever its position. Therefore, the particle motion, if perturbed by the wave, keeps a constant direction. The particles with an energy W larger than $e E_0 / k$ are, therefore, *untrapped* by the wave. The other particles, with $e E_0 > k W$, can move through a portion of space limited by $\|k x\| < \arccos (k W / e E_0)$, they are *trapped* by the wave. In the limit $W \sim 0$, the trapped particle velocities oscillate with the frequency ω_B. For larger energies, the bounce frequency is derived through the use of Jacobi elliptical functions.

The motion of untrapped particles is based on Eq. (1.143) and on the new variables $\xi = k x / 2$, $w = k W / e E_0$. For untrapped particles, $w > 1$. In terms of ξ and w, the first invariant equation becomes

$$(d_\tau \xi(\tau))^2 = \alpha^2 (1 - \beta \sin^2 \xi) \tag{1.144}$$

where $\alpha^2 = e E_0 k (1 + w)/2m = \omega_\mathrm{B}^2 (1 + w)/2$ and $\beta = 2/(1 + w) < 1$. It can be integrated,

$$\alpha t = \int_{k x(0)/2}^{k x(t)/2} \frac{d\xi}{\sqrt{1 - \beta \sin^2 \xi}} = F(k x(t)/2, \beta) - F(k x(0)/2, \beta) . \tag{1.145}$$

The integral is the elliptic integral of the first kind (usually met in physics textbooks when dealing with the oscillations of the pendulum). It reciprocal is the Jacobi amplitude am $= F^{-1}(t, \beta)$, and

$$kx(t) = 2\text{am}(\alpha t + F(kx(0)/2, \beta)) \,. \tag{1.146}$$

For trapped particles, a new variable ζ is defined by $\sin\zeta = \beta \sin\xi$. It is valid for a limited domain of values of ξ, compatible with the fact that the particles are trapped and that the values of $\|\xi\|$ are bounded. Then, Eq. (1.144) becomes

$$(d_\tau \zeta(\tau))^2 = \alpha^2 \left(1 - \frac{1}{\beta}\sin^2\zeta\right) \,. \tag{1.147}$$

Its solution is

$$\alpha t = F\left(\arcsin\left(\beta \sin kx(t)\right), \frac{1}{\beta}\right) - F\left(\arcsin\left(\beta \sin kx(0)\right), \frac{1}{\beta}\right) \,. \tag{1.148}$$

We can see that the trapped particles have a nonharmonic motion, and their frequency depends on their amplitude β. As we have seen, for a vanishing amplitude, the frequency is ω_{B}. For larger amplitudes, the frequency is lower. Consequently, the phases characterizing each individual motion mix, hindering coherent collective behavior of the trapped particles. This point, which goes beyond the motion of isolated particles, is discussed more extensively in Section 6.3.2.

2
Plasma Descriptions and Plasma Models

2.1
Distribution Function and Moments

2.1.1
From Individual Particles to Kinetic Description

When discussing the evolution of a gas in fluid mechanics, one is accustomed to describing the variations in space and time of only three variables: its density ρ, its velocity $\boldsymbol{u}$, and its pressure P. In most collisional cases, this description is indeed sufficient and can lead to "closed" theoretical models; whenever these three variables are known in the initial state of the gas, they are fully determined in its further evolution and can, therefore, be predicted via adequate modeling. This is true when the collisions are dominant just because the individual velocities of all particles, in this case, are bound to remain equal, on average, to the same "fluid" velocity. On the contrary, this property cannot be guaranteed in general when the collisions are not dominant. In the total absence of collisions or in a weakly collisional medium, the evolution of the system can depend on the initial microscopic state, that is, on the velocities of each individual particle, even if the three above fluid variables are the same in the initial condition.

In a full description, each particle is characterized by its position in the phase space $\xi = (\boldsymbol{r}, \boldsymbol{v})$, which is a 6D space. Investigating the detailed trajectories of all the individual particles in this space under the elementary electromagnetic binary forces is an extremely cumbersome N-body problem. Instead of solving it, one generally considers much simpler statistical descriptions. The following steps are used to obtain these approximate descriptions:

1. Go from a discrete description to a continuous one by replacing each point-particle by its "particle density". The N particle positions are then described by a single function of N variables, each of them being a 6D vector (therefore, a total of $6N$ scalar variables). The evolution of the system is then described by a differential equation with respect to space, velocity, and time of this function.
2. Introduce a probabilistic assumption based on the hypothesis that the particles are indiscernible. This consists of considering that the density of probability f_N

Collisionless Plasmas in Astrophysics, First Edition. Gérard Belmont, Roland Grappin, Fabrice Mottez, Filippo Pantellini, and Guy Pelletier.

for having any particle at positions $(\xi_1, \xi_2, \ldots, \xi_N)$ in phase space is just the sum of all particle densities corresponding to particle 1 in ξ_1, particle 2 in ξ_2, and so on, added with all the other possible arrangements of the particles in the same position set. The differential equation ruling the function of N variables f_N can be deduced from the preceding and is called the *Liouville* equation.

3. Integrate the preceding result to obtain "reduced" distribution functions, which are functions of only $p < N$ variables, $f_p(\xi_1, \xi_2, \ldots, \xi_p)$ and which are the probabilities for having any particle at each of p positions. The time and space variations of each function f_p is ruled by a differential equation, but this equation always involves a functional of order f_{p+1}. This set of equations, which can be obtained from the Liouville equation, therefore, relates the f_p all together. It is called the BBGKY hierarchy, from the names of the physicists who established it in the 1930s in the spirit of the Boltzmann works: Bogoliubov, Born, Green, Kirkwood, and Yvon (see for instance [5]. These theoretical calculations can be found in many textbooks of statistical physics and we will not repeat them here. We will focus here only on the last equation of this hierarchy, known as the "*kinetic equation*" ruling the variations of the one-point distribution function f_1, which is the probability of finding any particle at one point ξ. We will simply call it hereafter the distribution function and note it $f(\xi)$, or even $f(\boldsymbol{v})$ (the "velocity" distribution function) when no spatial dependence is implied.

The general form of the kinetic equation, as it comes directly from the BBGKY hierarchy, is as follows:

$$\partial_t f + \boldsymbol{v} \cdot \nabla f + \frac{(N-1)}{\mathcal{V}} \int d\xi_2 \boldsymbol{a}_{12} \cdot \nabla_{v1}(f_2) = 0\,. \tag{2.1}$$

Note that the distribution function that is considered is a number density in phase space. The mass density of course derives from it by just multiplying by the mass m of each particle. In this equation, $f_2(\xi_1, \xi_2)$ is the two-point distribution function, and $\nabla_{v1}(f_2)$ its derivative with respect to $\boldsymbol{v}_1$. The acceleration $\boldsymbol{a}_{12} = \boldsymbol{F}_{12}/m$ corresponds to the force $\boldsymbol{F}_{12}$, for instance, the electrostatic one, acting on a particle at position ξ_1 due to the particles located at position ξ_2. $\mathcal{V}$ is the volume of integration in real space.

As shown just above, the kinetic equation is a differential equation ruling the variations of the one-point distribution f and involving an integral of the two-point distribution f_2. As long as this last term is not known, the kinetic equation alone cannot be used for any practical application. Therefore, it needs to be replaced by approximate expressions as a function of f itself. Before any approximation, its expression can be usefully transformed by using:

$$f_2(\xi_1, \xi_2) = f(\xi_1) f(\xi_2) + g_2(\xi_1, \xi_2)\,. \tag{2.2}$$

This decomposition corresponds to the usual notion of conditional probability in statistics: the probability to have particles in (ξ_1, ξ_2) is equal to the product of the probabilities to have one particle in ξ_1 and one in ξ_2 if the two events are independent, but a correction has to be added whenever there are correlations between the two events. $g(\xi_1, \xi_2)$ is, therefore, the term which allows introducing the notion of two-point correlation, that is the notion of collision at the heart of this book. This decomposition allows rewriting the kinetic equation under the following form, which is the most commonly used:

$$\partial_t f + \boldsymbol{v} \cdot \nabla f + \boldsymbol{a} \cdot \nabla_v f = -\frac{(N-1)}{\mathcal{V}} \int d\xi_2 \boldsymbol{a}_{12} \cdot \nabla_{v1}(g_2) \,. \tag{2.3}$$

Under this form, the macroscopic acceleration $\boldsymbol{a}$ is an integration of the two-point acceleration $\boldsymbol{a}_{12}$ on all the positions ξ_2:

$$\boldsymbol{a} = \frac{(N-1)}{\mathcal{V}} \int d\xi_2 \boldsymbol{a}_{12} f(\xi_2) = \langle \boldsymbol{a}_{12} \rangle_{\xi_2} \,. \tag{2.4}$$

The RHS term of Eq. (2.3) is called the "collision term", while the term with $\boldsymbol{a}$ is the "collective term". It is still necessary to express the collision term as a function of f to be able to use the kinetic equation. There is no universal form for this expression and various approximate ones can be used. Depending on the approximation made, the kinetic equation is then named Vlasov, Boltzmann, Fokker–Planck, and so on. The first one is the simplest since it just consists of assuming that the collision term is negligible, that is the medium is collisionless. The Boltzmann term mimics the collisions of hard spheres (which is more suited for neutral gas than for plasmas as it will be shown in Chapter 4), Fokker–Planck assumes the collision term has the form of a diffusion, and so on.

It is worth noting, finally, that there is another route to get the kinetic equation. It allows one to get this equation directly, which concerns the one-point distribution function, without going as above through all the less-reduced distributions, from the Liouville equation and via the BBGKY hierarchy. This approach is due to Klimontovich [6]. If one considers the distribution function f as the number of particles in a small cell of the phase space, this number varies when the particles move in phase space: their velocities make the particles explore the gradients in real space and their accelerations, corresponding to the force field due to all the other particles, make them explore the gradients in velocity space. The difference between "collective" and "collision" fields comes in when computing the effective field viewed by each individual particle. The first part is an average on very large scales of the binary fields, while the "collisions" correspond to the differences from this average.

This gives an equation of the same form as above, even if the collision term does not have the same expression. We will not enter into the mathematical details of the Klimontovich approach here, but all the notions that are presented in the next section follow directly from this approach.

The Maxwellian (or Maxwell–Boltzmann) distribution is of particular importance in kinetic theory, because, as we will see, it is the natural form toward which any distribution tends, under homogeneous and stationary conditions, due to collisions.

It is defined by:

$$f(\boldsymbol{v}) = a e^{-b(v-v_0)^2} . \tag{2.5}$$

The three constants a, b and v_0 can take any value, but they are obviously related to the macroscopic quantities deriving from the three first moments (see next section): density n, fluid velocity u and thermal velocity V_{th}.

$$\begin{aligned} b &= \frac{1}{2V_{\text{th}}^2} \\ a &= \frac{n}{(\sqrt{2\pi}\, V_{\text{th}})^3} \\ v_0 &= u\ . \end{aligned} \tag{2.6}$$

Note that the thermal velocity is often replaced by the "temperature" defined by $T = mV_{\text{th}}^2$. Note also that the distribution of the modulus of the velocity, in the isotropic case, derives from by the preceding by: $f(v) = 4\pi v^2 f(\boldsymbol{v})$.

2.1.2
Kinetic Description and First Order Moments

In plasmas as in ordinary collisional fluids, the physical space can be subdivided into small cells of volume $d^3\boldsymbol{r}$, so that the total number of particles N of the gas is divided in subsets $d^3N = nd^3\boldsymbol{r}$, where the number density n depends on the cell, through its central location $\boldsymbol{r}$. The mass density ρ is just given by $\rho = nm$ whenever all particles have the same mass m. The number of particles in each cell is presumed large enough so that the histogram can be correctly approximated by a continuous function (even if the passage from a discrete to a continuous description is mathematically far from obvious). In each spatial cell, the particles are considered indiscernible and the information about their velocities $\boldsymbol{v}$ is retained only in the form of their statistical density of probability. This can be understood in separating each spatial cell into velocity cells with numbers of particles $d^6N = fd^3\boldsymbol{r}d^3\boldsymbol{v}$ where f is called the "velocity distribution function" (vdf). Once again, taking the limit of infinitely small cells, each of which still contain a large number of particles, the vdf can also be viewed as a continuous function $f(\boldsymbol{v})$ of the velocity. Of course, it is, as the density, also a function of $\boldsymbol{r}$ and t.

How is this kinetic description related to the above fluid description? From its very definitions, it is clear that the number density n is just the integral of the distribution function $f(\boldsymbol{v})$ over the velocity space.

$$n = \int f(\boldsymbol{v}) d^3\boldsymbol{v} . \tag{2.7}$$

It is the zero order moment of $f(\boldsymbol{v})$. The mass density, if all particles of the considered population have the same mass m is of course given by: $\rho = nm$.

The fluid velocity $\boldsymbol{u}$ is just the mean velocity of the particles in each cell:

$$\boldsymbol{u} = \langle \boldsymbol{v} \rangle_v = \frac{1}{n} \int \boldsymbol{v} f(\boldsymbol{v}) d^3\boldsymbol{v} . \tag{2.8}$$

It is, therefore, simply related to the first order moment of $f(\boldsymbol{v})$. Here, and in the following, the notation $\langle\cdot\rangle_v$ represents an average over velocity space, with the $f(\boldsymbol{v})$ weighting. The momentum density of the considered population is given by $\rho\boldsymbol{u}$.

All the higher order moments of $f(\boldsymbol{v})$ can be defined in the same manner, even if, beyond order 3 ("heat flux"), they no longer correspond to usual parameters in fluid mechanics. From the second order moments and beyond, it is more useful to consider the moments centered around the mean velocity $\boldsymbol{u}$ than the simple moments, since it is the natural way of making their definitions independent of the reference frame. It is done by introducing the variable $\delta\boldsymbol{v} = \boldsymbol{v} - \boldsymbol{u}$ instead of $\boldsymbol{v}$. In the fluid frame, where, by definition $\boldsymbol{u} = 0$, this variable is just the particle velocity since one has then: $\delta\boldsymbol{v} = \boldsymbol{v}$.

The centered moment of second order is called the pressure tensor $\boldsymbol{P}$. It is defined as:

$$\boldsymbol{P} = \rho\,\langle\delta\boldsymbol{v}\delta\boldsymbol{v}\rangle_v = m\int \delta\boldsymbol{v}\delta\boldsymbol{v}\, f(\boldsymbol{v})d^3\boldsymbol{v}\,. \tag{2.9}$$

The pressure tensor is a $3 * 3$ tensor since the velocity is a 3D vector. Each component (i, j) can be understood as the mean value of the flux of momentum $m\delta v_i$, transported at the velocity δv_j. What is the meaning of this tensor and how is it related to the usual scalar pressure P known in fluid dynamics?

The scalar pressure can be derived from the general pressure tensor whenever the medium is isotropic. In this case, the distribution function is the same in all directions, which means that $f(\boldsymbol{v})$ depends on $\boldsymbol{v}$ only through its modulus v. In these conditions, the symmetries lead to a pressure tensor of the form $\boldsymbol{P} = P\boldsymbol{I}$, where $\boldsymbol{I}$ is the identity tensor. It, therefore, depends only on *one* scalar parameter P, which is the usual scalar pressure of the fluid dynamics.

In this isotropic case, another usual notation is $P = nmV_{\text{th}}^2$, so defining the thermal velocity V_{th}. It can also be noted that $P = nT$, so defining the parameter T that we will call "temperature" in this book. (It is given here in energy units. Its value in kelvin (K) is obtained by dividing it by the Boltzmann constant k_{B}). It is worth noting, however, that this definition of T differs, in general, from the thermodynamic notion of temperature, which is based on the notion of entropy. But the notion of entropy is mainly meaningful in the strongly collisional limit (see Chapter 4) and its extension to collisionless media would give rise to a notion of temperature which is rarely evoked. For anisotropic media, three thermal velocities and three temperatures can be defined, one for each diagonal term.

An important remark: the density of kinetic energy of the plasma (in the flow frame) is just half the trace of the pressure tensor, since it is:

$$1/2(P_{xx} + P_{yy} + P_{zz}) = \rho/2\left\langle\delta v_x^2 + \delta v_y^2 + \delta v_z^2\right\rangle_v\,. \tag{2.10}$$

In anisotropic media, the form of the pressure tensor can differ from the preceding in two ways: (i) the diagonal terms can be different from one another, and (ii) the nondiagonal terms can be different from zero. The consequences can be important.

Considering for instance a planar geometry $\nabla = \boldsymbol{e}_n\partial_n$, it is well known in isotropic media that the pressure force $\nabla\cdot\boldsymbol{P}$ is always parallel to the direction

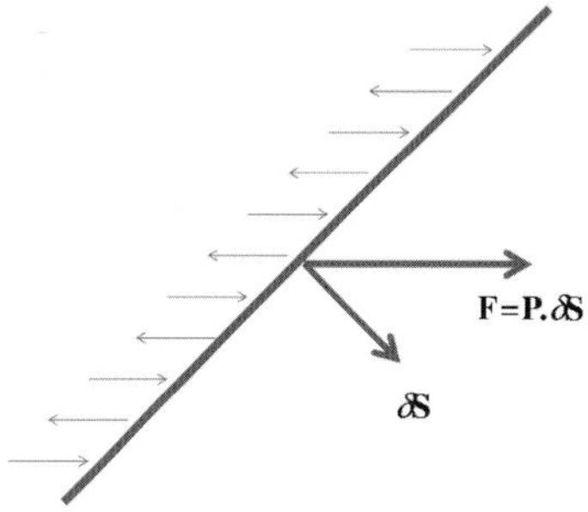

Figure 2.1 An example of anisotropic distribution. If all particles move along the same direction x (and $-x$), the force they exert on the surface is in the x direction. It is, therefore, not perpendicular to the surface.

$\boldsymbol{e}_n$ of the gradient. It can be different in an anisotropic one. This result can be told in another – more intuitive – way: the pressure force due to the particles located on one side of a surface δS is always normal to the surface if the medium is isotropic; it is not when the medium is anisotropic. A simple example of such a nonscalar pressure is given in Figure 2.1. One considers the particles located at a given time on the left side of the virtual surface drawn. It is supposed that, for some reason, all of them move in the same direction x, half with the velocity V_x and half with a velocity $-V_x$. Such a distribution is of course impossible in a collisional medium, but not in a collisionless one. With this "two beams distribution function", one can easily check that both parts of the population transport a positive momentum towards the right in the x direction. The force is, therefore, also always oriented along x, whatever the orientation of the surface. It is, therefore, not perpendicular to the surface in general (except, coincidentally, when this surface is normal to x).

In a magnetized plasma, the magnetic field is an obvious cause of anisotropy. We will see in Chapter 3 that, at large scale, the corresponding pressure tensor is still diagonal, but that its component $P_{\parallel}$ along the magnetic field is in general different from the two perpendicular components $P_{\perp}$, which are equal to each other. This is due to the "gyrotropic" symmetry of the distribution function in these conditions, which respect a cylindrical symmetry around $\boldsymbol{B}$ (related to the particle rotation around $\boldsymbol{B}$) but can depend differently on $v_{\parallel}$ and $v_{\perp}$.

The meaning of the three first moments that have just been defined is simple. If one considers for clarity an isotropic distribution function, that is, where $f(v)$ is a function of only the modulus v (or equivalently a 1D distribution function), the Figure 2.2 makes explicit these meanings with respect to the distribution function f. The density represents the surface under the curve $f(v)$, the fluid velocity u is the mean value of v, and the thermal velocity is the "width" of the curve, so characterizing the thermal spread of the particle velocities. It can be seen that these crude indications do not completely determine the form of $f(v)$. The third moment (see next subsection) is called the heat flux and it characterizes the asymmetry of the distribution function with respect to the mean velocity. Higher-order moments are related to finer and finer details of the distribution.

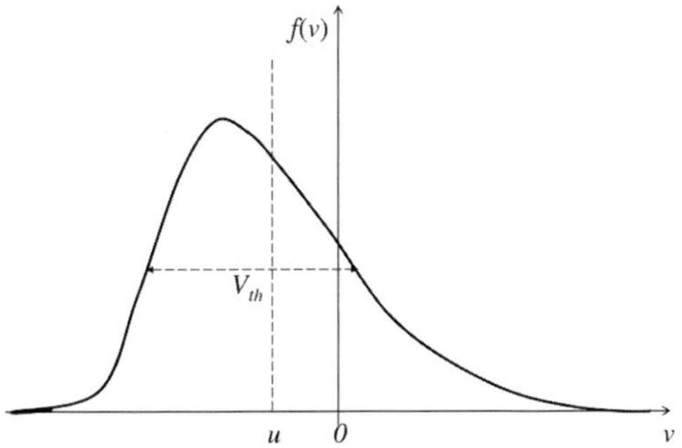

Figure 2.2 1D distribution function. The fluid velocity u and the thermal velocity V_{th}, related respectively to the first and second moment of the distribution function, characterize the mean value and the width of the curve. Higher moments characterize finer details of the distribution shape. The third moment ("heat flux") in particular characterizes the asymmetry.

2.1.3
Higher-Order Moments

All higher moments can be defined in almost the same way as the orders 0, 1, and 2 that we just described. The general definition of the moment of order p is:

$$\mathcal{M}_p = \rho \langle \delta \boldsymbol{v}^p \rangle_v = m \int \delta \boldsymbol{v}^p f(\boldsymbol{v}) d^3 \boldsymbol{v} \,, \tag{2.11}$$

where the power p must be understood as the tensor product of p successive vectors. The moment of order p is, therefore, a tensor of dimension $(3{*}3\ldots{*}3)$, with p times the number 3. These quantities are more and more cumbersome – and less and less used – when p increases. They depend in general on $(p+1)(p+2)/2$ independent parameters. There are not 3^p parameters, because all terms with transpositions such as $\delta v_x \delta v_y^2 \delta v_z^3$ and $\delta v_y^2 \delta v_x \delta v_z^3$ are trivially identical. It is nevertheless a large number above $p = 3$.

In any case, as for the second order tensor, the p order moments actually depend on much less independent parameters whenever particular symmetries can be invoked. If a full isotropy is verified in particular (which can be achieved in a fully collisional equilibrium), this number can easily be estimated:

1. all "odd" components (that is, with odd powers for at least one component, for instance, $\delta v_x \delta v_y^2 \delta v_z^3$) are zero;
2. many of the nonnull components are equal to each other (for instance, $\delta v_x^2 \delta v_z^4$ and $\delta v_x^4 \delta v_y^2$).

This leads to only $(p + 2)(p + 4)/24$ scalar parameters in this isotropic case. These results are general, for any distribution function f. If f is known, for instance, Maxwellian, all components of all moments can actually be expressed as functions of only *one* parameter, which is the thermal velocity V_{th}.

The "heat flux" tensor $\boldsymbol{Q}$ is the third order moment and characterizes, as outlined just above, the asymmetry of the distribution function with respect to the mean velocity value. It is a (3*3*3) tensor and so depends, in general geometry and general distribution functions, on 10 independent parameters. But reduced information can often be sufficient for special applications, which deserve some comments:

1. When considering energy transport (see Section 2.1.2), the full heat flux tensor is not needed, but only its "half-trace" with respect to the two last subscripts. For this reason, one is led to define the "heat flux vector" $\boldsymbol{q}$, which derives from the full heat flux tensor $\boldsymbol{Q}$ in the following way:

$$\begin{aligned} q_x &= 1/2(Q_{xxx} + Q_{xyy} + Q_{xzz}) \\ q_y &= 1/2(Q_{yxx} + Q_{yyy} + Q_{yzz}) \\ q_z &= 1/2(Q_{zxx} + Q_{zyy} + Q_{zzz}) \,. \end{aligned} \tag{2.12}$$

From the basic definition of the third moment $\boldsymbol{Q}$, it appears that each component of this vector represents nothing but the flux of kinetic energy (in the flow frame) $m/2(\delta v_x^2 + \delta v_y^2 + \delta v_z^2)$ in the direction of $\delta \boldsymbol{v}$.
2. When there is a gyrotropic symmetry, which is verified at large scales in a magnetized plasma, the distribution becomes symmetric with respect to the two directions perpendicular to the magnetic field, x and y, and the tensor simplifies. It then depends on only two independent parameters, $Q_{\parallel}$ and $Q_{\perp}$, and its only nonnull components are these ones:

$$\begin{aligned} Q_{zzz} &= q_{\parallel} \\ Q_{xxz} = Q_{xzx} = Q_{zxx} &= Q_{\perp} \\ Q_{yyz} = Q_{yzy} = Q_{zyy} &= Q_{\perp} \,. \end{aligned} \tag{2.13}$$

One can check that, with this symmetry, the heat flux vector is:

$$\boldsymbol{q} = \begin{pmatrix} 0 \\ 0 \\ Q_{\parallel} + 2Q_{\perp} \end{pmatrix} . \tag{2.14}$$

This shows that, in agreement with the gyrotropic hypothesis, the kinetic energy is transported only in the direction parallel to the magnetic field. It involves however both components $Q_{\parallel}$ and $Q_{\perp}$ of the heat flux tensor. This emphasizes that the subscripts $\parallel$ and $\perp$ just refer, for $\boldsymbol{Q}$, to the nature of the kinetic energy transported, related to the parallel or perpendicular components of the velocity, but not to the direction of the transport, always parallel.

2.1.4
Moments for a Mixture of Populations

On some occasions, a given population has to be considered as a mixture of several subpopulations, each with different characteristics. It is quite usual in ordinary gases; the air, for instance, can be considered as a mixture of several components:

nitrogen, oxygen, and so on, and the total pressure is usually said to be the sum of the partial pressures of the different components. In a plasma, beyond these trivial questions of composition (components with different masses), it can often also be useful to separate the global population of a given species into different subpopulations with different distribution functions. Are the moments additive in general as for the above notion of partial pressures in a collisional gas? The answer is "no"; all moments are, by definition, additive when they are not centered, but the centered moments defined above are not. Therefore, beyond order 1, it is important to keep in mind that, due to the expansion of the $(\delta \boldsymbol{v})^n$ terms, the additive quantities are not $\boldsymbol{P}$, $\boldsymbol{Q}$, and so on, but:

$$\boldsymbol{P}^* = \rho \langle (\delta \boldsymbol{v} + \boldsymbol{u})(\delta \boldsymbol{v} + \boldsymbol{u}) \rangle_v = \boldsymbol{P} + \rho \boldsymbol{u}\boldsymbol{u}$$

$$\boldsymbol{Q}^* = \rho \langle (\delta \boldsymbol{v} + \boldsymbol{u})(\delta \boldsymbol{v} + \boldsymbol{u})(\delta \boldsymbol{v} + \boldsymbol{u}) \rangle_v = \boldsymbol{Q} + 3\{\boldsymbol{u}\boldsymbol{P}\}^\circ + \rho \boldsymbol{u}\boldsymbol{u}\boldsymbol{u}$$

and so on. The notation $\{-\}^\circ$ is explained in the Appendix A.1.

Why are these differences not usually considered in collisional media? The reason is that when different populations are in strong collisional interaction, they all always have approximately the same fluid velocity $\boldsymbol{u}$, so that the centering has no effect: the centered moments are additive as the noncentered ones.

2.1.5
Nontrivial Generalization of the Fluid Concepts

The most common intuitions about fluids come from the collisional ones. For this reason, the very notion of a fluid behavior is generally believed to be essentially linked to the existence of collisions. It is even often thought that a Maxwellian distribution (which is a consequence of collisions) is necessary to understand the behavior of the fluids and that it is a hidden hypothesis behind any fluid modeling. We will see that the moment equations, from which all the fluid equations derive except one (the "closure" equation) are fully general and independent of the existence of collisions, even when the physical understanding seems a priori counterintuitive.

- The notion of density is clearly independent of collisions.
- The notion of fluid velocity as defined from the first moment, is independent as well. Nevertheless, there is a difference with the usual intuition. In any case, the fluid velocity can be defined as a space average over all the particles in a small volume. In a collisional fluid, it can also be viewed as the time-averaged velocity of any particle while, in a collisionless one, it cannot. The difference can easily be understood by an image: let us imagine a wide highway. Whatever the density of the traffic, the flow rate of the highway in a given slice can always be calculated from a mean velocity which is a space average over the vehicles in the slice. If the traffic is very dense, close to traffic congestion, all vehicles will accelerate and brake at different times, but all will move, in a time average, with approximately the same velocity since each vehicle cannot pass the vehicle

just before. If the traffic is very far from saturation, on the contrary, each vehicle can go with its own mean velocity, some faster, others slower, than the space average.

- The notion of pressure is generally understood as related to the collisions of the particles of a gas on the walls of a container. The above definition is much more general. The pressure is the transfer of momentum across a surface, virtual or not. If we consider a virtual surface located in $x = 0$, with a normal along the x axis, the P_{xx} component is due to the particles crossing the surface in both senses, the $x < 0$ particles with a $v_x > 0$ velocity and the $x > 0$ particles with a $v_x < 0$ velocity. Both categories contribute positively to the pressure. If there is a container with a wall at the same place as the previous surface, only the $x < 0$ particles participate, but considering that they are specularly reflected on the wall, the part of the particle distribution with $v_x < 0$ is exactly the symmetric of the $v_x > 0$ part. In the collisional hypothesis usual for ordinary gases, the distributions are indeed always symmetrical and the two results are identical.

From all the previous examples, one can understand that the fluid macroscopic quantities can be defined in a quite general way from the moments of the distribution function, and that these definitions do not need any collision to be meaningful. Furthermore, one will see in the next section that these quantities, with their general definition, respect "moment equations", which are identical to – or generalizations of – the usual fluid equations (transport of mass, momentum, energy in particular).

Let us finally show on a simplistic example that, even if the definitions are general and the fluid equations verified (except one, refer to Section 2.2.3), the physics underlying the fluid behavior can be counterintuitive with respect to the common intuition, which is based on the collisional case. In Figure 2.3, we take particles in only two thick slices going in opposite directions. In Figure 2.3a, the particles have a density n_0, a velocity V_0 and a negligible thermal spread. In Figure 2.3c, the particles also have a density n_0 and a negligible thermal spread but their velocity is $-V_0$. The density is 0 everywhere else. Let us assume that there is no interaction at all between the particles, neither collective nor collisional. What happens at the time when the two slices meet? In the region in common (Figure 2.3b), the density rises from n_0 to $2n_0$, the velocity falls from $\pm V_0$ to 0 and the pressure rises from 0 to $P_{xx} = 2n_0 V_0^2$. In ordinary fluid mechanics, such a transformation of the kinetic energy from "convective" to "thermal" is always indicative of a nontrivial physical process, generally dissipative. Here, it is just the effect of a linear superposition, without any possibility of invoking such microscopic phenomena since there is no particle interaction. This underlies the importance, in the description, of centering the moments around the mean velocity. This may seem arbitrary when one is accustomed to the collisional case, but it is important to realize that the definitions from the moments are the only ones that are fully generalizable, even in a collisionless case. Other examples will be given hereafter in the Section 9.1 (toy models).

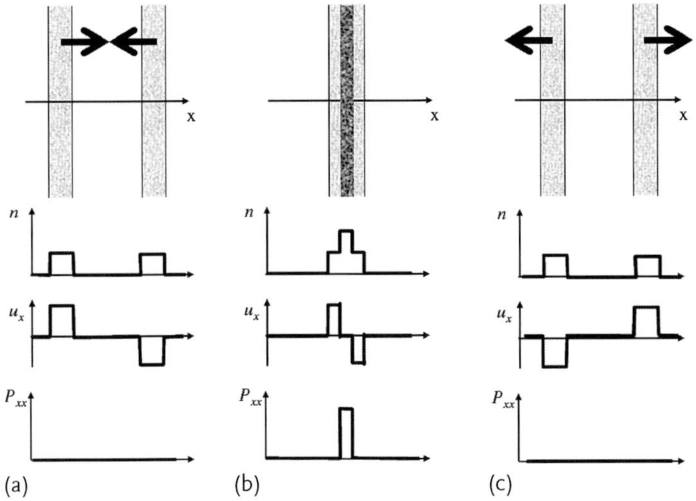

Figure 2.3 (a) When these two noninteracting clouds of particles meet and superpose, the pressure (second centered moment of the distribution function) increases suddenly, while the velocity (first moment) decreases. (b) This situation is quite different from the collisional media in which the collisions force the distribution function to return everywhere very quickly to Maxwellian (instead of remaining two-peaked as here in the superposition). (c) The transfer from convective to thermal energy occurs here in a perfectly reversible way, which is also quite different from the collisional case.

2.1.6 Fluid vs. Kinetic Description: An Example

Figure 2.4 shows experimental data from the Cluster mission (ESA). The electron distribution function $f(v_{\|})$, parallel to the magnetic field, exhibits a very remarkable form (generally considered to be unlikely, because of its instability), with two "beams" of opposite velocities. The spacecraft was in the Earth's magnetotail (see Chapter 1) during a perturbed period, such as those associated with magnetic substorms and auroras. The beams correspond to particles accelerated during the period and guided by the magnetic field.

If one has to model the evolution of such phenomena, one can think of using fluid theories and, therefore, describe the plasma only by its few moments ρ, $\boldsymbol{u}$ and P. If this is feasible, it would mean that the evolution of these three moments is the same as the evolution of the moments of a plasma with any different distribution, for instance, Maxwellian, provided only that these distributions share initially the same density, the same mean velocity (close to zero here) and the same thermal velocity (of the order of the velocity difference between the beams). It seems quite improbable and it is actually not strictly possible. Nevertheless, we will see that most of the equations ruling the evolution are conservation laws, fully independent of the form of the distribution function. This is generally sufficient to prevent the evolutions to strongly diverge one from the other. We will show in the next section that the whole fluid system is indeed fully independent of the distribution

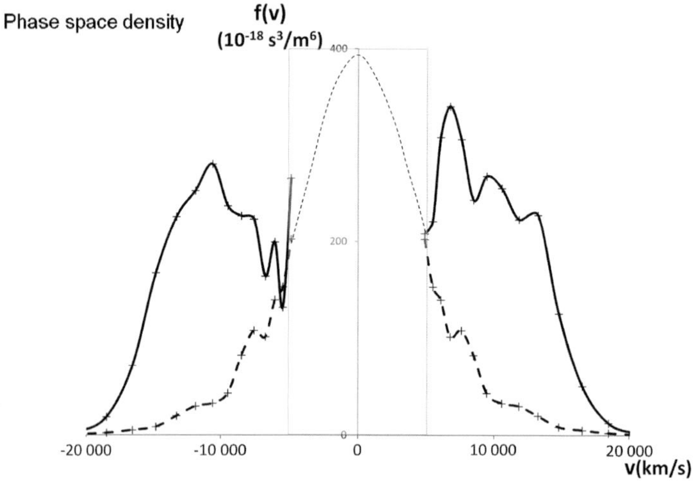

Figure 2.4 Electron distribution function in the Earth's environment, as measured by the PEACE experiment on board the ESA's Cluster mission spacecraft. The cut perpendicular to the magnetic field (dashed line) is here close to a Maxwellian, but parallel to the magnetic field (solid line), two additional electron beams with opposite velocities can be observed. These accelerated particles are bouncing between two mirror points of the terrestrial magnetic field. The shaded central part is not accessible to measurements and the Maxwellian shown with a light line is just indicative. The evolution of such a plasma can be different from the evolution of a Maxwellian plasma with the same three first moments: ρ, $\boldsymbol{u}$ and P, whereas most of the equations ruling this evolution are identical. Courtesy of D. Fontaine.

function, except one equation: the so-called closure equation, which is an approximate equation and through which all the differences arise. This closure equation is, therefore, of crucial importance in any fluid modeling.

2.2 From Kinetic to Fluid Equations

2.2.1 Moment Equations

All the moments defined above have their space and time variations related to each other by the so-called moment equations, from which the "fluid equations" derive (both are even identical for the first ones). We already know that the variations of the distribution function are ruled by the kinetic equation. As seen in Eq. (2.11) giving the definition of the moments, it is easy to derive from the kinetic equation all the equations of moment; one has to multiply the kinetic equation by $m\delta\boldsymbol{v}^p$ (or $m\boldsymbol{v}^p$ for orders 0 and 1) and integrate it with respect to velocity. Let us recall that the power p corresponds to a tensorial product.

For deriving the moment equations, one can start from the kinetic equation under the following form, where the expression of the collision term C is not specified:

$$\partial_t f + \boldsymbol{v} \cdot \nabla f + \boldsymbol{F}/m \cdot \nabla_v f = C \,. \tag{2.15}$$

Thanks to some adequate approximations, C is supposed to depend on $f(\boldsymbol{v})$ via the so-called collision operator (integral operator), which we will not discuss in this book since we are mainly interested by the collisionless limit. Let us, however, notice that it is homogeneous to f divided by a time. It is important to recall that this kinetic equation is written for any population of particles, with a single mass m and a single expression for the force $\boldsymbol{F}$, that is a single charge q if the electric force is considered. If several particle species are considered, one kinetic equation must be written for each of them. Beyond the collective force $\boldsymbol{F}$, the different populations can be coupled to each other via the collision term C. For the moments of order 0 and 1, the following results can be drawn:

$$\partial_t \rho + \nabla \cdot (\rho \boldsymbol{u}) = n \left\langle \nabla_v \cdot \boldsymbol{F} \right\rangle_v + \mathcal{C}_0 \tag{2.16}$$

$$\partial_t (\rho \boldsymbol{u}) + \nabla \cdot (\rho \boldsymbol{u}\boldsymbol{u} + \boldsymbol{P}) = n \left\langle \nabla_v \cdot (\boldsymbol{F}\boldsymbol{v}) \right\rangle_v + \boldsymbol{\mathcal{C}}_1 \,. \tag{2.17}$$

In each of these two equations, the first term of the RHS corresponds to the collective force, and the second one to the collision term. In the form given the first one has been expressed via an integration by parts, and the notation $\langle\cdot\rangle_v$ indicates an average over velocity space (weighted by the distribution function). The notation $\mathcal{C}_0$ stands for $m \int C d^3\boldsymbol{v}$. It represents the effect of the collision operator C for the evolution of the number of particles. In practice, as long as no chemistry is involved in the processes, the number of particles of a given population is constant and this integral cancels, due to the symmetries of the operator C. We will remain in this hypothesis afterward. The notation $\boldsymbol{\mathcal{C}}_1$ stands for $m \int C \boldsymbol{v} d^3\boldsymbol{v}$. It represents the effect of the collision operator C for the evolution of the global momentum. The symmetries also make this term cancel as long as only collisions internal to the given population are considered since the global population cannot exert forces on itself. We will, however, keep it since the collisions with the other populations can indeed exert such a force (see for instance the collisions of ions on electrons, which bring up the electron–ion friction called "resistivity").

In the most common hypotheses, these equations can be retained in the simplified form:

$$\partial_t \rho + \nabla \cdot (\rho \boldsymbol{u}) = 0 \tag{2.18}$$

$$\partial_t (\rho \boldsymbol{u}) + \nabla \cdot (\rho \boldsymbol{u}\boldsymbol{u} + \boldsymbol{P}) = n \boldsymbol{F}_0 + \boldsymbol{\mathcal{C}}_1 \,. \tag{2.19}$$

These two equations can be identified as the usual equations of fluid dynamics: continuity equation (conservation of the mass) and Euler equation (conservation of momentum). The simplifications that lead to them mainly derive from a hypothesis on the forces existing in the system; all of them have been supposed to verify $\nabla_v \cdot$

$F = 0$. This is actually justified for all the usual forces; it is obviously verified for all the forces independent of $\boldsymbol{v}$ (electrostatic, gravitational, and so on) and it can be checked to be also verified for the Lorentz force, which depends on $\boldsymbol{v}$ via $\boldsymbol{v} \times \boldsymbol{B}$. We restrict ourselves, here and in the following, to these two kinds of forces and write:

$$\boldsymbol{F} = \boldsymbol{F}_0 + q\delta\boldsymbol{v} \times \boldsymbol{B} \,. \tag{2.20}$$

In this notation, the mean part $q\boldsymbol{u} \times \boldsymbol{B}$ of the Lorentz force has been included in the term F_0 independent of the particle velocity.

For the orders p larger than or equal to 2, the same calculation can be done, except that $\delta\boldsymbol{v}$ must be used (centered moments) instead of $\boldsymbol{v}$. This makes the calculation less simple. One has to do the change of variable from $f(\boldsymbol{v}, \boldsymbol{r}, t)$ to $g(\delta\boldsymbol{v}, \boldsymbol{r}, t)$ before multiplying by $\delta\boldsymbol{v}^p$ and integrating (recalling that $\boldsymbol{u}$ depends on $\boldsymbol{r}$ and t). This leads to use this new form of the kinetic equation, better suited to the calculation of the centered moments:

$$\partial_t g + \boldsymbol{u} \cdot \nabla g + \nabla \cdot (\delta\boldsymbol{v} g) + [F/m - D_t \boldsymbol{u} - \delta\boldsymbol{v} \cdot \nabla \boldsymbol{u}] \cdot \nabla_v g = C \,. \tag{2.21}$$

The general equations for the centered moments of order $p \geq 2$ can be derived from it:

$$\begin{aligned} &\partial_t \boldsymbol{\mathcal{M}}_p + \nabla \cdot (\boldsymbol{u}\boldsymbol{\mathcal{M}}_p + \boldsymbol{\mathcal{M}}_{p+1}) - \nabla \cdot \boldsymbol{u}\boldsymbol{\mathcal{M}}_p + \rho D_t \boldsymbol{u} \cdot \langle \nabla_v (\delta\boldsymbol{v}^p) \rangle_v \\ &\quad + \rho\nabla \cdot \left\langle \nabla_v (\delta\boldsymbol{v}^{p+1}) \right\rangle_v \cdot (\boldsymbol{u}) = n \langle \boldsymbol{F} \cdot \nabla_v (\delta\boldsymbol{v}^p) \rangle_v + \boldsymbol{C}_p \,. \end{aligned} \tag{2.22}$$

The parenthesis around $\boldsymbol{u}$ before the equal sign means that the spatial derivative ∇ applies to $\boldsymbol{u}$ and not to the averaged quantity appearing between both (this position is necessary for the vector and tensor products). As previously, the notation $\boldsymbol{C}_p$ stands for $m \int C\delta\boldsymbol{v}^p d^3\boldsymbol{v}$ and represents the effect of the collision operator C for the evolution of the moment of order p. In the form given, the term $\nabla_v \cdot (\boldsymbol{F}\delta\boldsymbol{v}^p)$ has also been replaced by $\boldsymbol{F} \cdot \nabla_v (\delta\boldsymbol{v}^p)$, which corresponds to the already mentioned hypothesis $\nabla_v \cdot (\boldsymbol{F}) = 0$. The derivatives in velocity space can be written more explicitly by using the formulas:

$$\boldsymbol{A} \cdot \nabla_v (\delta\boldsymbol{v}^p) = \sum \delta\boldsymbol{v}^q \boldsymbol{A} \delta\boldsymbol{v}^{p-1-q} = \left\{\boldsymbol{A}\delta\boldsymbol{v}^{p-1}\right\}^\circ \,. \tag{2.23}$$

$\boldsymbol{A}$ can be any vector and the notation $\circ$ in the exponent denotes a sum on all the cyclic permutations of the indices, as defined in Appendix A.1. It can be checked that the resulting tensor is a symmetric one. Using the above formula, the general moment equation can be rewritten:

$$\begin{aligned} &\partial_t \boldsymbol{\mathcal{M}}_p + \nabla \cdot (\boldsymbol{u}\boldsymbol{\mathcal{M}}_p + \boldsymbol{\mathcal{M}}_{p+1}) + \left\{D_t \boldsymbol{u}\boldsymbol{\mathcal{M}}_{p-1}\right\}^\circ + \nabla \cdot \left\{\boldsymbol{\mathcal{M}}_p (\boldsymbol{u})\right\}^\circ \\ &\quad - \nabla \cdot \boldsymbol{u}\boldsymbol{\mathcal{M}}_p = \left\{\boldsymbol{F}_0/m\boldsymbol{\mathcal{M}}_{p-1} - \boldsymbol{\omega}_c \times \boldsymbol{\mathcal{M}}_p\right\}^\circ + \boldsymbol{C}_p \,. \end{aligned} \tag{2.24}$$

One can observe that the Lorentz force brings up the gyropulsation $\boldsymbol{\omega}_c = q\boldsymbol{B}/m$ in the result. Using Eq. (2.19), it can be observed also that the two terms $D_t \boldsymbol{u}$ and $\boldsymbol{F}_0$ can conveniently be grouped and replaced by a single term involving $\nabla \cdot \boldsymbol{P}$. We let the reader make this transformation when necessary.

At all orders, the collision operator C gives rise, generally speaking, to a collision term possibly acting on the evolution of the concerned moment. Its precise influence cannot be estimated without specifying the expression of C. Let us recall, however, that all these terms are zero in collisionless plasmas. Furthermore, some components of the moment equations are actually conservation equations, so that we know in advance that, due to symmetries of C, the corresponding integrals are zero, even in collisional plasmas. It is the case for the order 0 (conservation of the number of particles) and for order 1 (conservation of momentum) if one considers only the collisions internal to the considered population. (Other populations can actually exert a non-null collisional force on it; see the friction between electrons and ions giving rise to the resistivity). We will see in an upcoming chapter that some energy conservations also impose symmetries of C that lead to the cancelation of some collision terms at order 2.

The above general form of the moment equation is greatly simplified when the equation is restricted to the pure 1D case (for both space and velocity spaces) and in the absence of a magnetic field. In this case, the different dot products (contractions over two indices) are just ordinary products between scalars, all the tensors obtained by cyclic permutations of the indices are identical and their sum just amounts to a multiplication by the order p. The following result follows:

$$\begin{aligned} &\partial_t \mathcal{M}_p + \partial_x (u\mathcal{M}_p + \mathcal{M}_{p+1}) + p\left(D_t u \mathcal{M}_{p-1} + \mathcal{M}_p \partial_x u\right) \\ &= p\, F_0/m \mathcal{M}_{p-1} + \mathcal{C}_p \,. \end{aligned} \tag{2.25}$$

As a summary of this part, let us finally come back to the full magnetized and 3D case and put together explicitly the moment equations for orders 0 to 3 (full 3D), specifying the electromagnetic nature of the force $\boldsymbol{F}_0 = q(\boldsymbol{E} + \boldsymbol{u} \times \boldsymbol{B})$:

$$\partial_t \rho + \nabla \cdot (\rho \boldsymbol{u}) = 0 \tag{2.26}$$

$$q\partial_t(\rho \boldsymbol{u}) + \nabla \cdot (\rho \boldsymbol{u}\boldsymbol{u} + \boldsymbol{P}) = nq(\boldsymbol{E} + \boldsymbol{u} \times \boldsymbol{B}) + \boldsymbol{C}_1 \tag{2.27}$$

$$\partial_t \boldsymbol{P} + \nabla \cdot (\boldsymbol{u}\boldsymbol{P} + \boldsymbol{Q}) + \nabla \cdot \left[\boldsymbol{P}(\boldsymbol{u}) + (\boldsymbol{P}(\boldsymbol{u}))^T\right] = -\{\boldsymbol{\omega}_c \times \boldsymbol{P}\}^\circ + \boldsymbol{C}_2 \tag{2.28}$$

$$\begin{aligned} &\partial_t \boldsymbol{Q} + \nabla \cdot (\boldsymbol{u}\boldsymbol{Q} + \boldsymbol{R}) + \{D_t \boldsymbol{u} \boldsymbol{P}\}^\circ + \nabla \cdot \{\boldsymbol{Q}(\boldsymbol{u})\}^\circ - \nabla \cdot \boldsymbol{u}\boldsymbol{Q} \\ &= \{q/m(\boldsymbol{E} + \boldsymbol{u} \times \boldsymbol{B})\boldsymbol{P} - \boldsymbol{\omega}_c \times \boldsymbol{Q}\}^\circ + \boldsymbol{C}_3 \end{aligned} \tag{2.29}$$

...

$$\begin{aligned} &\partial_t \boldsymbol{\mathcal{M}}_p + \nabla \cdot (\boldsymbol{u}\boldsymbol{\mathcal{M}}_p + \boldsymbol{\mathcal{M}}_{p+1}) + \left\{D_t \boldsymbol{u} \boldsymbol{\mathcal{M}}_{p-1}\right\}^\circ + \nabla \cdot \left\{\boldsymbol{\mathcal{M}}_p(\boldsymbol{u})\right\}^\circ \\ &- \nabla \cdot \boldsymbol{u}\boldsymbol{\mathcal{M}}_p = \left\{q/m(\boldsymbol{E} + \boldsymbol{u} \times \boldsymbol{B})\boldsymbol{\mathcal{M}}_{p-1} - \boldsymbol{\omega}_c \times \boldsymbol{\mathcal{M}}_p\right\}^\circ + \boldsymbol{C}_p \,. \end{aligned} \tag{2.30}$$

In Eq. (2.28) giving the temporal evolution of the pressure tensor, the terms involving cyclic permutations $\nabla \cdot \left\{\boldsymbol{\mathcal{M}}_p(\boldsymbol{u})\right\}^\circ - \nabla \cdot \boldsymbol{u}\boldsymbol{\mathcal{M}}_p$ have been replaced by a sum of the tensor $\nabla \cdot \boldsymbol{P}(\boldsymbol{u})$ and its transposed since, in this case of an order 2 tensor, this expression is simpler.

As can be seen in the above equations, each moment equation relates the variations of the order p moment to those of the $p+1$ moment. This means that this infinite system cannot be closed at any finite order p without further information. A "fluid system" closed at order p contains $p+1$ variables, the moments from order 0 to order p. It must, therefore, contain $p+1$ equations: p moment equations from 0 to $p-1$, plus one "closure equation" relating the moment of order p to the moments of smaller order. This last "closure equation" derives from a well established expansion procedure in the strongly collisional case but it has to be justified, when it is possible, by different hypotheses in the collisionless or weakly collisional cases. It is important to recall that, in any case, the moment equations are "exact" equations as long as the kinetic equation itself can be considered as "exact". In a fluid system, only the "closure equation" depends on approximations that are to be discussed.

As an example, let us consider the 1D collisional case. The fluid description is then usually closed at order 3, which means that it contains four variables (n, u, P, Q) and that it contains three exact equations (continuity, momentum, energy); in addition, the heat flux is given by an approximate law such as the "Fourier law" (Eq. (2.38)). The same system can also sometimes be closed at order 2, the pressure being then expressed by some approximate "equation of state".

2.2.2 Lagrangian Form of the Moment Equations

For some uses, it is more convenient to put the system of moment equations under a Lagrangian form rather than under the above Eulerian form. This means that the operator ∂_t, which represents the temporal variation at a fixed position in space, can be conveniently replaced by the operator $D_t = \partial_t + \boldsymbol{u}\cdot\boldsymbol{\nabla}$, which represents the temporal variation when following the flow at its velocity $\boldsymbol{u}$. Using the continuity equation (2.26), the passage from the Eulerian form to the Lagrangian one (and vice versa) for any equation can be done via the formulas, valid for any quantity $\boldsymbol{A}$, scalar, vector or tensor:

$$\partial_t(\boldsymbol{A}) + \boldsymbol{\nabla}\cdot(\boldsymbol{u}\boldsymbol{A}) = \rho D_t(\boldsymbol{A}/\rho) \tag{2.31}$$

$$\rho D_t(\boldsymbol{A}) = \partial_t(\rho\boldsymbol{A}) + \boldsymbol{\nabla}\cdot(\rho\boldsymbol{u}\boldsymbol{A})\,. \tag{2.32}$$

The set of moment equations can, therefore, be put under the Lagrangian form:

$$D_t\rho + \rho\boldsymbol{\nabla}\cdot\boldsymbol{u} = 0 \tag{2.33}$$

$$\rho D_t\boldsymbol{u} + \boldsymbol{\nabla}\cdot\boldsymbol{P} = nq(\boldsymbol{E} + \boldsymbol{u}\times\boldsymbol{B}) + \boldsymbol{C}_1 \tag{2.34}$$

$$\rho D_t(\boldsymbol{P}/\rho) + \boldsymbol{\nabla}\cdot\boldsymbol{Q} + \boldsymbol{\nabla}\cdot\left[\boldsymbol{P}(\boldsymbol{u}) + (\boldsymbol{P}(\boldsymbol{u}))^T\right] = -\{\boldsymbol{\omega}_c\times\boldsymbol{P}\}^\circ + \boldsymbol{C}_2 \tag{2.35}$$

$$\begin{aligned}\rho D_t(\boldsymbol{Q}/\rho) + \boldsymbol{\nabla}\cdot\boldsymbol{R} + \{D_t\boldsymbol{u}\boldsymbol{P}\}^\circ + \boldsymbol{\nabla}\cdot\{\boldsymbol{Q}(\boldsymbol{u})\}^\circ - \boldsymbol{\nabla}\cdot\boldsymbol{u}\boldsymbol{Q}\\ = \{q/m(\boldsymbol{E} + \boldsymbol{u}\times\boldsymbol{B})\boldsymbol{P} - \boldsymbol{\omega}_c\times\boldsymbol{Q}\}^\circ + \boldsymbol{C}_3\end{aligned} \tag{2.36}$$

...

$$\rho D_t(\mathcal{M}_p/\rho) + \nabla \cdot \mathcal{M}_{p+1} + \left\{D_t \boldsymbol{u} \mathcal{M}_{p-1}\right\}^{\circ} + \nabla \cdot \left\{\mathcal{M}_p(\boldsymbol{u})\right\}^{\circ}$$
$$- \nabla \cdot \boldsymbol{u} \mathcal{M}_p = \left\{q/m(E + \boldsymbol{u} \times B)\mathcal{M}_{p-1} - \omega_c \times \mathcal{M}_p\right\}^{\circ} + \mathcal{C}_p \,. \qquad (2.37)$$

2.2.3
Fluid Equations: Necessity of a Closure Equation

In a fluid model, the plasma properties are described by the few first moments of each population, up to a maximum order p_{max}, which is often taken to be less than or equal to 3. It then involves at a maximum of four moments: mass density ρ, fluid velocity $\boldsymbol{u}$, pressure $\boldsymbol{P}$, and heat flux $\boldsymbol{Q}$. The moment of order 4 $\boldsymbol{R}$ is very rarely used in fluid descriptions and moments of higher orders almost never.

However, as seen in the above derivation, the moment equations actually form an infinite set, each relating the temporal evolution of the moment of order p to the spatial variations of the moment of order $p + 1$ and to the force fields. It is, therefore, impossible to get a closed system of fluid equations without introducing a "closure equation", which allows cutting this infinite system by giving an expression of the last moment $\mathcal{M}_{p_{max}}$ independent of the higher order moments. This equation cannot be a moment equation and must come from different physical hypotheses or approximations. For the system to be closed, it is also necessary that the force field can be expressed as a function of the p_{max} first moments, but this can generally be done without difficulty. The question of a closure equation is a much more difficult one.

A fluid model, therefore, always contains three different kinds of equations:

- A finite set of moment equations (for each population)
- A closure equation (for each population)
- A set of equations describing the evolution of the force fields.

The closure equation is a priori the only approximate equation of a fluid system since all the moment equations are "exact", that is, as exact as the kinetic equation from which they derive. Nevertheless, it must be noted that this is generally not completely true; most often, people actually don't use the full moment equations under their general tensorial form, but some simplified versions of them, based on symmetry hypotheses. As already emphasized, the full pressure tensor $\boldsymbol{P}$ can, for instance, be replaced by a scalar pressure P when the distribution function is isotropic. In strongly collisional media, as shown in the next section, one is so led to split the tensor into two parts: an isotropic one, described by a scalar pressure, and a slight departure from it, called the "viscosity tensor". The so-called Chapman–Enskog procedure provides a rigorous way of calculating this departure. It provides, for instance, the well-known form to the Navier–Stokes equation used in neutral fluids. In weakly collisional plasmas, such simplifications can of course not be used without care.

When there is no simple geometrical approximation available, one has to go back to the full tensorial form of the moment equations, which is complicated but feasible since the "exact" full equations are known. Nevertheless, the most basic limitation of any fluid theory remains, which concerns the existence and the nature of the closure equation. Several circumstances do justify simple closure equations:

1. *Strongly collisional case.* When the Knudsen number K_n (see Chapter 4) is small enough, that is, when the phenomena under study have scales large enough with respect to the mean free path, the same "Chapman Enskog" procedure also allows calculating a closure equation valid at first order in the K_n expansion. In a nonmagnetized plasma, this leads to the well-known "Fourier law":

$$q = -\kappa \nabla T \; . \tag{2.38}$$

   The coefficient κ is called "thermal conduction", and T is the temperature.
2. *Fast propagating perturbations.* For small perturbations propagating sufficiently faster than the thermal velocity of the population ($\omega / k V_{\text{th}} \gg 1$ if the characteristic temporal and spatial scales are noted ω^{-1} and k^{-1}), it can be shown that this population behaves adiabatically, meaning that its heat flux can be neglected everywhere and at any time (at the condition that the perturbation propagates in an equilibrium which itself does not involves any heat flux).
   The condition $\boldsymbol{Q} = 0$ can then be taken as a closure equation. Imposing this condition in the second order moment equation allows one to obtain the adiabatic equation as a function of the pressure and the density. If assuming a full isotropic pressure, the pressure tensor is characterized by a single scalar parameter P and the closure equation is:

$$D_t \left(P/\rho^{5/3}\right) = 0 \; . \tag{2.39}$$

   In the magnetized case, assume a gyrotropic symmetry is less restrictive and generally more realistic than assuming a full isotropy. The pressure tensor is then characterized by two parameters $p_{\parallel}$ and $p_{\perp}$. This leads to the so-called double-adiabatic closure equations, also known as "CGL" laws, from the names of their inventors (Chew, Goldberger, and Low, see [7]):

$$D_t (p_{\perp}/\rho B) = 0 \; . \tag{2.40}$$

$$D_t \left(p_{\parallel} p_{\perp}^2/\rho^5\right) = 0 \; . \tag{2.41}$$

3. *Slow propagating perturbations.* In the opposite hypothesis to the previous one ($\omega / k V_{\text{th}} \gg 1$), that is, for perturbations propagating much slower than the thermal velocity, the behavior of the population is not adiabatic, meaning that the heat flux is not negligible. In the fully isotropic hypothesis, it leads to an isothermal behavior:

$$\nabla(p/\rho) = 0 \; . \tag{2.42}$$

More generally, it can be shown to lead, in an anisotropic plasma, to:

$$\partial_z \left[T_{\parallel}\right] = 0 \tag{2.43}$$

$$\partial_z \left[\left(\frac{1}{T_\perp} - \frac{1}{T_\parallel} \right) B \right] = 0 \tag{2.44}$$

$$\partial_z \left[\frac{T_\perp}{\rho} \right] = 0 \,. \tag{2.45}$$

This shows that the anisotropy decreases along the field lines when B increases and vice versa. Various other closures can be proposed, which can be valid in various conditions (see examples in Chapter 9). Some of them simply aim at mimicking empirically the observed variations via simple relations, in particular, for the case of gyrotropic symmetries, the so-called polytropic closures:

$$D_t \left(p_\perp / \rho^{\gamma_\perp} \right) = 0 \tag{2.46}$$

$$D_t \left(p_\parallel / \rho^{\gamma_\parallel} \right) = 0 \,. \tag{2.47}$$

The values of $\gamma_\parallel$ and $\gamma_\parallel$ are then simply chosen in order to provide the best fit for a given phenomenon. The "toy models" presented in Section 9.1 will provide other examples of such empirical models.

Whenever the closure hypothesis can be based on a physical hypothesis, as in the three first examples above, it is worth understanding that it always comes in through arguments concerning the form of the distribution function perturbations that are allowed. For the fast perturbations in particular, the result can be understood going back to the Vlasov equation. It shows that, in this case, the particles are perturbed almost identically whatever their velocities so that the distribution function is almost displaced as a whole, the main change concerning the first moment $\delta \boldsymbol{u}$. The perturbations of all the other moments are small and almost no asymmetry can be brought: one can prove that, in this case, $\delta Q \ll P \delta u$, which guarantees that the heat flux perturbation can safely be neglected.

The crucial question of finding a relevant closure equation will find an exemplary illustration in Chapter 5, when comparing the fluid and kinetic treatments of the Langmuir wave.

2.2.4
Collisional Limit: Fluid Dynamics and Thermodynamics

The usual fluid dynamics, as well as the thermodynamics, with its well-known "state functions" have a limited scope of action, which is the "strongly collisional" case. This limit can be rigorously quantized thanks to the Knudsen number (see Chapter 4) and the application of the general results to this particular case can be done in the frame of the Chapman–Enskog procedure (already mentioned in Section 2.2.3). Thermodynamics, due to this strong collisionality hypothesis, is a theory of "quasi-equilibrium", the distribution functions being always close to Maxwellian. It is, therefore, not the topic of central interest for this book. Nevertheless, it is much more familiar to most physicists and can always be taken as a reference. The relations between collisions, entropy and damping in this context are approached in more details in Chapter 5.

The notions presented in this book aim at being "general", in the sense that they can be used as well in collisionless and in collisional limits. They are most often identical to those used in fluid dynamics and thermodynamics (which are a priori restricted to the collisional limit), but there are several differences that can be stressed:

- *Moments*
  The moments are defined in the same way in both domains, with one exception: what is called "pressure" in fluid dynamics and thermodynamics is not the full second order tensor that we call a "pressure tensor" here; it corresponds only to the scalar that characterizes its isotropic part $Tr(\boldsymbol{P})/3$. The full tensor is then usually called a "stress tensor", and the difference between this full tensor and its isotropic part is called the "viscosity tensor".
- *Temperature*
  In this book, the temperature is defined as $T = P/\rho$ (in energy units). It must, therefore, be considered just a convenient notation, which does not contain any additional information with respect to the moments ρ and P themselves. It is not a priori identical, in general, with the thermodynamical temperature $T = \partial U/\partial S$, which is based on the notions of internal energy and of entropy, and which is, therefore, a way of quantizing the role of the "disorderly" part of the energy. As these notions of "order" or "disorder" are not pertinent in the absence of collisions, we will avoid using them here.
  It is worth noting, however, that the two notions of temperature coincide in the case of an "ideal" gas, in which the definition $T = P/\rho$ can be considered an "equation of state". It is not just by chance. The condition of validity of the ideal equation of state is that the gas can be presumed rarefied enough so that the energy of interaction between the elements remains negligible with respect to the global kinetic energy of these elements. The elements can be protons or electrons as in a plasma; they can also be molecules in real gases. This condition is not in contradiction to the fact that the collisions are still presumed numerous enough to maintain the quasi-equilibrium state as the most probable one. The potential energy of interaction is always quite negligible with respect to the kinetic energy except in the close collisions, with impact parameters of the order of the Landau length (see Chapter 4), which are always extremely rare and don't impact the total energy at all.
- *State functions*
  In our context where the whole energy is the kinetic energy of the elementary particles (electrons and ions), the equations of energy can be deduced as shown above. The energy density and the energy flux appearing are:
  - *internal energy:* $U = 3/2P$ (no molecules and, therefore, no vibration–rotation energies)
  - *enthalpy:* $w = 5/2Pu$ (can be noted h in many contexts).

  The notion of entropy, and the consecutive notions of free energy and free enthalpy, can be defined as well, but we will not do it here because, if their defining formulas are relatively simple, their meaning and their use is much

less. The arguments about entropy will be developed in Chapter 4. However, as the entropy mainly aims at quantizing the probability of a state under the assumption of equiprobable microstates, we can already notice that this notion is obviously more suited for collisional plasmas than for collisionless ones. For a given density and energy, the collisions are supposed to make the system scan all the possible microstates, making one macrostate much more probable than the others. In the absence of collisions, any macrostate can remain stationary whatever its probability in terms of number of possible microstates.

2.3 Numerical Methods

Numerical simulation codes are based on a specific representation of a plasma defined through a choice of *fundamental data* that fully describes the plasma. These data are completed by a choice of equations governing its evolution, and of numerical methods to solve the equations.

The simulated plasma is supposed to evolve in a geometrical space (configuration-space or phase-space) whose geometry is described by a grid containing a finite (ideally very large) number of grid cells. A grid is constituted of points. We note $\boldsymbol{X}_j$ the points of a configuration-space grid, and $(\boldsymbol{X}_{j,k}, \boldsymbol{V}_{j,k})$ the points of a phase-space grid. The indices j and k can be made of 1, 2, or 3 integer numbers, according to the dimensionality of the code.

A numerical simulation involves the description of initial conditions (time $t = t_0$) and boundary conditions. Then, the simulation consists of a succession of time steps of duration Δt performed in order to compute the state of the plasma at times $t_n = t_0 + n\Delta t$.

In a regular grid, the points are regularly spaced. The space between the points of a grid corresponds to a distance (for a 1D grid), a surface (2D) or a volume (3D and more) that is called the grid cell.

2.3.1 Vlasov Codes

In a Vlasov code, the plasma is represented by the distribution functions f_s of all the particle species s, and the electromagnetic field. A distribution function is stored on the points of a grid that represents the phase space. It can be written as a function $f_s(\boldsymbol{X}_j, \boldsymbol{V}_k)$, where j, k is the index of the phase-space grid point, j for the position and k for the velocity (or impulsion). The electromagnetic field is stored on a subgrid that simply represents the configuration space. For instance, the electric field is $\boldsymbol{E}(\boldsymbol{X}_j)$ where j is the index of the configuration-space grid point.

For a 1D space with velocities (or impulsion) with one component, the corresponding phase space has two dimensions, and the configuration grid has one dimension. In the 3D general case, with the three components of the velocity/impulsion, the grid representing the phase space has six dimensions. Using computers, computing on a 2D phase space grid is very easy. Computing in a 6D space is still problematic. Therefore, Vlasov codes are seldom used for real 3D simulations.

A time step cycle comprises the following computations. Let us start the computation at time t_{n-1}.

Knowing the electromagnetic field on the grid, solve the Vlasov equation to iterate the value of f_s. Because the Vlasov equation can be written in the conservative form

$$f_s(\boldsymbol{x}(t_n), \boldsymbol{v}(t_n), t_n) = f_s(\boldsymbol{x}(t_0), \boldsymbol{v}(t_0), t_0) \tag{2.48}$$

a solution consists of propagating the value $f_s(\boldsymbol{X}_j, \boldsymbol{V}_k)$ for each couple j, k along the path that a particle of initial position $(\boldsymbol{X}_j, \boldsymbol{V}_k)$ in the phase space would follow. This stage of the computation, therefore, consists of solving the equation of motion many times. For instance, when the Vlasov code takes the full particle dynamics into account, the Lorentz force Eq. (1.108) is solved.

This path after a time Δt ends at a position $\boldsymbol{x}_{j,k}(t_n), \boldsymbol{v}_{j,k}(t_n)$, and the transported (invariant) value $f_s(\boldsymbol{X}_j, \boldsymbol{V}_k)$ is interpolated from this point toward the grid. The interpolation scheme can be characterized by an assignment function S and for any phase-space grid point j', k',

$$f_s(\boldsymbol{X}_{j'}, \boldsymbol{V}_{k'}) = \sum_i \sum_j S(\boldsymbol{x}_{j,k}(t_n) - \boldsymbol{X}_{j'}, \boldsymbol{v}_{j,k}(t_n) - \boldsymbol{V}_{k'}) f_s(\boldsymbol{X}_j, \boldsymbol{V}_k) . \tag{2.49}$$

If the assignment function S has a finite and small support (as is often the case), the number of assignment computations is much less than the number of phase-space grid points.

When the iteration of f_s is done at every phase-space grid point, the charge and current density on the configuration-space grid can be computed by a direct sum. For instance, the contribution of a species s of distribution f_s to the current density is

$$J(\boldsymbol{X}_j) = \sum_s \sum_k q_s \boldsymbol{V}_k f_s(\boldsymbol{X}_j, \boldsymbol{V}_k) . \tag{2.50}$$

Then the Maxwell equation is solved on the grid, and the following iteration can start.

2.3.1.1 Odds and Drawback of Vlasov Codes

Because they can associate phase-space grid cells for velocities, for which f_s is very small compared to its average value, Vlasov codes are appropriate to study phenomena, for which the tail of the distribution function plays an important role, as is the case with many plasma instabilities such as the "gentle bump in tail" instability (see Chapter 6).

If a Vlasov simulation is initiated with a uniform plasma, then the plasma stays strictly uniform, without any fluctuation. There is no noise. Consequently, Vlasov codes allow for a good control of the fluctuations level because the noise has to be introduced explicitly by the programmer in the initial conditions.

But, as we said, phase spaces have a lot of dimensions (up to six), and to remain within the fast storage capabilities of computers, it is necessary to work with a limited resolution in the phase space. This discourages simulations in the real six dimension phase space. But even with less dimensions, this limitation is problematic. It happens that in collisionless plasmas, the distribution in phase space generally becomes highly "filamented". This means that the iso-contours of f_s tend to develop very long and very thin filaments, as time goes by (see Figures 5.20 and 5.23 for instance). In these circumstances, there comes a time, not so large, when the size of the filaments is smaller than the resolution of the phase-space grid. Then the distribution f_s is artificially smoothed, and the simulation stops to give correct physical solutions. Practically, a reliable Vlasov simulation is stopped before the phase-space grid resolution limit is attained.

2.3.2
Particle in Cell Codes (PIC)

The plasma can be represented as an assembly of N particles P interacting through electromagnetic forces. The force on the particle P_i would result from its binary interaction with the $N - 1$ other particles. This is, the principle of N body simulations.

The fundamental data consists only of the positions $\boldsymbol{x}_i$ and velocities $\boldsymbol{v}_i$ of the N particles.

But this algorithm is costly because N^2 binary interaction terms must be computed at each time step, and they require the computation of all the interparticle distances. This algorithm is also very demanding when two particles are close to each other. In that case, as seen with Figure 4.2, the binary interactions are very intense over short time intervals. Taking them into account would require, at least for particles at close distances to each other, a very short time step. Moreover, because we deal with noncollisional plasmas, we are not interested by this kind of interaction.

Therefore, a method to compute the electromagnetic interaction that does not take the binary interactions into account in a "clean" way is necessary. The Particle In Cell (PIC) algorithm provides a way of doing that.

In terms of particles, the electromagnetic field $\boldsymbol{E}$, $\boldsymbol{B}$ associated with the collective interaction is computed on a grid (grid points at position $\boldsymbol{X}_j$), and the force applied to the particles simply results from the interpolation of this collective field at the position $\boldsymbol{x}_i$ of the particle. Then, two neighboring particles interact with almost the same electromagnetic field, and their binary interaction is completely neglected.

With a PIC code, the fundamental data is constituted of the N particles positions $\boldsymbol{x}_i$ and velocities $\boldsymbol{v}_i$ and the electromagnetic field $\boldsymbol{E}_j$, $\boldsymbol{B}_j$ on the G points of the grid. It is also useful (but not necessary to keep it all the time) to store the charge and current densities on each grid point.

The computation of the force requires NG operations, where N is the number of particles, and G the number of grid points. Because a grid cell contains a lot of particles ($N/G \gg 1$), this computation is much faster than the $O(N^2)$ computation of the N body problem.

Therefore, with PIC codes, the electromagnetic field is computed on a grid, and the particles move under the influence of this electromagnetic field. The plasma evolution is computed following the "plasma loop" scheme illustrated in Figure 1.1.

The electric field applied to a particle i of position $\boldsymbol{x}_i$ results from a B-spline interpolation from the grid associated with an assignment function S, as a sum performed over the grid cells j,

$$\boldsymbol{E}(\boldsymbol{x}_i) = \sum_j \boldsymbol{E}(\boldsymbol{x}_j) S(\boldsymbol{x}_i - \boldsymbol{X}_j) \,. \tag{2.51}$$

The magnetic field is interpolated in the same way.

With this local electromagnetic field, the Lorentz force acting on the particle is computed, and integrated in order to iterate the particle velocity and position. When the new position $\boldsymbol{x}_i$ and the velocity $\boldsymbol{v}_i$ of all the particles are computed, the charge density and the current density are interpolated onto the grid cells with the same assignment function S as before. (Otherwise, self-forces cause nonphysical oscillations of the simulated plasma). For a plasma with various species s, the charge density integrated over the grid cell j at position $\boldsymbol{X}_j$ results from a sum performed over the particles i

$$\varrho(\boldsymbol{X}_j) = \sum_s q_s \sum_i S(\boldsymbol{x}_i - \boldsymbol{X}_j) \,. \tag{2.52}$$

A similar formula is used for the assignment of the electric current density.

The Maxwell equations can now be solved. They are used to compute the electromagnetic field on each grid point. Generally, PIC codes are based on a regular grid (rectangular or cylindrical), and a finite difference method is used. Sometimes ("electrostatic codes") only the Maxwell–Gauss equation is solved (and the magnetic field, considered as constant, is not iterated).

Then, the iteration ends, after a series of numerical computations for diagnostics, and the next iteration can be done.

2.3.2.1 **Time Explicit PIC Scheme**

Here is a classical example of a time integration scheme used for the PIC simulations. Let us drop the index i characterizing the particle number, and let us note with an index n that the data is taken at time t_n. The equations of motion are

$$\boldsymbol{v}_{n+1/2} = \boldsymbol{v}_{n-1/2} + \Delta t \left[\frac{e}{m} \boldsymbol{E}_n + \frac{e}{2mc} (\boldsymbol{v}_{n+1/2} + \boldsymbol{v}_{n-1/2}) \times \boldsymbol{B}_n(\boldsymbol{x}_n) \right] \tag{2.53}$$

$$\boldsymbol{x}_{n+1} = \boldsymbol{x}_n + \Delta t \boldsymbol{v}_{n+1/2} \,. \tag{2.54}$$

We can notice that the position and the velocity are defined at different times, separated by an interval $\Delta t/2$. This time lag combined with the above equations allow for a second order scheme (the error depends only on $(\Delta t)^2$)) called “leap-frog”. When these equations are solved, the interpolation from the particle to the grid is done with Eq. (2.52) for the charge density ϱ_{n+1} and a similar one for the current density $\boldsymbol{J}_{n+1/2}$. Most often S is a piecewise linear function of the particle position relatively to the closest grid points (two points in 1D, four points in 2D, eight points in 3D). Then the Maxwell equations are solved. In order to improve the order of precision of the computations of the differential operators, the various components of the electromagnetic field and of the source terms are defined on slightly different position on the simulation grid. This is somehow the geometrical transposition of the “leap frog” method. The time marching equations are

$$\boldsymbol{E}_{n+1} - \boldsymbol{E}_n = c\Delta t \nabla \times \boldsymbol{B}_{n+1/2} - \mu_0 \Delta t \boldsymbol{J}_{n+1/2} \tag{2.55}$$

$$\boldsymbol{B}_{n+1/2} - \boldsymbol{B}_{n-1/2} = -c\Delta t \nabla \times \boldsymbol{E}_n \,. \tag{2.56}$$

The electric field at time t_{n+1} is directly given as

$$\boldsymbol{E}_{n+1} = \boldsymbol{E}_n + c\Delta t \nabla \times \boldsymbol{B}_{n-1/2} - c^2 \Delta t^2 \nabla \times \nabla \times \boldsymbol{E}_n - \mu_0 \boldsymbol{J}_{n+1} \,. \tag{2.57}$$

The magnetic field is deduced from it.

The problem with this algorithm is that it keeps absolutely all the physics, including the propagation of the light waves that have a too high frequency to interact with the plasma.

The smallest scale in a collisionless plasmas is typically the Debye length}?> λ_D. This implies that in an explicit code, every process involving the scale λ_D must be computed. Therefore, the sides of the grid cells must be equal or smaller, $\Delta x \leq \lambda_\mathrm{D}$. Because the particles typically propagate at the thermal velocity v_t, and because they participate in the charge density at every place along their path, the distance that they cover during a time step must be smaller than the length of a grid cell Δx. Therefore, we must ensure that

$$v_\mathrm{t} \Delta t < \Delta x \leq \lambda_\mathrm{D} \,. \tag{2.58}$$

Let us consider a light wave. Its phase velocity is c. It means that a perturbation of the electromagnetic field following this wave mode propagates at the velocity c. If the time resolution allows one to follow the evolution of that perturbation, and if the spatial resolution does not permit it, the explicit solver tends to provide nonphysical and possibly divergent solutions. Therefore, it is required that

$$c\Delta t \ll \Delta x \,. \tag{2.59}$$

These conditions are particular cases of the Courant–Friedrichs–Lewy condition (CFL condition), that arise with explicit time marching algorithms expressed with finite differences.

The CFL conditions can be very limiting constraints, and it is very useful to avoid some of them with the use of implicit time marching algorithms.

2.3.2.2 Time Implicit PIC Schemes

When simulating an electromagnetic plasma with an explicit particle in cell (PIC) code, several time scales must be taken into account. The aim of an implicit code is to solve the equations of evolution of a dynamical system without keeping the high-frequency fluctuations. Actually, implicitness must be treated with care; there is no universal theory on the convergence and stability of implicit schemes. In plasma physics, it can be useful to know exactly the nature of the high-frequency modes we get rid of. Then it is possible to get rid of the constraints imposed by the CFL conditions associated to the propagation of these modes.

With the direct implicit method, one can introduce a low pass filtering that damps the frequencies for which $\omega\Delta t \geq 1$.

For instance, to get rid of light waves, it is only needed to modify the Maxwell–Faraday equation. Instead of Eq. (2.56), one solves

$$\boldsymbol{B}_{n+1/2} - \boldsymbol{B}_{n-1/2} = -\Delta t \nabla \times \bar{\boldsymbol{E}}_n \quad \text{where} \quad \bar{\boldsymbol{E}}_n = \frac{1}{2}\left[\boldsymbol{E}_{n+1} + \bar{\boldsymbol{E}}_{n-1}\right] . \tag{2.60}$$

This method can be used as well to describe the motion of a particle guiding center using the slow time Eq. (1.124). The fast time oscillation is damped through the resolution of the following scheme

$$\begin{aligned} \boldsymbol{v}_{n+1/2} = \boldsymbol{v}_{n-1/2} + \Delta t \Big[& -\frac{e}{m}\boldsymbol{E}_n - \frac{\mu}{m}\nabla B_n(\boldsymbol{x}_n) \\ & - \frac{e}{2mc}(\boldsymbol{v}_{n+1/2} + \bar{\boldsymbol{v}}_{n-1/2}) \times \boldsymbol{B}_n(\boldsymbol{x}_n) \Big] \end{aligned} \tag{2.61}$$

where

$$\bar{\boldsymbol{v}}_{n+1/2} = \frac{1}{2}\left[\boldsymbol{v}_{n+1/2} + \bar{\boldsymbol{v}}_{n-3/2}\right] . \tag{2.62}$$

Other implicit schemes can be used. For instance time decentered schemes are made implicit by the time-discretization method, in which the intermediate time level is slightly decentered [8]. In the implicit moment method [9], the first moment equations are used jointly with particles to solve the electric field equation.

If the electric field is straightforwardly given in explicit codes, with implicit schemes, it is often combined with a linear operator. Therefore, a fairly large linear system must be solved.

2.3.2.3 Odds and Drawback of PIC Codes

PIC codes require less storage capabilities than Vlasov codes. They are not limited by the resolution in the velocity space, since there is no grid associated with that space. But, because the number of particles in a PIC code is much lower than in a real plasma, the artificially high level of thermal fluctuations reduces the signal-to-noise ratio. Therefore, PIC codes are not appropriate (unless a huge number of particles is used) for the simulation of low noise plasma, and of low amplitude phenomenons.

2.3.3 Perturbative PIC Codes

In collisionless plasma physics, the perturbative (δf) particle in cell (PIC) codes are an improvement of the usual PIC codes. They allow a low numerical noise by simulating only the electromagnetic field due to the perturbation (δf) of an equilibrium distribution f_0 of particles. This method permits one to reach a better signal-to-noise ratio, as long as the perturbed distribution δf remains small in comparison of the equilibrium distribution .

The distribution function f for each species which evolves in time according to the Vlasov equation:

$$D_t f = \partial_t f + \boldsymbol{v} \partial_x f + \boldsymbol{\gamma}(\boldsymbol{x}, \boldsymbol{v}) \partial_v f = 0 \tag{2.63}$$

where $f = f(t, \boldsymbol{x}, \boldsymbol{v})$ and $\gamma(\boldsymbol{x}, \boldsymbol{v}) = q/m(\boldsymbol{E} + \boldsymbol{v} \times \boldsymbol{B})$ is the acceleration due to the electric $\boldsymbol{E}$ and magnetic $\boldsymbol{B}$ fields. But it could be any other collective force compatible with the Vlasov equation. If we assume that the distribution f differs from an equilibrium distribution f_0 (corresponding to the $\boldsymbol{E}_0$ and $\boldsymbol{B}_0$ equilibrium fields, and $d_t \delta f_0 = 0$) by a weak perturbation δf, the Vlasov equation can be expressed as a time derivative of the perturbation only. For the equilibrium part:

$$\partial_t f_0 + \boldsymbol{v} \partial_x f_0 + \boldsymbol{\gamma}(\boldsymbol{x}, \boldsymbol{v}) \partial_v f_0 = \boldsymbol{\gamma}_1 \partial_v f_0 \tag{2.64}$$

where the acceleration γ is split in a stationary acceleration due to the equilibrium fields γ_0 and an acceleration γ_1 for the perturbed fields $\boldsymbol{E} - \boldsymbol{E}_0$ and $\boldsymbol{B} - \boldsymbol{B}_0$. Because the equilibrium function f_0 is in equilibrium with γ_0, it does not have to be recomputed at each time step. The evolution of the perturbation is given by:

$$D_t \delta f = -\boldsymbol{\gamma}_1 \partial_v f_0 \,. \tag{2.65}$$

The f distribution is the distribution of the particles for a given species.

The g macroparticle distribution function is:

$$g(\boldsymbol{x}, \boldsymbol{v}) = \sum_i \mathcal{S}(\boldsymbol{x}, \boldsymbol{x}_i, \boldsymbol{v}, \boldsymbol{v}_i) = \sum_i S(\boldsymbol{x} - \boldsymbol{x}_i)\delta(\boldsymbol{v} - \boldsymbol{v}_i) \tag{2.66}$$

where $\boldsymbol{x}_i$ and $\boldsymbol{v}_i$ are the position and the velocity of the ith macroparticle, and $\mathcal{S}(\boldsymbol{x}, \boldsymbol{x}_i, \boldsymbol{v}, \boldsymbol{v}_i) = S(\boldsymbol{x} - \boldsymbol{x}_i)\delta(\boldsymbol{v} - \boldsymbol{v}_i)$ is the interpolation function, and

$$\int S(\boldsymbol{x} - \boldsymbol{x}_i) d\boldsymbol{x} = 1 \,. \tag{2.67}$$

A statistical weight w_i is defined for each particle:

$$w_i = \frac{\delta f}{g}(\boldsymbol{x}_i, \boldsymbol{v}_i) \,. \tag{2.68}$$

For a Klimontovich distribution of the macroparticles (where S is a Dirac distribution) the evolution of the g distribution is a solution of the Vlasov equation

(Eq. (2.63)), thus the f/g ratio remains constant. It can be shown that with other forms of the interpolation function, this property can also be assumed as a correct approximation. Then the evolution of the weight is deduced from the perturbation evolution (Eq. (2.65)):

$$D_t w_i = -\frac{\gamma_1}{g}\partial_v f_0 = -\left(\frac{f}{g} - w_i\right)\frac{\gamma_1}{f_0}\partial_v f_0 \, . \tag{2.69}$$

The weighting of the particles is used when computing the charge and current densities. It can be deduced from Eq. (2.68) that when S is a Dirac distribution, $\sum_i w_i$ is a constant. With other forms of S, used in practice, this is approximately a constant, as is the total energy. (The variations of $\sum_i w_i$ are displayed in [10]. The numerical computation of this error can be used to improve the diagnostics of particles kinetic energy.)

An appropriate choice of the initial values of g and w_i for the simulation of a specific phenomenon can permit a very low signal-to-noise ratio. This ratio remains good as long as the amplitude of the perturbation δf remains low. Therefore, δf PIC codes are pertinent to the study of instabilities in their linear and weakly nonlinear phases. As an application in this book, the simulations of Chapter 5 have been performed with a perturbative PIC code.

2.4 Fluid Codes

We have described PIC codes, in which the trajectories of macroparticles are followed as precisely as possible, while the electric and magnetic fields are integrated on a fixed grid, using the fluid velocity and density. In these PIC codes, the macroscopic parameters, density, fluid velocity and pressure, are obtained by averaging the information that has been obtained by integrating the macroparticle trajectories. In contrast, the fluid codes allow obtaining directly the macroscopic parameters, so bypassing the microscopic step of trajectory integration. They consist of integrating the fluid system of equations, such as MHD, MHD–Hall, or bifluid, thus obtaining the first fluid moments (density, velocity, pressure) on a fixed grid. Of course, because particle trajectories don't have to be followed in detail (in particular the gyration around the magnetic field), the computational cost of these fluid codes is in general smaller than the PIC code with comparable accuracy and Reynolds. Also, at a fixed resolution (that is, a fixed number of grid points to describe the fields), a fluid simulation will in general be less noisy than a comparable PIC simulation, due to the fact that the fluid moments in the PIC code are computed on a limited number of macroparticles. This difference is the price to pay when using a PIC code, to encompass a larger number of physical effects.

2.5 Hybrid Codes

When the scales under studies are of the order of – or smaller than – the ion scales (inertial length or Larmor radius), a kinetic description must be used for the ions. Nevertheless, when these scales remain large enough with respect to the electron ones, the electron description can be simplified, allowing to save much computing time. The electrons can indeed be supposed in these conditions to ensure quasi-neutrality, so that their role only appears through the Ohm's law they provide through their momentum equation. In such a hybrid approach, PIC for ions and fluid for electrons, the electron description is fully determined by the choice of a fluid closure equation, most often chosen as polytropic. Other kinds of hybrid codes also exist (for instance fluid for ions and kinetic for electrons), to be applied in other circumstances, but they are less commonly used.

3
The Magnetized Plasmas

3.1
Ideal MHD

When the magnetic field has to be taken into account, the ideal magnetohydrodynamics (MHD) equation set is the simplest one that can be used to model the plasma variations. It is a mono-fluid system (no electron–ion distinction). It aims to model the large-scale variations, temporal and spatial (with respect to the microscopic scales such as the gyroperiods, the Larmor radii or the inertial lengths). Even for these scales, some of the approximations made for establishing it can be violated. As this model, and its numerous variants, is very widely used in astrophysics as in laboratory plasmas, it is of pivotal importance to know the conditions that must be met to make the results reliable for all these different applications.

In this perspective, we will first present the ideal MHD equation set and comment on some of its main properties. Then, we will go back to the more general plasma equations that have been presented in the previous chapters and show how the MHD system can be derived from them, and what the approximations made for this derivation are. This will allow us to understand what the limitations and the precautions for use are, especially when the medium is weakly collisional or fully collisionless. Beyond the validity domain of the ideal MHD, some extensions can still be called "nonideal" MHD models, but some others request more radical changes.

As for the moment equations already presented, we will see once again that, if the strongly collisional case is the only one for which a general and rigorous demonstration can be given, the applications of the MHD are much larger than this condition would suggest. Let us recall in particular that many astrophysical plasmas can be considered collisionless.

3.1.1
The Ideal MHD System

The ideal MHD system is the following:

Collisionless Plasmas in Astrophysics, First Edition. Gérard Belmont, Roland Grappin, Fabrice Mottez, Filippo Pantellini, and Guy Pelletier.

$$D_t\rho + \rho\nabla \cdot \boldsymbol{u} = 0 \tag{3.1}$$

$$\rho D_t \boldsymbol{u} + \nabla P = \boldsymbol{j} \times \boldsymbol{B} \tag{3.2}$$

$$\rho^\gamma D_t(P/\rho^\gamma) + (\gamma - 1)\nabla \cdot \boldsymbol{q} = 0 \tag{3.3}$$

$$\partial_t \boldsymbol{B} + \nabla \times \boldsymbol{E} = 0 \tag{3.4}$$

$$\nabla \cdot \boldsymbol{B} = 0 \tag{3.5}$$

where the vector variables $\boldsymbol{j}$ (current density) and $\boldsymbol{E}$ (electric field) are intermediate variables, which can be expressed as functions of the basic ones, $\boldsymbol{u}$ and $\boldsymbol{B}$, by:

$$\boldsymbol{j} = \nabla \times \boldsymbol{B}/\mu_0 \tag{3.6}$$

$$\boldsymbol{E} = -\boldsymbol{u} \times \boldsymbol{B} \,. \tag{3.7}$$

In the energy equation (3.3), one has a priori $\gamma = 5/3$. To keep some generality, the heat flux vector $\boldsymbol{q}$, appearing in this equation, has not been specified. It should be given by some "closure equation". It can be a nonideal term, but not necessarily. The simplest hypothesis of course just corresponds to the adiabaticone: $\boldsymbol{q} = 0$. In this case, this equation just becomes $D_t(P/\rho^\gamma) = 0$. The same polytropic form can also sometimes be kept with $\gamma \neq 5/3$ to close the MHD system in nonadiabatic situations. Such a closure is an ad hoc one, generally not justifiable theoretically. In collisional plasmas, the most usual form for $\boldsymbol{q}$ is Fourier's law: $\boldsymbol{q} = -\kappa\nabla T$, where kappa is the thermal conductivity. The term $\nabla \cdot \boldsymbol{q}$ then appears as a diffusion term, so showing its nonideal nature.

The above system can be viewed as an extension of the Navier–Stokes system: the main difference is the inclusion of the Laplace force in Eq. (3.2). This brings in the magnetic field $\boldsymbol{B}$, which requests new equations to determine the variations of this field. The Faraday equation (3.4), added with the ideal Ohm's law (Eq. (3.7) provides the needed equation, in conjunction with the divergence-free equation (3.5). The MHD system can indeed be introduced rather intuitively in this way, but we will nevertheless establish it more formally from the basic moment equations (which have themselves been demonstrated from the kinetic equation) because it is the only way for showing the assumptions necessary to make it valid, especially in the collisionless case.

Looking at the above MHD system, several obvious characteristics can be observed without further analysis. These characteristics correspond to simplifications, which will have to be discussed. The main ones are:

- it is "mono-fluid", which means that, although the plasma contains electrons and ions (with possibly several kinds of ions), it implies only one global density, one velocity and one pressure. We will see, when establishing the MHD system from the basic moment equations of the individual populations, that these single densities and velocities can be considered the ion ones (because the ions carry the mass) and the pressure as the sum of the electron and ion pressures

- it is a scalar model: the pressure tensor is reduced to a scalar parameter P. This simplification is fully justified in a strongly collisional medium, but it must certainly be questioned in a weakly collisional one.
- There is no RHS term coming from collisions (as the terms C_i in the moment equations of Chapter 2). This is consistent with the term "ideal" used for this model. The collisional interactions such as the resistivity (between ions and electrons) are thus supposed negligible.

Another important property appears when analyzing the ideal MHD system with more attention. As already mentioned, the electric field is not an intrinsic variable of this system: it is just an intermediate variable related to $\boldsymbol{u}$ and $\boldsymbol{B}$ via the ideal Ohm's law, and which can, therefore, easily be eliminated from the system. The important property to notice is that this electric field appears only via its curl, in Faraday's law. This means that the system can be entirely solved independently of the "longitudinal" part of the electric field (that is, the part corresponding to its divergence) and, therefore, of the charge density. Using the ideal Ohm's law itself does indeed provide one with a full electric field, including its longitudinal part, but this part is, in the ideal MHD case, a by-product, without feedback on the solution itself. If a generalized Ohm's law was used, differing from the ideal one by only a gradient term (for instance, an electron pressure gradient), nothing would change in the solution for $\boldsymbol{u}$, $\boldsymbol{B}$, ρ, and P. One should always keep in mind that the charge density IS NOT supposed to be zero in MHD. It can actually be ignored in the resolution, but it can always be computed anyway through the Maxwell–Gauss equation and Ohm's law (see also Chapter 1). Even if, as we will see, the establishment of the MHD system requires a condition called "quasi neutrality", it must be emphasized that this notion must not be confused with a condition of true neutrality.

3.1.2
The Ideal Ohm's Law

The ideal Ohm's law (3.7) used in the ideal MHD system has important consequences on the plasma evolution and deserves special comments. We will see later what its origin is and what its possible generalizations are, but let us first assume it is valid and describe these consequences. The first is that the electric field component $E_{\parallel}$ parallel to the magnetic field is zero; field lines are equipotential lines. This property leads to another one, called the "freezing-in theorem". In Chapter 1, it is shown that, whenever the parallel electric field is zero, one can define a "field line motion" at the magnetic velocity $\boldsymbol{v}_{\mathrm{m}} = \boldsymbol{E} \times \boldsymbol{B}/B^2$ and that, along this motion, the lines can be deformed in any manner but never lose their identity. This means that two points located on the same field line always remain on this same field line if both are moved with the local velocity $\boldsymbol{v}_{\mathrm{m}}$. This forbids any "magnetic reconnection" in which these magnetic connections would be changed. One can say that the lines are then "frozen" in the velocity field $\boldsymbol{v}_{\mathrm{m}}$ and transported by it. As the ideal Ohm's law indeed guarantees that $E_{\parallel} = 0$, it is a particular case of application of

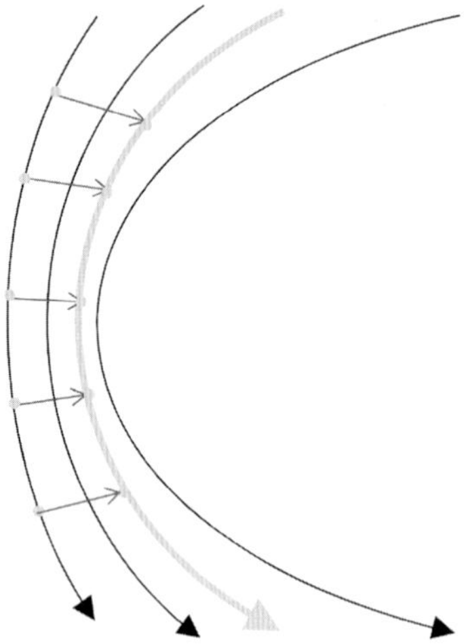

Figure 3.1 Freezing-in theorem. One considers points which are initially "magnetically connected", that is located on the same field line on the left-hand side of the figure. When all these points are moved at the velocity $\boldsymbol{v}_{\mathrm{m}}$, the resulting points are still magnetically connected, that is, located on a single field line if the freezing-in theorem holds. This allows defining the "field line motion", whatever the global magnetic field, stationary or not (in the figure, it is supposed stationary for convenience). In ideal MHD, one has $\boldsymbol{v}_{\mathrm{m}} = \boldsymbol{u}$, but the argument is the same in different contexts: in Hall-MHD for instance, one has $\boldsymbol{v}_{\mathrm{m}} = \boldsymbol{u}_{\mathrm{e}}$, and the topological constraint looks the same.

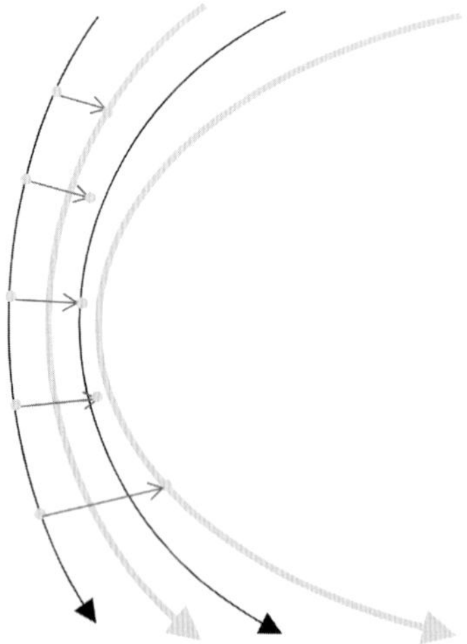

Figure 3.2 No freezing. If the freezing-in theorem is not valid (existence of a parallel electric field), the resulting points don't belong to a single field line. This prevents defining a field line motion. The upper point becomes connected to the first bold line while the lower one becomes connected to the second one. This slipping is actually quite negligible at large scale and can exist only in some localized regions where strong gradients exist. Even in these cases, the slipping has noticeable consequences only in special geometries.

the freezing-in theorem. One can see that, in this case, $\boldsymbol{v}_{\mathrm{m}} = \boldsymbol{u}$, so that the field lines are frozen in the plasma velocity field itself: the matter travels at the same velocity as the field lines. Figures 3.1 to 3.3 give illustrations of what if called freezing and what is called reconnection.

The existence of the magnetospheres (see Chapter 1) is an eloquent consequence of the freezing-in theorem. The solar wind, which is a flow of magnetized plasma coming from our star, transports with it the field lines of the solar magnetic field. These lines, while expanding with the flow, can take various forms due to the rotation of the Sun, combined with the variations – temporal and spatial – of the solar wind source. Nevertheless, as long as an ideal Ohm's law is verified everywhere (that is, as long as too short scales are not formed), these lines must keep their identity and no reconnection can occur along their motion. In particular, the solar field lines cannot reconnect with the lines of the planetary magnetic obstacles such as the terrestrial magnetic field. When approaching a planet with an intrinsic magnetic field, the solar lines are, therefore, forced to wrap around a boundary called "magnetopause" and to slip along it, while the planetary field lines are compelled to remain inside this boundary, in the so-called magnetosphere. Since the motion of the magnetic field lines is in this case identical to the plasma one, this is a quite important consequence of the freezing-in theorem that the solar wind

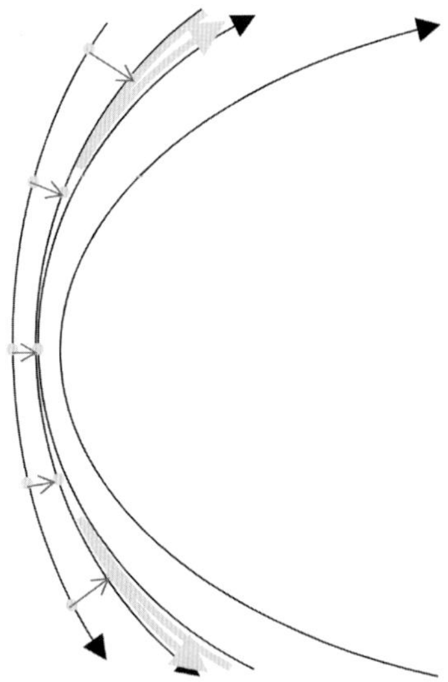

Figure 3.3 No freezing (2). In such a geometry ("X point"), strong gradients do exist and even a slight slipping indeed has strong consequences concerning the connectivity. One can see that the upper point becomes abruptly disconnected from the lower one and that the first one becomes connected to the highest lines in the figure, very far from the lines connected to the second one. This kind of phenomena, with a strong connectivity gradient, is called "reconnection". In this 2D picture, some field lines seem to cross each other at the X point. This is an effect, in general, of the projection (as long as the "guide field" in the third direction is not zero). It must in any case never be considered a characteristic feature of reconnection.

cannot penetrate the magnetosphere and the magnetospheric plasma cannot leak outside. For this reason, the magnetosphere is a tenuous bubble in a denser environment, with almost no exchange between both. There are actually exceptions to the magnetopause tightness: they are allowed when nonideal effects in the Ohm's law allow local defreezing of the field lines at some places, for instance, where the boundary thickness becomes sufficiently small. These exceptions can lead to brutal changes in the field line connections. They are then called "reconnection events".

It is worth emphasizing that the freezing-in theorem relies only on the ideal Ohm's law, without any connection with any other of the MHD equations. This means that it can remain valid, even when the other MHD equations become invalid, in particular when a more kinetic modeling is needed. Reciprocally, it can be violated in the frame of nonideal MHD models, where all equations are unchanged except the Ohm's law, which has to be replaced by some "generalized Ohm's law", as presented later.

3.2
Establishing the MHD Model

Equations (3.1)–(3.3) can be established from summations of the moment equations over the different populations, electrons and ions.

Equation (3.1) is exactly equal to the sum of the continuity equations, with, as explained in Chapter 2:

$$\rho = \sum \rho_p \tag{3.8}$$

$$\rho \boldsymbol{u} = \sum \rho_p \boldsymbol{u}_p \,. \tag{3.9}$$

The sum over the different populations p can actually be limited to only the ion populations i since the mass of the electron can be neglected in front of the ion ones.

Equation (3.2) can be obtained in the same way by the sum of the individual momentum equations of each population, but only under some approximations:

1. As already mentioned, the first approximation consists of supposing the scalar form for the pressure term. The validity of this approximation will be discussed later.
2. The second approximation consists of assuming:

$$P = \sum P_p \,. \tag{3.10}$$

As explained in Chapter 2, the moments of a mixture of populations are not additive in general because these moments, of order larger than or equal to 2, are centered moments while the additive ones are the noncentered ones. The departure is equal to $\rho u^2 - \sum \rho_p u_p^2 = \sum \rho_p \boldsymbol{u}_p \cdot (\boldsymbol{u} - \boldsymbol{u}_p)$. The electrons have generally a velocity different from the ions since the current density is generally not zero, but, once again, the approximation made in this case can be neglected because the electrons have a very small mass. On the contrary, when the plasma involves several ion species, this difference may have to be corrected whenever all the species do not have the same velocity $\boldsymbol{u}_i = \boldsymbol{u}$. This question can be ignored in strongly collisional plasmas because, in these cases, all species do move with the same fluid velocity with a very good approximation.

3. The last approximation made for writing the second equation consists of replacing $\sum n_p q_p(\boldsymbol{E} + \boldsymbol{u}_p \times \boldsymbol{B})$ by $\boldsymbol{j} \times \boldsymbol{B}$. This is clearly valid only if the term involving the electric charge density $\varrho\boldsymbol{E}$ can be neglected in front of $\boldsymbol{j} \times \boldsymbol{B}$. If one assumes that the longitudinal part of the electric field has the same order of magnitude as the transverse one or smaller, a dimensional analysis shows that neglecting ϱE is justified, providing that $\omega/k \ll c$ (ω and k characterize the temporal and spatial scales, supposing that the charge density and the magnetic field vary on the same scales). With the same hypothesis, it is worth noting that the longitudinal current j_l is negligible in front of the transverse one j_T, which justifies neglecting the displacement current $\partial_t(\epsilon_0 E)$ in the Ampère's law (3.6).

Equation (3.3) is the sum of the second order moment equations of all the populations, but once again subject to the approximation that all populations have very close mean velocities, except the electrons for which it does not matter much thanks to their small mass. With this approximation, the centered moments P and $\boldsymbol{q}$ can be summed as centered ones, as well as the quantities such as $P\boldsymbol{u}$. It is easy checking that, in the fully isotropic hypothesis chosen, the individual second order moment equation presented in its general form in Chapter 2 can indeed be put in the given simplest one:

$$\rho_p^\gamma D_t\left(P_p/\rho_p^\gamma\right) + (\gamma - 1)\nabla \cdot \boldsymbol{q}_p = 0\,. \tag{3.11}$$

An important warning must be given, however, concerning this Eq. (3.3): when assuming individual closure equations for each population, it does not provide, in general, a closure equation for the global system. Each individual closure does provide an expression of $\boldsymbol{q}_p$ as a function of ρ_p, $\boldsymbol{u}_p = \boldsymbol{u}$ and P_p. But the sum $\boldsymbol{q} = \sum \boldsymbol{q}_p$ of these expressions is not, unless all are linear, a function of the global variables $\rho = \sum \rho_p$ and $P = \sum P_p$. A trivial – and common –case where it works, however, is of course the case where all populations are adiabatic, since the sum of $\boldsymbol{q}_p = 0$ obviously gives $\boldsymbol{q} = 0$.

Equation (3.4) is Faraday's law and is without approximation. Similarly, the divergence-free equation, *Eq.* (3.5), is of course quite general.

In Eq. (3.6), the expression of $\boldsymbol{j}$ is approximated by neglecting the displacement current. This corresponds to neglecting the longitudinal part of the electric current with respect to its transverse part. As already mentioned, the range of validity of this approximation is the same as the range allowing one to neglect the ϱE term in Eq. (3.2): it corresponds to "phase velocities" ω/k small with respect to the velocity of light c.

Equation (3.7), finally, is the ideal Ohm's law, which is crucial in the MHD dynamics, as already mentioned, for the strong constraint (freezing-in theorem) that it introduces. This equation is no more than an approximation of the electron momentum equation. As seen in Chapter 2, this equation can generally be written:

$$E = -\boldsymbol{u}_e \times \boldsymbol{B} - \frac{1}{n_e e}\left[\rho_e d_t \boldsymbol{u}_e + \nabla \cdot \boldsymbol{P}_e\right] + \eta \boldsymbol{j} \,. \tag{3.12}$$

This equation can be called a "generalized Ohm's law". The specific form chosen here for the last term is the most usual form of a resistivity term, corresponding to the electron–ion collisions. It corresponds to the only term in the classical Ohm's law used in electrodynamics $\boldsymbol{E} = \eta \boldsymbol{j}$. It is actually the only common term between the Ohm's laws used in the two domains, which can be misleading since this term is even absent in a collisionless plasmas.

For reducing this generalized Ohm's law to its ideal form, the first approximation is to replace $\boldsymbol{u}_e$ by $\boldsymbol{u}$. This approximation is met when the current does not exceed a maximum value: $j \ll neu$, where $u \approx u_i$ is a typical order of magnitude of the ion velocity variations. Noting $\delta \boldsymbol{u}$ the velocity difference between ions and electrons, this current condition can be crudely written $\delta u/u \ll 1$; the velocity difference is much smaller than the global velocity variations.

Although this condition concerns basically the electric current (or the velocities), a condition on the charge density (or the number densities) can be deduced from it: $\varrho \ll nq$, which can be written as well $\delta n/n \ll 1$ if δn is the difference between electron and ion densities. This condition just tells that the departure from exact neutrality, which does exist, is much less than the total charge density of each species. This condition is called "quasi-neutrality". It should be kept in mind that it derives from a condition on the currents (which itself has no classical name). It is worth emphasizing also that quasi-neutrality is a rather loose condition; a quick numerical calculation indicates that the electric field corresponding to the total density would generally be a quite huge one. Any "normal" electric field is then small with respect to it. It is the reason why it would be a big mistake to confuse quasi-neutrality and neutrality, in spite of what the name may suggest. In any case, this condition allows important simplifications since it allows taking the same density n for electrons and for ions in the equations, whenever the difference $n_i - n_e$ itself is not implied.

A dimensional analysis shows that this first approximation is verified when the time and space scales verify: $\omega/k^2 \gg D$, where the variable D is $D = B/nq\mu_0$. This variable can be written in different manners, for instance, $D = V_A d_i$ if expressed as a function of the Alfvén speed and the ion inertial length (this variable is important in the theory of whistler mode, see Chapter 5). When these approximations are not verified, the corrective term $\boldsymbol{j}/nq \times \boldsymbol{B}$ must be added in the electric field expression, that is, in the Ohm's law. This term is called "the Hall electric field" since it is the cause of what is called the "Hall effect" in conductors.

The second hypothesis for reducing Eq. (3.12) to its ideal form consists of neglecting the three other terms of the RHS in front of $\boldsymbol{u} \times \boldsymbol{B}$. A dimensional analysis shows that the conditions are in general: $\omega/\omega_{ce} \ll 1$ for the first one and $k^2 R_e^2 \ll 1$ for the second (R_e being the electron Larmor radius). Concerning the last one, neglecting it just means that one assumes that the electron–ion collisions are negligible.

3.2.1
Large-Scale Conditions of Validity

It has already been mentioned that ideal MHD can be used only for large scales, temporal and spatial. Let us now try to find what the characteristic scales are that limit this validity. We have seen that two conditions are necessary for the ideal Ohm's law to be valid: $\omega/\omega_{ce} \ll 1$ and $k^2 R_e^2 \ll 1$. They just concern the electron dynamics; the time variations must be slow with respect their gyroperiod and the space variations must be large-scale with respect to their Larmor radius. These conditions are indeed necessary but far from sufficient. In the same Ohm's law, for neglecting the Hall electric field, one needs the condition $\omega/k^2 \gg D$. This is already more demanding. Considering, for instance, Alfvénic variations $\omega/k \approx V_A$, this condition leads to $kd_i \ll 1$: the scales must be larger than the ion inertial length d_i, which is generally much larger since it is an ion scale.

Apart from the above conditions, which are related only to Ohm's law, there are also two other assumptions, the validities of which have to be investigated:

- The pressure is assumed scalar, that is, the distribution functions are supposed fully isotropic. Such a hypothesis is certainly justified when the plasma is strongly collisional, that is, for $k\lambda_{mfp} \ll 1$ (λ_{mfp} being the mean free path), but there is actually no scale condition that could guarantee it in the opposite limit, for which the plasma can be considered as collisionless. Whatever the scale, the collisionless plasmas have no general reason to have fully isotropic distributions. If these distributions actually don't depart very much from full isotropy, it is for indirect reasons, in particular because of the kinetic instabilities that develop when the anisotropy is too strong.
  Nevertheless, at sufficiently large scales $\omega/\omega_{ci} \ll 1$ and $kR_i \ll 1$, the distribution always becomes gyrotropic, which means that the pressure tensor can be characterized by only two independent parameters: $P_\perp$ and $P_\parallel$. It can be proved by considering the moment equation for the pressure transport. In the adiabatic case, it is easy to see that $P_{xy} = P_{yz} = P_{zx} = 0$, and $P_{xx} = P_{yy}$ at zero order in ω/ω_{ci} (in a frame with $\boldsymbol{B}$ in the z direction). This is indeed the gyrotropic form of the pressure tensor. The condition in kR_i, even if it can be easily be guessed intuitively, is less easy to demonstrate; it comes through kinetic calculations showing that the finite Larmor radius effects (FLR) bring a contribution to the heat flux, which breaks the gyrotropy when kR_i is of the order or larger than unity.
  For these large scales, since full isotropy is never guaranteed in collisionless plasmas, but gyrotropy is, it would, therefore, be better to replace ideal MHD by an anisotropic MHD model, at the condition to have two closure equations instead of one. Such a theory exists for the adiabatic case; it is the CGL theory

(Chew, Goldberger, and Low). But beyond this simple adiabatic hypothesis, there is presently no general anisotropic MHD model.

- The model is mono-fluid, with all species with velocities sufficiently close one to another so that the centered moments (P, $\boldsymbol{q}$) can be added as the noncentered ones. In the plane perpendicular to $\boldsymbol{B}$, all species move at the same velocity $\boldsymbol{v}_{\mathrm{m}} = \boldsymbol{E} \times \boldsymbol{B}/B^2$ as soon as they all verify the above large-scale conditions $\omega/\omega_{\mathrm{cp}} \ll 1$ and $k R_p \ll 1$. The additivity is, therefore, guaranteed in these directions. In the direction parallel to $\boldsymbol{B}$, there is no general scale law guaranteeing the validity of the assumption. Nevertheless, it is not expected to lead to noticeable departure from MHD as long as the velocity difference $\boldsymbol{u}_p - \boldsymbol{u}$ remains small with respect to the thermal velocity of this population.

3.2.2
Departures from MHD: Multi-Fluid and Kinetic Effects

As we have just shown, ideal MHD has a limited range of validity, subject in particular to scale conditions. When some of the necessary conditions are violated, more refined models are needed, some of them remaining mono-fluid and being called "nonideal MHD models", others being multifluid or kinetic. What are these models and in which conditions is it necessary to use one or another?

1. If several ions are present, a global MHD description is possible if all the conditions are verified for all the individual species. The large-scale conditions, in particular, are imposed by the heavier ions. Otherwise, a multifluid is necessary.
2. If the collisions have to be taken into account, they can generally be introduced in the MHD system via terms such as $\eta \boldsymbol{j}$ in the Ohm's law. They lead to variants of the MHD system called "nonideal MHD" or specifically in this case, "resistive MHD".
3. If the condition $\omega/k^2 \gg D$ is not verified, the Hall electric field must be added in the Ohm's law. For Alfvénic variations ($\omega/k \approx V_{\mathrm{A}}$), it is a large-scale condition $k d_i \ll 1$, and $\omega/\omega_{\mathrm{ci}} \ll 1$. For variations with $\omega/k \ll V_{\mathrm{A}}$, neglecting the Hall current is still more restricted, and justified only to much larger scales. For describing the scales beyond these limits, the minimum model to be used is called the "Hall-MHD model". It is identical to the MHD model, except that the generalized Ohm's law must be used, which includes the Hall electric field. For cold plasmas ($\beta < 1$), this model has no other restriction, except when one reaches the electron scales; the mono-fluid description must then be abandoned and the minimum model becomes a bifluid model. Nevertheless, when the pressure has to be taken into account, the large-scale conditions (with respect to the ion scales) have to be verified to make the description valid, even for scales much smaller than the electron ones.
4. If the large-scale conditions $\omega/\omega_{\mathrm{ci}} \ll 1$ and $k R_{\mathrm{i}}$ are not verified, two main problems arise:

- The simple gyrotropic hypothesis is no longer justified, bringing in particular nondiagonal (viscous-like) terms.
- Finding a valid closure equation becomes increasingly problematic.

5. Closure. As already mentioned, the only circumstance where a simple closure can be applied is the adiabatic one, that is, when one has $\boldsymbol{q}_p = 0$ for all populations. A kinetic calculation is needed to assess in which conditions a population is actually adiabatic in a collisionless plasma. A condition can be shown, by this calculation, to be $\omega / k_{\|} V_{\text{th}\|p} \gg 1$. It can be characterized by the fact that the intrinsic temporal variations are faster than the spatial variations seen by a thermal particle in its motion along the magnetic field. The Vlasov equation then imposes that all particles of the considered population react almost identically to the collective field, so that the distribution is almost displaced as a whole. The main perturbation is then the fluid velocity perturbation while the density, pressure, and heat flux relative perturbations are much smaller. Calculating more precisely the pressure and heat flux perturbations further shows that the heat flux perturbation can even be neglected in front of the pressure one when $\omega / k_{\|} V_{\text{th}\|p}$ is large enough. When this condition is verified for all populations, a pure adiabatic closure is justified in the MHD system.
   There are indeed other circumstances for which closure equations for each population can be justified. The limit $\omega / k_{\|} V_{\text{th}\|p} \ll 1$, opposite to the previous one, can for instance lead also to simple closure equations. It can be an isothermal one if the population is isotropic and if the heat flux is initially zero. This can be physically explained: in these conditions, the thermal particles see field variations which are essentially due to spatial variations along $\boldsymbol{B}$ (which be considered as static at this temporal scale). Because they travel very fast along the magnetic field lines, the particles are then able to homogenize the temperature along them.
   The collisional case is clearly also a case where individual closure equations are valid, of the form $\boldsymbol{q} = -\kappa \nabla T$, with $T = P/\rho$ for each population.
   Nevertheless, it must be kept in mind that individual closure equations don't lead to global ones in general. As the usual individuals are generally not linear (except – trivially – $\boldsymbol{q} = 0$), they cannot be added and thus provide a closure equation involving the only global moments of the mixture. It can be checked, however, that this problem has no consequence if there is only one ion population and if the global pressure is carried by only one of the two existing populations, ion or electron.
6. Resonances. There are several kinds:

- Landau resonance, occurring when $\omega / k_{\|} V_{\text{th}\|p} \approx 1$. It is a resonance between the parallel electric field and the parallel velocity of the particles.
- Cyclotron resonances, occurring when $(\omega - n\omega_{\text{cp}}) / k_{\|} V_{\text{th}\|p} \approx 1$. It is a resonance between the perpendicular electric field and the perpendicular gyrating velocity of the particles.

- Bounce resonances, occurring at the bounce frequency for particles bouncing on a field line between two mirror points on both sides of a minimum of the magnetic field magnitude.
- Other resonances.

The existence of particle resonances makes finding pertinent closure equations particularly problematic. Close to a resonant velocity, particles with different velocities have actually very different reactions to the collective field. If this property leads to strong variations in the distribution function, it seems quite improbable that a description based on a small number of moments can be sufficient to model the system. This argument seems a matter of common sense. However, it must be qualified. As shown in Chapter 5 about Landau damping, even if a kinetic plasma has a quasi-infinite number of degrees of freedom, most of them actually give rise to very short-lived evolutions, so that most of the large-scale observable variations belong to a much smaller class, corresponding to the so-called "kinetic modes" in the linear case. The different particle behaviors close to the resonant velocity don't lead, for these long-lived modes, to any singularity in the distribution functions. They just lead to some damping, which is not a priori impossible to model in a fluid way. Approximate "Landau-fluid" models are indeed based on this principle.

3.3 Dimensional Analysis and Plasma Characteristic Scales

3.3.1 Dimensional Analysis: The General Methods

When investigating the validity range of the MHD model, several results have been presented in this chapter, which come from dimensional analyses. Let us now show how these analyses can be done and what the various characteristic scales of the plasma physics are that derive from them and the important dimensionless parameters which allow their comparison. Let us describe the method to obtain such analyses on a particular example, the momentum equation of MHD:

$$\rho D_t \boldsymbol{u} + \nabla \left(P + B^2/2\mu_0\right) = \boldsymbol{B} \cdot \nabla \boldsymbol{B}/\mu_0 \,. \qquad (3.13)$$

Using Ampère's equation, the Laplace force has been split here into a magnetic pressure term on the LHS and a magnetic tension on the RHS. The goal is now to compare the orders of magnitude of the variations of the different terms of the equation, in a given scale range characterized by ω and k (inverses of the temporal and spatial scales under study). For this purpose, each variable x is split into a part x_1 varying at the considered scale, plus a part x_0 varying at larger scales in time and space (which allows neglecting its derivatives) and a part x_2 varying at smaller scales. By this method, the equation ruling the variations at the considered scale appears similar to the one used in linear theory, when calculating the propagation modes in a plasma:

$$\rho_0 D_t \boldsymbol{u}_1 + \nabla(P_1 + \boldsymbol{B}_0 \cdot \boldsymbol{B}_1/\mu_0) = \boldsymbol{B}_0 \cdot \nabla \boldsymbol{B}_1/\mu_0 \,. \tag{3.14}$$

In this equation, the time and space derivatives of the order zero terms have been neglected because of their large-scale variations. Furthermore, some second order terms, which should appear due to the nonlinear terms such as $B^2/2\mu_0$ have not been introduced. The reason is that, even if conditions such as $x_2 \ll x_1$ of the linear theory are not needed, one supposes, however, that the second order terms are, at most, equal to the first order ones, so that neglecting them when a first order term does exist, can at most introduce a factor of 2 in the results, which can be ignored in the frame of such a dimensional analysis, even if it can actually correspond to measurable departures from the linear theory.

On the contrary, when the first order of a term is zero or close to zero, the nonlinear terms cannot be ignored. For these terms, the dimensional analysis will provide a result which depends on the fluctuation amplitude. It is important emphasizing that the method presented, used in such a way, can be used in systems where the nonlinear terms are of pivotal importance, as in turbulence theory. It is the case in particular when investigating the transfers of energy between waves; the above estimates of the nonlinear terms then give access to relations between the scales and the amplitudes, typically to spectra such as $u_1^2 = u_1^2(k)$.

A particular – and important – case where the nonlinear terms cannot be ignored has to be emphasized; when all linear terms tend to zero, the linear value of the inertial term also tends to zero, so that one must go an order further and change ω into something that can be noted $\omega \oplus \sum k u_1$. In the following, ω will, therefore, be reserved to the linear term (partial derivative: $\partial_t a \Rightarrow \omega u_1$), while the summation sign will correspond to the nonlinear terms brought by the total derivative. Writing $\boldsymbol{u} \cdot \nabla a \Rightarrow \sum k u_1 a_1$ just assumes that all the different terms u_1 and a_1 that are coupled nonlinearly have comparable scales ("local" interactions). This term is even considered dominant on the linear one in classical hydrodynamical turbulence theory.

The notation $A \oplus B$, used here and in all the next sections, is a symbolic one. It is just to recall that the two terms that have the orders of magnitude A and B are to be added in the considered equation. It must be kept in mind, however, that this addition has an order of magnitude which is not necessarily equal to $A + B$ since the two terms can have different signs and cancel each other.

Equation (3.14) is sufficient to the searched dimensionless analysis, as long as two dominant terms can be found, that is, far from some boundaries in the parameter space where three terms would have the same order of magnitude, allowing particular cancelations. To make it more explicit in a function of the temporal and spatial scales, let us now replace the time and space derivatives respectively by ω and $\boldsymbol{k}$, as one does for calculating the monochromatic plane waves, but forgetting all about the signs and the phases (no i in the calculations). Equation (3.14) can then be rewritten under the symbolic form:

$$\omega \rho_0 \boldsymbol{u}_1 = \boldsymbol{k}(P_1 \oplus \boldsymbol{B}_0 \cdot \boldsymbol{B}_1/\mu_0) \oplus \boldsymbol{B}_0 \cdot \boldsymbol{k} \boldsymbol{B}_1/\mu_0 \,. \tag{3.15}$$

For comparing two different terms to each other, they have then to be expressed as functions of the same fluctuation. This can be done by using any other equations available with the same rules. In the present example, it is easy to express all terms as functions of the vector $\boldsymbol{u}_1$. Applying the same method, one obtains:

$$\text{Continuity:} \qquad \rho_1 \Rightarrow \rho_0 \frac{\boldsymbol{k}}{\omega} \cdot \boldsymbol{u}_1 \tag{3.16}$$

$$\text{Polytropic closure:} \quad P_1 \Rightarrow \rho_1 c_s^2 = \rho_0 c_s^2 \frac{\boldsymbol{k}}{\omega} \cdot \boldsymbol{u}_1 \tag{3.17}$$

$$\text{Faraday/Ohm:} \quad \boldsymbol{B}_1 \Rightarrow \boldsymbol{B}_0 \cdot \frac{\boldsymbol{k}}{\omega} \boldsymbol{u}_1 \oplus \boldsymbol{B}_0 \frac{\boldsymbol{k}}{\omega} \cdot \boldsymbol{u}_1 \,. \tag{3.18}$$

The crudest estimation that can be done consists of fully ignoring the vector character of the different terms and just comparing the orders of magnitude of their modulus. It gives some kind of maximum contribution of each term. This is a useful step, even if it can be insufficient for special gradient directions or for special polarizations of the fluctuations, as it will be shown just after. It gives:

$$\omega \rho_0 u_1 = k(P_1 \oplus B_0 B_1/\mu_0) \oplus k B_0 B_1/\mu_0 \tag{3.19}$$

$$\rho_1 \Rightarrow \rho_0 \frac{k}{\omega} u_1 \tag{3.20}$$

$$P_1 \Rightarrow \rho_0 c_s^2 \frac{k}{\omega} u_1 \tag{3.21}$$

$$B_1 \Rightarrow B_0 \frac{k}{\omega} u_1 \,. \tag{3.22}$$

Putting all together, and dividing by $\omega \rho_0$:

$$u_1 = \frac{k^2 c_s^2}{\omega^2} u_1 \oplus \frac{k^2 V_A^2}{\omega^2} u_1 \oplus \frac{k^2 V_A^2}{\omega^2} u_1 \,. \tag{3.23}$$

One can observe that the two terms coming from the Laplace force, the magnetic pressure term and the tension term, have the same order of magnitude with this estimation, even if they actually correspond to quite different components.

The calculation, therefore, brings up two characteristic speeds: the Alfvén speed, defined by $V_A^2 = B^2/(\mu_0 \rho_0)$ and the sound speed, defined by $c_s^2 = \gamma P/\rho$ (with $\gamma \approx 1$). The latter speed is of the order of the ion thermal speed V_{thi} whenever $T_e \leq Ti$.

From the above dimensional analysis, one can conclude that the Lorentz force should always dominate the evolution when $\beta = c_s^2/V_A^2 < 1$ and that the spatial and temporal scales are then necessarily related by $\omega/k \approx V_A$. On the contrary, the pressure gradient force should always be dominant when $\beta > 1$, the scales being then related by $\omega/k \approx c_s$. In these conditions, the plasma behavior is not different from the behavior of an unmagnetized plasma.

One important and general property of the MHD model can be seen from this dimensional analysis; the calculation only brings up two characteristic speeds: the Alfvén speed, V_A and the sound speed c_s, but no characteristic scale arises, neither in time nor in space. This is indeed a general result for MHD, which can be seen directly on the differential equations of the system; the derivatives are homogeneous with respect to the two variables t and $\boldsymbol{r}$. The system is called "self similar", which means that, if one evolution is known, any variation α times larger and α time slower is also a solution, whatever α. This property is only true for ideal MHD and any departure from it indeed brings up characteristic scales, as shown in the next section.

Let us now underline that, if the above method is quite sufficient for many uses, several features, which may be important especially for particular polarizations, have been lost when forgetting the vector character of the equations. Let us consider, to emphasize this point, the particular case of purely parallel gradients. It can easily be shown that, whatever β is, these gradients can propagate at $\omega/k_\parallel \approx V_A$ or $\omega/k_\parallel \approx c_s$, only depending whether they concern perpendicular velocity fluctuations or parallel ones. The propagation is driven by the tension force in the first case (Alfvénic fluctuations) and by the pressure gradient force in the second one (sonic fluctuations). The fact that these results are fully independent of the β value is in contradiction to the above general conclusions. They indeed cannot be reached without looking more in detail into the vector polarizations.

Let us show how keeping the vector information allows us to get these more complete results. Separating the parallel and perpendicular components of Eqs. (3.15) to (3.18) yields:

$$\omega \rho_0 u_{1\parallel} = k_\parallel (P_1 \oplus B_0 B_{1\parallel}/\mu_0) \oplus k_\parallel B_0 B_{1\parallel}/\mu_0 \tag{3.24}$$

$$\rho_1 \Rightarrow \rho_0 \frac{k_\parallel}{\omega} u_{1\parallel} \tag{3.25}$$

$$P_1 \Rightarrow \rho_0 c_s^2 \frac{k_\parallel}{\omega} u_{1\parallel} \tag{3.26}$$

$$B_{1\parallel} \Rightarrow 0 \tag{3.27}$$

$$\omega \rho_0 \boldsymbol{u}_{1\perp} = k_\parallel B_0 \boldsymbol{B}_{1\perp}/\mu_0 \tag{3.28}$$

$$\boldsymbol{B}_{1\perp} \Rightarrow B_0 \frac{k_\parallel}{\omega} \boldsymbol{u}_{1\perp} \,. \tag{3.29}$$

This leads to:

$$u_{1\|} = \frac{k_{\|}^2 c_s^2}{\omega^2} u_{1\|} \tag{3.30}$$

$$u_{1\perp} = \frac{k_{\|}^2 V_A^2}{\omega^2} u_{1\perp} . \tag{3.31}$$

From these results, it is clear that the two kinds of velocity perturbations, parallel and perpendicular to the large-scale magnetic field, propagate independently, with characteristic velocities which are respectively c_s and V_A. The decoupling between the two kinds arises from the absence of magnetic compression $B_{1\|} = 0$ for such parallel gradients. For oblique gradients, the magnetic compression couples the two components $u_{1\|}$ and $u_{1\perp}$, so that they can be assumed to have the same order of magnitude.

To be more careful, the simple results obtained by ignoring the vector character of the variables should, therefore, be said to hold whenever the different components of the relevant vectors have the same order of magnitude. This rules out the special polarizations such as the previous one. In other circumstances, it will be important also to distinguish the "longitudinal" (divergent) components of the fluctuations and the "transverse" (nondivergent) ones. Concerning the electric field, it allows separating the roles of the electrostatic and the electromagnetic fluctuations. Concerning the velocity, it allows separating the compressible variations from the incompressible ones (all the classical dimensional analyses in incompressible turbulence neglect the first kind of fluctuations).

When the vector properties of the fluctuations are needed, it is clear that, because of the similarity of the formalisms, a good knowledge of the linear modes, their dispersions, and their polarizations greatly helps to answer the question and understand which terms are dominant and which can be neglected in the equations.

3.3.2
Temporal and Spatial Scales, Adimensional Numbers

3.3.2.1 Parallel Eletron Motion: Plasma Frequency and Debye Length

The most important role of the electrons in a plasma is to move rapidly along the magnetic field lines where they tend to limit – sometimes cancel – the parallel electric field via the space charge they can generate. This effect is more or less efficient, depending on two important characteristic scales of the plasma physics: the plasma frequency (temporal) and the Debye length (spatial). These scales indeed appear quite easily if we consider purely parallel gradients and write the parallel momentum equation of the electrons:

$$n_0 m_e \omega u_{e\|1} \oplus k P_{e1} \oplus n_0 e E_{\|1} = 0 . \tag{3.32}$$

The notation $\oplus$ has been defined in the previous section. This expression (3.32) is, of course, identical to the equation of an unmagnetized plasma. All terms can be expressed as functions of the electron and ion density fluctuations using the continuity equation for the velocity, a polytropic closure for the pressure and the Maxwell–Gauss equation for the electric field:

$$\omega^2 n_{\mathrm{e}\|1} \oplus k^2 V_{\mathrm{the}}^2 n_{\mathrm{e}1} \oplus \frac{n_0 e^2}{m_{\mathrm{e}}\epsilon_0}(n_{\mathrm{e}1} - n_{\mathrm{i}1}) = 0 \,. \tag{3.33}$$

Two characteristic scales so clearly appear in the equation, defined by:

$$\omega_{\mathrm{pe}}^2 = \frac{n_0 e^2}{m_{\mathrm{e}}\epsilon_0} \tag{3.34}$$

$$k_{\mathrm{De}} = \frac{\omega_{\mathrm{pe}}}{V_{\mathrm{the}}} \,. \tag{3.35}$$

Or, alternatively:

$$f_{\mathrm{pe}} = \omega_{\mathrm{pe}}/2\pi \tag{3.36}$$

$$\lambda_{\mathrm{De}} = \frac{1}{k_{\mathrm{De}}} = \frac{V_{\mathrm{the}}}{\omega_{\mathrm{pe}}} \,. \tag{3.37}$$

The first one is the "electron plasma frequency" and the second one the "electron Debye length". Introducing them, Eq. (3.33) can be rewritten as:

$$\frac{\omega^2}{\omega_{\mathrm{pe}}^2} \oplus k^2 \lambda_{\mathrm{De}}^2 \oplus 1 = \frac{n_{\mathrm{i}1}}{n_{\mathrm{e}1}} \,. \tag{3.38}$$

This equation involves four terms: three on the LHS concern only the electron motion, the fourth term on the RHS being the forcing by the ion charge density, which can be kept here as an unknown. The comparison of the different terms leads to the diagram shown in Figure 3.4. The only regions where the electron motion is not forced by the ion one are around the bold lines, which can be considered the "eigenmodes" of the electron motion. The three different lines correspond to the three different manners of finding two dominant terms among the three possible ones: electric field due to the electron space charge balanced by the electron inertia, $\omega \approx \omega_{\mathrm{pe}}$; electric field due to the electron space charge balanced by the pressure gradient force, $k \approx k_{\mathrm{De}}$; electron inertia balanced by the electron pressure gradient, $\omega/k \approx V_{\mathrm{the}}$. The first solution corresponds to the well known "Langmuir oscillation" at large scale, the second one to the "Debye screening" for the low frequency waves (which gives, for instance, sheaths in front of an antenna).

Conditions for quasi-neutrality The temporal and spatial scales are not necessarily limited to the vicinity of the above curves (electron "eigenmodes"). All the other regions of the (ω, k) plane may possibly be filled thanks to the ion forcing. The main term to be retained among the terms governing the electron motion is different in each region and indicated in bold. In the large-scale part of the plane $\omega \ll \omega_{\mathrm{pe}}$ and $k \ll k_{\mathrm{De}}$, the electron density fluctuation balances the ion one ($n_{\mathrm{e}1} \approx n_{\mathrm{i}1}$);

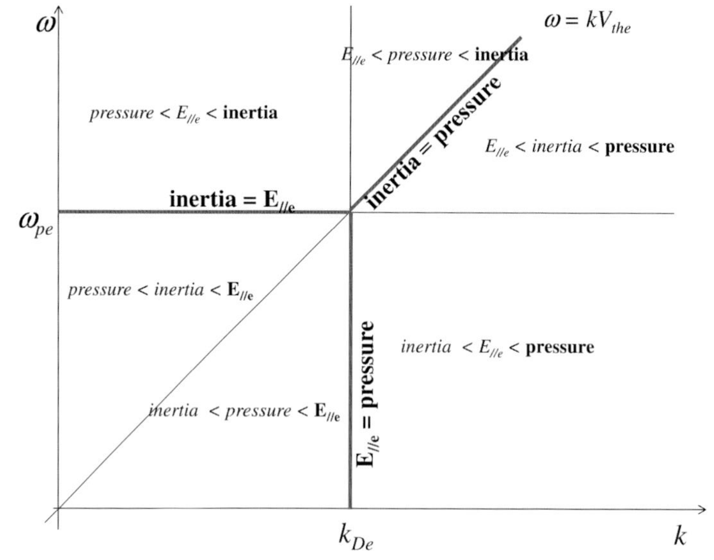

Figure 3.4 Dominant effects for the parallel electron motion at the different scales in the (ω, k) plane. The three forces compared are labeled as follows. Label "inertia", the inertia term $n_0 m_e D_t u_{e\|1}$. Label "pressure", the electron pressure gradient force. Label "$E_{\|e}$", the electrostatic force due only to the electron charge density. The dominant terms in bold are those to be retained in each scale range.

this corresponds to the well-known quasi-neutrality hypothesis used, for instance, in MHD. Outside this region, the electron contribution to the parallel electric field is minor; $n_{e1} \ll n_{i1}$, corresponding to a large departure from quasi-neutrality, and the electron forces to be retained are the inertia force or the pressure gradient one, as indicated.

The above demonstrations have been done for purely parallel gradients. They can indeed be extended to oblique ones without major changes in the results. It must be kept in mind, however, that, for strictly perpendicular gradients (or when the gradients depart from the perpendicular plane by less than m_e/m_i), the above conclusions no longer hold. Whatever the scale, there is then no possibility for the electrons to restore the quasi-neutrality by moving along the magnetic field lines. As shown above, the electron plasma frequency is one of the major characteristic scales in plasma physics. As it depends only on the electron density, it provides useful diagnosis tools for measuring this density. It is useful to have in mind its numerical values; having the electron density in cm^{-3}, the plasma frequency is, in kHz:

$$f_{pe} = 9\sqrt{n} \,.$$

In Table (3.1), the orders of magnitude for several typical media are given.

Changing m_e into m_i, an ion plasma frequency ω_{pi} and an ion Debye length λ_{Di} can be defined in the same way. The ion plasma frequency is much smaller than the electron one (mass ratio) while the ion Debye length has the same order

Table 3.1 Characteristic values of the plasma frequency in different media. The corresponding wavelength for an electromagnetic wave propagating in vacuum has been added for reference, because the plasma frequency acts as a cut-off frequency for these radiations. Some media correspond to partial ionization (for example discharge in a neon tube), others to fully ionized plasma (for example magnetospheric lobes).

Medium	Density (cm^{-3})	f_{pe} (kHz)	Corresponding wavelength in vacuum
Magnetospheric lobes	10^{-2}	1	300 km: radio waves
Solar wind	5	20	15 km: radio waves
Ionosphere (max)	5×10^5	6×10^3	50 m: radio waves
Solar corona	10^8	10^5	3 m: radio waves
Pulsar atmosphere	10^{12}	10^7	3 cm: μ waves
Neon tube	10^{12}	10^7	3 cm: μ waves
Fusion (tokamak)	10^{15}	3×10^8	1 mm: IR
Metal (conduction electrons)	10^{23}	3×10^{12}	0.1 µm: UV
Solar interior	10^{26}	10^{14}	3 nm: X-rays

of magnitude (temperature ratio) since it is independent of the mass. But these scales do not have the same physical importance concerning the above questions, in particular with respect to the quasi-neutrality conditions. When the notations ω_p or λ_D are used without subscript, it must be understood that the subscript e is implied. It is the case, for instance, for the "Debye sheaths". As explained in the chapter devoted to the notion of collision (Chapter 4), and as seen from the above calculations, the Debye length is a length related to the collective effects (electron density and pressure) and must not be attributed to an individual particle property.

Two last remarks have to be done at the end of this subsection:

- Some fluctuations don't imply compressibility, like the Alfvénic ones. As a consequence, they imply neither space charge (no electron or ion density change) nor parallel electric field, so that the above conclusions are not applicable for them.
- The diagram (Figure 3.4) is sufficiently general to be independent of the model considered, fluid or kinetic. One can nevertheless notice that, close to the line $\omega/k \approx V_{the}$, where the pressure effects are dominant, the Landau damping is effective, so that the exact treatment of this region demands a kinetic model. The same remark holds for the vicinity of the ion thermal velocity and around all the gyrofrequencies (cyclotron resonances).

3.3.2.2 Perpendicular Electron Motion: Electron Gyrofrequency and Electron Inertial Length

As seen above, the parallel electric field (and as a consequence the electric space charge) is mainly determined by the parallel motion of the electrons. Similarly, the perpendicular electric field is given by the perpendicular component of the same electron momentum equation, that is, the "generalized Ohm's law". Let us show what the characteristic scales are that correspond to this perpendicular motion. Considering strictly parallel gradients and forgetting about the vector character of $\boldsymbol{E}_\perp$ and $\boldsymbol{u}_\perp$, it gives:

$$\omega u_{\mathrm{e}\perp 1} \oplus \frac{e}{m_\mathrm{e}} E_{\perp 1} \oplus \omega_{\mathrm{ce}} u_{\mathrm{e}\perp 1} = 0 \,. \tag{3.39}$$

The first term corresponds to the perpendicular component of the inertial force; the second and third terms are the electric and magnetic parts of the Lorentz force. In the parallel gradient configuration chosen, the pressure gradient force has no perpendicular component if the pressure is isotropic. In more general hypotheses, this term should be estimated also. We will show it in the next subsection.

From this very first step, we can see that a temporal scale appears, given by the electron gyropulsation $\omega_{\mathrm{ce}} = e B_0/m_\mathrm{e}$. The gyrofrequency is just $f_{\mathrm{ce}} = \omega_{\mathrm{ce}}/2\pi$.

The perpendicular electric field is transverse with respect to the parallel gradients. It is, therefore, an induced electric field, related to the magnetic field variations, and not an electrostatic one as the parallel one. It can, therefore, be expressed as a function of $B_{\perp 1}$ via Faraday's law, which in turn can be expressed as a function of $j_{\perp 1}$ via Ampère's law. Since the current can be, by definition, expressed as a function of $u_{\mathrm{i}\perp 1}$ and $u_{\mathrm{e}\perp 1}$, one can express the different terms in function of $u_{\mathrm{e}\perp 1}$:

$$E_{\perp 1} = \frac{\omega}{k} B_{\perp 1} \tag{3.40}$$

$$B_{\perp 1} \oplus \frac{\omega^2}{k^2 c^2} B_{\perp 1} = \frac{1}{k} \mu_0 j_{\perp 1} \tag{3.41}$$

$$j_{\perp 1} = n_0 e (u_{\mathrm{i}\perp 1} - u_{\mathrm{e}\perp 1}) \,. \tag{3.42}$$

All together, this gives the dimensional equation:

$$u_{\mathrm{e}\perp 1} \oplus \frac{1}{k^2 d_\mathrm{e}^2} u_{\mathrm{e}\perp 1} \oplus \frac{\omega_{\mathrm{ce}}}{\omega} u_{\mathrm{e}\perp 1} = \frac{1}{k^2 d_\mathrm{e}^2} u_{\mathrm{i}\perp 1} \,. \tag{3.43}$$

Under this form, we have neglected the term $\omega^2/(k^2 c^2)$ with respect to unity, which amounts, as explained hereafter, to neglecting the displacement current. A new spatial scale so appears, the electron inertial length:

$$d_\mathrm{e} = \frac{c}{\omega_{\mathrm{pe}}} \,. \tag{3.44}$$

Even if this length implies the velocity of light c, it is worth noting that it is not at all related to the propagation of light since the displacement current has been neglected. This constant just appears because of the μ_0 of the Ampère's law, combined with ϵ_0 involved in the definition of ω_{pe}. Another characteristic parameter appears, which characterizes the ratio between the second and the third term, $D = \omega_{\mathrm{ce}} d_\mathrm{e}^2$. It is neither a temporal nor a spatial scale but a combination of both with a dimension

ω / k^2 like a diffusion coefficient. This new parameter can be seen to be independent of the particle mass, so that it can be written in many other different forms involving in particular the ion characteristic lengths: $D = \omega_{\text{ci}} d_{\text{i}}^2 = V_{\text{A}} d_{\text{i}} = V_{\text{A}}^2 / \omega_{\text{ci}}$.

As for the parallel motion equation, one can see that this perpendicular equation again leads to three electron "eigenmodes", which can exist independently of any forcing by the ions. There is one at $\omega \approx k^2 D$ and $\omega \ll \omega_{\text{ce}}$ and $k d_{\text{e}} \ll 1$, for which the electric force is balanced by the magnetic Lorentz force; one at $\omega \approx \omega_{\text{ce}}$ and $k d_{\text{e}} \gg 1$, for which the inertial force is balanced by the magnetic Lorentz one; and one at $k d_{\text{e}} \approx 1$ and $\omega \gg \omega_{\text{ce}}$, for which the inertial force is balanced by the electric one. In the limit of linear fluctuations, the first solution corresponds to the propagation of the so-called whistler eigenmode. The second one corresponds to the short scale asymptote of this mode at small scale, and the third one to the "electron skin depth", which is the limit of penetration of the higher frequency waves inside a medium.

Once again, the rest of the plane may possibly be filled by solutions which are not pure electron eigenmodes, but are forced by the ion dynamics. The resulting diagram is presented in Figure 3.5. It can be seen that, in the large-scale domain $\omega \ll k^2 D$ and $k d_{\text{e}} \ll 1$, the electron current density balances the ion one, $(j_{\perp \text{e}1} \approx j_{\perp \text{i}1})$. This weak current region is the equivalent, concerning the perpendicular velocity fluctuations, to the quasi-neutrality region found in the previous subsection, and which concerns the density fluctuations. When both conditions are met, one can make $n_{\text{e}} \approx n_{\text{i}}$ and $u_{\perp \text{e}1} \approx u_{\perp \text{i}1}$ in all the calculations that don't

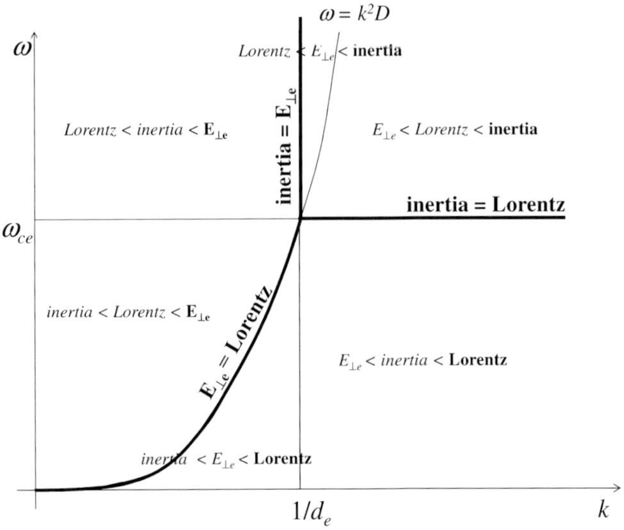

Figure 3.5 Dominant effects for the perpendicular electron motion at the different scales in the (ω, k) plane. The three forces compared are labeled as follows. Label "inertia": the inertia term $n_0 m_{\text{e}} D_t u_{\text{e}\perp 1}$. Label "$E_{\perp \text{e}}$": the force due to the induced electric field due only to the electron current. Label "Lorentz": the magnetic Lorentz force. The dominant terms in bold are those to be retained in each scale range.

rely explicitly on the differences. Outside this region, there exists a large departure from the weak current hypothesis. the electron contribution being minor for the current density and the corresponding induced electric field: $u_{\perp e1} \ll u_{\perp i1}$.

An important plasma parameter is the ratio f_{pe}/f_{ce}. It allows one to sort the previous effects, the parallel electric force (due to space charge) and the Lorentz one.

3.3.2.3 Electron Larmor Radius

In the above particular case of strictly parallel gradients, one important characteristic scale is missing, the electron thermal Larmor radius. This length is defined by:

$$R_{the} = \frac{V_{the}}{\omega_{ce}} . \tag{3.45}$$

From the point of view of the particle trajectories, it is just the gyration radius around the magnetic field of a thermal electron. It is thus quite intuitive that it should have some role in the collective plasma behavior. Nevertheless, this role is not as trivial as the previous ones to show through simple dimensional analyses. It did not appear in the parallel motion of the electrons since the magnetic field does not influence this motion. It did not appear either in the perpendicular motion because of the hypothesis of strictly parallel gradients and of an isotropic pressure; the pressure gradient is then strictly parallel itself and does not influence the perpendicular motion.

The Larmor radius does indeed appear when going beyond the above simplifying assumptions. Remaining in the parallel gradient hypothesis, it is necessary to abandon the hypothesis of an isotropic distribution function to see it. For an isotropic distribution function, as already mentioned, the divergence of the pressure tensor is always parallel to the large-scale magnetic field and, therefore, does not influence the electron's perpendicular motion. It is easy checking that this remains true even for anisotropic distributions, as long as these distributions remain gyrotropic. The nondiagonal terms are actually crucial to make the pressure play a role. Physically speaking, these effects come from particle "bunching" in the gyromotion around the magnetic field. In the considered case of parallel gradients, this bunching cannot be due to a spatial effect; it is temporal. A kinetic calculation shows that the nondiagonal terms are then on the order of ω/ω_{ce} times the diagonal ones, so that the pressure gradient term scales as $\omega/\omega_{ce} k^2 V_{the}^2/\omega^2$. This estimate can be compared to the previous ones, inertial (1), electric $(1/k^2 d_e^2)$ and Lorentz (ω_{ce}/ω). It is readily checked that the Larmor radius appears when comparing the gradient pressure term and the Lorentz one; both are equal when $k R_{the} = 1$.

Considering perpendicular gradients also leads to showing the role of the Larmor radius. The departure from gyrotropy is then due to a purely spatial effect. The so-called FLR (Finite Larmor Radius) effects appear in the kinetic calculation when integrating on the velocity gyrophase; this introduces a combination of Bessel functions whose arguments are $k R_e$.

The above considerations allow adding new boundaries in the (ω, k) plane of Figure 3.5. We let the reader do this exercise, the result being less clear and less universal than the simple ones that have been presented here.

3.3.2.4 Ion Characteristic Scales

The same dimensional analyses can be done for the momentum equation of any ion species. The same characteristic lengths can be defined, just replacing m_e by m_i and P_e by P_i. The method is similar, and the results can easily be guessed from the above ones. For the perpendicular motion:

- Comparing the gyrofrequency with respect to the considered time scale allows one to quantify the importance of the Lorentz force with respect to the inertial one.
- Comparing the inertial length to the considered spatial scale allows one to quantify the importance of the electric force with respect to the inertial one.
- Comparing the Larmor radius with respect to the considered spatial scale allows one to quantify the role of the pressure gradient force with respect to the Lorentz one.
- ...

3.3.2.5 Speed of Light

The current derives from the magnetic field by Ampère's equation. Generally speaking, it also depends on the electric field due to the so-called displacement current present in this equation. To estimate the importance of the second term, the electric field can be expressed as a function of the magnetic field by Faraday's equation for its transverse (nondivergent) component. On the contrary, its longitudinal (divergent) part is independent on the magnetic field and should be expressed via the Maxwell–Gauss equation as a function of the space charge.

Using the above methods of dimensional analysis and distinguishing the longitudinal and transverse parts leads to:

$$\frac{\mu_0}{k} j_{t1} = B_1 \oplus \frac{\omega^2}{k^2 c^2} B_1 \tag{3.46}$$

$$\frac{\mu_0}{k} j_{l1} = \frac{\omega}{k c^2} E_{l1} \,. \tag{3.47}$$

From Eq. (3.46), it is clear that the displacement current (second term of the RHS) can be safely neglected in front of the transverse particle current (first term) as soon as $\omega^2/(k^2 c^2) \ll 1$, that is, as the fluctuations propagate at speeds much smaller than the speed of light. This is verified whenever the electrons have enough time to react to the field, so creating the particle current by their motion, that is typically when $\omega \ll \omega_{pe}$. In these conditions, writing the Ampères's equation under the form $\mu_0 \boldsymbol{j} = \boldsymbol{\nabla} \times \boldsymbol{B}$ is sufficient to determine correctly the transverse part of the current. This approximation is called the "Darwin approximation". Its validity for the longitudinal part is, however, not guaranteed in full generality, even when the condition $\omega^2/(k^2c^2) \ll 1$ is fulfilled. Nevertheless, one can see that, in these conditions, one has $j_{l1} \ll j_{t1} E_{l1}/E_{t1}$, which means that even the longitudinal part of the displacement current can be neglected in Ampère's equation at the condition that E_{l1} is not too much larger than E_{t1}.

3.3.2.6 **Resistive Scale**

As seen in Chapter 1, taking into account the collision field can add corrective terms in all the fluid equations, in particular by introducing coupling between the different plasma populations. Let us take only one example of such an effect, the resistivity, that is, the friction force exerted by the ions on the electrons. When added to the above perpendicular electron momentum equation (generalized Ohm's law), the resistivity is written, without the pressure effects, as:

$$\omega u_{\mathrm{e}\perp 1} \oplus \frac{e}{m_\mathrm{e}} E_{\perp 1} \oplus \omega_{\mathrm{ce}} u_{\mathrm{e}\perp 1} \oplus \frac{\eta e}{m_\mathrm{e}} j_{\perp 1} = 0 \,. \tag{3.48}$$

Here we defined η as the resistivity, that is, the inverse of the conductivity, $\eta = 1/\sigma = E/j$. Note that the same notation η is sometimes used in the literature for labeling the "magnetic diffusivity", which is the same parameter as the resistivity, but divided by μ_0. Expressing again the electric field and the current density in function of $u_{\mathrm{e}\perp 1}$, one obtains:

$$u_{\mathrm{e}\perp 1} \oplus \frac{1}{k^2 d_\mathrm{e}^2} u_{\mathrm{e}\perp 1} \oplus \frac{\omega_{\mathrm{ce}}}{\omega} u_{\mathrm{e}\perp 1} \oplus \frac{\omega_\eta}{\omega} u_{\mathrm{e}\perp 1} = \frac{1}{k^2 d_\mathrm{e}^2} u_{\mathrm{i}\perp 1} \oplus \frac{\omega_\eta}{\omega} u_{\mathrm{i}\perp 1} \,. \tag{3.49}$$

A new temporal scale, related to the resistivity, is brought in, $\omega_\eta = \eta/\mu_0 d_\mathrm{e}^2$.

Once again, this effect introduces new boundaries in the (ω/k) plane which can allow completing Figure 3.5. In the case of weakly collisional plasmas, η is very small, and the resistive effect cannot be significant except for very small ω, that is on very large temporal scales.

3.3.2.7 **Reynolds Number and Magnetic Reynolds Number**

The magnetic Reynolds number is introduced in resistive MHD by analogy with the usual Reynolds number in hydrodynamics. In incompressible hydrodynamics, the velocity is purely transverse and its momentum equation is, when a viscous term and a forcing term are included:

$$\omega u_1 = \nu k^2 u_1 \,. \tag{3.50}$$

The parameter ν is the kinematic viscosity. It determines the temporal evolution as a function of the spatial scale by $\omega = \nu k^2$. This corresponds to a very large temporal scale for most of the usual values of ν, at least whenever k is not too large. In these conditions, one can no longer neglect the nonlinear terms with respect to the linear ones, so that one must replace ω by $\omega \oplus \sum k u_1$, leading to:

$$\omega + \sum k u_1 = \nu k^2 \,. \tag{3.51}$$

In these conditions, a diagram similar to Figures 3.4 and 3.5 can be built, allowing us to know which effects are dominant for all scales in the ω/k space. The three

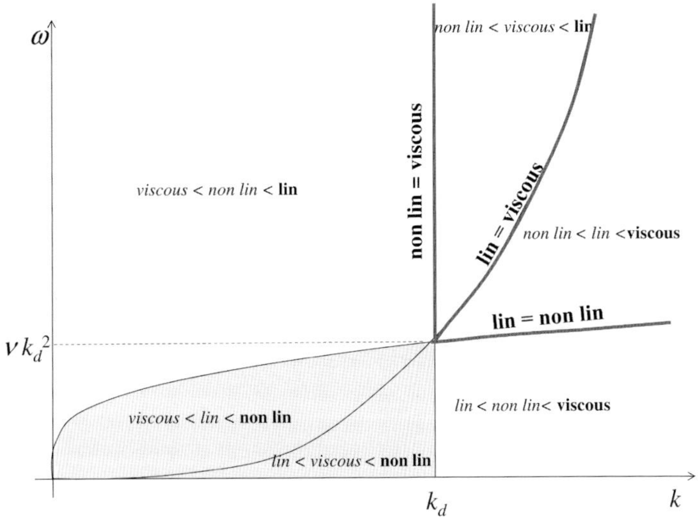

Figure 3.6 Diagram ω/k for the hydrodynamical case. The nonlinear inertial effects, scaled as ku_1, are dominant in the whole shaded zone and can balance each other. It is the so-called inertial range. Out of this range, the balance involves other forces, linear inertial (scaled as ω) or viscous (scaling as νk^2). The position of the vertical line $k = k_d$ depends on the turbulence spectrum since it corresponds to an equality $ku_1(k) = \nu k^2$ between the nonlinear terms and the viscous one (that is, a Reynolds number equal to unity)

effects to be distinguished are linear inertia, nonlinear inertia, and viscosity. The result is shown in Figure 3.6.

The Reynolds number aims at comparing the nonlinear terms $\sum k u_1$ to the viscous one νk^2. It is, therefore, defined as the ratio:

$$R = \frac{u_1}{\nu k} \,. \tag{3.52}$$

This number is of crucial importance in the case of hydrodynamic turbulence. It deserves special attention because the literature about this problem is immense. When it is larger than unity (small k's), and for sufficiently large time scales (small ω's), the nonlinear effects are dominant on the viscous and the linear ones and can balance each other. This defines the "inertial range" where the coupling of the given scale with the larger and smaller ones determines the equilibrium. On the contrary, for small scales (large k's), the viscous term is dominant and determines the evolution for the concerned "diffusion range". With this definition, the Reynolds number is a function of the considered spatial scale k. For stationary turbulence ($\omega_0 = 0$), which supposes some forcing at a large scale, the considered Reynolds number is generally the one corresponding to this large scale $k = k_0$.

Going now to nonideal MHD, the role of the nonideal term (for example resistive) is particularly important to investigate. The above study (3.3.2.6) has shown that this introduced a resistive time scale, characterized by ω_η, which allows completing the (ω, k) diagrams of Figure 3.4 or 3.5. The magnetic Reynolds number aims at comparing this nonideal term with the electric one. It is, therefore, the ratio:

$$R_{\mathrm{m}} = \frac{\omega}{\omega_\eta k^2 d_{\mathrm{e}}^2} \,. \tag{3.53}$$

Depending on the region of the (ω, k) plane considered, the symbol ω used here can correspond either to the linear term ω itself or to the nonlinear ones, which scale as $k u_{\mathrm{e1}}$. In turbulence theory, the region considered is always the region where the linear term is much smaller than the nonlinear ones. The typical case for which it is applicable is the Alfvénic turbulence in the quasi-perpendicular case $k_\perp \gg k_\parallel$. The expression of the Reynolds number is then:

$$R_{\mathrm{m}} = \frac{u_{\mathrm{e1}}}{\omega_\eta k d_{\mathrm{e}}^2} \,. \tag{3.54}$$

In the limit where u_{e1} has the same order of magnitude as the mono-fluid velocity u_1, one gets the classical expression:

$$R_{\mathrm{m}} = \frac{u_1}{\omega_\eta k d_{\mathrm{e}}^2} \,. \tag{3.55}$$

3.3.3
Dispersive and Dissipative Effects

We have shown here that the variations have to be characterized by two types of scales, ω (temporal) and k (spatial), and how the two types are related by the equations of the medium. It must be emphasized, however, that the same scale can correspond to two different behaviors, oscillatory fluctuations and secular ones. Typically, the same k can represent as well an oscillation as $\cos kx$ or a decrease as e^{-kx}. Similarly, the same ω can correspond to an oscillation as $\cos \omega t$ or a decrease as $e^{-\omega t}$. All combinations of both can exist, for instance, variations as $(\cos kx - \omega t)$ which are propagating waves, variations as $\cos \omega t e^{-kx}$ which represent a wave vanishing in space, or variations as $\cos kx e{-}\omega t$ which is a spatial oscillation damped in time. We can recognize examples of these different types of behaviors in Section 3.3.2.1. Considering for instance the boundaries drawn in Figure 3.4, one can see that the branch $\omega = \omega_{\mathrm{pe}}$ corresponds to a propagating wave, while the branch $k = \lambda_{\mathrm{D}}^{-1}$ corresponds to a spatial decrease (see the decrease of the electric field in the sheath that forms in front of a conductor, for any temporal excitation at small ω).

The information that would allow distinguishing the different kinds of variations has been lost in the above estimations by ignoring the phases, but it would have been kept in a full linear calculation. In these calculations, a real ω or a real k always represent an oscillation, while imaginary ones always represent a secular decrease. In this context, starting from a propagating wave, with both real ω and k, one can wonder whether the introduction of a new term modifies the real parts of ω and k, so modifying their relation (dispersive term), or whether it will add an imaginary part to ω or k, thus bringing some temporal or spatial damping (dissipative term).

The answer is in the linear calculation. Looking, for instance, to the Alfvén wave, the linear calculation shows that adding the Hall effect in the Ohm's law modifies the phase velocity of the wave and introduces a curvature in the (ω, k) relation (making the curve tend toward the $\omega = \omega_{ci}$ asymptote), but that no imaginary part, that is no damping, occurs. On the contrary, adding a resistive term in the Ohm's law weakly changes the dispersion but it introduces an imaginary part that is a damping.

Of course, the physical notion of dissipation is a well defined one, ruling the irreversible transfers from one form of energy to another one. Checking whether an additional effect brings dissipation or not should be viewed directly in the differential system of equations that models the system. Nevertheless, this direct investigation may be complicated so that testing whether the linear waves are damped or not often provides an efficient proxy.

3.3.4
Physical Importance of the Dimensionless Parameters

It is important to recall that *all* the physics of the phenomena occurring in a medium is ruled by the dimensionless parameters that characterize it, independently of the absolute values of the individual parameters. For illustrating this point, let us compare two media which seem a priori extremely different:

- The terrestrial magnetosheath. Located in front of the Earth in the solar wind flow, between the bow shock and the magnetopause boundary, it is a rather turbulent medium, with a density of about $50\,\text{cm}^{-3}$, the temperature is about 10^6 K, and the magnetic field about 20 nT.
- The plasma inside the tokamak reactor project for magnetic fusion ITER. The density is about $10^{14}\,\text{cm}^{-3}$, the temperature is about 10^8 K, and the magnetic field about 5 T.

At first sight, one could hastily conclude that the ITER medium is so much denser and hotter than the spatial one that nothing can be compared in the physics of the two media. Looking to the beta parameter, which tells whether the pressure effects are more important than the magnetic ones in determining the MHD plasma dynamics, one finds that $\beta \approx 10^{-2}$ in a tokamak and larger than unity in the magnetosheath. This means that, even if the two media are indeed different, the difference is not in the intuitive sense; the ITER plasma must be considered "cold" with regard to the magnetic field, while the magnetosheath is "hot". The coldness of the ITER plasma is even a mandatory condition in an experiment designed to confine the plasma, since a β larger than unity would correspond to a medium where the thermal effects are stronger than the magnetic confinement ones, which is the opposite of the searched goal. Other regions of the magnetosphere, closer to the Earth and of its intrinsic magnetic source, have β's less than unity, so being more directly comparable to the tokamak conditions. The so-called saw tooth instability is a MHD instability, which can break in a pseudoperiodic manner the magnetic configuration of a tokamak, so degrading its confinement. Quite similarly, the Earth's magnetotail undergoes brutal and quasi-periodic magnetic reconfigurations called "magnetic substorms", at the origin of the auroral precipitations. In spite of the huge differences between the orders of magnitude of the plasma parameters, the relative similarity of the dimensionless ones makes likely that a comparable physics underlies the two phenomena.

4
Collisional–Collisionless

4.1
Notion of Collisions in Plasma Physics

4.1.1
Coulomb Interaction: A Long Range Interaction

The common intuition about collisions mostly comes from our knowledge on neutral gases. In these media, the meaning of the word "collision" is then the usual one, the same as when one speaks of collisions between cars. It is not the case when speaking about collisions in plasmas. Because of the long range character of the Coulomb interaction prevailing in this case, the word is then used with an extended meaning, which could be misleading if some warnings were not given.

Usual gases are made of neutral atoms. These atoms are, of course, made of charged particles, but these charges are bound together so that the global charge of each of them is zero. The electrostatic long range interaction, in $1/r^2$, therefore, cancels, the effective interaction field decreasing much faster around each atom ($1/r^7$ for the interaction of two induced dipoles, cf. Van der Waals force). The interaction, therefore, only occurs at very short distance, which means that the particles don't interact with each other except when one particle comes extremely close to another, that is, when they "collide", in the usual sense of this term (see Figure 4.1). Between two collisions, each atom has a straight line trajectory or a gently curved one, under only the influence of external forces such as gravitation, without role of the other particles. During a collision, the interaction is short, and the two atoms abruptly exchange momentum and energy, changing drastically the direction of their trajectory. This kind of collision corresponds to the Boltzmann model, which is well imaged by the collisions of hard spheres like billiard balls.

In a fully ionized plasma, all the charged particles, positive and negative, are free, and the long range interactions, in $1/r^2$, can no longer be ignored. When the plasmas are sufficiently dilute, as they are in many space and astrophysical contexts, they are said to be "collisionless". It does not mean that each particle ignores the others, as one could think from the above hard sphere image; on the contrary, its trajectory is influenced by all the others, the closest and the furthest ones, via only the collective field they create, in an indiscernible manner. This collective field is

Collisionless Plasmas in Astrophysics, First Edition. Gérard Belmont, Roland Grappin, Fabrice Mottez, Filippo Pantellini, and Guy Pelletier.

Figure 4.1 Trajectory of a particle in a neutral gas. It is made of line segments between random and abrupt strong deviations.

indeed the only field that appears on the LHS of equations such as Maxwell–Vlasov or Maxwell–Boltzmann. It can take any value in agreement with these equations, that corresponds to the averaged "macroscopic" charge and current densities. It can in particular be zero when the plasma, at sufficiently large scale, is homogeneous and statistically neutral. When a plasma is said "collisional", which occurs when it is less dilute as in most of the laboratory experiments, it means that the description with only the collective field becomes insufficient and that some correction must be added to the RHS of the Maxwell–Boltzmann equations. If following a particle trajectory, it means that it is no longer determined by only the collective field, due to the other particles as a whole, all these particles being indiscernible, but that it becomes slightly sensitive to their exact spatial repartition, which is fluctuating in a random manner. In other words, the particle begins to be sensitive to the "granular" nature of the field sources, which can no longer be viewed as a pure continuous medium. The deviation with respect to the "collective" trajectory was, in the hard sphere model, null most of the time and strong during episodic and very short time, due to binary collisions. In a collisional plasma, it is progressive and "quasi-collective" in the sense that it involves a large number of remote particles.

Figure 4.2 is intended to illustrate how a particle feels the "granularity" of the plasma. It is done by the following numerical experiment: a large number of positive and negative particles is placed randomly in a cubic box of dimension L and the electric field is calculated along the x axis passing through the center of the box. This electric field is obtained by just adding at each point the elementary fields due to all individual particles. An equal number of positive and negative particles ($N = 10^7$ particles of charges respectively q and $-q$) has been distributed randomly in a uniform manner in the box and $\delta N = 0.05 N$ negative particles have been added randomly in a plane layer around $x = 0$ with a normal distribution of standard deviation $\sigma = 0.1\ L$ in x to make a slight departure from the neutrality at the macroscopic scales. The three components of the electric field have been calculated, but only the x component is plotted here. The macroscopic nonneutrality so created can clearly be observed in the form of a bipolar collective field that could be isolated by just smoothing the result. This collective field is exactly what would be

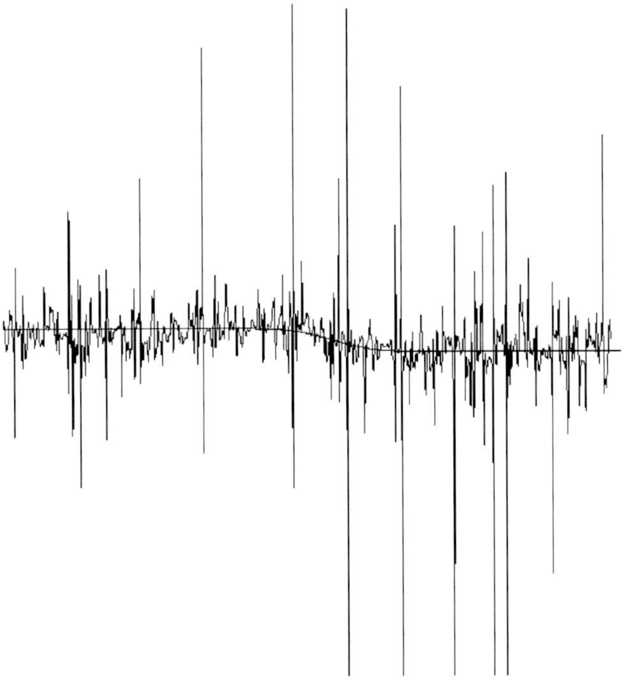

Figure 4.2 Electric field calculated on the x axis for a large number of steady particles randomly distributed in a square box (see text). The thin line shows the collective field alone. The difference between the actual field and the collective one can be labeled as "collision field". The largest scales, in spite of their small amplitudes, are the most efficient for deviating the particles, while the more evident short and high peaks have almost no effect on their trajectories.

found if the medium was modeled by a continuous space charge density $\rho = \delta n q$ with a profile $\rho(x)$ corresponding to the statistical repartition of the discrete particles. The thick line is obtained in this way. But Figure 4.2 shows that noticeable departures exist with respect to only this collective field. The first striking features are large spikes of very short lengths. These large spikes occur at all points where the x axis is very close to one particle. When looking to smaller variations, one can observe that they are more numerous and have a larger scale. The most visible ones still correspond to passages near individual particles, but less closely. When considering larger and larger scales, they are due to more and more remote particles, whose contributions overlap more and more. The largest scales correspond to the collective field for which, as already explained, all the particles intervene in an indiscernible manner, even the most remote of them.

What is the effect of this fluctuating "collision field" on a particle trajectory? It depends on its velocity. Let us consider a particle launched with a velocity v_0, in the field calculated for Figure 4.2. Such trajectories are displayed in Figure 4.3, showing that it depends noticeably on the initial velocity. The upper trajectories have been drawn for a "high" velocity particle (see next section for a precise definition of "high"), with several realizations for the random particle positions. One can see that all these trajectories are almost identical, and very close to the trajectory deriving from the collective field alone, calculated from the mean space charge density (dashed line). The slight bending of these trajectories is then only due to the collective field, and they can be considered as "collisionless" ones. In the lower part, the trajectories have been calculated for a smaller initial velocity (eight times smaller). The result is noticeably different; even if the trajectories don't look like the hard sphere ones seen in Figure 4.1, they also scatter far from the collective trajectory, due to the collision field, which has then enough time to be efficient. One can see that the deviation only involves large-scale curvatures, without any "strong deviation". The predominant role of the "almost collective" field, due to numerous remote particles, will be confirmed by the calculation of the next section. These particles actually provide smaller fluctuations than the closer ones, but their role is nevertheless dominant, since they are more numerous and they fluctuate for longer durations, due to their larger scale effect. The large spikes shown in Figure 4.2, which could be quite reminiscent of the usual concept of binary collisions, are actually not efficient for deviating the particles from the collective trajectory (that is, due to only the collective field); with the most usual velocities, their interaction time is too short. This confirms that the notion of collision in a fully ionized plasma is quite different from the usual binary image; it does not correspond to isolating strong deviations, but to a sum of many cumulative small effects due to overlapping peaks. If comparing Figures 4.1 and 4.3, one can say that the equiva-

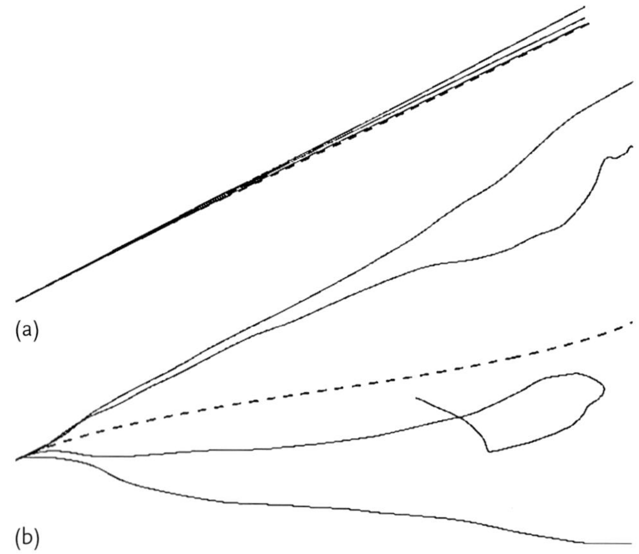

Figure 4.3 Trajectory of a charged particle in the electric field of Figure 4.2. (a) The trajectories correspond to a high velocity particle. (b) These trajectories correspond to a low velocity particle. The ones in (a) are almost fully determined by the collective field, while the others (b) are scattered by the collision field. In each case, the trajectory corresponding to only the collective field is plotted as a dashed line for reference.

lent "collision frequency" (for the slowest particle considered) is smaller, with the parameters used, in the second case than in the first one since getting a noticeable deviation demands about all the box dimensions in the second case while strong deviations were observed about 30 times in the first one. The increasing role of the collisions for decreasing particle velocities will also be confirmed by the calculation of the next section. However, one must keep in mind that the "collision field" is much more difficult to model than the fully collective one. In kinetic equations, it is responsible for the RH term, where different "collision operators" can be used for estimating its effect as a function of the distribution function (see Chapter 2). Let us outline finally that we only treat in this book the so-called Coulomb collisions. It means that the only interactions between particles we consider are those where the Coulomb force has no other effect than deviating the trajectories. It is the case of the fully ionized plasmas containing only elementary particles such as electrons and protons. In plasmas containing neutral particles and any kind of complex atoms, the "collisions" can also involve phenomena such as excitations, ionization, charge exchanges, and so on. In these media, one must estimate the probability of these phenomena (through various "cross sections"), which can in some instances be larger than the probability of getting a large deviation. When this is the case, it means that the strong and short peaks of the collision field are more efficient for these phenomena than for deviating the charged particles. This problem is beyond the scope of the present book.

4.1.2 Mean Free Path

The mean free path λ_{mfp} is defined as the length necessary, on average, for a particle trajectory to deviate noticeably from the "collective" one, that is, the trajectory due to only the collective field. Considering a collective trajectory which would be a straight line, one can typically estimate the mean free path as the length necessary to get a $\pi/2$ deviation from this initial direction. In the common hard sphere model, it is nothing but the mean distance between two close approaches (two "collisions"), since the trajectories are just straight lines between them and since the deviations, at each collision, are generally with a large angle. In a fully ionized plasma, it is quite different, as shown in Figure 4.3. Let us now estimate this mean free path. Consider a fully ionized plasma where the collective field is zero (that is, with a statistical neutrality at the macroscopic scales). When following a test-particle, its trajectory will, therefore, resemble one of those of Figure 4.3, except that the reference collective trajectory (dashed line) will be a straight line, say the x axis. As explained above, the actual trajectory is determined by the sum of fields created by all the individual particles of the medium. If the test-particle is launched at $x = -\infty$ along the x axis with a velocity v_0, we need to estimate the resulting deviation on a length L: $\delta\varphi(L)$ (for instance, between two abscissas $x = -L/2$ and $x = L/2$). The mean free path will then be the value of L giving a "large deviation", typically $\delta\varphi = \pi/2$.

Let us first calculate the elementary deviation $\delta\varphi(L)$ due to one single particle p_i at an abscissa $x = x_i$ and at a distance b_i from the x axis. The distance b_i is called the "impact parameter". The particle p_i is supposed immobile, which can always be supposed by choosing the right reference frame, as long a we do consider one single particle. This particle can be also presumed to be a repulsing or an attracting one, depending whether its charge has the same sign or not as the test-particle. From celestial dynamics and from the classical Rutherford calculation, the trajectory of a particle in a central force $F = k/r^2$ as the Coulomb force is a hyperbola. The total deviation between the two asymptotes is: $\tan(\delta\varphi_{\text{tot}}/2) = d_0/2b_i$. The characteristic length $d_0 = k/(1/2mv_0^2)$ is the so-called Landau length, which is the parameter by which the result depends on the initial velocity. It can be seen to be the distance from the fixed particle p_i where the potential energy is equal to the initial incident energy of the test-particle. It is, therefore, the closest approach distance for a frontal collision and it is actually an extremely short distance with all usual parameters (much smaller than the mean inter-particle distance). It can be viewed as the "size" of the particle for the electrostatic point of view. In the limit $d_0 \ll b_i$, the total deviation is small, and its expression can be simplified as:

$$\delta\varphi_{\text{tot}} = d_0/b_i \,. \tag{4.1}$$

The total deviation $\delta\varphi_{\text{tot}}$ is a priori not sufficient. Generally speaking, one would need the elementary deviation $\delta\varphi(L)$, even for distance L less than the distance necessary to go from one asymptote to the other. Unfortunately, if the total deviation has a simple expression, $\delta\varphi(L)$ does not, and can hardly be put under an analytical form. There is a limit, however, in which the complete calculation is not needed; it concerns the particles with impact parameters satisfying:

$$d_0 \ll b_i \ll L^* \,. \tag{4.2}$$

The upper limit L^* is equal to the half segment length $L/2$ whenever the particle p_i considered is not too close to the extremities of the segment. (Otherwise, it should be taken equal to $L/2 - \delta x_i$, where δx_i the shortest distance from the two extremities). In these conditions, which still leave one with a very large range of possible impact parameters since L is much larger than the mean interparticle distance, the test-particle undergoes a deviation $\delta\varphi(L)$ almost equal to the total deviation: $\delta\varphi_{\text{tot}}$. We will see that considering only the particle p_i verifying these conditions can be sufficient to find a correct estimate of the "global" deviation, due to all of them.

To estimate the global deviation, the above individual result has to be summed over all the surrounding particles. Considering all the particles that have the same impact parameter b will amount to considering a cylindrical shell around the axis x. If we assume the particles have a random distribution on this cylinder azimuthally, their deviations don't add linearly, but the squared deviations are to be added instead:

$$\langle\delta\varphi_b^2\rangle = \sum \delta\varphi_{bi}^2 = \delta\varphi_{bi}^2 \delta n(b) \,. \tag{4.3}$$

The number $\delta n(b)$ is the number of particles in the cylindrical shell of thickness δb. It can be estimated by $\delta n(b) = 2\pi n L b \delta b$, assuming a uniform local density n. The random hypothesis could be questioned, but it is clear that any coherent feature would result in a collective field, which we are not interested in. To focus on the question of collisions, we simply assume here that there is no mean charge density, so that the mean value of the deviation is zero. We thus focus only on the standard deviation of this deviation.

The total deviation is finally obtained by a summation over all the impact parameters. Replacing the discrete summations by integrals, we get:

$$\langle \delta\varphi^2 \rangle = 2\pi n L \int \langle \delta\varphi_b^2 \rangle \, b \, db \,. \tag{4.4}$$

If one retains the approximate expression (4.1) for $\delta\varphi_b^2$, this result is:

$$\langle \delta\varphi^2 \rangle = 2\pi n L d_0^2 \int \frac{db}{b} \,. \tag{4.5}$$

In order to take all the surrounding particles into account, all the impact parameters between $b = 0$ and $b = \infty$ must a priori be included in the calculation. But it does not mean that the approximate integral (4.5) has to be taken between these two limits; one can even see that this integral would actually diverge for both of them. There are two reasons for that: (i) the approximate expression (4.1) used for $\langle \delta\varphi_b^2 \rangle$ is valid neither for the too small nor for the too large values of b; (ii) calculating the deviation of a test-particle trajectory in a frozen field created by a fixed and uniform distribution of particles is not always sufficient. The plasma actually reacts self-consistently to all charge densities, even those which are statistical effects due to the finite number of discrete particles. This may change noticeably the above result at large scale, when this reaction has time to develop along the test-particle trajectory.

- Approximate value of $\langle \delta\varphi_b^2 \rangle$. Close to the lower limit $b = 0$, replacing the approximate expression (4.1) by the exact one, coming from $\tan(\delta\varphi_{\text{tot}}/2) = d_0/2b$ would obviously avoid the divergence since, for $b \ll d_0$, the angle $\delta\varphi_{\text{tot}}$ tends to π instead of infinity with the approximate expression. Another, simpler, way of doing a reasonable estimation is just to replace the $b = 0$ limit by a limit $b = d_0$, which does not change the result much. Concerning the upper limit, the correction would be a little more complicated; for the values of b larger than L, the deviation $\delta\varphi$ of a particle trajectory on the length L becomes smaller and smaller with respect to the total deviation $\delta\varphi_{\text{tot}}$ since the radius of curvature becomes smaller and smaller. The decrease of $\langle \delta\varphi_b^2 \rangle$ would then be as b^{-2} instead of b^{-1}. This does guarantee the convergence. The complete calculation is quite feasible, even if cumbersome. In any case, it is useless because of the second point, concerning the self-consistent reaction of the plasma. This effect actually takes place for scales larger than a scale λ_{sc} which is generally much shorter than L.
- Self-consistent reaction of the plasma. The above calculation has been done in the hypothesis of uniformly distributed particles, giving rise to a zero collective field. It allows one to consider the effects of all the diffusing particles as incoherent and add the squared deviations instead of the deviation themselves. However, this separation between collective and collision fields actually has some

limits. Defining the collective field as a spatial average over volumes containing many individual particles, the collision field appears to be due to the statistical differences from this average. One can understand that this collision field cannot be fully decoupled from the collective one. Any statistical difference indeed corresponds to a nonzero charge density and an associated electric field; this electric field acts on all the neighboring particles, which, in turn, react self-consistently and in a collective manner, tending to oppose the original electric field and bring the system back to neutrality (at least in a nonmagnetized plasma). This is the way the collision field is coupled to the collective one.

Because of the above self-consistent reaction, the collision field can no longer be considered constant in time during the whole time necessary for a test-particle to cross the length L. Above some characteristic scale λ_{sc} (sc as self-consistent), the plasma's self-consistent reaction always makes the field variations oscillating, which cancels the efficiency of the too remote particles for deviating the trajectory of the test-particle. Because of this, the deviation due to the particles with impact parameters $b > \lambda_{sc}$, instead of increasing as the square root of the number of particles, as supposed in the above calculation, actually decreases back to zero.

It is the reason why the above estimation, which neglects the self-consistent reaction of the plasma, must be cut at this upper scale. This conveniently avoids the cumbersome calculation of the upper contribution of the mean free path integral. In most conditions (especially in nonmagnetized plasmas), this upper limit will be shown to be the "Debye length".

Once the two limits of the integral (4.5) have been changed as explained to d_0 and λ_{sc}, the result can easily be found:

$$\delta\varphi_b^2 = 2\pi n L d_0^2 \ln \frac{\lambda_{sc}}{d_0} . \tag{4.6}$$

The mean free path is defined by a global deviation of the order of $\pi/2$. Its value is, therefore:

$$\frac{\lambda_{mfp}}{d} = \frac{\pi}{8 \ln \Lambda} \frac{d^2}{d_0^2} . \tag{4.7}$$

The result has been normalized to the mean interparticle distance $d = n^{-1/3}$. The parameter Λ stands for λ_{sc}/d_0, and $\ln \Lambda$ is called the "Coulomb logarithm". In almost all fully ionized plasmas, Λ is a very large number, typically between 10^4 and 10^{13}. However, since the dependence on Λ is only logarithmic, this number does not much influence the mean free path value (a precise determination of the upper boundary λ_{sc} is thus not a crucial issue). On the contrary, one can see that, whenever the interparticle distance is much larger than the Landau length, $d/d_0 \gg 1$, which is also the case for almost all usual plasmas, the mean free path is very much larger than the interparticle distance. This is consistent with the idea that what is called a collision in plasmas is actually the deviation resulting from a large number of weak interactions with remote particles.

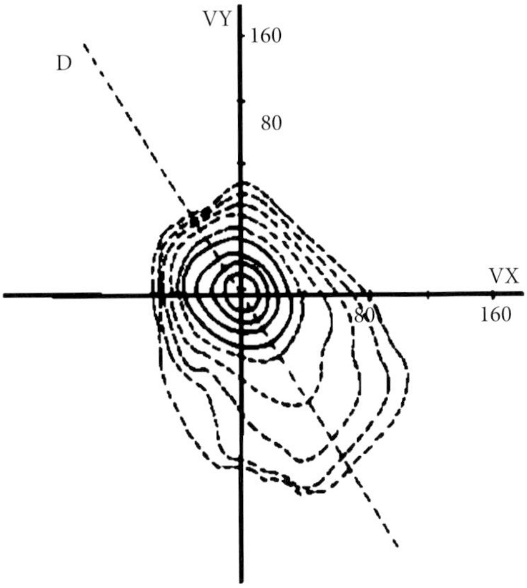

Figure 4.4 Role of collisions in the slow solar wind ($V_{sw} = 377$ km/s). The proton velocity distribution is measured by the Helios spacecraft. The 2D cut corresponds to the plane defined by the solar velocity (x axis) and the static magnetic field (dashed line). One can observe that the inner part is isotropized by collisions (small speeds) while the outer one is not (high speeds). From [11].

It is worth noting that the mean free path λ_{mfp} depends on the initial velocity of the test-particle via $1/d_0^2$ and therefore as $1/v_0^4$. This strong dependence on v_0 is also quite consistent with our first qualitative remarks: the slow particles are sensitive to the short electric peaks, whose amplitude is very strong (close particles), while the fast particles are only sensitive to the long lasting electric peaks, whose amplitudes are smaller (remote particles). This effect has visible consequences; a proton distribution measured in the solar wind, at about one astronomical unit, is shown in Figure 4.4. One can observe that the low energy particles have been isotropized and Maxwellianized by the collisions while the higher energy ones remain anisotropic, with higher velocities in the magnetic field direction.

A collision frequency can be deduced from the mean free path. A particle with a velocity v_0 is given by $\omega_{col} = v_0/\lambda_{mfp}$, that is, typically, for a thermal electron, $\omega_{col} = V_{the}/\lambda_{mfp}$.

4.1.3 The Debye Length and the Notion of Debye "Screening"

The most remote particles are the most efficient for scattering the trajectory of a test-particle because, in spite of their weaker field, they are more numerous and act on a longer distance. Nevertheless, as explained above, for scales larger than λ_{sc}, this effect is limited by the self-consistent reaction of the plasma, which prevents the too remote particles from acting coherently during the whole mean free path. To determine the effective value of λ_{sc}, one must, therefore, look at the self consistent motion of the electrons. This is done in Chapter 5 and it is shown that a charge separation, in a nonmagnetized plasma, is always oscillating at a frequency of the order of the plasma frequency $\omega \approx \omega_{pe}$ for all spatial scales larger than the Debye length λ_D. The definitions of the plasma frequency and of the Debye length are given in these chapters. This result can easily be proved in a nonmagnetized plasma. It can also be shown to be generally valid in a magnetized one, although the demonstration is less easy and even invalid when the gradients are too close to perpendicular to the static magnetic field.

Taking into account the oscillating nature of the large-scale electric field at ω_{pe}, a test-electron moving at V_{the} will never cross coherent peaks longer than $\lambda_D = V_{the}/\omega_{pe}$. For impact parameters $b > \lambda_D$, this means once again that the deviation of the particle during one individual peak will not be a complete one. Adding properly all these partial deviations does provide a rapidly converging integral, which makes negligible the contribution of this entire part $b > \lambda_D$ of the integral. This is the argument that allows choosing $\lambda_{sc} = \lambda_D$ as the upper boundary for the integral. If one introduces a large-scale body in a plasma (for instance, an antenna), with an imposed charge evolving slowly, it is easy to check that the electric potential always decreases exponentially away from the body with a characteristic scale λ_D. It is due to the self reorganization of the electron population, which makes a sheath and screens the body charge. This is the basic notion of Debye screening. This notion is, of course, applicable for large-scale bodies and its use for interpreting microscopic effects is actually perilous. It is, however, tempting to try to apply it to the individual ions and tell that the electric potential of each of them is a "screened Coulomb potential" $\Phi = qe^{-k_D r}/(4\pi\epsilon_0 r)$, instead of simply $\Phi = q/(4\pi\epsilon_0 r)$ in vacuum. This model is sometimes used for some applications and it is quite popular in plasma vulgarization. It indeed allows recovering several exact results of plasma physics. With respect to the previous section, it is clear in particular that screening the potential of each ion at λ_D will be equivalent, for the calculation of the mean free path, to cutting the integral at λ_D. Nevertheless, it must be understood that this simplistic view introduces a wrong image of the physics really involved. As explained above, the reason of the cutting at λ_D is a temporal effect. It cannot, therefore, be properly represented by any spatial and static image. This point is illustrated in Figure 4.5; supposing that the electrons have reorganized in such a way to screen the potential of one particular ion (the central one in this figure), it is clear that they cannot screen the potential of all the neighboring ions at the same time. For this obvious reason, the idea that the potentials of all the individual ions are screened simultaneously and in a quasi-static manner by some organization of the electrons is inconsistent.

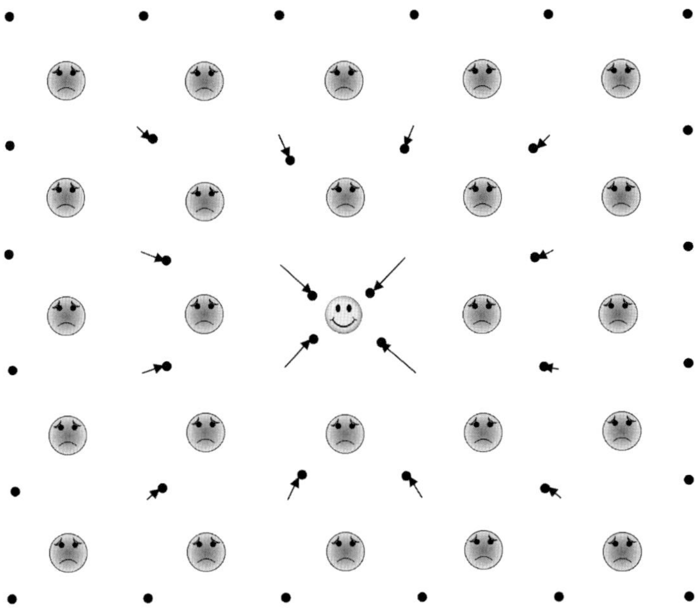

Figure 4.5 Notion of Debye screening for the individual ions, a traditional but misleading view. The figure shows that, if the electrons concentrate in a quasi-static manner around the central ion and are so able to screen its potential (happy smiley), they are obviously unable to screen at the same time the potential of all the other ions (unhappy smileys). The temporal variations cannot be ignored when interpreting why the deviation of one test-electron crossing the medium is insensitive to the ions located at more than one λ_D away from its trajectory.

4.1.4 Knudsen Number

The Knudsen number is the dimensionless parameter that allows quantifying the importance of the collisions for a given population with spatial variations at a scale k^{-1}. It is, therefore, the number:

$$K_n = k\lambda_{\text{mfp}} \,. \tag{4.8}$$

This standard spatial definition can be replaced by a temporal one when better suited to the study. For a phenomenon with a temporal time scale ω^{-1} and a collision frequency characterized by ω_{col}, one can introduce:

$$K'_n = \omega/\omega_{\text{col}} \,.$$

Considering the electron collisions, the first definition is better suited to "spatial" variations: $\omega/k \ll V_{\text{the}}$ and the second one for "temporal" variations $\omega/k \gg V_{\text{the}}$. Both are equivalent when $\omega/k \approx V_{\text{the}}$. To remain general, one must keep K_n as the largest value between the two.

When K_n is much larger than unity, the plasma can be said "collisionless". When it is much smaller than unity, the plasma is strongly collisional (for the given phe-

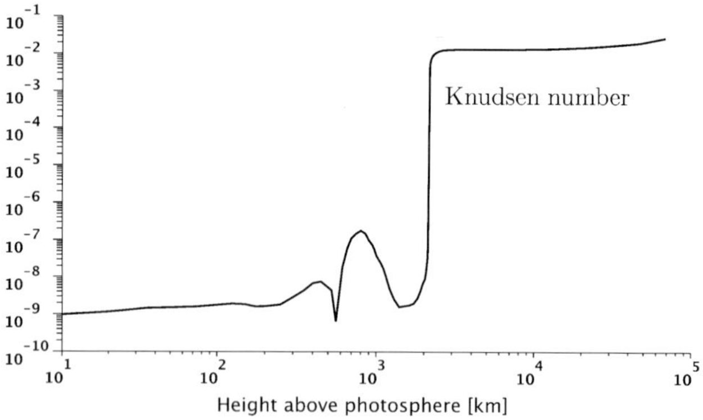

Figure 4.6 Knudsen number (Eq. (4.8)) in the corona and solar wind. The mean free path λ_{mfp} corresponds to the electron–electron collisions and the characteristic scale k^{-1} considered is the scale height of the temperature profile. The plasma goes rapidly from strongly collisional to weakly collisional in the range of altitudes considered. This figure was built from the model of solar atmosphere [12].

nomenon). The latter case is the most documented one since it corresponds to all the well-known theories of gas dynamics, hydrodynamics, thermodynamics, and all classical "fluid" theories. The former is less ancient, but still relatively developed, in particular in space physics and in magnetic fusion laboratory experiments. The case $K_n \approx 1$ is the most difficult since the collisions are then too numerous to be neglected, but not numerous enough to justify the usual fluid approximations. Figure 4.6 shows the collisional situation of the solar wind plasma. The collisions considered are the electron–electron ones, for a velocity $v \approx V_{\text{the}}$, the distances are relatively close to the Sun (photosphere, chromosphere and low corona) and the characteristic scale k^{-1} considered is the scale height of the temperature variations, as estimated from measurements. It can be seen that the solar wind is strongly collisional up to the chromosphere, but becomes much less collisional above the transition layer, tending to approach the difficult situation $K_n \approx 1$ away from it. Further into the solar wind (about the Earth's orbit, for instance), the Knudsen number can be estimated to be between 0.1 and 1. This is one of the reasons which make the solar wind modeling difficult.

The vicinity of the strong collisional case (small K_n) can be analyzed in a quite rigorous way by using the so-called Chapman–Enskog procedure. It allows us to derive, for instance, all the classical transport coefficients such as viscosity or thermal conduction. More generally, we have seen that the fluid theories relate the macroscopic variables ρ (mass density), $\boldsymbol{u}$ (fluid velocity), $\boldsymbol{P}$, pressure tensor, and $\boldsymbol{Q}$ (heat flux tensor). These four variables are related by three universal equations, based on conservation laws and therefore, independent of the character, collisional or not, of the medium. The fourth equation, called the "closure equation", and which is needed to get a predictive model, is a more approximative one, and is dependent on the collisionality via the symmetries of the distribution function. This last equation can be derived from the Chapman–Enskog procedure when K_n is small enough.

We will not explain this procedure in detail, but just give a general overview about its principles. Let us write the kinetic equation under the form $D_t(f) = C(f)$ of the kinetic equation, where $D_t = \partial_t + \boldsymbol{v}\cdot\nabla + \boldsymbol{F}/m\cdot\nabla_v$ is the total derivative in phase space and $C(f)$ is the collision integral operator. The procedure is a perturbative method with respect to the small parameter K_n. The order of magnitude of the RHS is $\omega_{\text{col}}\delta_f$, while the LHS is $\omega\delta_f$ if considering, for instance, "temporal" variations.

- at zero order in K_n, the kinetic equation is just: $C(f_0) = 0$. For realistic forms of the collision operator C, this just imposes that f_0 has a Maxwellian form. Fully neglecting the LHS terms just corresponds to the usual hypotheses of a complete thermodynamic equilibrium, no force and a uniform and stationary medium.
- at first order, the kinetic equation is: $f_1 C'(f_0) = D_t(f_0)$. This corresponds to the usual extension of thermodynamics to LTE (Local Thermodynamic Equilibrium). Provided that the collision operator C is known, this allows one to determine the closure equation and the transport coefficients.

Going to further orders is quite cumbersome and rarely practiced.

It must be kept in mind that this procedure, as all the deriving notions such as viscosity, are rigorously defined only in the case of sufficiently collisional media. It is an expansion in K_n and using it for K_n of the order of unity or larger is meaningless. Finding closure equations in the opposite limit of collisionless plasmas can, therefore, not be based upon this kind of expansion procedure. However, it can still be possible whenever special conditions impose some generic properties (for example, symmetries) to the distribution function. The adiabatic conditions resulting from scale conditions ($\omega/k \gg V_{\text{th}}$) are the most typical example of this possibility.

4.1.5
Plasma Regimes

The above sections have shown that the plasma properties, with respect to the collisions, depend on three main characteristic scales: the mean interparticle distance d, the Landau length d_0 and the Debye length λ_{D}. The first one is just a measure of the density ($d = n^{-1/3}$). The second characterizes the limit for a strong deviation by the electrostatic potential of a single particle. In the case of a repulsing potential, it can also been viewed as the closest approach distance in this potential. The third one is the largest distance to be taken into account when summing weak deviations to estimate the mean free path (the mean free path itself can be said to be a characteristic length, but we have seen that it results from the three other ones).

As already outlined, the plasma parameter Λ is the ratio between λ_{D} and d_0. It is huge in all standard plasmas, in space as in the laboratory and it is usually characterized by its logarithm $\ln \Lambda$, which is called the Coulomb logarithm. When the Coulomb logarithm is more than 10, the plasma is called "weakly coupled". It is the only case considered in this book. The weak coupling condition is even often called the "plasma approximation".

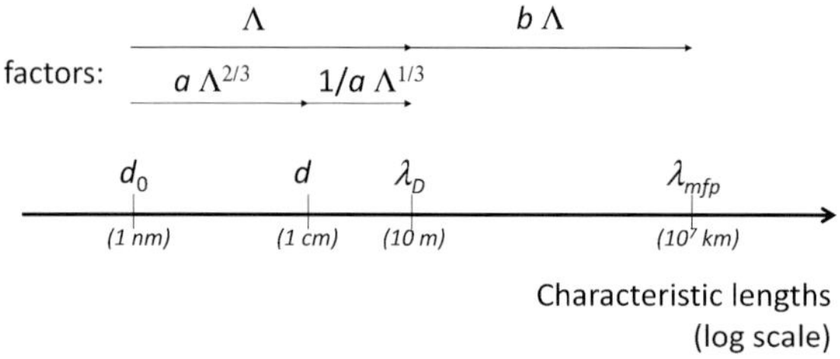

Figure 4.7 Collisions: the different characteristic lengths. Landau length (d_0), mean interparticle distance (d), Debye length (λ_D), and mean free path (λ_{mfp}). The scale is logarithmic. All ratios are huge in standard plasmas (weakly coupled) and determined by one single parameter Λ. The numerical values indicated in parentheses are given as an illustration of these huge ratios (with $\ln \Lambda = 23$ and $\lambda_D = 10$ m, which are close to the solar wind values). The coefficients are $a = (2\pi)^{1/3}$ and $b = \pi^2/(4 \ln \Lambda)$.

It is easy to check, from the very definition of these different characteristic lengths, that the ratio Λ rules all the ratio between them. This is shown in Figure 4.7. In the weak coupling standard situation, $\Lambda \gg 1$, one has $d_0 \ll d \ll \lambda_D$: the mean interparticle distance d is between the two limits, far from both the upper and lower limits. This means that the number of particles "in the Debye sphere" is very large ($\lambda_D^3/d^3 \approx \Lambda$) and the collision process is dominated by the weak deviations ($d^3/d_0^3 = \Lambda^2$). The mean free path λ_{mfp} is much larger than the Debye length ($\lambda_{mfp}/\lambda_D \approx \Lambda$), which means that the plasma can always be considered "collisionless" for all phenomena at the scale of the Debye length.

Beyond the case of weakly coupled plasmas, the coupling is said to be moderate for $2 < \ln \Lambda < 10$ and strong for $\ln \Lambda < 2$. These regimes can only be reached in plasmas extremely dense (Inertial Confinement Fusion), cold (antimatter and ultracold), or highly charged (dusty). When approaching closer to the lower limit, that is, Λ tending toward unity, the plasma is also said to become "correlated", to indicate that the strong collisions become nonnegligible. The mean free path then tends to the common limit: $\lambda_{mfp} \approx d \approx d_0$. Nevertheless, in these extreme situations, d has to be further compared with the "de Broglie" length, where the quantum effects cannot be neglected any more. When this limit is reached, the plasma is said "degenerate". It is the case, for instance, for the conduction electrons in a metal.

The distinction between "strong collisionality" and "weak collisionality" is only meaningful in the context of weakly coupled plasmas $\Lambda \gg 1$. It depends, in these conditions, upon the scale k^{-1} of the phenomena under study compared with the mean free path. As already mentioned, the plasma is said to be strongly collisional if $K_n = k\lambda_{mfp} \ll 1$ and weakly collisional if $K_n = k\lambda_{mfp} \gg 1$. The emphasis in this book is put on the weakly collisional, and even collisionless, case.

A last remark must be made concerning the noncollisionality. In the limit $K_n = k\lambda_{mfp} \ll 1$, one might be tempted to say that "if there is less than one single collision in a given scale, it means that there is no collision at all". This would be true for collisions of the hard sphere type. In the plasma case, if there is 0.1

collision, it means that a deviation does exist on all the particle trajectories, but is less than $\pi/2$.

4.2
Notion of Dissipation

4.2.1
Transfers of Energy and Dissipation

Two forms of energy coexist in a plasma: electromagnetic and kinetic. The kinetic energy itself is the sum over all the particle species of two different forms: the convective energy $\rho_s u_s^2/2$ (related to the fluid velocity u_s of the considered species) and the thermal energy (related to the centered second order moment P_s). For each of these energies, it is possible to write an "equation of energy", which rules its temporal evolution.

$$\partial_t \left(\epsilon_0 E^2/2 + B^2/2\mu_0\right) + \nabla \cdot (\boldsymbol{E} \times \boldsymbol{B}/\mu_0) = -\boldsymbol{j} \cdot \boldsymbol{E} \tag{4.9}$$

$$\partial_t(\rho u^2/2) + \nabla \cdot (\rho u^2/2\boldsymbol{u} + \boldsymbol{P} \cdot \boldsymbol{u}) = \nabla \cdot \boldsymbol{P} \cdot (\boldsymbol{u}) + n\,\langle \boldsymbol{F} \rangle \cdot \boldsymbol{u} \tag{4.10}$$

$$\partial_t(3/2P) + \nabla \cdot (3/2P\boldsymbol{u} + \boldsymbol{q}) = -\nabla \cdot \boldsymbol{P} \cdot (\boldsymbol{u}) + nX_F\,. \tag{4.11}$$

In the RHS terms of Eqs. (4.10) and (4.11), let us recall that the parentheses indicate that the differential operator ∇ is applied to the velocity and not to the pressure. Equation (4.9) directly derives from Maxwell equations. Equation (4.10) comes from the momentum equation (first order moment equation), dot multiplied by $\boldsymbol{u}$, with the help of the continuity equation. Equation (4.11) is half the trace of the centered second order moment equation. It must be kept in mind that all these equations are fully general and are valid in collisional plasmas as well as in collisionless ones.

For the sake of notation lightness, the subscripts s have been omitted in the two last equations: they must actually be verified by all species, ρ being their mass density, $\boldsymbol{u}$ their fluid velocity, and $\boldsymbol{P}$ their pressure tensor. The scalar P is one-third of the pressure tensor trace. The vector $\boldsymbol{q}$ is one-half of the trace of the third order moment $\boldsymbol{Q}$ (with respect to the two last subscripts). The force $\boldsymbol{F}$ acting on a particle of species s includes the electromagnetic force $\rho \boldsymbol{E} + \boldsymbol{j} \times \boldsymbol{B}$, due to both collective and "collision" fields (collisions with the same species particles and with the other ones), and possibly all other forces such as gravitation. The force $\boldsymbol{F}$ comes in Eq. (4.10) via its mean value (that is, average on the velocity space) and, in Eq. (4.11), via the term X_F, to be explained below. The mean value $\langle \boldsymbol{F} \rangle$ contains in particular the friction force due to the collisions between the considered species and all the others. It is the way the resistive dissipation comes in, since resistivity is a friction between ions and electrons.

By construction of Eq. (4.11), the general expression of the last term is $X_F = \left\langle \nabla_v \cdot \left(\boldsymbol{F}\delta v^2\right)\right\rangle$, where $\delta \boldsymbol{v} = \boldsymbol{v} - \boldsymbol{u}$, the symbol $\nabla_v \cdot (-)$ representing the divergence of

(−) with respect to velocity, and the brackets representing the averaging in velocity space. It is easy to check that the forces that do not depend on the velocity do not contribute to X_F since $\boldsymbol{F}$ can then be taken out of its expression, which simply becomes $2\boldsymbol{F} \cdot \langle \delta \boldsymbol{v} \rangle = 0$. This is the case of the electric force, the gravitational force, and all of the most common external forces. Furthermore, one can also check that the collective magnetic force, $q\boldsymbol{v} \times \boldsymbol{B}$, although it does depend on $\boldsymbol{v}$, depends on it in such a way that it does not bring any contribution either to X_F (the magnetic force does not work). On the contrary, when resistivity is present, this electron–ion friction does contribute to X_F: it is the Joule effect.

In each of the above equations, one can recognize that each energy varies in time, locally, for two different reasons:

- the divergence of an energy flux, which corresponds, when integrated over a finite volume, to the energy incoming through its boundaries
- one or several RHS terms, which are sources and sinks for the considered form of energy. They actually correspond to exchanges with the other forms of energy. Considering a finite volume, these terms don't correspond, as the previous one, to transfers of energy across the boundaries but to phenomena occurring inside the volume.

Of course, summing all the equations gives a purely "conservative" equation for the total energy, that is, an equation where all temporal variations are due to only the divergence of an energy flux, the sum of all the RHS terms being zero. The $\boldsymbol{j} \cdot \boldsymbol{E}$ term in particular appears as the sum of all the terms $nq\boldsymbol{E}$, that appear in Eq. (4.10), via the electric forces acting on the different species: $n \langle \boldsymbol{F} \rangle \cdot \boldsymbol{u} = nq\boldsymbol{u} \cdot \boldsymbol{E}$.

Dispatching the energy variations between the energy fluxes, which are on the LHS and the exchange terms, which are on the RHS, may seem arbitrary; both actually depend on the derivatives of $\boldsymbol{P}$ and $\boldsymbol{u}$. Nevertheless, the decomposition presented above is the only one that guarantees that the exchange term is frame independent since the derivatives of $\boldsymbol{u}$ have this property while $\boldsymbol{u}$ itself has not. It is worth noting also that the heat flux $\boldsymbol{q}$ is never a mean of exchange between the different forms of energy, it is just a spatial transfer of the thermal energy.

We shall now focus on the energy exchange terms and show that, depending on their characteristics, some can be described as "reversible" (nondissipative exchanges), and some as "irreversible" (dissipative exchanges).

4.2.2
The Concept of Dissipation in Collisional Fluids

In ordinary fluid dynamics, the collisions are the fastest process, and one is justified in considering an expansion in the small parameter $K_n = \lambda_c/\lambda$ (ratio between the mean free path and the scale of the gradients, called Knudsen number, see Section 4.1.4). As a result of this expansion, the pressure tensor (defined, as always in this book, as the second order centered moment) can be written as the difference of two terms: $\boldsymbol{P} = P\boldsymbol{I} - \boldsymbol{\sigma}$, where P is a scalar and $\boldsymbol{I}$ the identity tensor. The first term is isotropic and the second one is called the "viscosity tensor".

- The first term $P\boldsymbol{I}$ corresponds to the zeroth order in the collisional expansion, that is, to the homogeneous case. For a fluid made of elementary particles (typically protons and electrons), the scalar P corresponds to the internal energy $E_{\text{int}} = 3P/2$. The scalar P is simply called "pressure" in this context, the complete tensor $\boldsymbol{P}$ being then called "stress tensor" to make clear the difference. The variations of P can sometimes be determined by those of lower order moments by approximate closure equations such as $d_t(P/n^\gamma) = 0$, but this is not general; the collisional expansion does provide a closure equation, but for the heat flux $\boldsymbol{q}$, not directly for P.
- The second term $\boldsymbol{\sigma}$ can be estimated by the first order of the expansion. This viscous tensor has both diagonal and off-diagonal terms. They correspond to frictional forces opposed to velocity gradients, respectively along and perpendicular to the flow motion. One can indeed show that this component is a symmetrical tensor proportional to the velocity gradients. It is generally noted:

$$\boldsymbol{\sigma} = \mu \boldsymbol{D}_u + \zeta \nabla \cdot \boldsymbol{u} \boldsymbol{I} \tag{4.12}$$

with

$$\boldsymbol{D}_u = \nabla \boldsymbol{u} + (\nabla \boldsymbol{u})^T - 2/3 \nabla \cdot \boldsymbol{u} \boldsymbol{I} \ .$$

This form allows distinguishing the velocity variations that imply pure shears (first part) and pure compressions (second part). The second part is consistently ignored in all conditions when the fluid is quasi-incompressible. The coefficients μ and ζ are positive scalars, respectively called shear and volume viscosities.

When inserting this form of the pressure tensor, one can estimate the energy exchange term between Eqs. (4.10) and (4.11):

$$\begin{aligned} \nabla \cdot \boldsymbol{P} \cdot (\boldsymbol{u}) &= P \nabla \cdot \boldsymbol{u} - \nabla \cdot \boldsymbol{\sigma} \cdot (\boldsymbol{u}) \\ &= P \nabla \cdot \boldsymbol{u} - \mu \nabla \cdot \boldsymbol{D}_u \cdot (\boldsymbol{u}) - \zeta \left[\nabla \cdot \boldsymbol{u}\right]^2 \\ &= P \nabla \cdot \boldsymbol{u} - \mu/2 \sum \mathrm{D}^2_{u i j} - \zeta \left[\nabla \cdot \boldsymbol{u}\right]^2 \ . \end{aligned} \tag{4.13}$$

To find the last expression of the term with μ, one has to make use of the symmetry of the tensor $\boldsymbol{D}_u$ and the property that its trace is zero.

This result clearly shows that, contrary to the ideal term in P, the viscous terms, with μ and ζ, are strictly negative. This means that the viscosity always induces energy transfers in the same direction, from convective to thermal. This irreversible transfer is what is called "dissipation". It has important consequences, which can be observed in many fluid applications. For instance:

- A linear sound wave can be damped by viscosity, but never amplified,
- The heating of the flow when crossing a shock wave (see Figure 4.8) is always larger than adiabatic and never smaller. Irreversibility is indeed essential in shocks since, to be stationary, a shock must involve viscosity to oppose the indefinite steepening of the velocity profile due to nonlinear effects. Reversing the

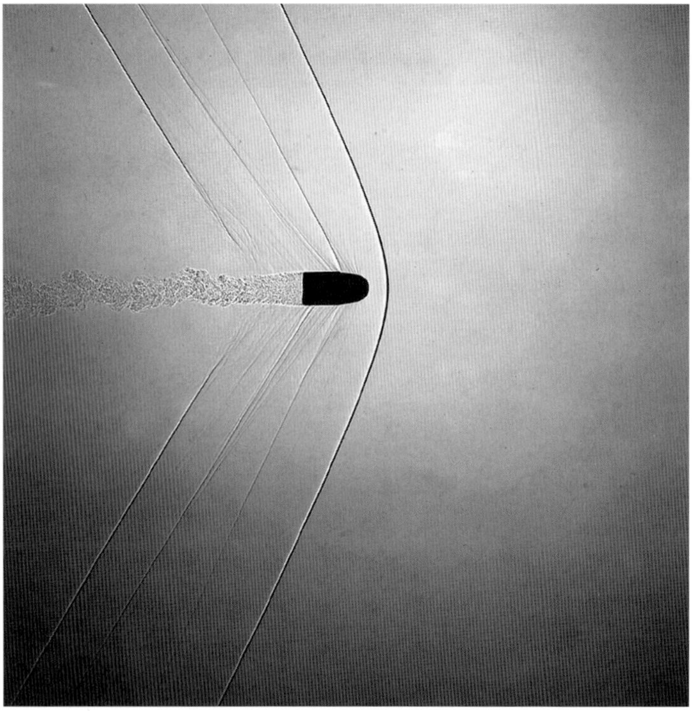

Figure 4.8 When a flow has a supersonic velocity with respect to an obstacle (here a bullet), its velocity decreases to zero at the obstacle's surface through one or several shock waves. These structures involve essentially irreversible phenomena since the shock layers are made stationary by the viscous terms inside them. The viscous terms are negligible everywhere else, where the gradients are not so high. Credit H. Edgerton.

flow velocity, that is, considering a flow evolving from sub- to supersonic}?>, would clearly not correspond to a stationary state since the viscous effects could not oppose the nonlinear flattening which would occur in this case (although it would verify as well the Rankine–Hugoniot conservation laws, see Chapter 6).

The general definition of "dissipation" which we propose here generalizes what has just been described in the above section: "*dissipation is an irreversible transfer of energy from one form to another*". In the usual collisional fluid case, the final form is always thermal, and the transition from convective (or magnetic) energy to thermal energy clearly corresponds to the passage from "coherent" to "incoherent". In the collisionless case, the classical notion of viscosity no longer holds, and the term $\boldsymbol{\sigma}$ is not necessarily related to velocity gradients. The distinction between convective and thermal energies is then not so clearly associated with the distinction between "coherent" and "incoherent", the velocity distribution functions being possibly quite different from shifted Maxwellians. The notion of reversibility itself is then not so clear and requires a more general definition.

4.2.3 Reversibility

The notion of reversibility (which is actually not restricted to the question of energy dissipation), is of major importance in plasma physics. The general definition can be expressed as: "*a physical system is said reversible if, for any of its evolutions between states A and B, it is possible to get the reverse one between B and A.*" How must one manage to get the reverse evolution when the system is reversible? How can one recognize that a given system of equations models – or not – a reversible physical system? These are the two main questions we shall answer in this section. But let us first give some simple illustrations to show that the notion of reversibility does possess a clear intuitive content.

1. Let us consider a single rubber ball bouncing on the floor. If the bounces are elastic, the bounces are perfectly symmetric in time and, if this motion is filmed, the resulting movie can be projected in the right sense or in the wrong one without the viewer noticing anything which would seem to violate the physical laws. On the contrary, if the bounces are not elastic, some energy is lost at each bounce and the bounce amplitude decreases systematically with time. When looking at the movie, seeing the ball bouncing higher and higher would be a clear indication that it is viewed in the wrong sense. This means that the viewer is quite aware that some motions are possible, but that the reverse motions are not, whatever the manner the balls are launched, and, therefore, that the system is irreversible.
2. Let us consider now the motion of several balls on a billiard board. This system is reversible as long as the balls have no friction on the board and as the collisions between them and with the edges are elastic. The film can be seen once again in any sense, and the motion of the balls will still look like possible ones. But if the friction is not negligible, or if the collisions are not elastic, the physics becomes irreversible; friction always makes the velocity of the balls decrease with time, and never the opposite. In the movie, if the balls are accelerated instead of being slowed down, one can easily guess that it is projected in the wrong sense.
3. Going away from these simple mechanical examples, one can find an infinite number of examples in everyday life where the notion of irreversibility is intuitively obvious and where some reverse evolutions appear clearly unlikely or even impossible. Seeing bank notes burning (Figure 4.12) and transforming into ashes is sad, but possible. Seeing the reverse evolution, that is, seeing ashes gathering and being transformed into bank notes is clearly much less likely. This example will be taken hereafter as an illustration for the notion of probability of an evolution, in connection with the concept of entropy.

Let us now try to be more specific and give a form more operational for the physicist to the general definition above.

We consider a system whose state is characterized, at each time t by a "state vector" $S(t)$ which is the set of variables that contains all the available knowledge on the system. In the mechanical example above of billiard balls, the state vector S is made of all the positions and velocities of the balls; in fluid mechanics, it is made of the three functions of space: density $n(\boldsymbol{r})$, velocity $\boldsymbol{u}(\boldsymbol{r})$, and pressure $P(\boldsymbol{r})$. In kinetic plasmas, it contains the distribution functions of all the species $f_s(\boldsymbol{v}, \boldsymbol{r})$ and the electromagnetic fields $\boldsymbol{E}(\boldsymbol{r})$ and $\boldsymbol{B}(\boldsymbol{r})$, and so on.

How can one define a "reverse" evolution? It is the evolution obtained by changing the sign of all variables whose definition is odd with respect to time, the space being yet unchanged. This means in particular that the sign of all velocities $\boldsymbol{v}$ must be changed, since $\boldsymbol{v} = d_t(\boldsymbol{r})$ is indeed odd in time, as well as for all the variables having an odd dependence on $\boldsymbol{v}$, such as the heat flux, for instance, in fluid mechanics. The forces remain unchanged in such a reversal. Considering the electromagnetic forces, it shows that the sign of $\boldsymbol{B}$ must be changed, but not the change of $\boldsymbol{E}$.

Thanks to this definition, the reversibility can the be tested as follows.

Suppose that, after having been started in a state S_i at time t_i, the system has reached a state S_f at time $t_f = t_i + \delta t$. Let us now reverse the final state, as described above. If the system is reversible, the system will then follow a backward evolution which will be the exact "mirror" of the forward one. It will in particular go back to its initial state after the same delay δt.

Let us go back to the example of the bouncing rubber ball: we suppose that it is dropped without initial velocity at time $t = 0$ at a height $z = z_0$, and that its velocity is measured at time δt. If the ball is launched again with the opposite velocity, what will happen? If the bounces are elastic, the motion of the ball will then go back to the same height with the same null velocity, after the same delay δt. If some energy is lost at each bounce, the motion will become slower and reach smaller heights. With the image in mind of a movie viewed forward or backward, the reversal of the final conditions exactly amounts to a "time reversal" if the system is reversible, but rapidly differs from it when some irreversibility is present.

In the billiard balls example, one can imagine that all the balls are organized initially in a central triangle, except one, which is initially launched with a fast velocity from one edge and hits the others. To test the reversibility of this system, one should stop it at some time t_f, measure all the velocities, and relaunch all the balls at the same positions, but with opposite velocities. If the result is the "reverse film", one returns to the triangle plus the isolated ball, and the system can be said to be reversible. If the result is different, it is irreversible.

A warning must be given about the image of the reversed movie, which can sometimes be misleading. Some evolutions can look unlikely, not because they are physically irreversible, but because we know that some initial conditions don't exist in nature. For instance, seeing a film where a river goes up to its source would clearly indicate that the film runs backward, whatever the physics of the flow, reversible or not. It is simply because we know that nobody can launch the flow from the sea with the necessary upstream velocities. The definition must of course remain insensitive to these considerations and no evolution must be rejected because of a particular selection of the initial conditions: all the reversals must be admitted. We will see hereafter that, in the Landau damping phenomenon, a universal damping is obtained for all usual perturbations of a distribution function, although the system is reversible. To reverse the phenomenon and obtain a locally increasing perturbation at these times, one should take the exact distribution function (which is quite particular) and reverse all the velocities. This is not feasible in reality, but this does not question the reversibility of the system (it is indeed feasible in simulation).

Finally, if the physical system evolution is ruled by a system of differential equations relating the state variables, is one able to recognize from this system of equations whether the physical system that it describes is reversible or not? In other words, what are the mathematical properties of the equations that guarantee the reversibility? If the variables are reversed as explained above in the system of equations, reversibility is guaranteed if this system remains unchanged at the condition of also changing t into $-t$ (the space variables still being unchanged). This is sometimes called "time reversal".

In the preceding example of billiard balls without friction and with elastic collisions, the equations are trivially reversible between the collisions since they are just $d_t(\nu) = 0$. The collisions with the billiard edges are also trivially reversible since the rule relating the velocities before and after the collision is simply $\nu_+ = -\nu_-$. The elastic collisions between hard spherical balls are indeed also ruled by reversible laws, even if these laws are a little more difficult to write down.

Let us look now from this point of view at the fluid and kinetic descriptions, which are the subjects of this book. It is easy to check that the usual collisional fluid equations (see Chapter 2) are reversible when they are "ideal", that is in the absence of viscosity, thermal conduction, and any other transport term of this kind. They become irreversible as soon as one of these terms is included, because all of them have a parity different from the ideal ones. It can be seen as well that a plasma described by a Vlasov equation (see Chapter 2) is reversible but that the collision term makes the other kinetic equations, such as the Boltzmann equation, irreversible.

4.2.4 Irreversibility and Damping

In a collisional fluid, as we have seen, the nonideal effects such as viscosity are a cause of irreversibility. They also bring damping in the propagation of waves, the latter phenomenon being a proof of the former. Let us describe what happens in this case. If one imposes an initial density perturbation in such a fluid, this perturbation propagates, due to the quasi-cyclic exchanges between convective and thermal energies (velocity and pressure). In the presence of viscosity, the calculation shows that these exchanges are not exactly in quadrature so that the exchanges are indeed dissipative, with a net transfer, on average, from convective to thermal. The damping of the wave amplitude results from this dissipation. How is this damping a proof of irreversibility? If changing t in $-t$ at time t_f and reversing the corresponding final state, it is clear that the system would indeed go back to its initial position if ideal propagation was alone. But, in the presence of viscosity, the amplitude of the wave decreases during the backward evolution as it decreased during the forward one, so that the system can certainly not go back to its initial state.

Does this mean that any damping always implies irreversibility, even in collisionless media? The answer is *no*, and the "Landau damping" (see Chapter 5) is the most famous counterexample for that. Let us consider a homogeneous Vlasov–Poisson electrostatic plasma and let us perturb it at some time $t = 0$ with a perturbation of the distribution function modulated in space. To fix ideas, let us consider a sinusoidal initial perturbation of the density with a single wave number k and let us consider the linear evolution only (although the Landau effect is a much more general effect). Let us suppose also that the zero order plasma is stable (for instance, with a Maxwellian distribution). It can be shown that the amplitude of the associated density perturbation, after some more or less long transient period, always eventually decreases at asymptotic times. This is Landau damping [13], and it is a statistical effect: the particles transport the initial perturbation at different velocities, and the corresponding phase mixing results in a damping for all macroscopic variables such as density. The eventual evolution, which is always a decrease and never an increase, seems a priori characteristic of an irreversibility. Yet, it is the result of the Vlasov equation, which is reversible. There is, therefore, an interesting paradox to solve.

The main difference between the two previous examples, collisional and Vlasov, can be understood by thinking about the notion of "ideal propagation". In collisional fluids, when the transport coefficients such as viscosity can be neglected, the system can be labeled as "ideal". The propagation is then characterized by a combination of a small number of modes (related to the number of moments retained), which is just a linear superposition of them as long as the perturbation has a sufficiently small amplitude. Each mode is undamped and the same holds for their superposition. When considering a pressure perturbation in air, for instance, the evolution of any initial perturbation is just the superposition of two sound waves, one in each sense (in 1D). The introduction of some viscosity makes an irreversible departure from this behavior for each of the individual modes, which become damped. The situation is different for a Vlasov plasma: the number of propagation modes tends to infinity when the number of velocities considered in the distribution function tends to infinity and, as long as one remains in the Vlasov frame, there is no "nonideal" term to add. The evolution then comes from the statistical phase mixing between the infinity of propagation modes. Two remarkable – and not trivial – features then characterize the asymptotic behavior: (i) it much resembles the fluid evolution, with its small number of modes (if a correct closure equation is chosen, see Chapter 2) and (ii) the departure from this fluid behavior is always in the sense of an amplitude decrease.

As shown on the sketch of Figure 4.9, the systematic decrease at asymptotic times is actually not in contradiction with reversibility in this case; as long as the distribution function $f(\boldsymbol{v}, \boldsymbol{r})$ is completely described, no irreversibility occurs. Starting for its value at time t_f, one can go back to its initial value by just changing $f(\boldsymbol{v})$ into $f(-\boldsymbol{v})$. In this backward motion, the wave amplitude then begins by increasing, as the phase mixing decreases, meets the initial condition at time $t = 2t_f$ and, if go-

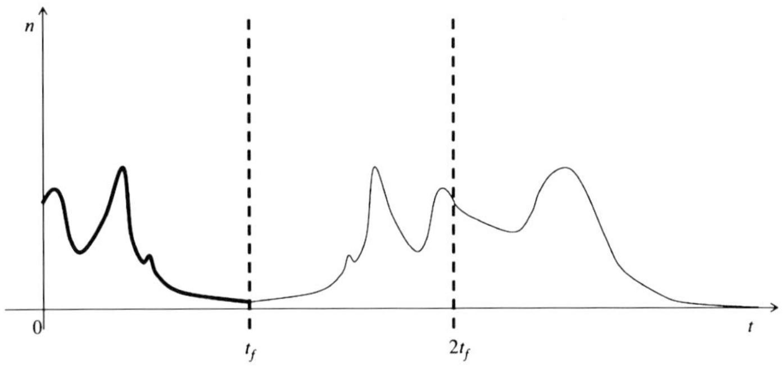

Figure 4.9 Sketch illustrating the possibility of a reversible damping (applicable to Landau damping). The density is plotted as a function of time from $t = 0$ to $t = t_f$. If the system is reversed at a large time t_f, the density returns to its initial value at $t = 2t_f$, but eventually decreases again at asymptotic times

ing on, decreases again when time tends towards $-\infty$, the phase mixing increasing again.

The amplitude decrease, at asymptotic times, that is, long after the initial condition, is actually associated with a statistical selection of the possible perturbed distribution functions, not to a dissipation at microscopic scale. Nevertheless, when looking to only the first moments such as density, this fine selection cannot be observed, and the phenomenon indeed appears as a dissipation.

A close comparison between fluid and kinetic theory is needed to understand the similar propagation and the different damping found by the two approaches. Without going into it, a comment can be made right now: since the fluid theories have a quite limited number of modes, it is clear that the process of damping by phase mixing cannot take place in this frame. Therefore, if one wants to make a fluid model account for the damping, it is necessary to include in it viscous-like processes, which are dissipative. It must be noted that the irreversibility then arises from the approximate description and not from the microscopical phenomenon itself. This leads to a quite interesting – and general – conclusion: the notion of irreversibility is description-dependent; reversible elementary physical processes lead to a more and more irreversibility when they are described with less and less accuracy. This point is developed in next section.

4.2.5 The Notion of Reversibility Depends on the Description

Figure 4.10 intends to illustrate the dependence of the notion of reversibility (and, therefore, of dissipation) on the accuracy of the description. In this simplistic toy model, one has N particles, without any interaction, neither at large distance (no collective field acting on the trajectories), nor at small ones (no collisions). The particles are put inside a square box with sides of length $l = 1$, and they only interact with the four sides of this box, the collisions being then elastic and specular. By these rules, the complete system is obviously and trivially reversible. The par-

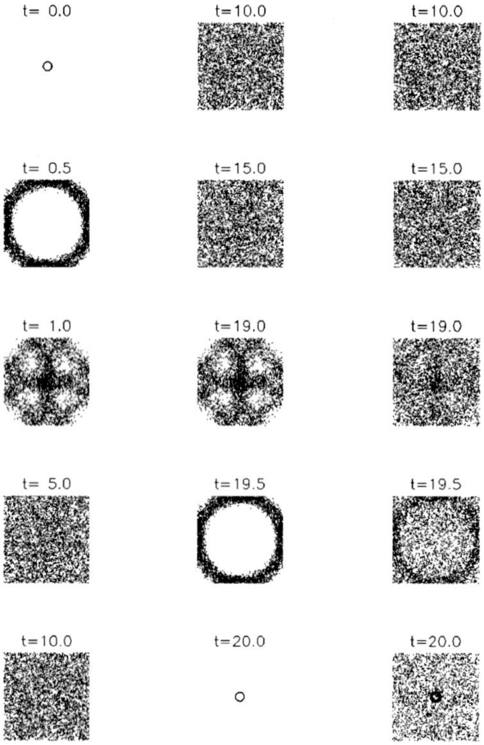

Figure 4.10 Dependence of the reversibility on the description accuracy. First column: Motion of independent particles launched in all directions from a central ∘ and reflecting specularly on the box edges. Second column: Exact "time reversal": each particle velocity at the final time is changed from $\boldsymbol{v}$ to $-\boldsymbol{v}$: the initial ∘ is exactly recovered. Third column: Only some statistical properties of the particle distribution are reversed: the initial ∘ is only partly recovered, and a clear irreversibility occurs.

ticles are initially distributed, at time $t = 0$ on a circle of diameter 0.1. They are then launched with velocities having random directions (uniformly distributed between 0 and 2π) and random magnitudes (normal distribution around $v = 1$ with $\sigma = 0.05$). The first column shows the particle positions at five different times between 0 and 10. It can be seen that the reflections on the edges, together with the small velocity dispersion rapidly mix the particles in such a way that they finally look randomly distributed in the whole box. In the second column, we make the numerical test of "inversing" time in the final state $t_f = 10$: we change the velocity $\boldsymbol{v}$ of each particle into $-\boldsymbol{v}$, and look at the subsequent evolution. This evolution is clearly observed, in the second column, to be the mirror of the previous one, the positions at time $t_f + \delta t$ of this backward evolution being exactly identical to the positions at time $t_f - \delta t$ of the forward one, up to retrieving the initial circle, the order progressively emerging from the disorder. The expected reversibility is, therefore, perfectly respected. In the third column, a different test is done: instead of reversing all the particle velocities, a finite resolution is imposed (64×64 pixels).

Inside each pixel, the particles are redistributed randomly (uniform distribution) and their velocities are also redistributed randomly (normal distribution), in such a way that the mean velocity in this pixel is reversed and the variance conserved. The consecutive evolution is not far from the preceding, but instead of going back exactly to the initial circle, the system now includes particles that have been spread in the whole box. Only reversing the first moments at the given scale has, therefore, introduced irreversibility in the system.

The same occurs when modeling a collisionless plasma with a fluid theory, that is, by only its first moments: density, velocity, and pressure. This description can, without contradiction, be irreversible while the Vlasov description of the phenomenon is reversible. It just means that one can go back to the initial state by changing $\boldsymbol{v}$ into $\boldsymbol{v}$ in the distribution function, but that one cannot go back to the initial state if one just reverses the mean velocity $\boldsymbol{u}$ into $-\boldsymbol{u}$, letting density and pressure be unchanged.

This is a general conclusion: the passage from a description to another less accurate description generally introduces irreversibilty. The passage from a Vlasov theory to a fluid one, with the notion of Landau damping, was a first example. Many other examples can be found as illustrations. Let us cite three of them, which are both classical and of great importance.

Fluid turbulence Let us consider an ideal fluid, without any of the transport coefficients that could lead to irreversible diffusion of matter, momentum, or heat. This medium can be subject to some large-scale motion, superposed to some turbulent agitation (see the downstream region of Figure 4.8, for instance). If one is interested in only the large scales, which would be the only relevant scales in a laminar motion, one can average the medium properties over this scale and characterize the turbulence through only its statistical properties, typically its spectrum. If there is a sufficient scale separation between the characteristic scales of the turbulence and the considered "large scales" (determined in the case of the bullet by the size of this obstacle), the average has to be done on a scale larger than the correlation length of the turbulence. So doing, the details, which are reversible at small scales, can no more be described, and the description becomes irreversible, implying diffusion terms traditionally called "anomalous": anomalous diffusion, anomalous viscosity, anomalous conductivity, and so on. The mixing effects belong to this category: if one takes an initial cloud of milk in a cup of coffee (Figure 4.11), the role of turning a spoon inside this cup is to create turbulence down to very small scales. The effect is to make a strong diffusion of the white fluid particles in the black ones, rapidly leading to an intricate mixing of both, with an apparently homogeneous creme color. It is clear that inversing the large-scale rotation, that is, turning the spoon in the reverse sense, would not lead to separating the milk and the coffee again. At this scale, there is indeed an irreversible diffusion. But if one could describe all the fine details of the turbulence, one would recover the initial situation by just "reversing the time". This is actually possible in simulation, whenever all the positions and ve-

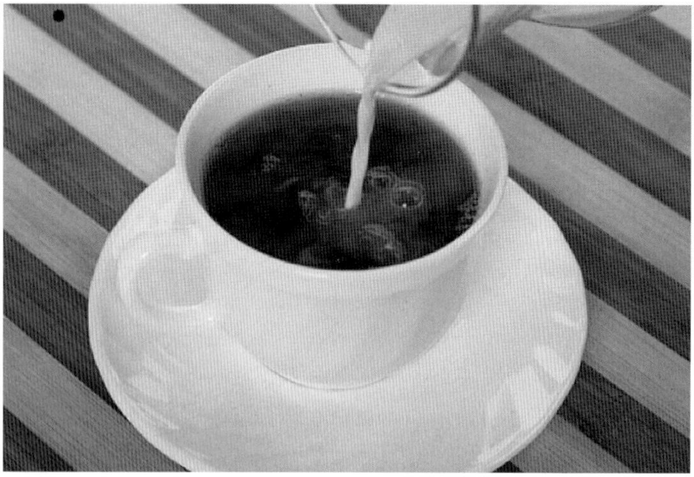

Figure 4.11 If the milk is first separated from the coffee, the turbulence created by the motion of a spoon will rapidly mix both liquids in a practically "irreversible" manner.

locities of the fluid particles are described, in the limits of the calculation accuracy. In this sense, the system would thus be reversible.

Collisions If one considers the equations of motion for a large number of individual particles in electromagnetic interaction, it is easy to check that these equations are reversible. However, when restricting the complete N particle description to the "kinetic" one, that is, with only the one-particle distribution function $f(\boldsymbol{v})$, the medium becomes irreversible if collisions are not negligible. This can be seen via the RHS of the kinetic equation (see Chapter 2), where the collision operator describes approximately the departure of the field from only its collective value. This second member is negligible only when the plasma is sufficiently dilute. The kinetic equation then becomes the Vlasov equation. In the strongly collisional limit, on the contrary, we know that the collisions introduce a viscosity coefficient (the "normal" one), which as a consequence transfers systematically the convective energy to thermal energy. Dissipation has, therefore, appeared when going from one description to the other and replacing the motion of all the individual particles by a statistical hypothesis describing them through only their velocity distribution function.

Heat flux In the context of irreversibility, and in connection with the problem of fluid closure equation, we make special mention about the heat flux modeling. The heat flux vector $\boldsymbol{q}$ that appears in Eq. (4.11) derives from the heat flux tensor $\boldsymbol{Q}$. By its very definition, it is an odd quantity with respect to $\boldsymbol{v}$, and, therefore, a reversible term. In strongly collisional plasmas, however, this term can be modeled as $\boldsymbol{q} = -\kappa \nabla T$. This term is even in $\boldsymbol{v}$, and, therefore, is irreversible. This paradox comes, once again, from a degradation of the information. Because of collisions, the heat flux is supposed to evolve on scales much shorter than the scales under study, and so to relax very rapidly toward the most probable state. As these short scale variations are not described, this is the origin of the irreversibility. If one would change $\boldsymbol{v}$ into $-\boldsymbol{v}$, the instantaneous heat would indeed change its sign, but it would go back very quickly to the $\boldsymbol{q} = -\kappa \nabla T$ value. Since this fast evolution is not described, the system is clearly irreversible.

The heat flux is not a term of exchange between two forms of energy: it is just a spatial transfer of only the thermal energy. However, when this term brings irreversibility in the system, it must be kept in mind that all the variables behave in an irreversible manner. It is true in particular for the term $\nabla \cdot \boldsymbol{P} \cdot (\boldsymbol{u})$ (work of the pressure force) that appears in the RHS of Eq. (4.11): it means that the exchange of energy from convective to thermal energy may very well be irreversible even when the viscosity is negligible, that is, when the term of exchange itself is not the cause of the irreversibility. This remark can be important when investigating the dissipation in media such as solar wind (see Chapter 9).

In all the previous examples, the simplification introduced when going from a description to a less accurate one, consists of a probabilistic assumption. In the toy model used in the introduction, one supposes, for instance, that under some spatial resolution, the particles can be supposed randomly distributed in phase space (position and velocity) at the final time, and their description is then reduced to their mean velocity and "temperature" (velocity variance). The same holds when going from kinetic to fluid description, and the notion of "fluid particle" is the direct consequence of this statistical hypothesis about the particles included. In the case of Landau damping, the asymptotic damping is actually reversible, as already explained, and due to the selection of particular distribution functions at these times. But in the same way, irreversibility appears as soon as a finite resolution is introduced in velocity, replacing the exact particle distribution in a given pixel of the phase space by a uniform random distribution. As the exact evolution of the initial perturbation actually forms extremely sharp modulations in velocity of the distribution function, it is easy understanding that the reversibility is rapidly lost when losing these fine structures, so making it impossible to go back in time.

The importance of probabilistic arguments for irreversibility has indeed been well-known for a long time in statistical physics, where it is called the "reversibility" or "Loschmidt" paradox. It is commonly used for explaining what is called the *time arrow*, which makes the distinction between past and future (see, for instance, R. Feynmann in his lessons about "The character of physical law"). Beyond fluid or plasma physics, it can be imaged by an extreme example (Figure 4.12): when a bank note is burning, this phenomenon, which seems to be one of the most irreversible possible, can indeed be fully compatible with reversible microscopic laws at the level of the atomic interactions. Obtaining the reverse phenomenon, that is, building a bank note from its ashes, is, therefore, not strictly impossible: it is only highly improbable (very highly improbable, unfortunately!). It is even so improbable that, when doing physics, it is quite mandatory to concentrate on only the most probable states and disregard all the countless others.

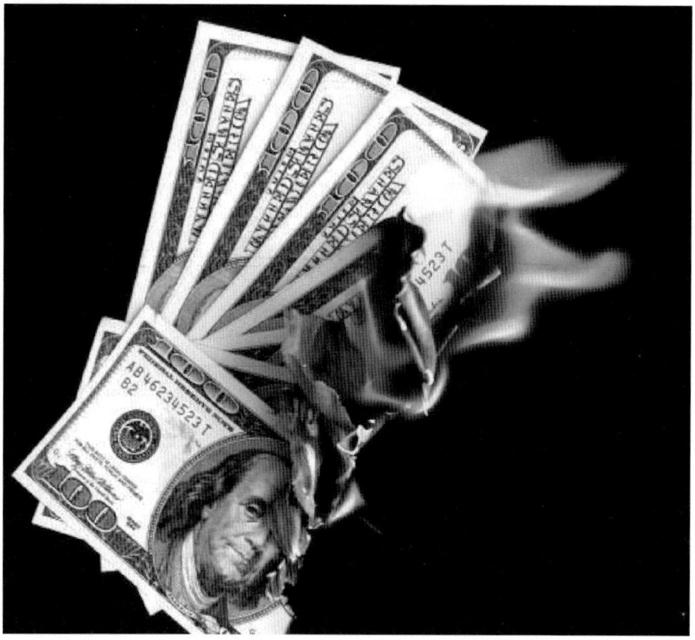

Figure 4.12 If all the phenomena involved in burning are reversible at the most microscopic scale possible, which is reasonable, the opposite phenomenon is theoretically possible: building a bank note from its ashes. It is simply very highly improbable. The notion of entropy allows evaluating this probability.

In the frame of the probabilistic assumptions that allow simplifying the description of a medium, the entropy is the quantity that allows quantifying the incompleteness of the description and how much the most probable state is more probable than the others.

4.2.6 Entropy

As shown above, the irreversibility increases when going from a given description to a degraded one, that is when considering only "macrostates", each of them corresponding to a lot of "microstates", generally assumed to be equiprobable. Whenever such probabilistic arguments can be justified, one needs to know which macrostate is the most probable and how to quantify this probability. The estimation depends on the involved levels of description. We will just do it in the simplest example, when the macrostate is defined, at each time, by the distribution function $f(\boldsymbol{v}, \boldsymbol{r})$ for each species. This level indeed corresponds to a quite fine description, but one must be aware that it is still not, by far, the complete description of the system. It is just the number of particles in all elementary volumes of the 6D space $(\boldsymbol{v}, \boldsymbol{r})$, while the complete description would demand to know the position and the velocity of *each* of these individual particles ($6N - D$ space). We will simply assume that all the particle arrangements that correspond to the same distribution function are equiprobable. This is actually a huge simplification since all particles in the same elementary volume δV_i centered at the point $(\boldsymbol{v}_i, \boldsymbol{r}_i)$ of the phase space are considered indiscernible instead of being followed individually. Calling $\delta n_i = f_i \delta V_i$ the number of particles in such an elementary volume of the phase space, one can evaluate the effect of the simplification done by calculating the number of "microstates" (that is considering all particles as discernible) corresponding to the same "macrostate" (that is with indiscernible particles inside each elementary volume). Considering for simplicity a discretized phase space, this calculation can be done as follows:

- Estimate the total number of possible microstates in the total volume: it is $N!$ if N is the total number of particles.
- Estimate the number of microstates that are equivalent from the macrostate point of view: it is the product $\delta n_1! \delta n_2! \delta n_3! \ldots$, where the $\delta n_i!$ are the number of particles in the different elementary volumes i.
- The number of microstates states N_f corresponding to the same f is then just the ratio between the two preceding numbers: $N!/(\delta n_1! \delta n_2! \delta n_3! \ldots)$.

For the usual numbers of particles considered, which are very large, this number N_f is huge for the most probable macrostate and the importance of the other microstates is comparatively extremely low: as it can be guessed from the central limit theorem, the relative width of the microstate distribution decreases with the number δn particles in each cell of the phase space as $\delta n^{-1/2}$, which actually corresponds to an extremely peaked distribution. This shows the very large probability of observing the only most probable macrostate f rather than all the other macrostates (at least when all the microstates are equiprobable). This result can be generalized and provides a mathematical basis for explaining the very weak probability of seeing a burned bank note coming back to its initial form.

The so called statistical entropy, which coincides in this case with the thermodynamical notion of entropy, is just the logarithm of this number (with, possibly, arbitrary additive and multiplicative constants). Thanks to the logarithm, it is an "extensive" function, that is, additive when two populations are added. Its usual form is obtained by replacing all factorials thanks to the Stirling formula, valid for large numbers:

$$S = \log(N!) - \sum \log(\delta n_i!) \approx N \log(N) - \sum \delta n_i \log(\delta n_i) . \tag{4.14}$$

Disregarding all the additive and multiplicative constants, which are all unimportant for testing the variations of entropy, one can replace δn_i by f_i, ignore the term $N \log(N)$ and get:

$$S = -\sum f_i \log(f_i) . \tag{4.15}$$

And finally going back to the continuous space, one gets the usual formula for the entropy:

$$s = -k \int d^3 v\, f \log(f) . \tag{4.16}$$

In thermodynamics, the constant k added in front is the Boltzmann constant (in order to get a quantity with an energy unit). Note that, in the above formula, the integration is done over only the $\boldsymbol{v}$ space, and not the $\boldsymbol{r}$ one, so that the function s defined in this way is actually, not the total entropy, but its spatial density. Knowing the mathematical expression for entropy, the state observed can easily be determined, as soon as it can be supposed to be the most probable one: it is just the state that maximizes the entropy, taking into account the constraints imposed by the physical system. For instance, for a steady equilibrium state with given energy and density, the well-known Maxwellian distribution is readily retrieved in this way.

On can summarize the essential ideas developed in this section by saying that irreversibility – and thus dissipation – comes from a degraded information on the system via a probabilistic argument. As soon as all the microstates, which are "hidden variables", are supposed equiprobable, the macrostate observed can be supposed to be just the most probable one, and it can be determined as the state which maximizes the entropy, that is, the number of microstates not described. The irreversibility comes in at the moment when the information reaches the microstate fine description, and where it is lost since it is then just replaced by the "reasonable" equiprobable hypothesis.

Of course, the hypothesis that all microstates are equiprobable is pivotal to make the entropy concept useful. In strongly collisional plasmas, and with a kinetic description, it is a quite reasonable hypothesis as soon as the collisions can be supposed sufficiently frequent to grant a homogeneous mixing of all the particles that belong to a small volume in phase space. But it must certainly be questioned when the number of collisions decreases, and it is certainly wrong for a pure collisionless plasma: let us suppose, for instance, that the system is initialized in some equilibrium macrostate, Vlasov equation shows that it will then stand identical, that is, with the same distribution functions, independently of any consideration about them, probable or not.

The Landau effect [13], cited earlier, is a good plasma illustration of these concepts, when it is applied to another level of description. In this case, the "complete" description considered is the Vlasov one. In the absence of collisions, a description at this level is indeed exact, without going to the description of all individual particle positions, which is useless. As explained, the exact evolution of the full distribution function is then fully reversible. Nevertheless, when looking to only the first moments, or equivalently to a "coarse-grained" distribution function, the evolution becomes irreversible, the degraded information on the system coming from this coarse-graining. It is worth noting that any limitation of the velocity resolution, as small as it can be, has always the effect of bringing this irreversibility at some time: if the perturbed distribution function is supposed monotonous in $\boldsymbol{v}$ and gently modulated in $\boldsymbol{r}$, always becomes more and more sharply modulated in $\boldsymbol{v}$ when the time increases (see Chapter 5). A statistical entropy can also be calculated in this case, but its expression must be adapted to the different situation.

5
Waves in Plasmas

There is a large variety of propagation modes for linear waves in a homogeneous plasma, and especially in a magnetized one. The dispersion and the polarization of these modes are all different and depend on

- the zero order parameters of the medium in which the waves propagate. The dimensionless parameters such as β or ω_{pe}/ω_{ce} can help distinguish the different regimes and simplify the description of the medium and of its perturbations.
- the scale range (spatial and temporal) considered. Comparing, for instance, the phase velocity v_φ with respect to the thermal velocity of the different species determines what can be easily treated in the frame of a fluid modeling and what needs a kinetic description.
- the propagation angle of $\boldsymbol{k}$ with respect to $\boldsymbol{B}_0$ in the case of a magnetized plasma or, more generally, of any direction of anisotropy.

This makes the "zoology" of plasma modes quite abundant and complex. There are many textbooks entirely devoted to this question of plasma waves and the present book will not duplicate them. We will just describe a few for illustrating the general principles to be used when calculating the linear modes and point out those which are the most commonly used in astrophysics. Since MHD is indeed the most commonly used description for the magnetized plasma, we will first describe the MHD modes and their properties. Then, we will present the high frequency (electron) waves because they are closely related to the well-known vacuum electromagnetic waves (light, radio, and so on) and can be used for remote measurements of very distant astrophysical objects. These waves generally have phase velocities much larger than the thermal velocities (not far from c), so that a simple cold plasma approximation is generally sufficient to their description. We will further restrict this description, for the sake of simplicity, to two propagation angles: strictly parallel and strictly perpendicular. Finally, we will make special mention of the so-called whistler mode, which extends the MHD scale range toward higher frequencies and which has interesting properties and applications. Among the numerous other existing wave modes, the famous "Langmuir wave", which exists in a nonmagnetized plasma, is not presented in this first part, but it is presented in the next one, as the simplest illustration of the so called Landau effect or "collisionless damping".

Collisionless Plasmas in Astrophysics, First Edition. Gérard Belmont, Roland Grappin, Fabrice Mottez, Filippo Pantellini, and Guy Pelletier.

After the first "zoological" part about the wave modes, the rest of the chapter is devoted to the questions of damping and instabilities of the plasma waves. These questions are important from a fundamental point of view. It will be shown that they occur in a quite different manner depending whether the medium is collisional or not. It is interesting for instance to note that a damping always occurs, in a collisional medium, via the dissipative terms added explicitly in the fluid description, while it appears in the kinetic case via a choice of accessible initial conditions, the Vlasov equation describing the evolution being strictly reversible. We will see also that, in a collisionless medium, instabilities can occur due to peculiarities of the velocity distribution function, even when the medium is homogeneous.

5.1 MHD Waves

The linearized ideal MHD equations are:

$$\frac{\partial \rho_1}{\partial t} = -\rho_0 \nabla \cdot \boldsymbol{v}_1 \tag{5.1}$$

$$\frac{\partial \boldsymbol{B}_1}{\partial t} = -\boldsymbol{B}_0 (\nabla \cdot \boldsymbol{v}_1) + (\boldsymbol{B}_0 \cdot \nabla) \boldsymbol{v}_1 \tag{5.2}$$

$$\rho_0 \frac{\partial \boldsymbol{v}_1}{\partial t} = -\nabla p_1 - \frac{1}{\mu_0} \boldsymbol{B}_0 \times (\nabla \times \boldsymbol{B}_1) \ . \tag{5.3}$$

Considering the closure equations

$$p = f(\rho) \quad \text{and} \quad C_s^2 = f'(\rho) \tag{5.4}$$

and Eq. (5.3), the mass velocity perturbation is given by

$$\rho_0 \frac{\partial^2 \boldsymbol{v}_1}{\partial t^2} = -C_s^2 \nabla \left(\frac{\partial \rho_1}{\partial t} \right) - \frac{1}{\mu_0} \boldsymbol{B}_0 \times \left(\nabla \times \frac{\partial \boldsymbol{B}_1}{\partial t} \right) \ . \tag{5.5}$$

With Eq. (5.1),

$$\rho_0 \frac{\partial^2 \boldsymbol{v}_1}{\partial t^2} = +C_s^2 \rho_0 \nabla (\nabla \cdot \boldsymbol{v}_1) - \frac{1}{\mu_0} \boldsymbol{B}_0 \times \left(\nabla \times \frac{\partial \boldsymbol{B}_1}{\partial t} \right) \ . \tag{5.6}$$

These equation are Fourier transformed. Contrarily to the case of a cold multicomponent plasma, we base the dispersion equation on the plasma velocity and not on the electric field perturbation, because as we have seen in Section 1.3.10.1, the electric field perturbation has a low order of magnitude. We find

$$\left(C_s^2 + V_A^2\right) \boldsymbol{k}(\boldsymbol{k}\cdot\boldsymbol{v}_1) - V_A^2 \boldsymbol{k}_\parallel (\boldsymbol{k}\cdot\boldsymbol{v}_1) + \boldsymbol{v}_1 \left(-\omega^2 + V_A^2 k_\parallel^2\right) - V_A^2 k_\parallel \boldsymbol{k} v_{\parallel 1} = 0 \ . \tag{5.7}$$

In matrix form with an equilibrium uniform magnetic field $\boldsymbol{B}_0 = (0, 0, B_0)$ and $\boldsymbol{k} = (k_x, 0, k_z) = (k_\perp, 0, k_\parallel)$, this is equivalent to

$$\begin{pmatrix} -\omega^2 + V_A^2 k^2 + C_s^2 k_\perp^2 & 0 & C_s^2 k_\parallel k_\perp \\ 0 & -\omega^2 + V_A^2 k_\parallel^2 & 0 \\ C_s^2 k_\parallel k_\perp & 0 & -\omega^2 + C_s^2 k_\parallel^2 \end{pmatrix} \begin{pmatrix} v_{1x} \\ v_{1y} \\ v_{1z} \end{pmatrix} = \begin{pmatrix} 0 \\ 0 \\ 0 \end{pmatrix} . \tag{5.8}$$

Nontrivial solutions are possible if the determinant of this equation is null. This happens if

$$\omega^2 = k_\parallel^2 V_A^2 \tag{5.9}$$

or

$$\omega^2 = \frac{k^2}{2} \left[V_A^2 + C_s^2 \pm \sqrt{(V_A^2 + C_s^2)^2 - 4 V_A^2 C_s^2 \cos^2\theta} \right] \tag{5.10}$$

where $\boldsymbol{k} = (k \sin\theta, 0, k \cos\theta)$. Equation (5.9) defines the Alfvén wave, also called the shear Alfvén wave. Equation (5.10) defines the fast and slow magnetosonic waves, also called fast and slow compressional Alfvén waves.

The magnetosonic waves are characterized by the dispersion Eq. (5.10). When the plasma is dominated by the magnetic pressure, that is, $\beta \ll 1$ or $C_s \ll V_A$, a first order development in C_s/V_A gives

$$\omega^2/k^2 = V_A^2 + C_s^2 \cos^2\theta \quad \text{and} \quad \omega^2/k^2 = C_s^2 \cos^2\theta$$

for, respectively, the fast and the slow modes. When the plasma is dominated by the particle pressure, that is, $\beta \gg 1$ or $C_S \gg V_A$, the dispersions become:

$$\omega^2/k^2 = C_s^2 + V_A^2 \cos^2\theta \quad \text{and} \quad \omega^2/k^2 = V_A^2 \cos^2\theta \ .$$

When $C_s = V_A$, $\omega^2/k^2 = V_A^2(1+|\sin\theta|)$ and $\omega^2/k^2 = V_A^2(1-|\sin\theta|)$. In any case, the fast mode has the largest phase velocity and the slow mode has the smallest one. The shear Alfvén and the slow modes do not propagate at right angles from the magnetic field. In parallel propagation, the Alfvén wave has a dispersion identical to the fast one for $C_s \ll V_A$ and to the slow one for $C_s \gg V_A$ (the three modes have the same dispersion relation for $C_s = V_A$).

5.1.1 Polarization of the MHD Waves

The dispersion equation does not provide the full description of a linear wave. One may also need to know its polarization. The polarization of the three MHD waves is summarized in Tables 5.1 and 5.2. We can notice that the shear Alfvén waves are

Table 5.1 Alfvén wave polarization.

Variables	Shear Alfvén wave function of v_1	function of B_1
v_1	$(0, v_{1y}, 0)$	$\left(0, -V_A \dfrac{B_{1y}}{B_0}, 0\right)$
B_1	$\left(0, -B_0 \dfrac{v_{1y}}{V_A}, 0\right)$	$(0, B_{1y}, 0)$
E_1	$(-B_0 v_{1y}, 0, 0)$	$(B_{1y} V_A, 0, 0)$
J_1	$\left(\dfrac{ik_z B_0}{\mu_0} \dfrac{v_{1y}}{V_A}, 0, -\dfrac{ik_x B_0}{\mu_0} \dfrac{v_{1y}}{V_A}\right)$	$\left(-\dfrac{ik_z B_{1y}}{\mu_0}, 0, \dfrac{ik_x B_{1y}}{\mu_0}\right)$
ϱ_1 (charge density)	$-i\epsilon_0 B_0 k_x v_{1y}$	$i\epsilon_0 k_x V_A B_{1y}$
ρ_1 (mass density)	0	0
group velocity	$v_g = (0, 0, V_A)$	$v_g = (0, 0, V_A)$
Poynting flux	$S = S_z = \frac{B_0^2}{\mu_0 V_A} v_{1y}^2$	$S = S_z = \frac{V_A}{\mu_0} B_{1y}^2$

Table 5.2 Compressional Alfvén waves polarization.

Variables	Magnetosonic wave = compressional Alfvén wave function of v_{1x}
v_1	$\left(v_{1x}, 0, \dfrac{k_x k_z C_s^2}{\omega^2 - k_z^2 C_s^2} v_{1x}\right)$
B_1	$\left(-B_0 \dfrac{k_z v_{1x}}{\omega}, 0, B_0 \dfrac{k_x v_{1x}}{\omega}\right)$
E_1	$E_1 = (0, B_0 v_{1x}, 0)$
J_1	$\left(0, \dfrac{ik^2}{\omega} \dfrac{B_0}{\mu_0} v_{1x}, 0\right)$
ϱ_1 (charge density)	0
ρ_1 (mass density)	$\rho_0 \dfrac{\omega}{\omega^2 - k_z^2 C_s^2} k_x v_{1x}$

associated with a charge density perturbation ($\varrho \neq 0$), but not with a compression of the plasma ($\rho = 0$). On the contrary, the magnetosonic waves are purely electromagnetic ($\varrho = 0$), but they are associated with a plasma compression ($\rho \neq 0$). None of these waves carry an electric field in the direction parallel to the ambient magnetic field $\boldsymbol{B}_0$; therefore, they are not able to accelerate particles in this direction of high mobility.

The magnetic field line motion is particularly simple with shear Alfvén waves. The field lines are defined by $dy/B_y = dz/B_z = dz/B_0$. In Fourier space dy/dz becomes $ik_z y$. With the shear Alfvén waves, $ik_z y = B_{1y}/B_0 = -k_z v_{1y}/\omega$, or equivalently, $-i\omega y = v_{1y}$. The partial time derivative $-i\omega y$ of the coordinate y of the magnetic field line is the magnetic field line velocity. The other components of the velocity and magnetic field perturbation are null. Therefore, with the shear Alfvén waves, the magnetic field lines move with the same velocity as the plasma.

5.1.2
Application: Alfvén and MHD Waves in the Earth's Magnetosphere

The global configuration of the magnetospheres is subject to frequent changes. For instance, near the inner boundary of the Earth's magnetic tail, local changes of the structure called *dipolarizations* occur during substorms, in a phase that is called the *onset*.

These local changes of the magnetic field and plasma structures can be propagated elsewhere through MHD waves. A part of the changes with plasma compression can be propagated by fast MHD waves. Another part of the changes with compression can be carried by slow waves and the part without compression can be carried away from the dipolarization locus by Alfvén waves. The Figure 5.1a shows the direction of the group velocity of the MHD waves as a function of their direction of propagation, given by the angle θ between $\boldsymbol{k}$ and the local magnetic field. The fast waves can propagate in a large range of directions and, consequently, their energy tends to decrease with distance as r^{-2}. On the contrary, the perturbations carried by the slow and Alfvén waves propagate mainly along the magnetic field. Therefore, with a quasi-unidirectional propagation, their intensity tends to remain constant in that direction. Therefore, Alfvén and slow waves can be seen far from their place of emission, approximately along the same field lines. For instance, Alfvén waves can be observed in the auroral zone, at only a few hundreds of kilometers above the ionosphere, on magnetic field lines connected to the inner boundary of the plasma sheet, a few minutes after substorm onsets, that is, after dipolarization events.

With the measurements made aboard spacecrafts, Alfvén waves can be identified through the ratio of the perturbations of the electric and of the magnetic field. With ideal MHD Alfvén waves, this is V_A (see Table 5.1). Actually, for shorter wavelengths (a quantity that cannot be measured with a single spacecraft), the theory of Section 5.3.2 must be applied. In this context, the noncompressional Alfvén waves are decoupled into the left-hand and the right-hand side of the polarized waves. Figure 5.2 shows that their E/B ratio can be notably different from V_A. Then, to be certain of the Alfvénic nature of these waves, it is also necessary to check that

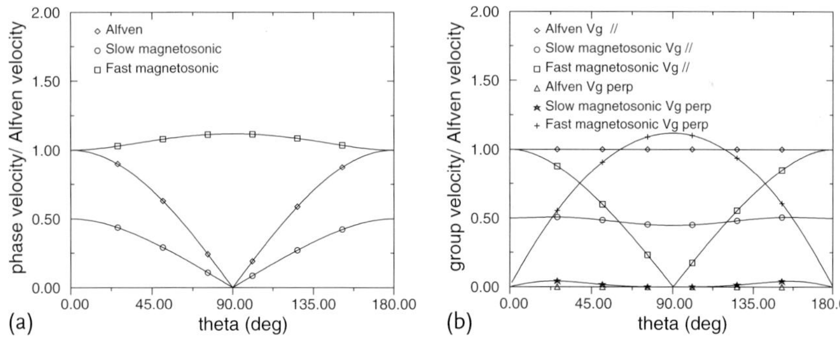

Figure 5.1 MHD propagation velocities as a function of the propagation angle, for $C_s = 1.5 V_A$. (a) Modulus of the phase velocities. (b) x and z components of the group velocities.

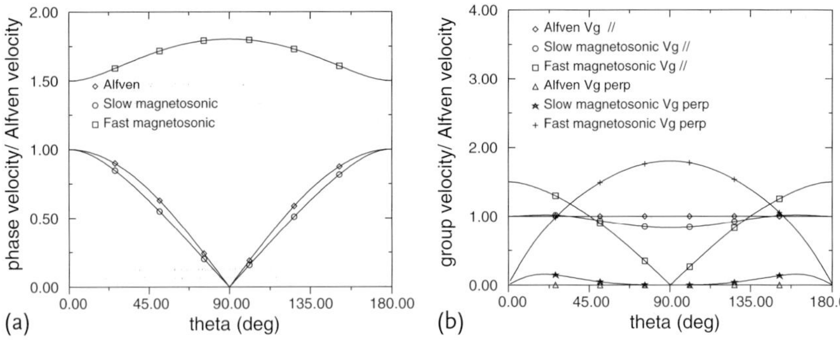

Figure 5.2 MHD propagation velocities as a function of the propagation angle, for $C_s = 0.5 V_A$. (a) Modulus of the phase velocities. (b) x and z components of the group velocities.

they carry no plasma compression. This can be done through the use of density measurements.

The Poynting flux of Alfvén waves packets (see Table 5.1) measured in the Earth's auroral zone can carry enough energy to permit the acceleration up to a few keV of a flux of electrons sufficient to allow for auroral display, even for energy conversion rates (from waves to particles) of only a few percent. Therefore, these wave are not only the signature of distant plasma changes, they are also an efficient means of propagating important fluxes of energy along the magnetic field lines.

5.2 Transport Induced by Waves

The presence of waves (or more generally, fluctuations) in a fluid may change the average state of the fluid, by playing the role of an effective pressure. A first example is that of turbulent fluctuations in a neutral fluid. Consider the advection term $(u \cdot \nabla) u$ in the Navier–Stokes equations. We write the velocity field u as the sum of a mean value $\langle u \rangle$ and of a fluctuating field δu with zero time average $\langle \delta u \rangle = 0$. Then the *average* value of the advection term will be

$$\langle u \cdot \nabla u \rangle = \langle u \rangle \cdot \nabla \langle u \rangle + \langle \delta u \cdot \nabla \delta u \rangle \ . \tag{5.11}$$

The fluctuations will thus in general provide a net contribution to the global balance of forces that constrain the mean flow. Chandrasekhar proposed [14] that in a turbulent fluid the term $\langle \delta u \cdot \nabla \delta u \rangle$ may in some conditions be replaced by the gradient of a term of the form $\nabla \left\langle \delta u^2 \right\rangle$. The effective total pressure would then become the sum of the ordinary pressure, plus a "turbulent pressure": $P + \rho \left\langle \delta u^2 \right\rangle$ which could expel the plasma from the regions where fluctuations are more important. This scenario was proposed to explain why turbulent interstellar matter could resist self-gravitational collapse. This requires assuming everywhere statistical isotropy for the velocity field, so that the tensor $\left\langle u_i u_j \right\rangle$ is zero except for diagonal terms which are positive definite. In practice, however, this is not realized: there are always special locations where this is not true, and where turbulence, instead of resisting, actually favors gravitational collapse. In a sense, this case is an example of (local) conversion of random flows into ordered (local) flows, inflows, or outflows.

A second example is that of acoustic waves. Can a localized source of acoustic waves "push" the fluid away from the source, thus exerting a wave pressure analogous to the radiative pressure? Note that acoustic waves propagating away from a source do not have net velocity, as they are characterized by a velocity fluctuation along the wave vector which is zero on average. However, acoustic eigenmodes do carry momentum. More precisely, consider acoustic waves, in a fluid globally at rest, escaping a source located at the origin. Examine the right-hand side of the Ox axis with $x > 0$; the x component of the average momentum density is

$$\langle \rho u_x \rangle = \langle \delta\rho \delta u_x \rangle = \rho_0 \delta u_x^2 / c_s \tag{5.12}$$

as the eigenmode propagating rightward satisfies $\delta u_x / c_s = \delta\rho/\rho_0$ (c_s is the sound speed). Hence, if there is any way to damp the amplitude of the waves as they propagate, due to, for example, viscous dissipation, then the waves should give back their momentum to the fluid. However, this does not necessarily give rise to a net acceleration or deceleration of the fluid. In the case of (nonmagnetic) stellar winds for instance, it can be shown that acoustic waves escaping from the surface actually lead to a net deceleration of the wind [15].

5.2.1 Alfvén Wave Pressure

We examine now the case of Alfvén waves, which are particularly relevant for astrophysical flows, in view of the ubiquitous presence of magnetic fields. Consider for simplicity plane waves (with velocity and magnetic field fluctuations depending on the x coordinate) with circular polarization, that is, nonlinear Alfvén waves with constant magnetic pressure, propagating in a plasma with a uniform mean magnetic field $B_0 \| Ox$. The wave is characterized by its fluctuating field δB which is perpendicular to the mean field. The excess magnetic pressure, or wave pressure P_w due to Alfvén waves is

$$P_w = \delta B^2 / (2\mu_0) \,. \tag{5.13}$$

It is directly associated with the fluctuating velocity: $\delta u = \pm \delta B / \sqrt{\rho}$, depending on the direction of propagation.

The Alfvén wave pressure phenomenon is illustrated here by considering successively two configurations: (i) a 1D homogeneous plasma, (ii) a stratified 3D plasma flow, namely a solar wind model.

5.2.1.1 Alfvén Pressure in a Homogeneous Plasma

We consider a 1D interval $-\pi \le x \le \pi$ with density unity and a mean magnetic field $B_0 = 1$ along the Ox axis, so that the Alfvén speed is $V_A = 1$ (see the "Alfvén speed" definition in Appendix A.1). To generate the Alfvén wave, we impose a periodic force $F \propto (0, \cos(\omega t), \sin(\omega t))$ on the two perpendicular components of the velocity u_y and u_z. This forcing is confined within a very small interval around the center of the domain at $x = 0$. As a result, two waves with circular polarization (that is, constant magnetic pressure $(B_y^2 + B_z^2)/(2\mu_0)$) start propagating in two opposite directions along the Ox axis. The frequency is $\omega = 5$, so that the wavenumber of the progressing waves is $k = \omega/V_A = 5$. The temperature of the plasma is such that $T/m = 1$, with a polytropic index $\gamma = 1.4$, implying a sound speed $c_s \simeq 1.18$.

Two distinct wave fronts are seen to propagate (see Figure 5.3): that of the Alfvén wave itself, represented in the figure by the magnetic pressure front, which proceeds at the Alfvén speed, and, slightly ahead of that, a localized pulse of parallel velocity that proceeds at the sound speed. The figure shows the wave pressure $P_w = (B_y^2 + B_z^2)/(2\mu_0)$, the gas pressure P, the total pressure $P + P_w$, and the parallel velocity U_x, at three successive times $t = 0.8, 1.6, 2.4$, respectively, as solid, dotted, and thick lines. The steady peak of gas pressure in the center of the domain ($x = \pi$) reveals the extent of the forced region.

Because of a small viscosity, the magnetic wave pressure decreases slowly starting from the forced region at the center. The region behind the two pulses of parallel velocity has no or negligible velocity. In this example, the acceleration by the wave pressure is thus a transient phenomenon: the injection of Alfvén waves at the center of the domain just sends two acoustic pulses, but does not generate any long lasting flow.

In a second experiment, we increase the viscosity by a factor of 20 (note that replacing viscosity by resistivity gives similar results). The results are shown in Figure 5.4. Because of the increased viscosity, the wave pressure now develops a

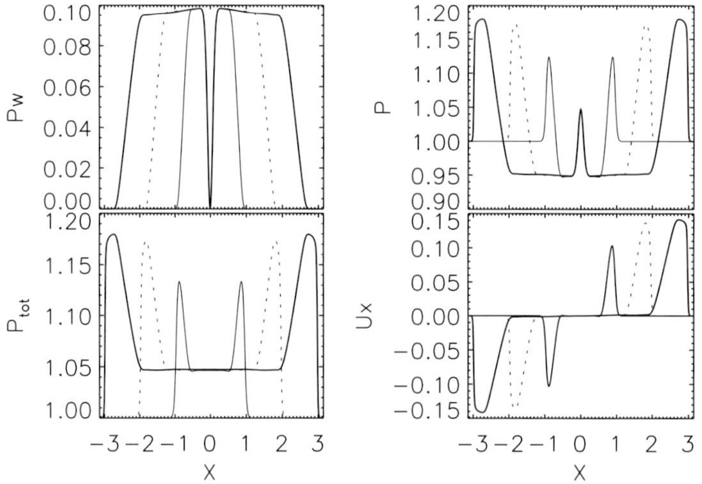

Figure 5.3 Pressure waves generated by forcing Alfvén plane waves with circular polarization in the middle of the domain ($x = 0$). Case with small viscosity. The mean magnetic field is parallel to Ox axis, the Alfvén speed is $V_A = 1$, and sound speed is $c_s = 1.18$. Wave magnetic pressure $P_w = (B_y^2 + B_z^2)/(2\mu_0)$, gas pressure P, total pressure $P_{tot} = P + P_w$, parallel velocity u_x are shown at times $t = 0.8$, 1.6 and 2.4, the time $t = 0$ corresponding to the beginning of the forcing.

much larger gradient in the region filled with Alfvén waves, behind the wave front. As a consequence, a quasi-steady outflow develops behind the acoustic pulse. The outflow region expands with the Alfvén waves. This is a simple example of Alfvén waves pushing the plasma out of the region where they propagate. We have considered here a particular value of the plasma β, close to unity. The outflow profile and its evolution does actually vary, depending on the plasma β, but the phenomenon remains general.

5.2.1.2 **Alfvén Wave Pressure in the Solar Atmosphere and Solar Wind**

Flow acceleration by Alfvén wave pressure may be most effective in stellar atmospheres and stellar winds [16]. The Alfvén wave pressure is thought to be an important ingredient particularly in generating the fast solar wind (with speeds larger than 700 km/s). Actually, in this case, dissipation (turbulent or not) is not really necessary, as the stratification with the distance to the Sun directly generates a nonzero gradient of the wave pressure, which adds to the gas pressure gradient to accelerate the wind up to large distances from the Sun.

Before giving a numerical example, let us briefly explain how the wave pressure gradient is related to stratification. Assume for simplicity spherical symmetry. The wind flow and the magnetic field are thus purely radial. Constancy of the magnetic flux ($\nabla \cdot B = 0$) along a radial magnetic line is expressed as

$$BA = \text{const} \tag{5.14}$$

where $A \propto r^2$ (r being the heliocentric distance) is the section of both a magnetic tube or a *flow tube* as well. (Note that the argument can be extended to more general topologies than pure spherical symmetry and thus use more general expressions

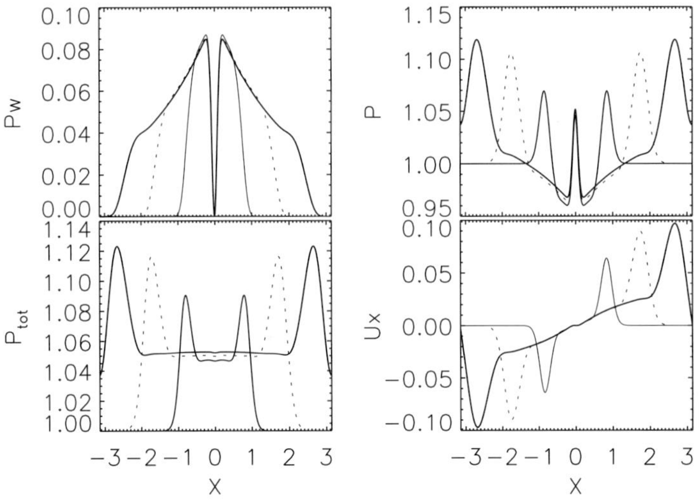

Figure 5.4 Pressure waves and quasi-steady flow formed by forcing Alfvén waves with circular polarization in the middle of the domain ($x = 0$). Case with larger viscosity. Same caption as previous figure.

of $A(r)$). Constancy of the mass flux along a flow line reads:

$$\rho u A = \text{const} . \tag{5.15}$$

Simple analytic forms can be derived either (i) close to the Sun where the atmosphere is quasi-static (ii) far from the Sun where the flow is supersonic and superalfvénic. First consider the quasi-static region (i). The *energy flux* of the waves is constant there in the high-frequency (WKB) limit:

$$(1/2)\left(\delta B^2/\mu_0\right) V_A A = \text{const} . \tag{5.16}$$

Using the constancy of the mass flux equation (5.15) and of the magnetic flux equation (5.14), Eq. (5.16) gives for the wave pressure

$$P_w = \delta B^2/(2\mu_0) \propto 1/(A V_A) \propto \sqrt{\rho} . \tag{5.17}$$

Alternatively, in the superalfvénic region far from the Sun ($u > V_A$), the eigenfrequency ω in the plasma frame is (slowly) changing as the wave propagates, so that the invariance of the wave energy flux is to be replaced by the invariance of the flux of the wave *action* $\delta B^2/(2\mu_0\omega)$, where the frequency ω is the Alfvén frequency in the wind frame. The wave action flux reads

$$\frac{\delta B^2}{2\mu_0}(u + V_A) A/\omega = \text{const} . \tag{5.18}$$

The frequency ω is related to the (distance-independent) frequency ω_0 in the absolute solar frame by $\omega = \omega_0 V_A/(u + V_A)$, so that Eq. (5.18) becomes

$$\frac{\delta B^2}{2\mu_0}(u + V_A)^2 A/V_A = \text{const} . \tag{5.19}$$

In the distant superalfvénic solar wind, the group velocity reduces to $u + V_A \simeq u$, which finally gives

$$\frac{\delta B^2}{2\mu_0} u^2 A/V_A = \text{const} . \tag{5.20}$$

Using the constancy of mass and magnetic fluxes (Eqs. (5.15)–(5.14)), this reads $\delta B^2 \propto \rho V_A/u \propto \rho^2 A V_A$ or, as well

$$P_w = \frac{\delta B^2}{2\mu_0} \propto \rho^{3/2} \propto r^{-3} . \tag{5.21}$$

The r^{-3} law in the RHS comes from the wind speed being quasi-uniform at long distances, so that the mass flux conservation Eq. (5.15) becomes $\rho \propto 1/A \propto r^{-2}$.

The wave pressure (Eqs. (5.17) and (5.21)) thus scales as $\rho^{1/2}$ and $\rho^{3/2}$, respectively, close and far from the Sun. Since the density decrease with distance is quasi-exponential in the quasi-static region close to the Sun, this leaves room for a large increase of wave pressure in the atmosphere, even taking into account the large temperature increase of the corona, which opens the possibility of a substantial contribution of waves to the acceleration of the wind.

The spherical symmetry, as well as the radial assumption for the magnetic field, rarely holds for real astrophysical objects. To illustrate how the wave pressure acts in more complex geometries, we now consider a numerical model of the solar wind in which Alfvén waves are injected starting from the base of the hot corona [17]. In these calculations, two main approximations are made. First, the part of the magnetic field due to internal currents of the Sun is assumed dipolar (the total magnetic field is thus the sum of this dipolar field and of the magnetic field induced by the flow in the atmosphere and wind). Second, the thermal effects have been eliminated from the model which integrates the equations of isothermal MHD. These approximations still allow us to look at how the wave pressure accelerates the flow in a complex magnetic field structure.

The numerical domain is a spherical shell between 1.8 and 16 $R_\odot$. The temperature of the corona is assumed to be 1.3 MK. A monochromatic wave of about 150 km/ s in rms amplitude is injected at 1.8 $R_\odot$, which is comparable to observed upper bounds [18]. The wave is monochromatic, with period equal to 20 min.

The figure (Figure 5.5) represents in a meridian the magnetic field lines (a), speed isocontours (b), and snapshots of (i) wind speed radial profiles shown in the top part of (c) at two latitudes (equator and 45°), with and without waves; (ii) the profiles of the azimuthal component (u_φ) at the same two latitudes shown in the bottom part of (c).

Consider the wind speed profile in Figure 5.5c (top). The small amplitude profile corresponds to the equatorial profile of u_φ, the large amplitude profile to the latitude 45°. The two profiles with negligible velocity in the interval $[2, 4]$ $R_\odot$ corresponds to the equator, the other two profiles to the 45° curves. In each case, there is a substantial velocity gain, of the order of 100 km/ s, due to the Alfvén wave pressure. Note that no wind comes directly from the equatorial regions, since the magnetic field there forms closed loops which trap the plasma and forbids any systematic steady outflow. In these equatorial regions, the magnetic field becomes open only at distances larger than about 5 $R_\odot$. This remark explains both the absence of radial velocity at the equator at smaller distances, and also the very low level of Alfvén waves at all distances at the equator: strictly, no Alfvén waves should survive in this region where the mean field is either undefined or not at all directed along the radial.

Large oscillations show up in the radial velocity at large latitude (Figure 5.5c, top). These are acoustic-like waves which propagate together with (and are generated by) the Alfvén waves. These waves come from the fact that the Alfvén waves, although circularly polarized when starting to propagate from the surface, lose this property during propagation, which drives pressure fluctuations in the form of forced acoustic waves.

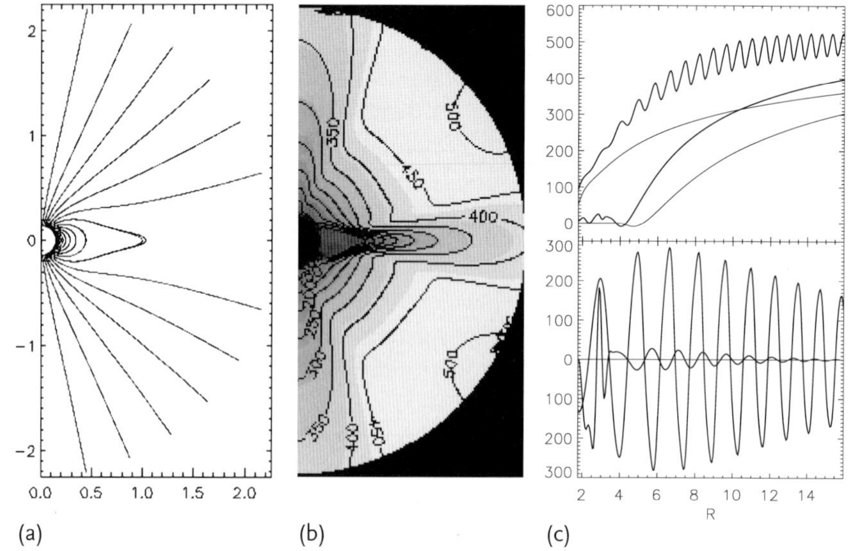

Figure 5.5 Additional acceleration of the solar wind by Alfvén waves using an isothermal axisymmetric MHD model. (a) Unperturbed magnetic field lines. (b) Radial velocity U_r in km/s, averaged during one Alfvén period. The slow polar wind speed is due to the absence of Alfvén wave injection there. (c, upper panel) Radial velocity U_r vs. distance r at equator and 45° latitude (faster speed), with Alfvén waves (thick lines) and without waves (solid lines). (c, lower panel) Azimuthal velocity U_φ at equator and 45° latitude (higher values). (From [17])

5.3 High-Frequency Waves

5.3.1 Cold Plasma Model

We consider a cold multicomponent plasma (species s), where the equilibrium densities n_{0s} and the velocities $\boldsymbol{v}_{0s}$ are uniform. The hypothesis that the different populations can be considered "cold" should have to be checked a posteriori by controlling the phase velocity found to be larger than all the thermal velocities. It allows one, when valid, to avoid any question about the generally difficult choice of a closure equation. Following the procedure given in Section 1.3.4, the cold plasma dielectric function $\boldsymbol{\epsilon}$ comes from the linearized equations of continuity and momentum of a cold plasma:

$$\begin{aligned} 0 &= \partial_t n_{1s} + \boldsymbol{v}_{0s} \nabla \cdot n_{s1} + n_{0s} \nabla \cdot \boldsymbol{v}_{s1} \,, \\ n_{0s} m_s d_t \boldsymbol{v}_{s1} &= n_{0s} e_s (\boldsymbol{E}_1 + \boldsymbol{v}_{0s} \times \boldsymbol{B}_1 + \boldsymbol{v}_{s1} \times \boldsymbol{B}_0) + n_{1s} (\boldsymbol{E}_0 + \boldsymbol{v}_{0s} \times \boldsymbol{B}_0) \,, \\ 0 &= n_{0s} (\boldsymbol{E}_0 + \boldsymbol{v}_{0s} \times \boldsymbol{B}_0) \,. \end{aligned} \tag{5.22}$$

The last equation comes from the equation of motion and relates the equilibrium quantities. The density n_{0s} being uniform, it shows that the perpendicular velocities are equal for all the species. For simplicity, we now consider a frame of reference where these velocities, as well as the equilibrium electric field are null. Then, the equilibrium velocities are parallel to $\boldsymbol{B}_0$. After a Fourier transform,

$$\begin{aligned}(\omega - \boldsymbol{k}\cdot\boldsymbol{v}_{0s})n_{s1} &= n_{0s}\boldsymbol{k}\cdot\boldsymbol{v}_{s1}\,,\\ -i(\omega - \boldsymbol{k}\cdot\boldsymbol{v}_{0s})m_s\boldsymbol{v}_{s1} &= e_s(\boldsymbol{E}_1 + \boldsymbol{v}_{0s}\times\boldsymbol{B}_1 + \boldsymbol{v}_{s1}\times\boldsymbol{B}_0)\,. \end{aligned} \tag{5.23}$$

Because the velocities are parallel to $\boldsymbol{B}_0$, we can write $\boldsymbol{k}\cdot\boldsymbol{v}_{0s} = k_{\|}v_{0s}$. From these and the Fourier transform of the Ampère equation,

$$\boldsymbol{v}_{s1} = \frac{ie_s}{\omega_{ds}m_s}\left[\boldsymbol{E}_1 + \frac{1}{\omega}\boldsymbol{v}_{0s}\times(\boldsymbol{k}\times\boldsymbol{E}_1)\right] + \boldsymbol{v}_{s1}\times\left(\frac{ie_s}{\omega_{ds}m_s}\boldsymbol{B}_0\right)\,. \tag{5.24}$$

where $\boldsymbol{\omega}_{ds}$, $\boldsymbol{\omega}_{cs}$ and $\boldsymbol{\omega}_{ps}$ are a Doppler shifted frequency, the cyclotron and the plasma frequencies associated with each species,

$$\omega_{ds} = \omega - k_{\|}v_{0s} \quad \text{and} \quad \boldsymbol{\omega}_{cs} = \frac{e_s\boldsymbol{B}_0}{m_s} \quad \text{and} \quad \omega_{ps}^2 = \frac{n_{0s}e_s^2}{\epsilon_0 m_s}\,. \tag{5.25}$$

Numerically, the corresponding frequencies f (in Hz) for electrons are $f_{pe} = 9\sqrt{n_e}$ and $f_{ce} = 28\times10^9 B$ where n_e is in m^{-3} and B in T. From Eq. (5.24),

$$\begin{aligned}\boldsymbol{v}_{s1} = & \frac{ie_s}{\omega_{ds}m_s}\frac{1}{1-(\omega_{cs}/\omega_{ds})^2}\left(\boldsymbol{I} - i\frac{\boldsymbol{\omega}_{cs}}{\omega_{ds}}\times - \frac{\boldsymbol{\omega}_{cs}}{\omega_{ds}}\frac{\boldsymbol{\omega}_{cs}}{\omega_{ds}}\right)\\ & \cdot\left(\boldsymbol{E}_1 + \frac{1}{\omega}\boldsymbol{v}_{0s}\times(\boldsymbol{k}\times\boldsymbol{E}_1)\right)\end{aligned} \tag{5.26}$$

where $\boldsymbol{I}$ is the unity tensor and $\boldsymbol{\omega}_{cs}\boldsymbol{\omega}_{cs}$ the dyadic square product of $\boldsymbol{\omega}_{cs}$ (this is a tensor). The contribution of the species s to the electric current perturbation is

$$\boldsymbol{J}_{1s} = n_{0s}e_s\boldsymbol{v}_{1s} + n_{1s}e_s\boldsymbol{v}_{0s}\,. \tag{5.27}$$

It can be written with a conductivity tensor $\boldsymbol{\sigma}$, that after some algebra appears to be

$$\boldsymbol{J}_{s1} = \boldsymbol{\sigma}_s\cdot\boldsymbol{E} = i\epsilon_0\omega\boldsymbol{M}_s\boldsymbol{N}_s\cdot\boldsymbol{E} \tag{5.28}$$

where

$$\boldsymbol{M}_s = \begin{pmatrix} \frac{\omega_{ds}}{\omega}\frac{\omega_{ps}^2}{\left(\omega_{ds}^2-\omega_{cs}^2\right)} & i\frac{\omega_{cs}}{\omega}\frac{\omega_{ps}^2}{\left(\omega_{ds}^2-\omega_{cs}^2\right)} & 0\\ -i\frac{\omega_{cs}}{\omega}\frac{\omega_{ps}^2}{\left(\omega_{ds}^2-\omega_{cs}^2\right)} & \frac{\omega_{ds}}{\omega}\frac{\omega_{ps}^2}{\left(\omega_{ds}^2-\omega_{cs}^2\right)} & 0\\ \frac{k_x v_{0s}\omega_{ps}^2}{\omega\left(\omega_{ds}^2-\omega_{cs}^2\right)} & -i\frac{k_x v_{0s}\omega_{cs}\omega_{ps}^2}{\omega\omega_{ds}\left(\omega_{ds}^2-\omega_{cs}^2\right)} & \frac{\omega_{ps}^2}{\omega^2} + \frac{k_z v_{0s}\omega_{ps}^2}{\omega\omega_{ds}} \end{pmatrix} \tag{5.29}$$

$$\boldsymbol{N}_s = \begin{pmatrix} 1-\frac{v_{0s}k_z}{\omega} & 0 & \frac{v_{0s}k_x}{\omega}\\ 0 & 1-\frac{v_{0s}k_z}{\omega} & 0\\ 0 & 0 & 1 \end{pmatrix}\,. \tag{5.30}$$

The propagation tensor $\boldsymbol{P}_{\mathrm{E}}$ defined in Eq. (1.76), also noted ϵ in the literature, is

$$\boldsymbol{P}_{\mathrm{E}} = \epsilon_0 \left(\boldsymbol{I} - \sum_{\mathrm{s}} \boldsymbol{M}_{\mathrm{s}} \boldsymbol{N}_{\mathrm{s}} \right) . \tag{5.31}$$

When the velocities $v_{0\mathrm{s}}$ are null, it can be expanded in the following way

$$\boldsymbol{P}_{\mathrm{E}} = \epsilon_0 \begin{pmatrix} \kappa_{11} & -i\kappa_{12} & 0 \\ i\kappa_{12} & \kappa_{11} & 0 \\ 0 & 0 & \kappa_{33} \end{pmatrix} \tag{5.32}$$

where

$$\kappa_{11} = \kappa_{22} = 1 - \sum_{\mathrm{s}} \frac{\omega_{\mathrm{ps}}^2}{\left(\omega^2 - \omega_{\mathrm{cs}}^2\right)}$$

$$\kappa_{12} = -\kappa_{21} = + \sum_{\mathrm{s}} \frac{\omega_{\mathrm{ps}}^2 \omega_{\mathrm{cs}}}{\omega\left(\omega^2 - \omega_{\mathrm{cs}}^2\right)}$$

$$\kappa_{33} = 1 - \sum_{\mathrm{s}} \frac{\omega_{\mathrm{ps}}^2}{\omega^2} . \tag{5.33}$$

The first two coefficients can be written as

$$\kappa_{11} = \kappa_{22} = \frac{1}{2}(R + L) \quad \text{and} \quad \kappa_{12} = -\kappa_{21} = \frac{1}{2}(R - L) \tag{5.34}$$

with

$$R = 1 - \sum_{\mathrm{s}} \frac{\omega_{\mathrm{ps}}^2}{\omega(\omega + \omega_{\mathrm{cs}})} \text{ and } L = 1 - \sum_{\mathrm{s}} \frac{\omega_{\mathrm{ps}}^2}{\omega(\omega - \omega_{\mathrm{cs}})} . \tag{5.35}$$

From Eq. (1.77), with the same conventions for the orientation of the magnetic field and of the wave vector, the dispersion relation is

$$\begin{vmatrix} \omega^2\kappa_{11}/c^2 - k_{\parallel}^2 & -i\omega^2\kappa_{12}/c^2 & k_{\perp}k_{\parallel} \\ i\omega^2\kappa_{12}/c^2 & \omega^2\kappa_{11}/c^2 - k^2 & 0 \\ k_{\perp}k_{\parallel} & 0 & \omega^2\kappa_{33}/c^2 - k_{\perp}^2 \end{vmatrix} = 0 . \tag{5.36}$$

5.3.2
Parallel Propagation

When $k = k_{\parallel}$, Eq. (5.36) is decoupled in two independent equations

$$\kappa_{33} = 0 \quad \text{and} \quad \kappa_{11} + s\kappa_{12} = \frac{k^2c^2}{\omega^2} , \tag{5.37}$$

where $s = \pm 1$. The first case corresponds to "plasma waves"; their dispersion equation is $\omega^2 = \omega_{\mathrm{p}}^2$. In the second case, the dispersion equation is

$$\frac{k^2c^2}{\omega^2} = 1 - \sum_s \frac{\omega_{ps}^2}{\omega(\omega - s\omega_{cs})} \ . \tag{5.38}$$

Remembering that D is the determinant of the propagation matrix $\boldsymbol{P}_E$, Eq. (1.76) implies $E_z = 0$, or equivalently $E_\parallel = 0$. Therefore, unlike electrostatic waves, $\boldsymbol{k} \cdot \boldsymbol{E}_1 = 0$. The Faraday equations then tells us that $\boldsymbol{E}_1 \cdot \boldsymbol{B}_1 = 0$; the electric and magnetic perturbation vector fields are perpendicular to each other. With a single ion species and $\boldsymbol{v}_0 = 0$, Eq. (5.38) the dispersion equation is

$$\omega^4 - s\omega^3(\omega_{ce} + \omega_{ci}) + \omega^2 \left[\omega_{ce}\omega_{ci} - k^2c^2 - \left(\omega_{pe}^2 + \omega_{pi}^2\right)\right] - s\omega \left[(\omega_{ce} + \omega_{ci})\, k^2c^2 + \omega_{pe}^2\omega_{ci} + \omega_{pi}^2\omega_{ce}\right] - k^2c^2\omega_{ce}\omega_{ci} = 0 \ . \tag{5.39}$$

(Keep in mind that $\omega_{ce} < 0$.) When the frequency ω is much larger than the ion plasma and cyclotron frequencies,

$$\frac{k^2c^2}{\omega^2} = 1 - \frac{\omega_{pe}^2}{\omega(\omega \pm \omega_{ce})} \ . \tag{5.40}$$

The case of the positive (negative) square root corresponds to a right-handed (left-handed) polarization. The cut-off frequencies, that is, for $k \to \infty$, are called the left and right cut-off frequencies,

$$\omega_G = \frac{1}{2}\left(+\omega_{ce} + \sqrt{\omega_{ce}^2 + 4\omega_{pe}^2}\right) = \frac{1}{2}\left(-|\omega_{ce}| + \sqrt{\omega_{ce}^2 + 4\omega_{pe}^2}\right) , \tag{5.41}$$

$$\omega_D = \frac{1}{2}\left(-\omega_{ce} + \sqrt{\omega_{ce}^2 + 4\omega_{pe}^2}\right) . \tag{5.42}$$

The right-hand wave then enters into resonance with the electron gyrofrequency. A computation based on Eq. (5.38) taking the ion frequency into account would show that the left-handed polarized wave enters into resonance with the ion gyrofrequency.

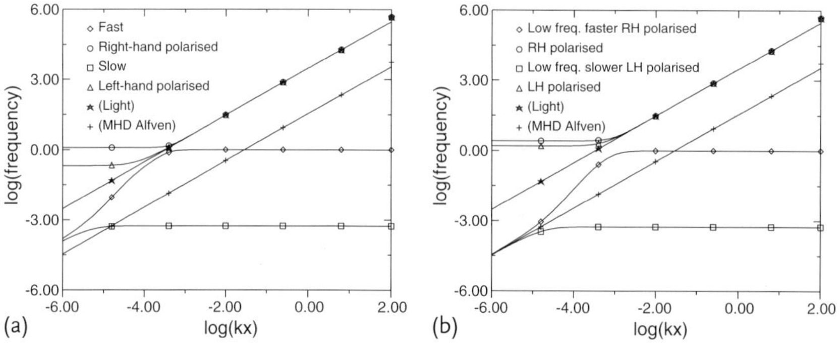

Figure 5.6 Dispersion relation for waves in parallel propagation in a cold hydrogen plasma. The wave number ($k = k_\parallel$) is normalized by the electron inertial length $x = d_e$, the frequency by the electron gyrofrequency ω_{ce} vs. k. (a) For $\omega_{pe}/\omega_{ce} = 0.5$. (b) For $\omega_{pe}/\omega_{ce} = 2$.

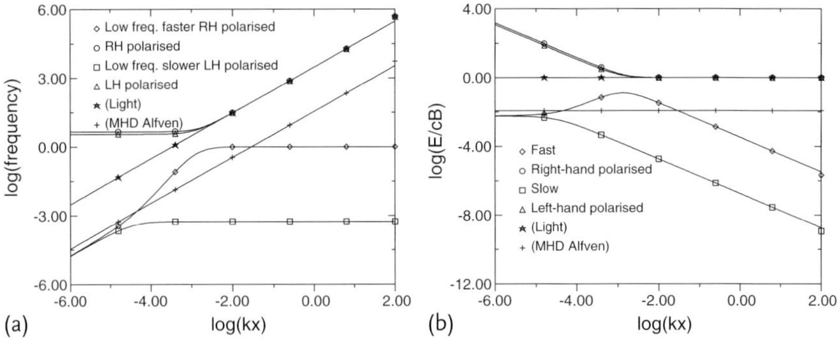

Figure 5.7 Waves in parallel propagation ($k = k_{\parallel}$) in a cold hydrogen plasma where $\omega_{pe}/\omega_{ce} = 4$. (a) Dispersion ω/ω_{ce} vs. k. (b) Electric to magnetic field ratio vs k.

The dispersion equation (5.39) is a fourth order polynomial, with four roots. The corresponding wave dispersion is shown in Figures 5.6 and 5.7. Among these roots, the two with the lowest frequency correspond to the right-hand side and left-hand side circularly polarized Alfvén waves.

5.3.3
Perpendicular Propagation: Ordinary and Extraordinary Waves

When $k = k_{\perp}$, the dispersion equation is split into two independent parts

$$\frac{k^2c^2}{\omega^2} = \kappa_{33} \quad \text{or} \quad \kappa_{11}\left(\kappa_{22} - \frac{k^2c^2}{\omega^2}\right) + \kappa_{12}^2 = 0\,. \tag{5.43}$$

The first equation defines ordinary waves. Considering the total plasma frequency $\omega_p^2 = \sum_s \omega_{ps}^2$, it is simply

$$\omega^2 = \omega_p^2 + k^2c^2\,. \tag{5.44}$$

For $\omega \gg \omega_p$, it coincides with light waves; with a relation of dispersion $\omega \sim kc$, it does not "see" the plasma. On the contrary, for $\omega \sim \omega_p$, in spite of an almost infinite phase velocity, the group velocity is null: the wave propagation is stopped by the plasma. This wave does not exist for $\omega < \omega_p$.

With several plasma populations, the resolution of the second equation is tedious. For a single ion species plasma and null $\boldsymbol{v}_{0s}$ velocities, using Eq. (5.34) and neglecting Zm_e/m_i in comparison to unity,

$$\frac{k^2c^2}{\omega^2} = \frac{RL}{S} = \frac{\left(\omega^2 + \omega\omega_{ce} + \omega_{ce}\omega_{ci} - \omega_{pe}^2\right)\left(\omega^2 - \omega\omega_{ce} + \omega_{ce}\omega_{ci} - \omega_{pe}^2\right)}{\left(\omega^2 - \omega_{lh}^2\right)\left(\omega^2 - \omega_{uh}^2\right)}, \tag{5.45}$$

where ω_{uh} and ω_{lh}, respectively, are the upper and the lower hybrid resonant frequencies

$$\omega_{\mathrm{uh}}^2 = \omega_{\mathrm{pe}}^2 + \omega_{\mathrm{ce}}^2 \quad \text{and} \quad \frac{1}{\omega_{\mathrm{lh}}^2} = \frac{1}{\omega_{\mathrm{ci}}^2 + \omega_{\mathrm{pi}}^2} + \frac{1}{|\omega_{\mathrm{ci}}\omega_{\mathrm{ce}}|} .$$

This is a sixth degree equation relatively to ω. This allows possibly six possibilities of wave propagation. This simple example shows the complexity of wave analysis, even in the simple case of cold and uniform plasmas. The problem is made even more complex when other ion species are introduced. An abundant amount of literature is devoted to wave propagation in plasmas [19]. It is very often quite useful to solve the wave equations numerically. For instance, far beyond the simplifying assumption of cold plasmas, the WHAMP program [20] computes wave dispersion and polarization in homogeneous plasmas composed of many species, with finite (possibly anisotropic) temperatures.

5.3.4
Application: Plasma Cut-offs and Limits to the Radio Astronomy

Radio waves propagated into a vacuum naturally convert into ordinary waves when they come into a plasma. With an increasing electron density, as long as the requirements for Eq. (1.82) are met, the waves adopt a path that is analogous to those shown on the left-hand side of Figure 1.6, and that include wave reflection before reaching the location where their frequency equals the local plasma frequency.

As the Earth is surrounded by the plasmas of the ionosphere and of the magnetosphere, low-frequency waves from the ground can be reflected back upon the ionosphere. The characteristic cut-off is the plasma frequency, that is, proportional to the square root of the electron density. As the densest plasma around the Earth is in the ionosphere (at an altitude of approximately 250 km), the reflection of waves occurs in the ionosphere. The ionospheric cut-off corresponds in general to a frequency close to 10 MHz. This frequency fluctuates according to the fluctuation of the ionospheric density. The cut-off frequency is lower near the poles, at night time, and when the solar activity is low.

The amateur radio frequency bands include channels between 1.8 and 28 MHz (medium and short waves). The waves where $f < 10$ MHz are currently reflected by the ionosphere and they allow long range, and even interhemispheric communications. In the range 14–28 MHz, long-range communications are possible occasionally, according to solar activity. The other wave bands used by amateur operators are well above the ionospheric cut-off and do not allow ionospheric reflection. Waves below the ionospheric cut-off are also used with radars sounding the ionospheric activity. For instance, the Super Dual Auroral Radar Network (SuperDARN) is an international radar network comprising 18 radars in both hemispheres. They operate in the high-frequency (HF) bands between 8 and 22 MHz. The radars, based on "coherent diffusion" by the ionsopheric plasma measure the Doppler velocity of plasma density irregularities in the ionosphere. The choice of their frequency is obviously based on the possibility of wave reflection by the ionosphere. Similarly, the EISCAT (Europe), HAARP (USA), Sura (Russia), and so, on radar networks, based on a somewhat different principle ("ionospheric incoherent diffusion"), operate at frequencies below the ionospheric cut-off.

Reciprocally, the ionospheric cut-off isolates radio-astronomers from distant radio emissions, and also from the powerful terrestrial radio emissions. Indeed, the auroral zone of the Earth is a strong radio emitter in the kilometric range of wavelengths. The emitted waves are called Auroral Kilometric Radiations (AKR). Their sources have altitudes of a few thousand kilometers. The waves frequency $f \sim 250\,\text{kHz}$ is close to the electron gyrofrequency in the source regions. The AKR is triggered by bunches of accelerated electrons connected to the auroral activity and to magnetospheric substorms, and the total emission power can reach 10^7 W. Because these waves cannot reach the ground (except occasionally in polar regions [21]) they were discovered during the space era, thanks to measurements aboard space probes. On the contrary, similar emissions above Jupiter, where the local electron gyrofrequency ($f < 40\,\text{Mhz}$) can be above the ionospheric cut-off, are studied (for $f > 10\,\text{MHz}$) with radio telescopes on ground, operating in the decametric range of wavelengths.

A radio-telescope installed on the far side of the moon (protected from terrestrial low frequency emissions) would allow us to observe much below the Earth's ionospheric cut-off. The low frequency cut-off would then be imposed by the solar wind density. Generally speaking, the solar wind density varies as $n = n_0 r^{-2}$ where r is the distance to the Sun. When r is expressed in astronomical units (AU), $n_0 = 5 \times 10^6\,\text{m}^{-3}$, and $\omega_{\text{pe}} = 9\sqrt{5}\text{r}^{-1}\,\text{kHz}$. At 1 AU, the cut-off frequency is about 20 kHz, this is the lowest frequency allowed for the observations of remote radio waves in the vicinity of the Earth. A radio-telescope on the far side of the moon would allow observations down to $f = 20\,\text{kHz}$.

Of course, at larger distances to the Sun, lower frequency waves can be observed. For instance, 2–3 kHz signal, well above the background noise was observed aboard the Voyager 1 and 2 spacecrafts when they reached a distance of 7–10 AU. The 2–3 kHz indeed correspond to the local cut-off frequency of the solar wind.

5.3.5
Application: The Dispersion of Radio Waves from Pulsars

The name "pulsar", a contraction of "pulsating star", comes from the pulsed character of their radio emissions. The radio waves are observed over a broad bandwidth including all the atmospheric windows, from decametric to decimetric waves. The envelope of the signal repeats with a period (typically in the range 1 ms–10 s) that corresponds to the spin period of the neutron star. The pulsed character of the signal appears on every radio frequency, but the signal arrives later in the lower frequency channels than in higher-frequency ones, as can be seen in Figures 5.8 and 5.9. This can be explained by the dispersion of ordinary waves in a plasma. Actually, the interstellar medium is not completely empty; it rather contains a low density plasma. The radio waves emitted by the pulsar then propagate mostly as ordinary waves, and the pulse shape is propagated at the ordinary mode's group velocity, that from Eq. (5.44) is

$$v_g = \frac{kc^2}{\omega} \sim c\left(1 - \frac{\omega_p^2}{2\omega^2}\right) . \quad (5.46)$$

The approximation on the right-hand side is valid when the wave frequency is large compared to the plasma frequency, $\omega \gg \omega_p$. The time delay between the pulse emission and its reception over a distance L is

$$T = \int_0^L \frac{dl}{v_g} \sim \int_0^L \frac{dl}{c}\left(1 + \frac{\omega_p^2}{2\omega^2}\right) = \frac{L}{c} + \frac{e^2}{2c\epsilon_0 m_e \omega^2}\int_0^L n_e dl . \quad (5.47)$$

The term L/c is the free space travel time, and the last term is proportional to the column density of plasma crossed by the radio wave. Radio-astronomers call *dispersion measure* the integral $\mathrm{DM} = \int_0^L n_e dl$. The delay is then

$$t = \mathrm{DM}/(2.41 \times 10^{-4} f^2) \quad (5.48)$$

where t is in seconds, and the frequency f in MHz. Then, a short pulse measured on a receiver of bandwidth B is stretched out to a length $\tau = 8.3 \times 10^3\,\mathrm{DM}\,f^{-3}B$ measured in seconds. The dispersion of the pulses is then used as an indication of the amount of interstellar plasma crossed by the wave, and, therefore, as an indicator of the distance to the pulsar.

5.3.6 Application: Faraday Rotation in the Interstellar Medium

Figures 5.6 and 5.7 and the corresponding existence of different cut-off frequencies for left and right handed waves imply that a linearly polarized wave in a plasma will see its polarization plane turn by an angle $\Delta\Psi$ as long as it propagates along the ambient magnetic field. It can be shown that this property is true also for oblique propagation. The difference between the left and the right cut-off frequencies is due only to the component of the magnetic field along k; therefore, a measure of the rotation of the polarization plane can be used to gain information about the intensity of the magnetic field into the plasma. More precisely, the rotation of the polarization plane is

$$\Delta\Psi = \mathrm{RM}\lambda^2 \quad (5.49)$$

where λ is the wavelength, and RM is called the *rotation measure*, and is defined (in SI) as

$$\mathrm{RM} \sim 2.7 \times 10^{-13} \int n_e(l) B_{\parallel}(l) dl \quad \mathrm{rad\,m^{-2}} \quad (5.50)$$

where n_e is the electron density, and $B_{\parallel}$ is the line of sight magnetic field strength, and the integral is taken over the entire path from the source to the observer. Practically, the Faraday effect can be shown with multiwavelength measurement and Eq. (5.49) provides RM. This effect is stronger with low frequency waves.

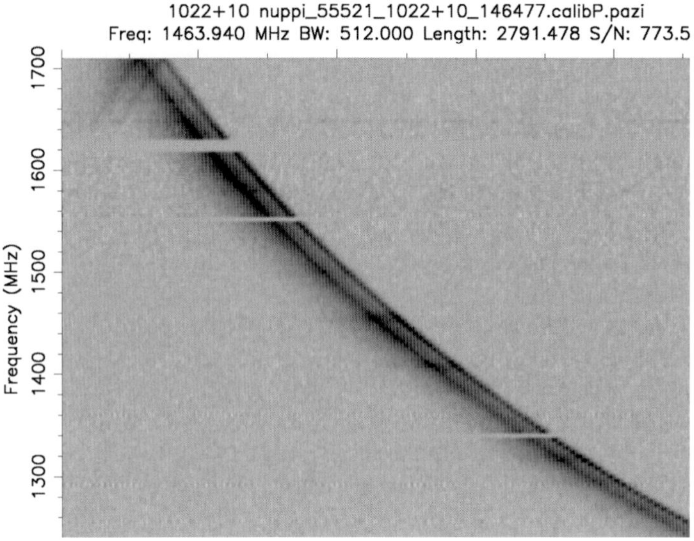

Figure 5.8 Phase-frequency spectrum of the pulsar J1022+1001. The ordinate is the emission frequency and the abscissa is the time of one period of the pulsar $P = 16.453$ ms. The spectrum is taken in the GHz range, measured on 21 November 2010, with the Nançay Radio Telescope. The fact that the signal arrives later at lower frequencies is a result of the dispersion. The fit of this curve with Eq. (5.48) indicates a dispersion measure DM $= 10.246\,\text{pc}\,\text{cm}^{-3}$. Credit: Cognard I., Station de Radioastronomie de Nançay, LPC2E/CNRS and Université d'Orléans.

The closest plasma providing Faraday rotation is those of the ionosphere. Because of the rather unpredictable fluctuations of the ionospheric density, the polarization plane of linearly polarized waves reflected upon the ionosphere is unpredictable too, and this can raise a problem when simple dipole antennas are used. (Because they are very sensitive to Ψ.)

Hopefully, the Faraday effect can be used in astronomy to evaluate the order of magnitude of interstellar magnetic fields. This is especially true when the rotation measure (depending on $n_e B_{\parallel}$) is combined – as for pulsars – with the dispersion measure DM (depending only on n_e) or when the density can be estimated through a different method. In astrophysical sources, synchrotron radiation is largely polarized, and constitutes a good signal for rotation measure estimates.

Faraday measurements in the Milky Way conduct to an average galactic magnetic field about 0.1 nT.

5.4 Whistler Mode

As shown in this chapter, there are a finite number of linear propagation eigenmodes in fluid theories. Similar modes can generally be retrieved by kinetic mod-

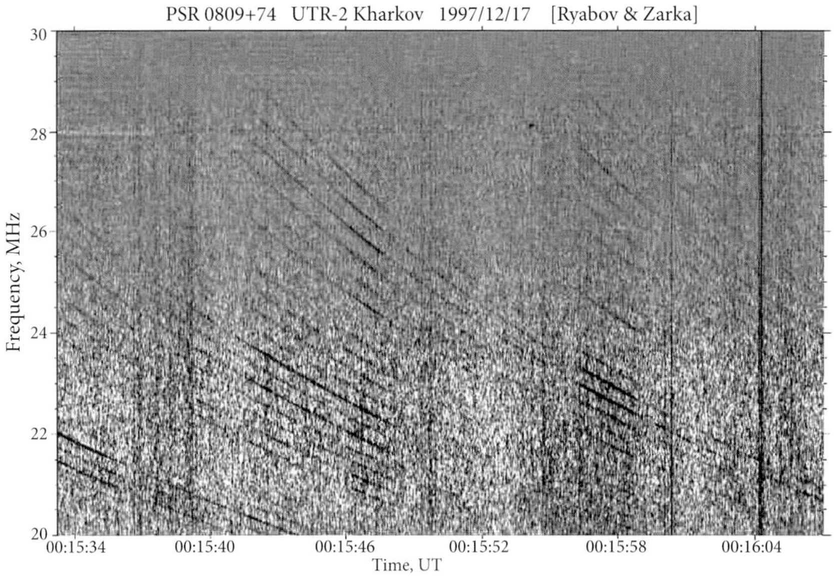

Figure 5.9 Time-frequency spectrum of the pulsar PSR 0809+74, measured in the 20–30 MHz range with the Kharkov UTR2 radio telescope, on 17 December 1997. Because the range of frequencies is much lower than in Figure 5.8, and in spite of a small dispersion measure (DM = 5.75 pc cm^{-3}), the time delay is much more pronounced. For a frequency range of only 10 MHz, it is larger than the pulsar period (P = 1.2922 s) by a factor exceeding 20. Credit: P. Zarka, CNRS-Observatoire de Paris.

eling (see Section 5.6) with more or less noticeable changes, such as damping. In each scale range (spatial and temporal), the modeling of these modes can be more or less simplified, due to the dominance of some effect on the others (see Chapter 3). The case of the whistler mode is an example of such simplifications and it deserves special attention because this mode has many interesting properties that have applications in space plasmas.

The whistler mode is the only mode that exists in a magnetized plasma in the "electron MHD" range, that is, for temporal scales limited by $\omega_{ci} \ll \omega \ll \omega_{ce}$. In this range, the ions can be considered steady because of their large mass, and the electron motion can be just determined by an Ohm's law $\boldsymbol{E} = -\boldsymbol{u}_e \times \boldsymbol{B} - \nabla P_e / ne$ because of their small mass. It is quite noticeable that neither the ion nor the electron inertial comes into the description of this range: all phenomena are fully independent of m_i and m_e.

The linear modes in this range are determined by only two equations: the Faraday/Ohm's equation and the Ampère's equation, taking into account that the whole current is carried by the electrons since the ions can be supposed steady. After linearization, it is:

$$\frac{\boldsymbol{B}_1}{B_0} = \frac{1}{\omega} [\boldsymbol{e}_z \boldsymbol{k} - \boldsymbol{e}_z \cdot \boldsymbol{k} \boldsymbol{I}] \boldsymbol{u}_{e1} \tag{5.51}$$

$$\boldsymbol{u}_{e1} = -i \frac{B_0}{n_0 e \mu_0} \boldsymbol{k} \times \frac{\boldsymbol{B}_1}{B_0} . \tag{5.52}$$

In this system, $\boldsymbol{e}_z$ is the unit vector along $\boldsymbol{B}_0$. Note that the pressure tensor does not appear in the equations, indicating that the scale range under study is independent of the temperatures. This is a priori not fully general: it is valid only when the curl of $\nabla \cdot \boldsymbol{P}_{e1}$ is null. Nevertheless, it is generally well satisfied in the scale range under study. It would be important to look with more detail if investigating the continuation of the whistler mode for very oblique propagation and large k, when the condition $\omega_{ci} \ll \omega$ is no longer satisfied.

Choosing the reference frame such that $\boldsymbol{e}_z = (0, 0, 1)$ and $\boldsymbol{k} = (k_\perp, 0, k_\parallel)$ one can easily solve explicitly the system in $\boldsymbol{u}_{e1}$ and obtain:

$$\begin{pmatrix} 1 & i\frac{k_\parallel^2 D}{\omega} & 0 \\ -i\frac{k^2 D}{\omega} & 1 & 0 \\ 0 & -i\frac{k_\parallel k_\perp D}{\omega} & 1 \end{pmatrix} \cdot \boldsymbol{u}_{e1} = 0 . \tag{5.53}$$

The constant D is defined, as in Chapter 3, by $D = B_0/n_0 e \mu_0$. Introducing artificially the electron or ion masses in the formula, this can be written as well as a function of the inertial lengths and gyrofrequencies: $D = \omega_{ce} d_e^2 = \omega_{ci} d_i^2 = V_A^2/\omega_{ci} = V_A d_i$.

The dispersion is obtained, as usual, by canceling the determinant, which provides:

$$\omega = \pm k k_\parallel D . \tag{5.54}$$

Reporting this dispersion in the above matrix provides the polarization: the whistler velocity perturbation is proportional to $\boldsymbol{u}_{e1} = (\pm k_\parallel, ik, \mp k_\perp)$. It means that this perturbation is transverse to $\boldsymbol{k}$, and, therefore, that, in the hypotheses done, the whistler mode is incompressible. Its other important properties are:

- For strictly parallel propagation $k = k_\parallel$, the group velocity along the magnetic field is proportional to k. This means that an initial localized perturbation will become dispersed along the propagation, the shortest wavelength (larger frequencies) arriving at any given point earlier than the longer ones. This observation of "decreasing tones" is at the origin of the name of the mode: if a lightning strike in the atmosphere emits such a wave, this one can be observed far from the source (in the magnetosphere or on the ground in the opposite hemisphere) and it sounds as a decreasing "whistle" (the signal is actually in the audible frequencies, so that one can listen directly to it, via a loudspeaker).
- For a propagation close to the parallel propagation, but not strictly parallel, there is a ratio on the order of $2k_\parallel/k_\perp$ between the parallel and perpendicular group velocities. The consequence is that the mode is well guided along the magnetic field, which explains why it can indeed be observed far from its source: if the group velocities were equal in all directions, the wave would experience a geometrical decrease in $1/r^2$, which would much limit the possibilities of observations at large distance.

- The group velocity, via D, also depends on the local density and magnetic field. This leads a whistler wave propagating along an inhomogeneous medium (assuming that these zero order gradients occur on scales much larger than the wavelength, not to invalidate the above treatment) to have a varying velocity, and possibly to reflect. This occurs, for instance, for the above lightening whistlers, which can reflect back and forth several times between the two hemispheres before becoming completely unobservable. Excited artificially, the whistler waves can also be used to perform remote measurements of the ionospheric density profile.

The above dispersion equation is valid only in the scale range between ion and electron gyrofrequencies. When prolongated toward low frequencies, it merges with a MHD mode, which is generally the fast magnetosonic mode (so becoming compressible). At high frequency, the electron inertia makes it tend toward an asymptote, which is $\omega = \omega_{\rm ce}$ whenever the angle is not too close to perpendicular (the ion and electron inertia would have then to be taken into account at the same time). The asymptotic solution is electrostatic and, because it depends on the propagation angle, it is called "the resonance cone". Figure 5.10 shows these two continuations in the parallel propagation case. It is to be noted that the low frequency continuation in this case just transforms the previous dispersion relation into $\omega = \pm k^2 D(1 + \omega_{\rm ci}/\omega)$, which make apparent the connection with the MHD mode since $D\omega_{\rm ci} = V_{\rm A}^2$.

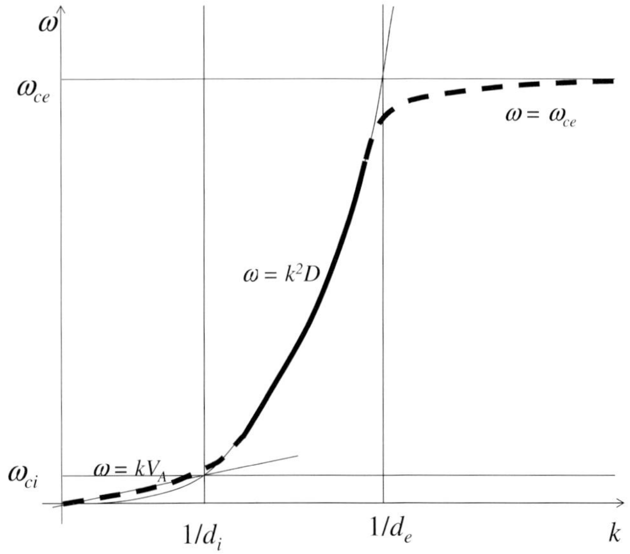

Figure 5.10 Dispersion relation of the whistler mode for a propagation parallel to the magnetic field. The bold line corresponds to the domain of validity of the "electron MHD" approximation used in the present calculation.

5.5
Collisional Damping in Fluid Theories

Fluctuations of any nature (temperature, velocity, magnetic field, etc.) tend to become damped by nonreversible processes involving interparticle collisions. At the level of a fluid description, damping is always associated with a rise of the total entropy in the considered system. At the microscopic level damping and entropy are equally intimately linked as long as collisions are the driving factor of the damping mechanism. However, as illustrated in a later section of this chapter, fluid oscillations can be damped even in a rigorously collisionless plasma.

5.5.1
Dissipative Effects and Entropy

A dissipative process is a process occurring in a nonhomogeneous thermodynamic system implying the irreversible transformation of some form of energy (bulk, internal, potential, magnetic, etc.) into another. In a fluid, the dissipation of temperature fluctuations through conduction, the dissipation of velocity fluctuations through viscosity, the dissipation of a gravitationally driven surface wave through viscosity or the dissipation of magnetic field fluctuations through electric resistivity are examples of dissipative processes. In principle, the final form of energy for a dissipative process is the internal energy U and the final state of the system a homogeneous one characterized by a unique temperature T. In classical thermodynamics, a dissipative process is associated with a positive variation of the entropy $ds > 0$ and a rise of the internal energy according to the fundamental thermodynamic relation

$$d U_{\text{diss}} = T ds \, . \tag{5.55}$$

It is worth remembering here that a positive variation of the internal energy is not necessarily associated with a dissipative process and a growth of entropy. It can be an adiabatic process where the internal energy varies according to the prescription of the first law of thermodynamics

$$d U_{\text{rev}} = -P dV = m \frac{P}{\rho^2} d\rho \tag{5.56}$$

where ρ is the density, P the pressure and $V = m/\rho$ is the volume occupied by a molecule of mass m. From the continuity equation $\partial_t \rho = -\nabla \cdot (\rho \boldsymbol{u})$ and the equation of motion $D_t \boldsymbol{u} = -\nabla P/\rho$, it follows that the variation of the sum of the internal and kinetic energy within a fixed volume V is given by

$$\begin{aligned} \partial_t \int\limits_V \left(n U_{\text{rev}} + \rho \frac{u^2}{2} \right) dV &= - \int\limits_V \nabla \cdot \left[\rho \boldsymbol{u} \left(\frac{u^2}{2} + \frac{w}{m} \right) \right] dV \\ &= - \int\limits_{\partial V} \rho \boldsymbol{u} \left(\frac{u^2}{2} + \frac{w}{m} \right) \cdot \boldsymbol{dS} \end{aligned} \tag{5.57}$$

where $w = U_{\text{rev}} + P/n$ is the enthalpy per mass, ∂V the surface of the volume V and $\boldsymbol{dS}$ the outwards pointing surface element. Let us now consider an infinite monochromatic, propagating wave, of wavelength λ and angular frequency ω. We assume the coordinate system to be oriented in such a way for the wave vector $\boldsymbol{k}$ to be aligned with the x axis so that we can write the field fluctuations as

$$\boldsymbol{u}, \delta P, \delta\rho \propto \exp[i(kx - \omega t)] \,. \tag{5.58}$$

This wave travels from left to right through the volume V represented in Figure 5.11 with a phase velocity ω/k. For a linear wave (for instance, a sound wave) the internal energy of the fluid increases or decreases locally following Eq. (5.56).

The variation of the energy within the volume can be computed at any given time using Eq. (5.57). If the box length L_x along the propagation axis x is taken to be a multiple of the wavelength $L_x = n\lambda$, the variation of the internal energy is zero due to the vanishing of the surface integral in Eq. (5.57) which can be easily verified by decomposing the integration over all surfaces of the volume

$$\begin{aligned}\int_{\partial V} \boldsymbol{F}\cdot d\boldsymbol{S} = {} & F(x,t)(-A_x) + F(x+L_x,t)A_x \\ & + \int_{A_y} \boldsymbol{F}\cdot d\boldsymbol{S} - \int_{A_y} \boldsymbol{F}\cdot d\boldsymbol{S} \\ & + \int_{A_z} \boldsymbol{F}\cdot d\boldsymbol{S} - \int_{A_z} \boldsymbol{F}\cdot d\boldsymbol{S} \\ & \propto e^{i(kx-\omega t)}\left(1 - e^{ikL_x}\right) \,.\end{aligned} \tag{5.59}$$

The system being invariant under translations along the y direction, the two surface integrals over A_y cancel each other identically on the right-hand side of Eq. (5.59).

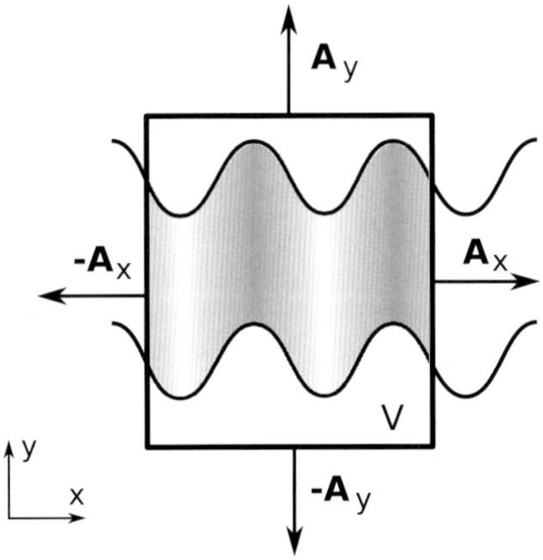

Figure 5.11 A nondissipative propagating wave modifies locally and temporarily the internal energy of the fluid according to Eq. (5.56). The time averaged internal energy within a fixed volume V is unaltered as the total density flux through all its faces is zero.

The same argument does obviously apply for two surface integrals over A_z. The two terms with A_x do also cancel identically if the box length L_x is an integer multiple of the wavelength λ. For L_x not a multiple of λ the integral (5.59) will be a zero average oscillating function of time.

Let us now consider the less trivial case of a pure nonreversible variation of the internal energy of a fixed volume of the fluid. The latter can be computed using Eq. (5.55) and the continuity equation leading to the following expression (5.55)

$$\begin{aligned}\partial_t(n U_{\mathrm{diss}}) &= n\partial_t U_{\mathrm{diss}} + U_{\mathrm{diss}}\partial_t n \\ &= -\nabla\cdot(n\boldsymbol{u}\,U_{\mathrm{diss}}) + nT\left(\partial_t s + \boldsymbol{u}\cdot\nabla s\right) .\end{aligned} \tag{5.60}$$

The general energy equation, including both reversible and dissipative effects can now be written by merging Eq. (5.57) (not considering the volume integration) and Eq. (5.60), viz

$$\partial_t\left(nU + \rho\frac{u^2}{2}\right) = -\nabla\cdot\left[\rho\boldsymbol{u}\left(\frac{u^2}{2} + \frac{w}{m}\right)\right] + nT D_t s \tag{5.61}$$

where the internal energy is now the sum of reversible and nonreversible contributions $U = U_{\mathrm{rev}} + U_{\mathrm{diss}}$ and the enthalpy is consistently defined as $w = U + P/n$. From Eq. (5.61) we deduce that the variation of the energy content brought into a volume V by nonreversible processes is given by

$$\int_V nT D_t s dV . \tag{5.62}$$

It should be emphasized that a moving fluid parcel is said to behave adiabatically if its entropy content ns obeys the conservation equation

$$\partial_t(ns) + \nabla\cdot(ns\boldsymbol{u}) = nD_t s = 0 \quad \text{(adiabatic)} . \tag{5.63}$$

In this case the variation of the parcels internal energy is a reversible one. As we shall see in Section 7.1, the viscous friction which tends to dissipate velocity gradients in the fluid, as well as heat conduction which tends to dissipate temperature gradients, are examples of such nonreversible processes. In the case of the infinite monochromatic wave of Figure 5.11 with temperature and velocity fluctuations, both viscous and conductive dissipation may operate. We can verify that dissipation in the volume is necessarily accompanied by a rise of its entropy content. We do already know that the reversible fluctuations do not modify the average internal energy in the volume V and consequently consider only the variations of the internal energy due to dissipation. Thus, using the identity $d(nU) = ndU + Udn$ and setting $dU = Tds$ we obtain

$$\frac{1}{nU}\partial_t(nU) = \frac{T}{U}\partial_t s + \frac{1}{n}\partial_t n . \tag{5.64}$$

Integrating this equation over the volume V and assuming that the total mass of fluid in the volume does not change in time (no sources or sinks) we do finally obtain

$$\partial_t \int_V \ln(nU)dV = \int_V \frac{T}{U}\partial_t s dV > 0 \,. \tag{5.65}$$

The reason for the time derivative of the integral on the left of Eq. (5.65) be positive is that dissipation increases the internal energy nU in the volume V at the expense of other forms of energy. Equation (5.65) also implies that the temporal variation of the entropy is also a growing function of time $\partial_t s > 0$, as both T and U are positively defined quantities.

5.5.2
Dissipation and Collisions

From the microscopic point of view dissipation is intimately related to collisions. Given the distribution function $f(t, \boldsymbol{x}, \boldsymbol{v})$ to describe the detailed state of the fluid and assuming that the total number of particles in the system is conserved, the equation describing the time evolution of the system is a continuity equation in the 6D space $(\boldsymbol{x}, \boldsymbol{v})$:

$$\partial_t f + \nabla_x \cdot \boldsymbol{j}_x + \nabla_v \cdot \boldsymbol{j}_v = 0 \,. \tag{5.66}$$

where $\boldsymbol{j}_x$ and $\boldsymbol{j}_v$ are the microscopic currents in real space and in velocity space, respectively. In the absence of collisions the currents are conveniently approximated by $\boldsymbol{j}_x = \boldsymbol{v} f$ and $\boldsymbol{j}_v = \dot{\boldsymbol{v}} f$, that is, by the local values of f times the six-dimensional velocity. Thus, in a noncollisional fluid, the currents entering or leaving the six-dimensional volume Ω in Figure 5.12 do only depend on the values of f close to the volume boundary $\partial\Omega$.

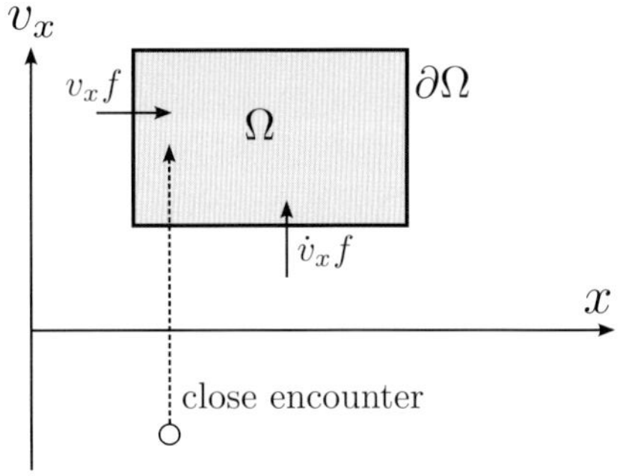

Figure 5.12 Collisional and collisionless currents through the boundaries of a fixed volume in phase space. In a noncollisional system the changes of f inside a fixed phase space volume Ω only depends on the currents $j_x = v_x f$ and $j_{v_x} = \dot{v}_x f$ defined locally near its boundary $\partial\Omega$. In the event of a collision (for example, due to a close encounter with another particle) a particle may instantly modify its velocity v_x by a large amount. If the number of such collisions is not negligible nonlocal contributions to the current j_{v_x} must be retained to compute the variations of f inside Ω, that is, more than just the first term of the development (5.67).

However, in a collisional system, particles can change their velocity instantly and enter into the volume Ω even if before the collision they are located at a large distance from $\partial\Omega$. For example, in the case of a close encounter, a particle may reverse velocity as illustrated in Figure 5.12. In order to account with more or less accuracy for such collisions, the nonlocal structure of the distribution function f has to be included in the description of the $\boldsymbol{j}_v$ current. The mathematical technique to describe the nonlocal structure of the distribution function is a Taylor expansion which allows us to describe the i component of the current as

$$j_{v,i} = a_i f + b_{ij}\partial_{v_j} f + c_{ijk}\partial^2_{v_j v_k} f + \dots \tag{5.67}$$

A development is not required for the $\boldsymbol{j}_x$ current as collisions do not modify the physical position of a particle. In (5.67) the collisionless model corresponds to the case where $a_i = \dot{v}_i$ and all higher-order coefficients of the development are set to zero leading to the collisionless equation

$$D_t f = \partial_t f + \dot{\boldsymbol{x}} \cdot \nabla_x f + \dot{\boldsymbol{v}} \cdot \nabla_v f = 0 \tag{5.68}$$

where $\dot{\boldsymbol{v}} = F(t, \boldsymbol{x}, \boldsymbol{v})/m$ is the acceleration of the particles in the external force field F. For a plasma, where particles interact via the Coulomb potential, retaining the first two terms of the development of $\boldsymbol{j}$ allows us to recover the Fokker–Planck equation which is a convenient approximation in the case where trajectory deviations due to collisions are mostly weak and close encounters with strong "jumps" in velocity space are rare. This is the case in weakly coupled plasmas where the Coulomb logarithm λ^{-1} (with $\lambda = \ln \Lambda$) is much larger than unity as for example in the solar wind where λ is generally larger than 15, but not in the moderately coupled plasma of the solar convective zone where λ may be less than 2. The third term in Eq. (5.67) represents a λ^{-1} order correction to the Fokker–Planck model. Symbolically, the generalization of Eq. (5.68) including collisions can be written as

$$D_t f = \left(\frac{\delta f}{\delta t}\right)_{\mathrm{c}} \tag{5.69}$$

where $(\delta f/\delta t)_{\mathrm{c}}$ is a collision operator eventually containing zero and higher order velocity derivatives of the distribution function f.

5.5.2.1 Entropy in the Collisionless Limit

For a given distribution function f the quantity $f d^3x d^3v$ represents the number of particles in the infinitesimal six-dimensional volume $d^3x d^3v$ surrounding the phase space point $(\boldsymbol{x}, \boldsymbol{v})$. The number f is, therefore, proportional to the probability of finding an arbitrary particle of the system in the corresponding volume element. Following Gibbs, we may then use the statistical definition of the entropy of the system as

$$S_{\mathrm{G}} = -k \int f \ln f d^3x d^3v \,. \tag{5.70}$$

In the case of a collisionless system obeying Eq. (5.68) it is easy to verify that because of $D_t f = 0$, the entropy must be constant:

$$D_t S_G = -k \int D_t f (1 + \ln f) d^3 x d^3 v = 0 . \tag{5.71}$$

With a nonzero collision operator the Gibbs entropy is no longer conserved.

5.5.3
Strongly Collisional Systems

In a system where collisions are frequent, in the sense that the time between subsequent collisions for a typical particle is short with respect to the characteristic time scales for the evolution of the macroscopic fluid quantities, the distribution function $f(\boldsymbol{x}, \boldsymbol{v})$ is expected to depart only weakly from a Maxwell–Boltzmann distribution. In a system where the dominant variations are spatial rather then temporal it is expected that the distribution function is everywhere close to a local Maxwell–Boltzmann distribution if the mean free path λ between successive collisions for a typical particle is small with respect to the characteristic spatial variation scales of some pertinent macroscopic fluid quantity.

Let us consider a strongly collisional system with temperature gradients along the x axis. Under such circumstances we may expand the distribution function around the local Maxwell–Boltzmann distribution f_0 as follows

$$f(v_x, v, x) = f_0(v, n, T) \left[1 + v_x D \left(\frac{v}{v_T} \right) K_T(x) + \mathcal{O}\left(K_T^2\right) \right] \tag{5.72}$$

where $K_T(x) \equiv \lambda \partial_x (\ln T)$ is the thermal Knudsen number which is by definition $|K_T| \ll 1$ in a strongly collisional case. The implicit function $D(v/v_T)$ can be computed applying the Chapman–Enskog formalism to the kinetic Eq. (5.69) after specification of the collision operator on its right-hand side. It has been first computed for a plasma by Spitzer and Harm (1953) using a Fokker–Planck collision operator $(\delta f/\delta t)_c$. These authors found that the associated conductive heat flux $q = \int dv^3 (m/2) v_x v^2 f$ is given by the widely used expression for the electron flux in a collisional plasma

$$q_{e,\,SH} = -2.26 n \frac{(kT)^{3/2}}{m_e^{1/2}} \lambda \partial_x (\ln T) . \tag{5.73}$$

From the fluid point of view it appears immediately that the conductive flux is associated with an entropy flux. For example, for a nonmoving fluid with a temperature gradient, the entropy inside a fixed volume V increases or decreases depending on whether the conductive flux flows in or out from the volume according to

$$\int_V n T \partial_t s = -\int_V \nabla \cdot \boldsymbol{q} = -\int_{\partial V} \boldsymbol{q} \cdot d\boldsymbol{S} . \tag{5.74}$$

From a microscopic point of view, assuming a temperature gradient along the x direction, the conductive heat flux into the volume V corresponds to exchanging a particle of energy $3/2kT_0$ from inside the volume with a particle of energy $3/2k(T_0 + \lambda \partial T/\partial x)$ from outside the volume. The variation of the entropy being associated with the variation of the internal energy, each particle increases the entropy inside the volume V by an amount

$$\delta s = \frac{\delta U}{T} = \frac{3}{2} k_{\mathrm{B}} K_T \,. \tag{5.75}$$

Of course, this amount is negative if the temperature increases when crossing the boundary of the volume from outside to inside.

5.5.4
Heat Conduction: From Collisional to Collisionless

It has been pointed out by various authors that the transport coefficients calculated using an approximation of the local distribution function of the type of Eq. (5.72) does fail at relatively small values of the Knudsen number $K_T > \mathcal{O}(10^{-3})$. The small Knudsen number failure for the transport coefficients stems from the fact that the Fokker–Planck mean free path grows as $(v/v_T)^4$. By comparison, the mean free path is independent of velocity in a system of hard spheres, a rough approximation of a gas of neutral molecules. Thus, in a plasma with Maxwellian velocity distribution, the number of particles at three times the thermal velocity is approximately 10^{-4} smaller than the number of particles at thermal velocity but carry a 27 times larger energy flux. The transport of energy in the plasma may, therefore, be essentially collisionless despite particles at thermal velocity being dominated by collisions. In addition, the reduced number of collisions for the high-energy particles potentially allows for strong departures from isotropy and an overcontribution to the heat flux from the high-energy parts of the distribution function.

Let us consider a plasma confined between two thermostats at temperature T_0 and T_1 as shown in Figure 5.13. Each time a particle hits a thermostat it is injected back into the system following the velocity distribution function of a nondrifting Maxwell–Boltzmann distribution at the given temperature (T_0 or T_1). The number of the particle in the system is, therefore, constant and once a stationary state has been reached the fluid velocity must be zero in the whole system independently on the collisionality of the system.

Given that $T_1 > T_0$ a heat flux is expected to flow from right to left. In the strongly collisional limit, that is, if the density in the system is high enough for the temperature variation $\delta T/T$ being small over a distance of the order of the mean free path, the electron heat flux is given by the collisional expression (5.73). As in a stationary state energy cannot accumulate in the system and if no energy is brought to the electrons by other species (that is, protons) the energy equation reduces to

$$\partial_x q_{\mathrm{e,SH}} = 0 \,. \tag{5.76}$$

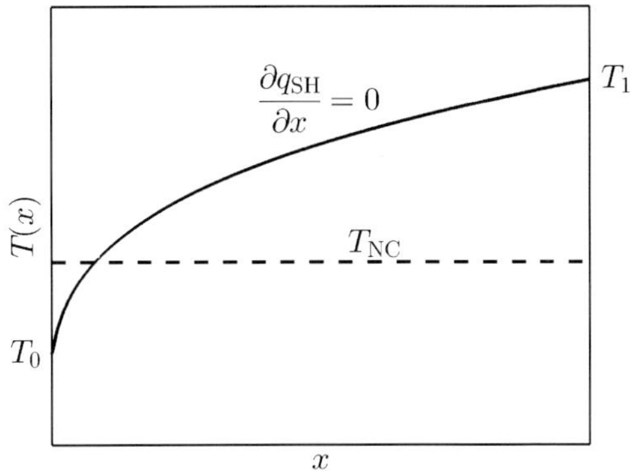

Figure 5.13 Plasma confined by two impermeable thermostats. Temperature profile $T(x)$ for a plasma between two thermostats at temperature T_0 and T_1 in the strongly collisional limit (continuous line) and in the collisionless limit (dashed line). The experiment illustrates the difficulty to support a static temperature gradient in the collisionless limit.

Using the Fokker–Planck approximation for the mean free path of electrons $\lambda \propto T^2/n$ it follows from Eq. (5.76) that the temperature profile must satisfy the differential equation $T^{5/2}\partial_x T = \text{constant}$. The corresponding temperature profile for the case where $T_1 = 4T_0$ is plotted in Figure 5.13. We note that in the collisional case the conductive flux $q_{e,SH} \propto -T^{5/2}\partial_x T$ is independent of density so that the heat flux per particle is inversely proportional to density. As suggested by Eq. (5.73), the heat flux transported by a given species of particles is inversely proportional to the square root of its mass, so that high-mass species are expected to be less efficient in transporting the energy through thermal conduction compared to low mass species.

5.5.4.1 **Collisionless Limit**

In the collisionless limit and in the absence of any external field (gravitational field, electrostatic field, and so on), particles reflected by one of the thermostats in Figure 5.13 travel at constant velocity to the opposite thermostat. In stationary conditions, a self produced electrostatic field E may still exist in the system with the consequence that the pressure of the protons and electrons must vary spatially with the opposite sign to ensure total pressure balance $\partial_x(P_e + P_p) = 0$. We may exclude the possibility of a nonzero electric field by noting that in the case of only one species and no external field, the static solution implies $\partial_x P = 0$, that is, a constant pressure. Indeed, given that individual species tend to settle at constant pressure and since the superposition of such constant pressure populations is a constant pressure plasma with $E = 0$.

In the absence of forces the velocity distribution function of both electrons and protons in the system must be identical in all parts of the system as $D_t f = u\partial_x f = 0$. Thus, the part of the distribution function $f(v_x > 0)$, corresponding to positive velocities $v_x > 0$ is given by the conditions imposed on the left boundary, at temperature T_0, while the part corresponding to negative velocities $f(v_x < 0)$

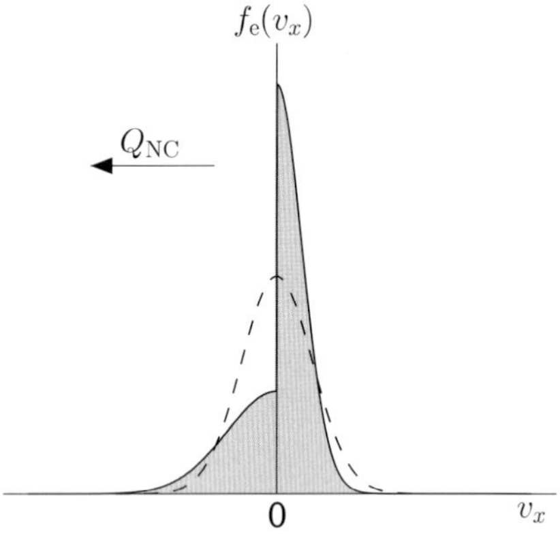

Figure 5.14 Distribution function in the collisionless limit. In the collisionless limit the velocity distribution function between the two thermostats of Figure 5.13 does not depend on position. Assuming particles hitting a thermostat are instantly reinjected into the system following a Maxwellian distribution at temperature T_0 or T_1, the distribution of the v_x component must be the sum of these two Maxwellians. The relative amplitude of the two Maxwellians is determined by the constraint of zero average velocity $\int_{-\infty}^{\infty} v_x f(v_x) dv_x = 0$. According to Eq. (5.77) such a distribution implies a strong energy flux form the hot to the cold thermostat.

is given by the conditions imposed at the right boundary, at temperature T_1. In stationary conditions the flux of particles going from the left to right must balance the flux of particles from right to left. Thus, if n_0 is the density of the particles streaming away from boundary 0 and n_1 the density of particles streaming away from boundary 1 the condition $n_0\sqrt{T_0} = n_1\sqrt{T_1}$ must hold. The velocity distribution function corresponding to the case $T_1 = 4T_0$ is shown in Figure 5.14. Also plotted is the Maxwellian distribution with the same total density $n = n_0 + n_1$ and pressure $P = P_0 + P_1 = nk_B T$ with $T = (T_0 T_1)^{1/2}$.

Thus, despite that fact that the distribution has no spatial gradient and zero mean velocity, it has a strong collisionless conductive flux q_{NC} from the hot to the cold thermostat. Computing the third moment of the distribution function $q_{NC} = \int 0.5 m v_x v^2 f(v^2) d^3 v$ we obtain:

$$q_{NC} = \left(\frac{8}{\pi}\right)^{1/2} \frac{n k_B^{3/2}}{\sqrt{m}} \left(T_0\sqrt{T_1} - T_1\sqrt{T_0}\right) . \qquad (5.77)$$

It is interesting to note that unlike the collisional flux q_{SH}, the collisionless flux Eq. (5.77) is proportional to density, meaning that the flux per particle does not change with density. Thus, while in the collisional case the heat flowing from the hot to the cold thermostat does not change with the number of carrier, in the collisionless case the heat flux can be increased indefinitely by increasing the number of carrier (provided the plasma remains collisionless).

5.5.5
The Thermoelectric Field: Another Consequence of Collisions between Ions and Electrons

We note that in the collisional case a thermoelectric field E_T associated with inter-species collisions is normally present in a configuration of the type of Figure 5.13. In the static limit, the electric field is required to balance the thermal force R_p exerted by the electrons colliding with the ions. For a proton–electron plasma the equilibrium of forces requires

$$
\begin{aligned}
-\nabla p + neE_T &= -R_\mathrm{p}\\
-\nabla p - neE_T &= R_\mathrm{p}
\end{aligned}
\tag{5.78}
$$

where we have used the fact that momentum conservation implies that the thermal force on the electrons and protons must compensate exactly, that is, $R_\mathrm{e} = -R_\mathrm{p}$, and where we have assumed that the system is large enough with respect to the Debye length for quasi-neutrality $n = n_\mathrm{e} = n_\mathrm{p}$ to hold and collisions to be capable of ensuring equal electron and proton pressure. In order to compute R_p the collision operator for electron–proton collisions must be specified on the right-hand side of Eq. (5.69) applied to the protons. A rough estimate of the thermal force can nevertheless be obtained by considering that the momentum per time unit transmitted to a proton is the difference between the momentum received from electrons on its left (sign $-$) and the electrons on its right (sign $+$). Since at a given temperature the characteristic velocity of protons is 43 times smaller than the characteristic velocity of an electron, we consider a proton at rest as in Figure 5.15.

Noting that in stationary conditions the number of electrons per time unit passing the proton from the right must equal the number of electrons passing the proton from the left, that is, $n_\mathrm{e}^- v_\mathrm{e}^- = n_\mathrm{e}^+ v_\mathrm{e}^+ = n_\mathrm{e} v_\mathrm{e}$, we must have

$$
\frac{R_\mathrm{p}}{n} \sim \frac{m_\mathrm{e} v_\mathrm{e}^-}{\tau_\mathrm{ep}^-} - \frac{m_\mathrm{e} v_\mathrm{e}^+}{\tau_\mathrm{ep}^+} = m_\mathrm{e} n_\mathrm{e} v_\mathrm{e} \left(\frac{1}{n_\mathrm{e}^- \tau_\mathrm{ep}^-} - \frac{1}{n_\mathrm{e}^+ \tau_\mathrm{ep}^+} \right) . \tag{5.79}
$$

This equation illustrates the fact that the frictional force arises due to the spatial variation of the product of the electron proton collision time and the density. In the

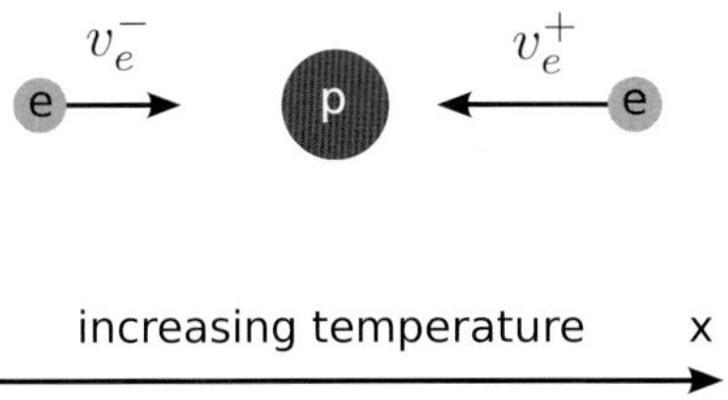

Figure 5.15 Given a temperature gradient in a collisional plasma a thermal force R_p operates on the protons due to a difference in the velocity of the electrons coming from regions with different temperatures. Note that R_p is in the direction of the temperature gradient. In static conditions the frictional force is balanced by an electric field E_T.

Fokker–Planck approximation the product $n_e \tau_{ep}$ is only weakly dependent on the plasma density (via the Coulomb logarithm), but varies with temperature as $T^{3/2}$ suggesting that the frictional term is proportional to the derivative of the temperature. The electrons hitting the test proton transporting the characteristics of the plasma at a distance on the order of the mean free path λ_{ep} on each side we approximate the terms in parenthesis on the right-hand side of Eq. (5.79) by a first order Taylor expansion

$$n_e^{\pm} \tau_{ep}^{\pm} \approx n_e \tau_{ep} \pm \lambda_{ep} \partial_x (n_e \tau_{ep}) \,. \tag{5.80}$$

Generalizing the collisional time to $n_e \tau_{ep} \propto T^a$ so that $a = 3/2$ for a plasma and $a = -1/2$ for hard sphere interactions, we finally obtain

$$\frac{R_p}{n} \sim 2 \frac{m_e v_e}{\tau_{ep}} \frac{a \lambda_{ep}}{T} \partial_x T = 2a \partial_x T \,. \tag{5.81}$$

Subtracting one equation form the other in (5.78) we do finally obtain an estimate of the thermoelectric field

$$E_T \sim -\frac{a}{e} \partial_x T \,. \tag{5.82}$$

This equation shows that in a weakly coupled plasma where $a = 3/2$ the thermoelectric field is directed downwards from the temperature gradient. More rigorous computations using Eq. (5.69) and a Fokker–Planck collision operator gives $e E_T = -0.71 \partial_x T$.

5.6 Collisionless Damping

5.6.1 Number of Eigenmodes: Fluid vs. Kinetic

A fluid system is composed of N moment equations, one closure equation, and a number m of Maxwell equations. Working in Fourier space ω, k, let us assume a linear perturbation with a given spatial variation k, and look for the possible temporal evolutions characterized by ω.

The N moment equations are first order in time. If the closure equation and the Maxwell equation used (for instance, the Gauss equation) are not differential with respect to time, the global system is, therefore, of a global order N, corresponding to a polynomial dispersion equation of degree N also. There are, therefore, N solutions (N "modes") in ω. If the system is stable, all the solutions are real. If not, there are complex conjugate solutions. If the closure equation or the Maxwell equation used are differential with respect to time, this increases the number of modes of a few units.

Taking the example of a sound wave in a neutral collisional medium, one uses two moment equations (density and momentum), an algebraic closure equation (polytropic), and zero Maxwell equations. There are, therefore, two solutions:

two sound waves propagating in the two opposite senses. Taking the example of electromagnetic waves in vacuum, one uses zero moment equations, zero closure equations, and two Maxwell equations (Faraday and Ampère): this again leads to two waves in the two opposite senses. Considering the MHD system closed with an adiabatic condition $q_1 = 0$, we let the reader check that, under a few additional hypotheses, it leads to six modes (three in each sense).

Considering now a kinetic problem, one has to use a kinetic equation for each population (typically a Vlasov equation in the collisionless case), instead of a set of a limited number of moment equations. This means, with a naive (nonmathematical) vocabulary, that there are an "infinite" number of variables $f(v)$ corresponding to the infinite number of values of v. The Vlasov equation is, as the moment equations, a first order equation with respect to time. Extrapolating the above arguments, there should, therefore, be an infinite number of solutions, that is, an infinite number of modes corresponding to the infinite number of degrees of freedom.

In full generality, this is true. If a medium gives rise, for instance, to two modes when modeled in a fluid theory, the kinetic modeling of the same medium gives rise to an infinity of modes and nothing allows a priori to distinguish two of them as being a kinetic counterpart of the previous two fluid modes. In this sense, it seems that considering the kinetic treatment of waves just as an improvement of the fluid treatment of them is quite hopeless. Nevertheless, we will see that the notion of a "kinetic mode" is, however, meaningful, adding strong restrictions in the distribution functions acceptable. In common words, one can say that "the kinetic nature has an infinite number of degrees of freedom, but for statistical reasons to be explained, it does not make use of all this possible freedom, restricting itself to a small number of modes, generally close to the fluid ones". This important physical point is at stake for any kinetic treatment of waves, and it must be kept in mind for understanding the fine mathematical methods that are needed for it. It is shown in next section on the simple example of the Langmuir wave.

5.6.2
A Simple Example: The Langmuir Wave, from Fluid to Kinetic

The Langmuir wave is the simplest example of a plasma wave. It is often taken as the paradigm of all the plasma physics and it allows us in particular to show clearly the relationship between fluid and kinetic treatment in the absence of collisions. It is an electrostatic wave which occurs at sufficiently high-frequency so that it concerns only the electrons, the ions being considered motionless. The basic theory concerns plane waves in an unmagnetized plasma. It is then a purely 1D problem since the velocity perturbations as well as the electric field ones are parallel to the direction of the gradient.

The physics of the Langmuir waves is based on three main equations: the first two fluid equations of the electron population (density and momentum), which tell how these particles react to the electric field, and the Maxwell–Gauss equation, which tells how the electric field is created by the electron charge density:

$$D_t n_e + \partial_x (n_e u_e) = 0 \tag{5.83}$$

$$n_e m_e D_t u_e + \partial_x P_e = -n_e e E \tag{5.84}$$

$$\partial_x E = -n_e e/\varepsilon_0 \,. \tag{5.85}$$

The P_e notation stands for the P_{exx} component of the pressure tensor $\boldsymbol{P}_e$ in the direction x of the gradient. Linearizing and specifying the search of monochromatic solutions (one single ω and one single k), the same system is written as (with $v_\varphi = \omega/k$):

$$u_{e1} = v_\varphi \frac{n_{e1}}{n_{e0}} \tag{5.86}$$

$$v_\varphi u_{e1} = \frac{P_{e1}}{m_e n_{e0}} + \frac{e}{m_e}\frac{E_1}{ik} \tag{5.87}$$

$$E_1 = -\frac{1}{ik}\frac{n_{e0} e}{\varepsilon_0}\frac{n_{e1}}{n_{e0}} \,. \tag{5.88}$$

These equations are obtained by assuming that the zero order distribution is homogeneous and stationary and that we are in a reference frame where $u_0 = 0$ and $E_0 = 0$.

These three equation are quite general, and they are valid as well in kinetic as in fluid theory. Nevertheless, this system is incomplete since it has only three equations while it involves four variables: n_{e1}, u_{e1}, P_{e1}, and E_1. In fluid modeling, a closure equation is needed to make it solvable.

5.6.3 Fluid Treatment of the Langmuir Wave: Choice of a Closure

5.6.3.1 Polytropic Closure

Lets us first assume that the pressure perturbation can be expressed as a function of the density one by a polytropic law: $P_{e1} = \gamma m_e V_{the}^2 n_{e1}$. Whenever this assumption is verified, one can then easily eliminate two of the three remaining variables, for instance, u_{e1} and E_1, and obtain:

$$v_\varphi^2 \frac{n_{e1}}{n_{e0}} = \left(\frac{\omega_{pe}^2}{k^2} + \gamma V_{the}^2\right)\frac{n_{e1}}{n_{e0}} \,. \tag{5.89}$$

It is clear from this equation that, apart from the trivial solution $n_{e1} = 0$, all the monochromatic waves must satisfy the dispersion equation:

$$\omega^2 = \omega_{pe}^2 + \gamma k^2 V_{the}^2 \,. \tag{5.90}$$

This is the fluid result for the Langmuir wave, also called the "Bohm and Gross expansion" of its dispersion equation. The plasma frequency which appears in this result is given by:

$$\omega_{\mathrm{pe}}^2 = \frac{n_{\mathrm{e}0} e^2}{m_{\mathrm{e}} \varepsilon_0} . \tag{5.91}$$

The coefficient γ is here arbitrary. Its value can be fixed by more precise theories, either fluid or kinetic.

The fluid system used is here composed of two first order equations with respect to time so that, in agreement with the property outlined in the Section 5.6.1, there are, therefore, two solution branches (for a given k): one positive and the other negative. These two "eigenmodes" are identical except that they propagate in two opposite directions. For any real k, these waves propagate with a real ω, that is, without any damping.

In reality, as we will see, the full kinetic calculation shows that waves can indeed propagate with a dispersion very close to this one, but only for small k's. When the pressure term (responsible for the $\gamma k^2 V_{\mathrm{the}}^2$ term) gets an order of magnitude equal or larger than the electric one (responsible for the ω_{pe}^2 term), the above solution no longer exists: the kinetic solutions that are physically relevant are then damped, and their dispersion becomes different, as shown in Figure 5.16.

This corresponds to a scale condition $k\lambda_{\mathrm{D}} \ll 1$ for the above result to be valid, the spatial parameter λ_{D} being the Debye length, given by $\lambda_{\mathrm{D}} = V_{\mathrm{the}}/\omega_{\mathrm{pe}}$. Where does this condition of validity come from?

It comes necessarily from the only assumption made in this calculation, that is, from the closure equation. Performing the full kinetic calculation, we will show that the polytropic closure equation chosen is indeed valid, with a value of γ equal to 3, when the above condition on k is met, and that it is always violated for larger k's. How does the fluid system need to be modified to improve its result and make it more general? As the polytropic law can appear as a technical and rather unjustified closure, one can think of using more physical assumptions, such as the adiabatic one, or increase the order of the closure, so using a larger number of exact fluid equations, or look for any other mean of finding relevant closure equations, coming for instance from kinetic calculations.

5.6.3.2 **Adiabatic Closure**

A modest change can be made to try improving the above result: it is to close the system at order three instead of two. One has then to add the transport equation for the pressure to the above system. As shown in Chapter 2, this equation is written in full generality, with the same notation:

$$v_\varphi P_{\mathrm{e}1} = q_{\mathrm{e}1} + 3 n_0 m_{\mathrm{e}} V_{\mathrm{the}}^2 u_{\mathrm{e}1} . \tag{5.92}$$

The simplest closure equation at order 3 is the adiabatic one: $q_1 = 0$. As the heat flux q is a measure of the asymmetry of the distribution function, it is likely that the adiabatic assumption can be justified in some particular instances. Even if we

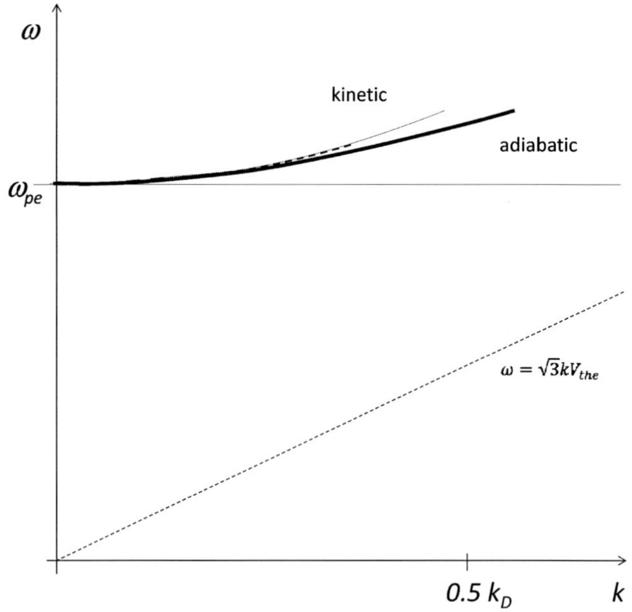

Figure 5.16 The Langmuir mode dispersion. The fluid adiabatic model provides the curve in solid line, corresponding to an undamped propagation. The experiments, as the kinetic calculation, indicate a slight departure from this solution concerning the real part of the frequency (dashed line) but an important difference in the imaginary part (damping), which becomes very large at large k. The other lines are given for reference: $\omega = \omega_{\rm pe}$ (thin solid line) is the cut-off frequency and $\omega = \sqrt{3}kV_{\rm the}$ (thin dashed line) is parallel to the asymptote of the fluid adiabatic result.

don't know a priori to what extent this simple hypothesis can be verified in the real – kinetic – system, we can easily investigate its consequences.

It is clear from the above equation that the adiabatic hypothesis leads to two kinds of solutions:

- $v_\varphi = 0$, which is called the "entropy mode" and is not directly related to the Langmuir mode, or
- $P_{\rm e1} = 3m_{\rm e}V_{\rm the}^2 n_1$

The first solution (entropy mode) corresponds to a stationary mode, in which the relation between $P_{\rm e1}$ and $n_{\rm e1}$ can be arbitrary. It is called an "entropy mode", and is far from the Langmuir mode. The second solution is identical to the above polytropic one, just specifying the value of gamma: $\gamma = 3$. The two above Langmuir modes are, therefore, recovered in this way. These results once again confirm the usual rule: a closure at order 3 leads to three eigenmodes: two Langmuir modes (identical to the preceding ones) and one entropy mode.

The adiabatic result is valid as long as the adiabatic hypothesis is valid, that is, as long as the heat flux perturbation can be neglected in front of $v_\varphi P_{\rm e1} \approx 3m_{\rm e}V_{\rm the}^2 u_{\rm e1}$ in Eq. (5.92). This is of course not universal. As no restriction of the type $k\lambda_{\rm D} \ll 1$

appears in the frame of this adiabatic calculation, one can guess that it is the condition of validity of the adiabatic assumption itself. The kinetic calculation will prove that the adiabatic assumption is indeed met for v_φ sufficiently smaller than V_{the}, but is violated for larger phase velocities, which is indeed equivalent.

Concerning the value of gamma coming from Eq. (5.92), one must note that the value $\gamma = 3$ appearing here is actually dimension dependent. This value does correspond to the collisionless 1D problem we consider, where the perturbation pressure tensor is 1D because all velocity evolution is in the same x direction. If the perturbation was 3D, for instance, if the medium was sufficiently collisional to isotropize rapidly the thermal part of the velocity perturbations, the γ value would be 5/3. More generally, if the pressure perturbation is made isotropic in a subspace of dimension D, the thermal energy (that is, half the trace of the pressure tensor perturbation in this subspace), can be written $D\,P_{\text{e1}}/2$, and the corresponding polytropic index is $\gamma = 1 + 2/D$.

5.6.3.3 A Special Closure

Let us consider the following special closure at order three, which is quite illuminating for understanding the following kinetic calculations and the notion of a "kinetic mode":

$$q_{\text{e1}} = P_{\text{e0}} v_\varphi \left(\frac{P_{\text{e1}}}{P_0} - 3\frac{n_{\text{e1}}}{n_0} \right) . \tag{5.93}$$

One can see that this closure involves v_φ, which means, coming back from Fourier space to the real one, that it is a differential closure equation involving time and space derivatives. This closure in Eq. (5.93) is not arbitrary. It will be shown that it is an unavoidable relation whenever a purely monochromatic solution is considered, with one single ω and one single k (both real), if one restricts oneself to symmetric zero order distribution functions, without odd moment.

The kinetic calculation of the next section also shows that, if one looks for pure monochromatic solutions, not only for the macroscopic moments, but also at the microscopic level of the distribution function perturbation, this hypothesis imposes that the corresponding perturbation $f_1(v)$ is singular in $v = v_\varphi$ (asking for the use of mathematical appropriate tools such as the theory of distributions). This is a crucial property, which will lead us to loosen the monochromatic hypothesis.

Similar closure equations could be found at higher orders, with the same hypotheses. The result would then depend on the form of the zero order distribution function, but this calculation is actually useless, since it would lead to exactly the same solutions.

Trying to solve the fluid system when it is closed with Eq. (5.93), one can check that the determinant of the system is zero and that the system is verified irrespective of the value of v_φ. The reason is that Eq. (5.93) is actually just the combination of the fluid equations of order 1 and 2 (Eqs. (5.83) and (5.92)). The full system (with this closure) is, therefore, always satisfied whenever the initial fluid system (without closure) is satisfied, without any condition on v_φ.

This surprising result deserves some comments: any pulsation ω can be associated, when the closure equation is valid, with any k without contradiction. This means that there is no dispersion relation in this case or, equivalently, that there is an infinity of modes, without any trace of the above adiabatic Langmuir mode. The heat flux perturbation, as the perturbations of all the other moments, is just a known function of the linear amplitude of the linear wave (for instance n_{e1}) as soon as v_φ is known.

In front of this result, one may ask: is it possible – and realistic – to excite modes with any ω and any k, even far from the Langmuir mode? The answer cannot be found inside the fluid framework itself: the kinetic theory is needed for that. And the answer is: "any of these solutions are mathematically possible, but not physically realistic in most experimental configurations". As already mentioned, this closure equation can be obtained as soon as everything is strictly monochromatic, which implies that a part of the distribution perturbation is carried by a Dirac distribution. In theory, and even in simulation, it is indeed possible to perturb only one monokinetic slice of the distribution function: the above calculation is then quite relevant and it is then possible to choose freely any ω and any k, the propagation at $v_\varphi = \omega/k$ being then guaranteed, without damping and without any restriction. In this limit, one can say that there is actually an infinity of modes and that the neighborhood of the above adiabatic Langmuir mode is not favored in any manner. Nevertheless, whenever an initial perturbation is created by a macroscopic field, as in almost all real physical experiments, the perturbation of the distribution function is necessarily smooth, without any singularity in $\delta(v - v_\varphi)$. This imposes that the initial distribution can never be one single mode among the purely monochromatic modes that have just been found. If the initial perturbation is decomposed over these singular ones, it must imply a large number of them so that the resulting distribution perturbation $f_1(v)$ can be smooth. The summation over the different values of v_φ then leads to a wave packet, which generally implies damping along time. This damping then comes from the phase mixing between the different components, which each propagate at a different v_φ. Some of these wave packets have a strong damping, other have a weaker one. We will show that the vicinity of the adiabatic mode corresponds to the weakest damping. It is the reason why the fluid notion of adiabatic Langmuir mode does not fully disappear from the kinetic results as it first seems to do.

The reason why, for smooth initial conditions, the vicinity of the adiabatic conditions corresponds to the weakest damping will also appear in the kinetic calculation: it will show that, in the conditions $v_\varphi \gg V_{\text{the}}$ where this mode can be observed, it corresponds to the mode that can be determined by fully ignoring the quasi-resonant particles and taking into account only the nonresonant ones. This "nonresonant mode", then identical to the adiabatic one, is real, that is, without damping. The damping and more generally the possibility of obtaining an infinity of other modes, fully depends on the role of the quasi-resonant particles. It can actually be easily understood that there are in general very few particles with velocities close to the resonant velocity in the given conditions for v_φ (think of a Maxwellian distribution). As the number of quasi-resonant particles weights their effect in determining the mode, one can guess that, smoothing their singular behavior will only introduce a small damping, while all the other modes will present strongly singular signatures close to the resonant velocity.

At the end of this section, we have, therefore, introduced the important notion, which will be demonstrated in the next section, that it is the restriction brought on the admissible initial conditions at the microscopic level of the distribution function that is the physical basis of the notion of "kinetic mode". The property of "collisionless damping", or "Landau damping" (Section 5.6.4), of these modes is also a consequence of this restriction. When this condition is not imposed, there is indeed an infinity of modes, with singular distribution functions, regardless of whether or not these modes are close to the adiabatic conditions.

5.6.3.4 **Other Closures**

We have mentioned that closing the system at order two by a polytropic law leads to a solution which is approximately valid for $k\lambda_D$ much smaller than unity. We have now seen that closing at order three with an adiabatic condition adds a different mode, but does not make the description of the Langmuir ones closer to the kinetic solutions (with smooth distribution functions). Could a closure at an higher order improve this result?

Closing the system with higher-order closures would add more new eigenmodes to the system. If the closure is not differential, one would get N modes for a closure at order N. If it is physically relevant, two of them should correspond to the Langmuir modes we are looking for, and one may expect that the description of these modes is improved. Two difficulties, however, arise for performing this program: (i) if the adiabatic condition is a well understood physical one, there is nothing comparable at higher orders. One can even say that the larger the order, the more difficult the guess of a relevant closure; (ii) if the goal is to get closer to some "kinetic mode", this mode has to be defined and calculated before inferring the closure equation.

A typical example can be given for illustrating the first difficulty: let us try to impose that a large odd order moment is unperturbed. This might be viewed as an extension of the adiabatic hypothesis, which would be just the $p = 3$ particular case. The calculation shows that the corresponding result is actually irrelevant. The solutions found close to the "Langmuir modes" are actually more distant from the kinetic solutions than the adiabatic ones, and still without damping, and even more seriously, most of the other $N - 2$ new solutions are complex conjugate solutions in ω: this would make the system unstable if used in a numerical code. All of this clearly shows that such an arbitrary closure cannot be a good approximation to model the medium.

Satisfactory closure equations can indeed be obtained, at any order, using the results of the kinetic calculations. Of course, just reproducing the same result as the kinetic one for a mono-k wave would be useless, since the kinetic calculation has to be done first. The goal is actually to get a closure equation under the form of a differential equation in real space, which should be valid for any variation in space of the initial perturbation. This is the principle of the so-called Landau fluid codes, a version of which also exists in the MHD context [22, 23].

From the kinetic calculation, the perturbed distribution f_1 is known as a function of the electric perturbation E_1, this relation involving ω and k. All the perturbed moments can be deduced as well. Effective closure equations can be found in this manner at any order. The closure equation at order p can be taken directly as this expression of the perturbed moment of order p in function of E_1, whatever p is. Using the moment equation system up to order $p-1$, many other equivalent closure equations can also be chosen, where the electric field has been excluded and replaced by various appropriate combinations of the lower order moments. Some orders and some combinations are better suited than the others to get the best accuracy in the results, but we will not present these details here.

The main difficulty comes when going back from Fourier to real space. It is feasible by inverse Fourier transform only at the condition that the closure can be put under the form of ratios of polynomials in v_φ. As mentioned, some possible closures are simpler than the others, but the kinetic result always implies complex transcendental functions that can never be put exactly under this form. This difficulty is approximatively solved by replacing these exact functions by their "Pade approximants" (which are ratios of polynomials), with a small number of poles. When used with small amplitude perturbations, the Landau fluid models allow reproducing quite closely the kinetic solutions with the same macroscopic initial conditions. When used in weakly nonlinear systems, they generally also provide results that are approximately correct, depending on the nature of the nonlinear effects, the order chosen for the closure (all the moment equations are nonlinearly exact), and the number of poles taken in the Pade approximants.

5.6.4
Kinetic Treatment of the Langmuir Wave: Landau Damping

5.6.4.1 In Search of a Monochromatic Kinetic Solution

The simplest adiabatic description of the plasma is sufficient to derive the main properties of the Langmuir mode when $k\lambda_D \ll 1$. For shorter wavelengths, different closure equations have to be used to describe the departure from this simple modeling, and particularly to show the appearance of a kinetic damping. We have seen that appropriate closure equations are able to mimic these kinetic properties, at least in the linear case, but that the kinetic calculation has to be done first in order to derive these appropriate closures. Let us now do this kinetic calculation.

The same work, linearization, Fourier transform, and so on, has now to be done directly on the Vlasov equation itself instead of done on a limited number of its moments. This first step readily leads to:

$$(v - v_\varphi) f_1 = a_1 \quad \text{with} \quad a_1(v) = f_0' \frac{e}{m} \frac{E_1}{ik} = f_0' \frac{\omega_{pe}^2}{k^2} \frac{n_{e1}}{n_0} . \tag{5.94}$$

In this equation, f is the electron distribution function, and f_0' is the derivative of its unperturbed part with respect to velocity. The perturbed electric field E_1 has been replaced by its value in function of the density perturbation n_{e1} thanks to the Maxwell–Gauss equation. This gave the plasma frequency ω_{pe}.

In order to simplify the notation and to better show the dimensions, we will hereafter use a dimensionless distribution function $\tilde{f}_0(\xi) = f_0(v)/f_{00}$, with $\xi = v/\sqrt{2}V_{the}$ and $f_{00} = n_0/\sqrt{2\pi}\,V_{the}$. With this notation, a Maxwellian distribution is just: $\tilde{f}_0(\xi) = e^{-\xi^2}$. Using it, a_1 can be rewritten as:

$$a_1(v) = \frac{\tilde{f}'_0}{2\sqrt{\pi}}\frac{\omega_{pe}^2}{k^2 V_{the}^2} n_{e1} \,. \tag{5.95}$$

The problem will be completely solved when eliminating the unknown function $f_1(v)$ by imposing that its integral $\int f_1 dv$ is equal to n_{e1}. This integro-differential character of the problem makes its specific difficulty. To express f_1 as a function of n_{e1} and doing the integration, one must divide by $v - v_\varphi$, which is zero in $v = v_\varphi$. The general solution of Eq. (5.94), using the mathematical theory of distributions, is:

$$f_1 = \text{p.v.}\left\{\frac{a_1(v)}{v - v_\varphi}\right\} + \delta(v - v_\varphi)b_1 \,. \tag{5.96}$$

The notation p.v. stands for "principal value" and means that all the moments of this part of the distribution function will have to be calculated with the principal value of the integrals, so cutting them symmetrically on a small interval around the pole $v = v_\varphi$. The second term (Dirac distribution) indicates that a part of the density can be carried by a monokinetic perturbation localized at $v = v_\varphi$. The Vlasov equation is a first order differential equation in time. The existence of two terms in its solution is the natural translation, in Fourier space, of the fact that this solution in real space is the sum of two terms: a particular solution (principal value) plus the general solution of the equation without its RHS term (delta). The use of the distributions theory in this context has been introduced by [24].

The result shows the central role played by the "resonant velocity" $v = v_\varphi$, which makes quite particular the interaction between the electric field and the distribution perturbation in its vicinity. This characteristic feature of the kinetic calculation comes from the fact that, whatever the phase velocity v_φ, there are always particles with this velocity in the distribution. Nevertheless, equivalent features can be found in ordinary mechanics.

5.6.4.2 **Specific Mathematical Property of the Kinetic Treatment**

Why do the above cautions, which led us to use the concept of mathematical "distributions", seem to be needed in only the kinetic treatment? Why could they be ignored in the fluid treatment or more generally in any problem involving a finite number degrees of freedom, for instance, for a set of coupled oscillators? This important question deserves some comments: understanding the notions of kinetic mode and of kinetic damping mainly follows from it.

Considering the Vlasov equation alone, without considering its coupling with the Maxwell Gauss equation, one should first observe that it is a first order differential equation with respect to time. The electric field, in the RHS term is then a forcing term. As already mentioned, the solution contains two parts: the particular solution of the complete equation, with its RHS term, and a general solution of the LHS term alone. The first part of Eq. (5.96) corresponds to a particular solution forced by the electric field, while the second one corresponds to the ballistic trajectories of the particles, in the absence of any electric field. For a given electric field variation, the ratio between the two parts just depends on this initial conditions on $f_1(v)$.

It is to be noted that the two parts of f_1 are singular in $v = v_\varphi$, in the sense that, whatever E_1, the first one diverges when v tends to v_φ while the second corresponds to a finite density concentrated in a single point $v = v_\varphi$. This is an avoidable consequence of the considered monochromatic variation assumed for $f_1(v)$.

In order to better understand this characteristic feature of the Vlasov equation, Figures 5.17 and 5.18 intend to show what happens when the number of degrees of freedom increases and even tends to infinity. The first figure shows a small number (four) of coupled oscillators made of four weights, each related to the other and to external walls by five springs. The spring constants K are all supposed identical, but the masses are all different, so that we have indeed four oscillators with different

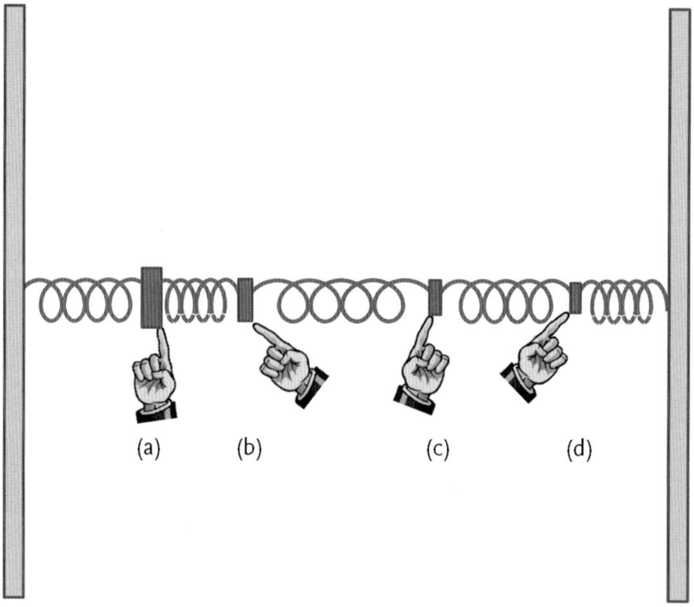

Figure 5.17 Mechanical system with a small number of resonant frequencies: four coupled oscillators made with masses and springs, with increasing resonant frequencies from left to right with decreasing masses (a–d). The temporal evolution of this system has four eigenmodes and each of them can be excited individually by choosing the four initial positions of the masses (with the four fingers drawn). The third oscillator (c) is supposed here to have a resonant frequency close to the excited eigenmode. It, therefore, demands larger displacements, with different signs on both sides. The fluid calculations belong to this mathematical class of problems.

resonant frequencies ω_n. This means that there are actually eight monochromatic eigenmodes for the whole system, distinguishing ω and $-\omega$, that is, distinguishing the left and right directions for the mass motions. For the sake of clarity, the weights are supposed ordered by masses, the values of ω_n^2 being ordered as well, but in the reverse sense.

For an arbitrary initial condition, for instance, displacing only one single mass, the eight modes will be excited simultaneously, which will result in a complex evolution of the system, clearly nonmonochromatic. To excite only one single eigenmode, one must give eight initial conditions corresponding to it. Starting with zero velocities, this simply amounts to fixing precisely the four positions of the four masses, as indicated by the fingers on the figure. If the eigenmode so excited has a frequency close for instance to the resonant frequency of the third mass, a small analytical calculation shows that the amplitude of the displacement of this mass

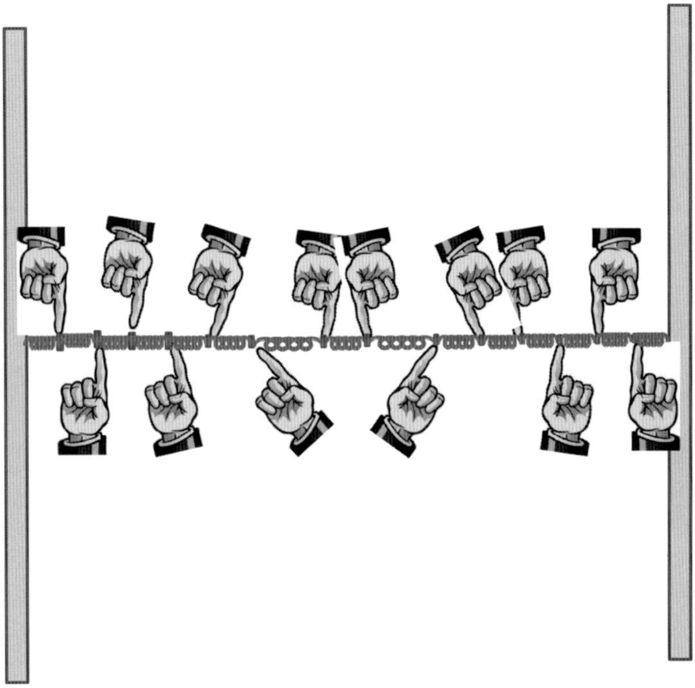

Figure 5.18 Mechanical system with a large number of resonant frequencies: same as Figure 5.17, but the number of masses and springs increases, together with the number of fingers necessary for selecting one eigenmode in the initial condition. The initial displacements also demand more and more accuracy because they vary on smaller and smaller distances around the resonant oscillator. The kinetic calculation belongs to this class of mathematical problems, in the limit of a continuous set of resonant frequencies. It would then demand an infinite accuracy to select one single (undamped) eigenmode. With a limited accuracy, a packet of eigenmodes is inevitably excited, leading to damped solutions.

has to be larger than the others, and that the senses of displacement have to be reversed on both sides, where the resonant frequencies are smaller and larger.

In this first example, four fingers are, therefore, enough to select a known eigenmode. Figure 5.18 shows what happens when the number N of oscillators increases. The number of eigenmodes then increases, and the number of fingers necessary to excite only one single mode increases at the same time. This makes the selection more and more difficult when N increases and when one has a limited number of fingers to control the initialization. This effect is still enhanced by the fact that any given eigenfrequency becomes increasingly close to one of the spring resonant frequencies since, if one assumes a constant range for the set of resonant frequencies, their density increases with the number of oscillators.

This makes larger and larger the gradient of displacements around the resonant oscillator, this displacement being large on both sides close to the resonant velocity, but with different signs, and decreasing rapidly further away. You must have very small fingers to succeed to excite this only mode. If the available fingers are too thick to initialize differently close oscillators, this has no big consequences far from the resonant oscillator, but in its vicinity, it is different: the result is unavoidably to excite a wave packet instead of a single monochromatic wave. This gives a faithful picture of what kinetic damping is: a material inability to excite a purely monochromatic mode.

When the number of degrees of freedom tends to infinity, the number of initial conditions tends toward infinity as well, and any global eigenmode becomes necessarily infinitely close to one of the resonant frequencies. It is exactly what happens with the Vlasov equation. And this makes necessary the special treatment presented just above for describing the vicinity of the resonant velocity. One could theoretically excite any real eigenmode, without any damping, with proper initial conditions, but except in simulation, this is generally not possible for two reasons: (i) one can generally control only a small number of macroscopic parameters, electric field, density, velocity, pressure, ..., and not the detailed form of the distribution function (one does not have a "sufficient number of fingers"); (ii) the very peaked perturbations of the distribution around the resonant velocity, for most eigenmodes, makes it impossible to describe them, even approximately, through a limited number of moments. Initializing the system with a perturbation without such a peaked distribution necessarily excites a bunch of eigenmodes, resulting in the kinetic damping.

5.6.4.3 **Resolution**

As mentioned, the calculation will be finished when one will have imposed that the integral of f_1 is equal to n_{e1}. This gives:

$$n_{e1} = I_1 + b_1 \tag{5.97}$$

$$\text{with} \quad I_1(v_\varphi) = \text{p.v.} \int \frac{a_1(v)}{v - v_\varphi} dv \, . \tag{5.98}$$

The second term is an arbitrary constant b_1. The first one, with a principal value, is a classical mathematical transform applied to the function a_1, called a "Hilbert transform". It can be noted as $I_1 = \pi\mathcal{H}(a_1)$. From the definition of a_1, which depends on v only through the derivative of the dimensionless zero order distribution function f_0', this can be rewritten as:

$$I_1 = \sqrt{\pi}\frac{\omega_{\text{pe}}^2}{2k^2 V_{\text{the}}^2} n_{\text{e1}} \mathcal{H}(\tilde{f}_0') \,. \tag{5.99}$$

In the following, we will note F_0 the Hilbert transform of $\tilde{f}_0$. It is a simple property of the Hilbert transform that the transform of $\tilde{f}_0'$ is then just F_0', so that we will use Eq. (5.99) under the form:

$$I_1 = \sqrt{\pi}\frac{\omega_{\text{pe}}^2}{2k^2 V_{\text{the}}^2} n_{\text{e1}} F_0' \,. \tag{5.100}$$

The most classical results concern the particular case of Maxwellian zero order distributions ($\tilde{f}_0 = e^{-\xi^2}$ with the dimensionless notation). The function F_0 is then a known function, tabulated, and with well-studied mathematical properties. It can be noted using various classic function names:

- $F_0(\xi) = -\text{w}_\text{i}(\xi)$, where w is the "Faddeeva function", or "complex error function", and w_i its imaginary part;
- $F_0(\xi) = 1/\sqrt{\pi}\text{Z}_\text{r}(\xi)$, where Z is the "Fried and Conte function" or "plasma dispersion function" and Z_r its real part;
- $F_0(\xi) = -e^{-\xi^2}\text{erfi}(\xi)$, where erfi is the "imaginary error function".

The function Z is the most commonly used in this context of plasma physics.

Keeping our general name F_0, which does not need to specify the form of the distribution function f_0, Eq. (5.97) becomes:

$$b_1 = D_\text{p} n_{\text{e1}} \tag{5.101}$$

$$\text{with} \quad D_\text{p} = 1 - \sqrt{\pi}\frac{\omega_{\text{pe}}^2}{2k^2 V_{\text{the}}^2} F_0' \,. \tag{5.102}$$

The subscript p in D_p is for reminding us that it comes from the principal value part of the solution.

Historically, the first solution of this Vlasov–Gauss problem had been derived by Vlasov himself. It was almost the above one, but with only the principal value part, that is, without the constant b_1 coming from the Dirac part of the above general solution, which describes the ballistic motion of the particles. It provided him thus with a dispersion equation $D_\text{p} = 0$, which is indeed a valid particular solution of the monochromatic type. But, as all of them, this particular solution, with its divergent behavior around the resonant velocity $v = v_\varphi$, can hardly be excited by any usual experimental means. We will see, however, that this solution is close to the physical "kinetic solution" that we will derive hereafter, when $F_0'(\xi)$ is small, which occurs when ξ is much larger or much smaller than unity (that is when v_φ is much larger or much smaller than V_{the}).

The complete solution of Eq. (5.101) is not less difficult than the Vlasov solution to be excited by a reasonable initial condition, but it is quite different. Since b_1 is arbitrary, the equation can be verified for any values of ω and k, and there is, therefore, no dispersion relation. This means that, by choosing adequately the initial distribution perturbation $f_1(v, x)$, it is possible to get any monochromatic evolution ω that one wants. This infinity of solutions fits quite well with the fact that the Vlasov equation has an infinite number of degrees of freedom. The original Vlasov solution was just one of them.

If we finally reject all the solutions $f_1(v)$ that are singular, with a divergence at $v = v_\varphi$ (described above thanks to mathematical distributions, principal value, and Dirac), we must abandon the search of solutions that are purely monochromatic at the microscopic level. For defining what is called a "kinetic mode", two approaches are possible:

1. Solve an initial value problem.
2. Look for "eigenmodes".

In both cases, the selection comes form the class of functions accepted, excluding the singular ones. We will present here the arguments in the case of stable systems, speaking of damping rather than growth rate, but the calculation is actually valid in both cases. The specific question of kinetic instability will be treated in the next section.

The original work of Landau [13], in 1946, used the first approach. The selection criterion amounts there to imposing that the initial functions $f_1(v)$ have the following property: their mathematical continuation in the complex plane has no pole (at least in the lower half-plane). This obviously rejects the above monochromatic solutions, since we have seen that these solutions have a pole on the real axis at $v = v_\varphi$, but it is actually a much more restrictive condition. Even if it may seem nonintuitive under this mathematical form, the condition can be understood: if $f_1(v)$ has a pole which is not exactly on the real axis, but with a small imaginary part, $\tilde{v}_\varphi = v_{\varphi r} + i v_{\varphi i}$, it is easy checking that this situation would correspond to a function $f_1(v)$ without divergence for v real, but still with a strong signature in the vicinity of $v = v_{\varphi r}$. The function $1/(v - \tilde{v}_\varphi)$ has indeed a large modulus for $v \approx v_{\varphi r}$, with a fast phase change from $-\pi$ to $+\pi$ when going from smaller to larger values. This is quite reminiscent of the problem of an oscillator: when forcing it close to its resonant frequency, there is an infinite resonance if the oscillator is nondissipative, but a similar odd resonance curve, without infinite divergence, when dissipation is present. Extending this argument to larger and larger imaginary parts leads to excluding all poles in the complex plane in order to exclude any particular signature, even weak, in the initial condition of $f_1(v)$. Landau's calculation shows that, when this condition is verified for the initial condition, whatever its detailed form, a dispersion equation can be derived, which is universal.

The dispersion equation derived in this way is complex, which allows determining complex solutions $\omega = \omega_r + i\omega_i$ (assuming, for instance, a given real k). The less damped solution determines the asymptotic behavior of the system, which is, therefore, also universal. The transient period that precedes it is a mixture of all the other solutions of the same dispersion equation, and it depends on the detailed initial condition on $f_1(v)$ (within the "no pole condition" restriction). The less damped mode (generally called "Landau mode") is generally close to the fluid monochromatic mode, but it nevertheless involves a damping rate, that is, a negative imaginary part of the frequency.

Landau's calculation Eq. (5.6) can be found in many textbooks. We will not reproduce it here. It makes use of a Laplace transform in time, instead of the usual Fourier transform. This allows one to introduce directly the initial condition in the calculation and to use integration contours with complex ωs for the inverse transform, so avoiding the mentioned problem of decomposing the regular initial condition on a basis of singular solutions. In this method, the calculation requires some cleverness in deforming the integration contours in the complex plane and handling the notion of analytic continuation (in particular for the function $f(v)$ continued in the complex plane of the variable v). This relative difficulty may sometimes blur the physical basis of the method: the selection of perturbations without complex poles must not be taken as a secondary detail, just necessary for mathematical correctness, but as the fundamental physical assumption. Without this assumption, there is no universal behavior at asymptotic times and there is an infinity of solutions, corresponding to the infinite number of degrees of freedom, which makes the notion of kinetic mode meaningless.

We shall develop a little bit more the second approach, as done in [25]: looking for "kinetic eigenmodes". The dispersion relation that we will derive is identical to the one established by Landau, but without treating the full initial value problem. Keeping close to Landau's calculation, we shall adopt here the following definition for a kinetic eigenmode: it is a solution where the macroscopic variables as n_{e1} (or E_1) are

- null for $t < 0$
- and varying as $n_{e1} = \hat{n}_{e1} e^{-i\omega t} = \hat{n}_{e1} e^{-ikv_\varphi t}$ for $t > 0$,

$\omega = \omega_r + i\omega_i$ being complex, with a negative imaginary part (damping). At the microscopic level, as we will see, this does not imply that the distribution function has the same mathematical form: $f_1(v)$ is not monochromatic. This form for n_{e1} can be built as a wave packet of the pure monochromatic solutions calculated above. Using Laplace transform tables, it can be seen that this superposition can be written:

$$n_{e1} = -\frac{\hat{n}_{e1}}{2i\pi} \int \frac{e^{-ikv'_\varphi t}}{v'_\varphi - v_\varphi} dv'_\varphi \,. \tag{5.103}$$

The variable v'_φ is real, but v_φ is complex, so that the integral has no divergence, contrary to Eq. (5.98). The exponential decrease, in any such wave packet, clearly comes from the phase mixing between the monochromatic components. Finding the distribution function corresponding to such a wave packet just amounts to applying the same linear superposition to the elementary functions

f_1 of all the monochromatic components. This elementary function comes from Eqs. (5.95), (5.96) and (5.101):

$$f_1 = n_{e1}\left[D_p\delta(v - v_\varphi) + \frac{\omega_{pe}^2}{k^2}\,\text{p.v.}\left\{\frac{f_0'/n_0}{v - v_\varphi}\right\}\right] . \tag{5.104}$$

In this notation, f_0, f_1 and n_{e1} are the usual dimensioned quantities, and D_p is dimensionless, as defined in (5.102). Applying the above linear superposition to these distributions, one gets the general expression for the perturbed distribution function $f_1(v)$ that corresponds to this kind of time behavior:

$$f_1(v) = -\frac{\hat{n}_{e1}}{2i\pi}\frac{1}{v - v_\varphi}\left[D_p e^{-ikvt} - 2i\pi\frac{\omega_{pe}^2}{k^2}\frac{f_0'(v)}{n_0}\left(e^{-ikv_\varphi t} - 1/2e^{-ikvt}\right)\right] . \tag{5.105}$$

We don't present here the detail of this calculation (the integrations to be done are of the same kind as the Landau ones). Let us just mention that the function $f_1(v)$ found is no longer singular for any real v whenever the imaginary part of v_φ is not zero. One can notice, furthermore, that $f_1(v)$ does not imply only the complex frequency that characterizes the mode $\omega = kv_\varphi$ and which is imposed for the macroscopic quantities, but also frequencies kv, which are different for the different velocities. This corresponds to the ballistic motions of the particles and this is important for the notion of phase mixing giving rise to the decrease of the macroscopic quantities. Let us finally outline that $f_1(v)$ is so found for any value of v_φ, without any notion of dispersion equation at this stage. To find a dispersion equation, one still has to add a selection condition. Let us take the same as Landau's one, that is, the absence of a pole of $f_1(v)$ in the complex plane at time $t = 0$. For this time, Eq. (5.105) provides:

$$f_1(v) = -\frac{\hat{n}_{e1}}{2i\pi}\frac{1}{v - v_\varphi}\left[D_p - i\pi\frac{\omega_{pe}^2}{k^2}\frac{f_0'(v)}{n_0}\right] . \tag{5.106}$$

For arbitrary values of v_φ, it is clear that this function has in general a complex pole at $v = v_\varphi$. The only way of suppressing this pole is by making the numerator null for the same value $v = v_\varphi$, leading to:

$$D_p - i\pi\frac{\omega_{pe}^2}{k^2}\frac{f_0'(v_\varphi)}{n_0} = 0 . \tag{5.107}$$

It must be noted that imposing the absence of a complex pole at any time t instead of imposing it at $t = 0$ would lead to the same result, since the numerator has the same value at any time for $v = v_\varphi$, except for the propagation coefficient $e^{-i\omega t}$. This could be guessed in the logic of an eigenmode search: if there is a pole at $t = 0$, the same pole is present at any time. The Vlasov solution $D_p = 0$ thus corresponds to retaining only the first part of this solution, that is, taking only the principal value part into account. One can see that adding a more physical description of the distribution perturbation close to the resonant velocity adds an imaginary part to the dispersion equation which, in turn, brings imaginary parts (damping) in the solutions.

This result of Eq. (5.106) is exactly the same as the result obtained by Landau using the first approach. The less damped solution of this equation has a damping rate that we will note γ_L and that is called "Landau damping rate".

5.6.4.4 Landau and Non-Landau Solutions: Numerical Simulation

The above results, which may seem a little abstract, can be made more concrete by looking at numerical simulation data. The results of Figures 5.19–5.23 illustrate this point. They come from [25]. The simulation used is a "perturbative PIC" simulation (see Section 2.3), which allows to get a low noise level in a fully kinetic simulation.

Figure 5.19 displays the electric energy density, in logarithmic scale, when initializing the system with a sinusoidal density perturbation, the electron density being everywhere Maxwellian with random particle positions and velocities. The straight line shows the theoretical decrease rate $2\gamma_L$ predicted by Landau's theory. Apart from a small oscillation due to an interference with the residual wave propagating in the reverse sense (due to an improper initialization), one can see that the decrease follows quite accurately this theoretical prediction.

Figure 5.20 shows the perturbed distribution function $f_1(v)$ at different times. The increasingly short oscillations of $f_1(v)$ with respect to v is due to the basic ballistic effect (the terms in e^{-ikvt}), as it easily be understood from Figure 5.21. Instead of the cuts at a given position x_0 shown in Figure 5.20, this figure is the

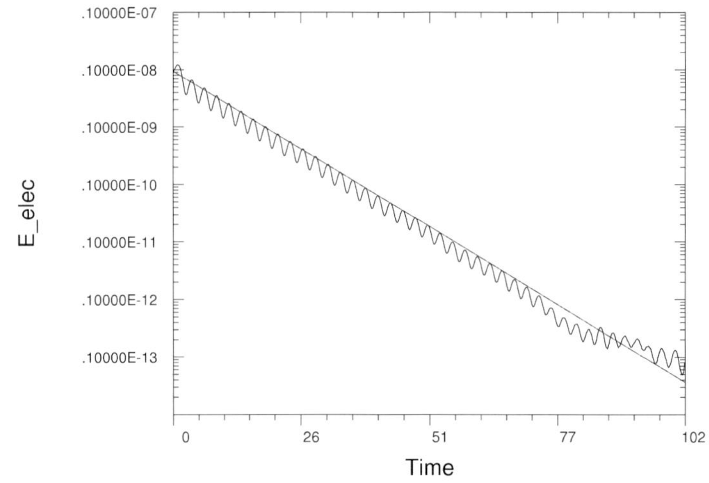

Figure 5.19 Electrostatic energy decrease due to Landau damping in a standard case, that is, with an initial density perturbation varying sinusoidally with x but without special initialization for $f_1(v)$ (simply Maxwellian here). The line indicates the theoretical $-2\gamma_L$ rate.

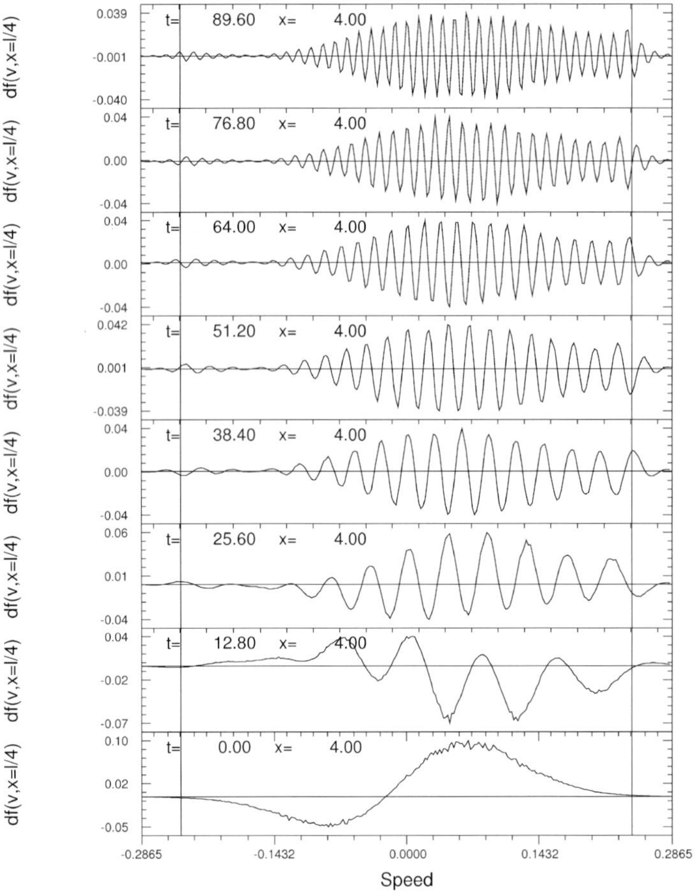

Figure 5.20 Perturbed distribution functions $f_1(v)$ in the same case as in Figure 5.19 at a given position x and at different times. No resonant feature is observed near the resonant velocity (vertical line), which could be a signature of the particle's resonant behavior.

full 2D display of the function $f_1(x, v)$ in phase space. As particles go faster toward the right-hand side in the upper part of the figure and in the opposite direction in the lower part, any perturbation is increasingly sheared by this effect. The simple structure observed at $t = 0$ becomes more and more inclined, resulting in the oscillating cuts of Figure 5.20. This dominant effect progressively masks the global oscillation, forced by the electric field, which propagates with the positive velocity v_φ.

In Figure 5.20, the vertical line on the right shows the resonant velocity v_φ (the velocity $-v_\varphi$ is also drawn for symmetry). It clearly appears that no signature is present at this velocity, contrary to what should occur if there was a resonant behavior in $1/(v - v_\varphi)$. This would occur only if a monochromatic mode had been

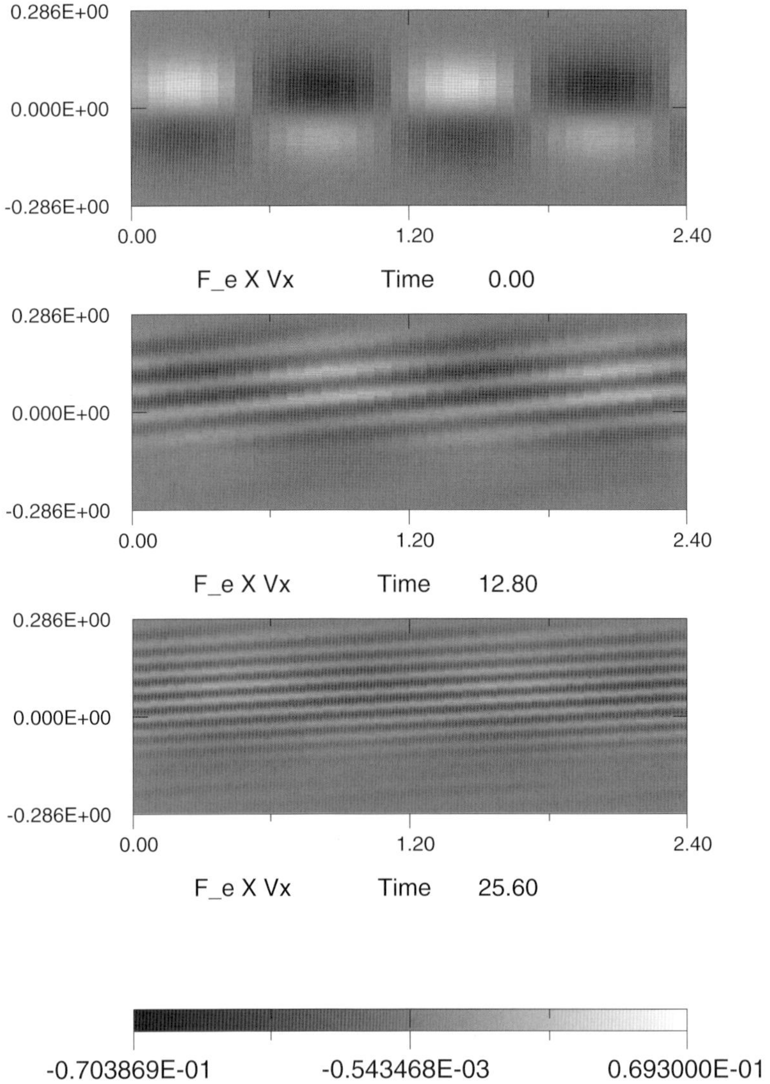

Figure 5.21 Phase space (x, v) at different times t for the same case as in Figures 5.19 and 5.20 (Landau damping). The "ballistic effect" manifests itself by the increasing shear of the distribution function. This explains the increasingly short oscillations appearing in the distributions shown in Figure 5.20, which are just cuts at a given x of this 2D plot.

excited by the initial condition, but it is clear that the initial form of the perturbation $f_1(v)$ ($t = 0$, lowest panel) does not involve any signature of this kind.

The above-mentioned figures illustrate what "always" happens in a plasma, whenever the initializations of $f_1(v)$ are smooth, without special preparation of the perturbation phase. Nevertheless, to make the theory more explicit, let us show that in a simulation, where any initialization is possible, "non-Landau solutions"

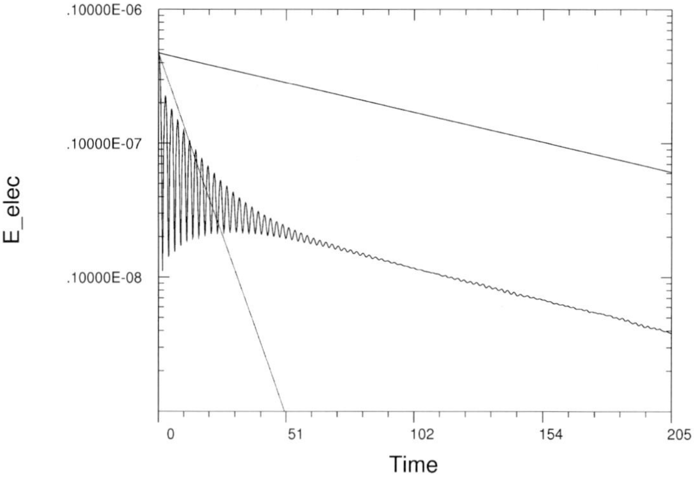

Figure 5.22 Same as Figure 5.19, but with a special initialization of $f_1(v)$ with a complex pole implying a signature at $v = v_\varphi$.

can exist, with an asymptotic behavior different from any solution of Landau's dispersion equation. This asymptotic behavior is then just fixed by the position of the initial distribution pole in the complex plane.

Figure 5.22 shows an example where the initial perturbation of the distribution possesses a pole in the complex plane. As this pole has an imaginary part smaller, in absolute value, than γ_L, it fully determines the asymptotic behavior. The straight line at $2\gamma_L$ is the steepest line, while the other one corresponds to the imaginary part of the pole. The energy clearly follows the second line (one can forget about the transient part, which is larger than the previous one because it is technically more difficult to excite only one single eigenmode in this case). Figures 5.23 and 5.24 show the corresponding perturbations $f_1(v)$. It is clear, in this case, that a peak signature is present at $v = v_\varphi$ at time $t = 0$, and that this signature goes on existing afterward, just experiencing a sinusoidal oscillation in space and time, corresponding to the mode propagation.

Initializing a non-Landau solution demands giving the right values of $f_1(x, v)$ for all x and for all v, all being different. It is clear that such an initialization cannot occur by chance, even if the peak was smaller and wider (corresponding to a pole further from the real axis, that is, a larger damping). Even if extremely unlikely in natural conditions, these simulations confirm that an infinity of "non-Landau" modes can be excited by this method, and that the Landau solution is only one particular solution between them.

These illustrations show, as the preceding theoretical part, that the Landau damping is due to a restriction in the accepted distributions functions. The full Vlasov equation, without this restriction, is reversible, in agreement with the possibility of any real ω in this case. This same restriction can also be recovered by considering a "coarse grained" velocity space, which provides a limitation in the gradients

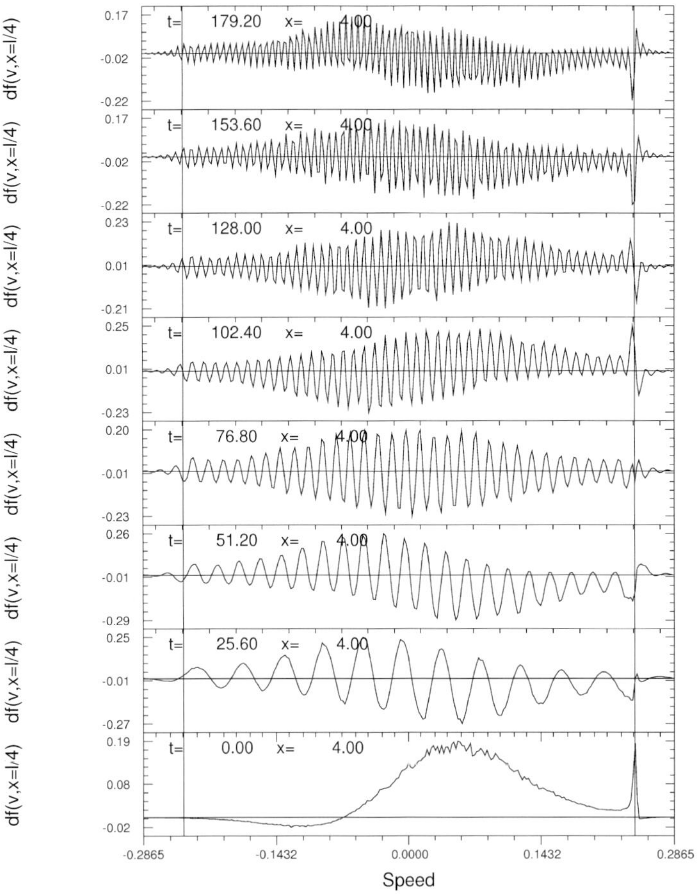

Figure 5.23 Same as Figure 5.20 for the special initialization of Figure 5.22. A resonant signature is observed near the resonant velocity (vertical line).

of $f(v)$ and thus prevents any singularity in this space. The restriction is then a result of the incompleteness of the description. This is in agreement with the fact that the damping is an irreversible effect and that any irreversibility, as explained in Chapter 4, is always related to an incompleteness of the description when the microscopic effects are reversible.

It is to be noted that the above linear calculations can be continued by nonlinear ones, taking into account the nonnegligible energy of the wave perturbation with respect to the energy of the zero order medium. This brings important phenomena such as particle trapping and quasi-linear diffusion. These phenomena are presented in Chapter 6. It must be kept in mind however that the basic notions of kinetic mode and of kinetic damping are not essentially linked to these nonlinear phenomena: they are based on the linear notions explained above.

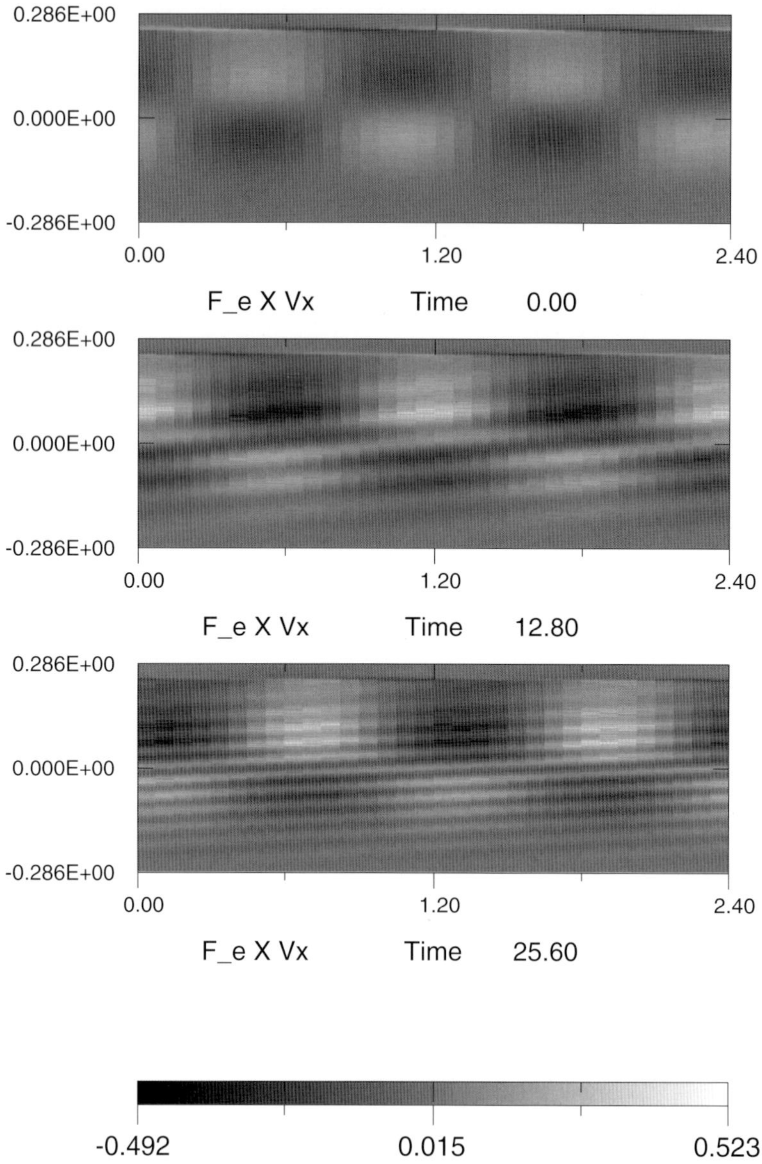

Figure 5.24 Same as 5.21, for a non-Landau solution. One can clearly see the permanent feature at $v = v_\varphi$ and its modulation with x. It is the signature of the resonant particle behavior, which exists for all solutions except the Landau one.

5.6.4.5 Energy Conservation and Kinetic Damping

There are a number of textbooks where arguments about the energy conservation are used to demonstrate the Landau effect in a way supposed to circumvent the mathematical difficulties and give a more physical and a more intuitive image of the Landau effect. These arguments are actually to be used with much caution. As for any problem of mechanics, the energy conservation can indeed be used instead of the momentum equation to reach the same results. Nevertheless, several remarks have to be made:

- The energy arguments just provide a possible shortcut to describe the individual particle trajectories. Considered alone, they can, therefore, explain neither what a kinetic mode is nor why it is damped. All the important ingredients concern collective effects (absence of resonant signature) and are thus independent of the way the individual trajectories are calculated.
- All solutions, Landau and non-Landau, of course respect energy conservation. The energy conservation must not be understood as the cause of Landau damping.
- Energy is a squared variable and using it in a linear calculation requires very meticulous care.
- The role of the resonant particles can be misunderstood when looking at these arguments too superficially. The decrease of the electrostatic energy is actually not due to an increase of the amplitude of f_1 at the resonant velocity, but due to the phase properties of the particles on both sides of it. As emphasized above, it is even this absence of the singular behavior of f_1 around v_φ which defines what the kinetic mode is. A more detailed calculation shows that, if one notes $\mathcal{E}$ the global energy exchange, (i) the strictly resonant particles don't participate at all to this exchange; (ii) the particles close to the resonant velocity (on both sides) have a participation equal to $2\mathcal{E}$, due to this phase effect; (iii) the nonresonant particles, further from the resonant velocity, have a contribution in the opposite sense equal to $-\mathcal{E}$, due to an amplitude variation of f_1.
- Some classical illustrations of the Landau effect can also be misleading. The pictures of small boats, or surfers, going up and down on sea waves is nothing but an illustration of energy conservation. This, as explained above, is quite insufficient to explain the Landau effect whenever the collective aspects are missing. Many classical illustrations also show the particle phase space with trapping effects, or distribution functions with quasi-linear effects in the vicinity of the resonant velocity. These effects do indeed exist, and the nonlinear treatment of the Landau effect (which is not treated in the present chapter) is indeed quite interesting. But one must not think that the role of the resonant particles lies in these nonlinear effects: the basic Landau effect is a linear one.

In spite of these various reservations, the energy arguments can be used, however, to present the Landau damping phenomenon in a short and simple way. In linear theory, there are two first order terms governing the kinetic energy increase at the expense of the electrostatic energy: one is associated with the ballistic transport of the initial perturbation f_1 (integrated over velocity), the other is due to the forcing of f_0 by the electric field E_1. The non-Landau solutions are associated with this first kind of energy transport. If one admits that, for the reasons of accessibility of the solutions that have been discussed above, the kinetic mode is characterized by the absence of signature in $f_1(v)$ at $v = v_\varphi$, this first contribution can be neglected, the phase mixing making null the corresponding integral. The second kind of energy transport term corresponds to the Landau solutions. It usually corresponds to an increase of the kinetic energy because, in distributions such as Maxwellian, the slope is negative at $v > 0$, so that there are more particles at $v < v_\varphi$ than at $v > v_\varphi$. As the first category of particles gains energy while the second one looses energy, the net balance is indeed a gain of energy for the global particle population. More details about the energy arguments can be found in [26].

5.6.5
Other Types of Kinetic Damping

All effects of kinetic damping, in the case of electrostatic phenomena (E_1 parallel to k) can be called Landau damping, whatever the kind of wave involved and the nature of the particles involved. The example of the Langmuir wave is the simplest one, and the effect then concerns only the electrons. But it is easy replacing it by any other kind of electrostatic wave, for instance, a sound wave, and it can concern as well the ions and the electrons, depending whether the propagation velocity is close to the ion thermal speed or to the electron one. The effect can exist as well in magnetized plasmas. In all these cases, the calculation concerning the Vlasov equation is fully identical, the only change is the coupling with Maxwell equations and with the other particle populations. Instead of being expressed as a function of the electron density by Maxwell equation in the case of a Langmuir wave, it can, for instance, be expressed as a function of the electron pressure gradient in the case of a sound wave.

Landau damping is not the only possible kinetic damping. Similar phenomena can occur for any possible resonance in the particle trajectories.

In a magnetized plasma, the most important resonance, after Landau's one, is the cyclotron resonance. While Landau resonance concerns the parallel motion of the particles forced by the parallel electric field, the cyclotron resonances concern the perpendicular motion of the particles, with its cyclotron gyration, forced by the perpendicular electric field. While the resonance condition is $\omega - k_\| v_\| = 0$ for Landau resonance, it is $\omega - k_\| v_\| = \pm n\omega_c$ for the cyclotron one (with n = any integer).

Other resonances can occur as well, as the "bounce resonances" for particles bouncing between two mirror points (in a magnetosphere for instance). All these effects can give rise to kinetic damping.

5.7 Instabilities

5.7.1 Real Space Instabilities: Fluid Treatment

In gas dynamics or in any medium where a fluid treatment is considered as valid, for instance, in MHD, one has a relatively small number of variables which are functions of space and time, for instance, the density $n(\boldsymbol{r}, t)$, the velocity $\boldsymbol{u}(\boldsymbol{r}, t)$, the pressure $P(\boldsymbol{r}, t)$, the magnetic field $\boldsymbol{B}(\boldsymbol{r}, t)$, and so on. These variables are related by the same number of fluid equations. Linearizing these equations and looking for monochromatic solutions in ω and k, one can establish a dispersion equation (see Chapter 1). For a given k, if all the solutions in ω are real, the system is stable and the different solutions provide the different modes of propagation in the medium. If there is at least a pair of complex conjugate solutions, the solutions ω with positive imaginary parts correspond to exponentially growing perturbations and the medium is unstable.

Such instabilities generally occur when the unperturbed medium (order zero in the linearization process) is nonhomogeneous and/or nonstationary. Several examples of such instabilities are famous and can exist in neutral media as well as in plasmas. For a given layer, surface waves can occur, for instance, in the following cases:

- the Kelvin–Helmholtz instability, when there is a shear across the layer for the tangential velocity;
- the Rayleigh–Taylor instability when, across the layer, there is a force normal to it with a gradient in the sense opposite to the force itself (for the gravity force, this is the case for a heavy medium on a light one);
- the tearing instability (in MHD), when there is a shear of the tangential magnetic field, especially when the normal magnetic field is zero.

In circumstances when the monofluid description is insufficient, a kinetic treatment is generally needed or, – at least – a multifluid one (see Example 5.7.4). We will see that instabilities can then occur independently of any gradient in real space. The evolution of the system then tends nonlinearly to modify the different moments, but without involving any gradient or any motion in real space.

5.7.2 Velocity Space Instabilities: Kinetic Treatment

In the kinetic frame, one has to solve kinetic equations (Vlasov equations in the collisionless case) instead of a limited set of N fluid equations relating N variables. Each population is, therefore, described then by an infinite set of variables $f(v)$ (for all values of v) instead of the previous limited number N.

The same instabilities due to nonhomogeneities and/or nonstationarities in the zero order medium can be described in this kinetic formalism as well, but this is not the most interesting thing: there are now instabilities due to the shape of the distribution function $f_0(v)$ in velocity space, even in homogeneous and stationary media. The nonlinear tendency is then to modify the shape of $f_0(v)$. We will show in this chapter that a distribution function with two peaks is unstable while a distribution with one peak is stable.

The above calculation has shown that, in a plasma, there is no possible propagation with a real ω, whenever the distribution functions are perturbed in a standard manner (that is, whenever the perturbation $f_1(v)$ has no pole when continued in the complex plane, this condition being valid at any time $t > 0$ if it is imposed at $t = 0$). In a stable plasma, this leads to the conclusion that any wave is always damped, corresponding to $\omega_i < 0$.

Nevertheless, not one of the arguments given have ever specified explicitly that the plasma was stable. The results obtained should, therefore, be valid also when the plasma is unstable, so leading to solutions with $\omega_i > 0$. Keeping the hypothesis that the zero order distribution is homogeneous and stationary (with $u_0 = 0$ and $E_0 = 0$), let us try to find out what the conditions on f_0 are that can make the plasma unstable. The "Landau effect", which is generally considered in the stable case is then called the "inverse Landau effect".

5.7.3
Weak Kinetic Effects

The dispersion Eq. (5.107) is indeed valid in full generality, in an unstable plasma as in a stable one. Even if the calculation may seem different in the two cases in the initial value problem of Landau (contour deformation useful in one case and not in the other), the physics is actually the same in both cases, as it appears more clearly in the eigenmode approach. The question is only to solve this dispersion equation and to investigate when it gives rise to solutions with $\omega_i < 0$ and when it leads to solutions with $\omega_i > 0$. It is difficult to give general answers to this question, but it is possible to do it in the particular case of weak kinetic effects, that is, when the term $\varepsilon_k = \pi(\omega_{\mathrm{pe}}^2/k^2) f_0'(v_\varphi)/(n_0)$, which comes from the assumed absence of resonant signature in $f_1(v)$ at $v = v_\varphi$, is small enough so that a perturbative method can be used.

This allows us to write the dispersion equation under the form:

$$D_{\mathrm{p}} = i\varepsilon_k \tag{5.108}$$

k is supposed fixed and both D_{p} and ε_k are functions of v_φ. We now further assume that the introduction of ε_k brings a small departure from the solution $v_{\varphi p}$ of $D_{\mathrm{p}} = 0$. Using an expansion around this value leads to:

$$\delta v_\varphi D_{\mathrm{p}}' = i\varepsilon_k \,. \tag{5.109}$$

Where D'_p represents the derivative of D_p with respect to v_φ. We thus finally obtain:

$$\delta\omega = ik\frac{\varepsilon_k}{D'_\text{p}} . \tag{5.110}$$

The function D_p has been defined in (5.102). It derives from the zero order distribution function through an integral form with a principal value (Hilbert transform). Coming back to the Langmuir mode at $v_\varphi \gg V_\text{the}$ ($\xi \gg 1$), let us check that this formula does allow us to find the expected result that the Landau effect, for a Maxwellian distribution, leads to a small damping. In this case, the asymptotic expansion of the Fried and Conte function (see Section 5.6.4.3) shows that:

$$D_\text{p} \approx 1 - \frac{1}{k^2\lambda_{De}^2}\frac{1}{2\xi^2 - 3} \tag{5.111}$$

$$D'_\text{p} \approx \frac{1}{\sqrt{2}V_\text{the}}\frac{1}{k^2\lambda_{De}^2}\frac{4\xi}{(2\xi^2-3)^2} . \tag{5.112}$$

One can see that D'_p is odd in ξ (defined in Section 5.6.4.1), and, therefore, in v_φ: it is positive for a positive v_φ and negative for a negative v_φ. At the same time, $f'_0(v_\varphi)$, and, therefore, ε_k has the opposite sign: the Maxwellian has a negative derivative for the positive velocities and vice versa. This shows that the departure from the real solution of $D_\text{p} = 0$ is purely imaginary and negative: the kinetic effects indeed lead to a small damping, as expected.

Another example can easily be derived from this one, but leading now to an instability: it is sufficient to add to the above Maxwellian distribution another one, but much smaller in amplitude and drifted at a velocity $V_\text{D} \gg V_\text{the}$ far in the tail of the first Maxwellian (see Figure 5.25). The second population is generally called a "beam", to emphasize its drift velocity, (but it does not imply any geometrical limitation as the word "beam" could also invoke). This configuration is generally called a "gentle bump in the tail". Thanks to the small amplitude of the second

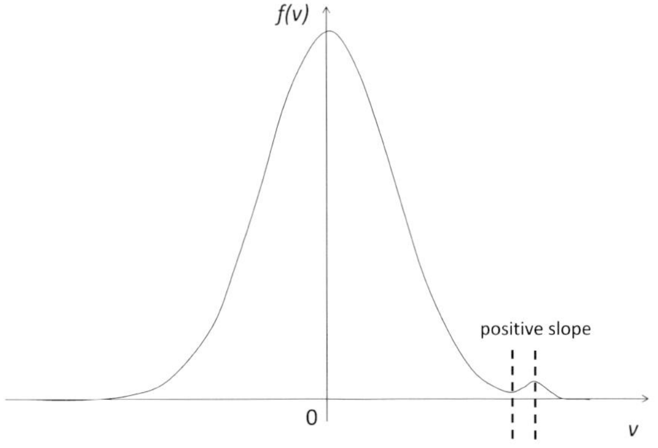

Figure 5.25 Maxwellian distribution with a gentle bump in the tail (small beam). The presence of a positive slope somewhere on the $v > 0$ semiaxis makes the plasma unstable.

population, one can consider that it does not play a significant role in the value of D_p, which is an integral. On the contrary, thanks to its position far in the tail, this second population is able to introduce a positive value of the derivative f_0' in a limited range of v_φ slightly below the drift velocity V_D. The Landau effect, in this range of phase velocities therefore results in a positive ω_i, that is a growth rate. This is what is called the "inverse Landau effect".

Although the above demonstration is very limited (weak kinetic effects, Maxwellian distributions), the result can be considered as general: a monopeaked distribution is stable, and a two (or more) peaks distribution is unstable.

5.7.4
An Example: The Two-Stream Instability

This example further develops the preceding one concerning the two-peaks distributions. This will allow distinguishing two types of instabilities in velocity space: the multifluid type and the inverse Landau type.

Let us consider two beams of equal amplitudes and opposite drift velocities $-V_D/2$ and $V_D/2$ (Figure 5.26). The two beams will be considered also with equal temperatures, and their common thermal speed called V_{the}. We will show how the result then depends on the value of this temperature.

Let us first consider the case when the relative drift is much larger than the thermal speed $V_D \gg T_{the}$ (bold line of Figure 5.26). In this case, one can easily predict that a two-fluid method, with polytropic hypotheses for each beam (with $\gamma = 3$), is quite well adapted. One has, therefore, to use exactly the same method as shown above (Figure 5.16), with the only difference being that the zero order densities for the two populations a and b are $n_{a0} = n_{b0} = n_{e0}/2$ and the zero order velocities are $u_{b0} = -u_{a0} = V_D/2$. The total derivatives for each beam, therefore,

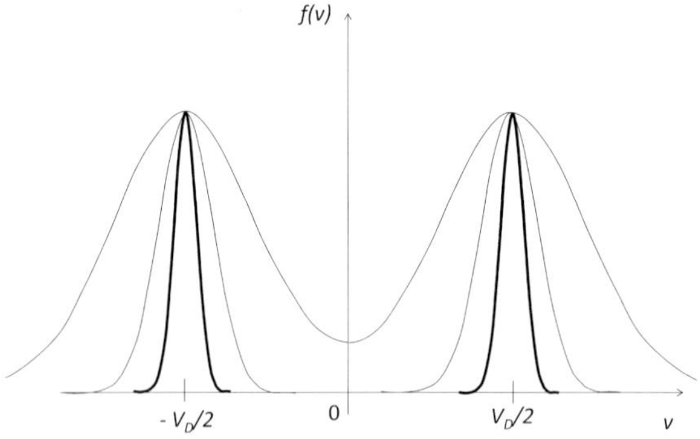

Figure 5.26 The zero order distributions leading to the two-stream instability. The bold curve corresponds to a small temperature for each beam, which provides essentially the same results as completely cold beams. The thin curves correspond to higher temperatures. The highest temperatures no longer lead to a multifluid instability. Nevertheless, the resonant effects can still be effective to make the system unstable.

bring the two pulsations: $\omega_a = \omega + k V_D/2$ and $\omega_b = \omega - k V_D/2$. This leads to:

$$\left(\omega_a^2 - 3k^2 V_{\text{the}}^2\right) \frac{n_{a1}}{n_0} = -ik\frac{e}{2m} E_1 \qquad (5.113)$$

$$\left(v_{\varphi b}^2 - 3 V_{\text{the}}^2\right) \frac{n_{b1}}{n_0} = \frac{e}{2m} \frac{E_1}{ik} \qquad (5.114)$$

$$-ik\frac{e}{m} E_1 = \omega_{\text{pe}}^2 \left(\frac{n_{a1}}{n_{e0}} + \frac{n_{b1}}{n_{e0}}\right) . \qquad (5.115)$$

The first equation is the combination of the two first fluid equations for the population a; the second equation is similar for population b; and the third one comes from the Maxwell–Gauss equation. Solving in E_1 leads to the dispersion equation:

$$\frac{1/2}{\omega_a^2 - 3k^2 V_{\text{the}}^2} + \frac{1/2}{\omega_b^2 - 3k^2 V_{\text{the}}^2} = \frac{1}{\omega_{\text{pe}}^2} . \qquad (5.116)$$

Using the definitions of ω_a and ω_b, this equation appears as biquadratic in ω. This simplicity comes from the symmetry chosen (equal densities, opposite velocities) and allows for an analytical solution. This solution is given in Figure 5.27.

The solutions $\omega = \pm\omega_{\text{pe}}$ still exist at small k, but they are modified for larger k, due to the two velocity drifts of the beam components (which act as a thermal spread on the global plasma) as well as to the temperatures of the beams themselves. These solutions remain real, however. Most interestingly, other solutions appear at smaller ω; some of these new solutions happen to be complex conjugates (null real part and opposite imaginary parts), which correspond to a plasma instability. For too hot beams, however, this instability disappears.

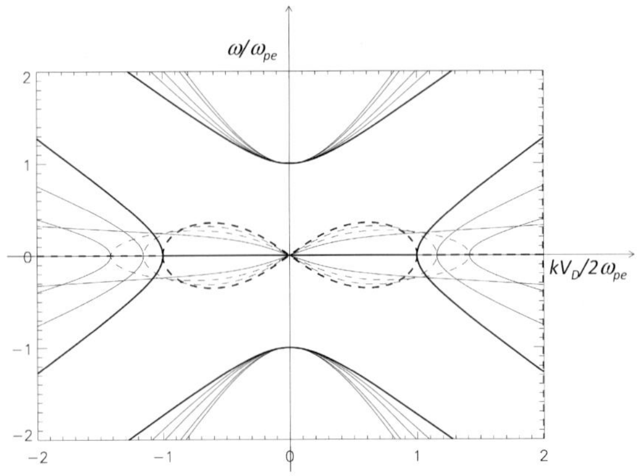

Figure 5.27 Solutions of the fluid dispersion equation for the two stream instability. The full lines are the real parts, the dashed lines are the imaginary ones. Notice that, for some values of k, some real solutions happen to be replaced by a pair of complex conjugate ones (instability). The bold curves correspond to $V_{\text{the}} = 0$, the other ones to $V_{\text{the}}/V_D = 0.14$, 0.20, 0.29, 0.32. The hottest solutions have no instability.

This is an example of the "multifluid" type of instability. It always results in solutions which are complex conjugate, the solution with the positive imaginary part being the clue for an instability. Solving the real part of the dispersion equation $D_p = 0$ has exactly the same characteristics, even if the function D_p, which comes from the kinetic treatment can be slightly different from the above polytropic dispersion equation. However, it must be observed that, in this case, the instability does not come from the kinetic term in f_0' coming from the resonant particles, but from the complex conjugate solutions of a real equation. It occurs whenever the separation between the two peaks of the distribution function is larger than their thermal spread because, in this case, the wave can develop with a phase velocity which is in between, so being in the fluid (adiabatic) case for both components. It can be seen in Figure 5.27 that the real waves develop at phase velocities larger than $V_D/2$ or smaller than $-V_D/2$ and that the unstable ones occur between these two values, with a null real phase velocity. No waves occur with phase velocities close to the beam ones.

When the temperature of the two beams increases, there is less and less room for the waves to develop between the two beams without experiencing a significant f_0'. Considering again the limit of a weak departure from the fluid solution due to the resonant effects, the calculation is exactly the same as previously for all the waves that are real in the limit $D_p = 0$. It can be checked that, with the symmetry assumed (and contrary to the above case of a small beam in the tail of a large Maxwellian), the sign of D_p' is everywhere such as leading to damping and never making these real modes unstable. Concerning the purely imaginary waves, D_p' and f_0' have to be calculated for an imaginary value of the argument. It can be checked that the two complex conjugate solutions are modified in the sense of increasing their imaginary part; therefore, making the unstable solution more unstable and the damped solution less damped, and so breaking the conjugation.

The above result can be considered as general: two complex conjugate solutions are characteristic of an instability of the "multifluid" type, that is due to the solution of $D_p = 0$ while a departure from this conjugation is characteristic of an instability due the resonant particles, via f_0', and combinations of both types are always possible.

6
Nonlinear Effects, Shocks, and Turbulence

6.1
Collisionless Shocks and Discontinuities

6.1.1
Nonlinear Propagation, Discontinuities, Jumps

6.1.1.1 Discontinuities

In plasmas, as in hydrodynamics, nature often tends to form thin interface layers separating media with different properties. It is the case, for instance, upstream of a supersonic jet, where a "shock wave" forms, separating the unperturbed steady, atmosphere from the gas in contact with the plane, which is hotter and denser, with a velocity approaching that of the plane. It is the case as well upstream of a boat, where the level of the sea is unperturbed until a similar "bow shock" close to the bow of the boat, or upstream of a magnetized planet like the Earth where the incident flow is the solar wind and the obstacle the planetary magnetic field.

These thin layers will be called here "discontinuities," as it is commonly done, although this name is not used then in its mathematical sense: the layers are generally quite continuous and can be well resolved. Their only important characteristic is to be "thin", which means that they can be considered locally 1D, the gradient along their normal being much larger than the tangential gradients (and than the normal gradient outside, but this hypothesis is less critical).

These thin layers can be formed occasionally due to particular boundary conditions or initial conditions, but the main mechanism leading to them is the "nonlinear steepening". The notion of nonlinear steepening belongs to the more general notion of "nonlinear propagation" in hyperbolic systems, which is a full-fledged branch of mathematics. We will not develop here these theories (for more details, see books such as [27]), but just summarize some important results and give an illustration. The reader is referred to the specialized literature for more details.

6.1.1.2 Nonlinear Propagation

The two main points to remember are:

1. For 1D variations, the notion of linear mode can be generalized into the notion of a "simple wave". Each linear mode has then a nonlinear version, which prop-

Collisionless Plasmas in Astrophysics, First Edition. Gérard Belmont, Roland Grappin, Fabrice Mottez, Filippo Pantellini, and Guy Pelletier.

agates in the same sense, at approximately the same speed, but gets deformed along the propagation instead of remaining unchanged in its own frame. The small amplitude limit of the simple wave is of course the linear wave.

2. The calculation of the simple waves is based on the notion of characteristic variables and characteristic equations.

If the original system is made of first order differential equations with respect to time and to space, it can generally be written under the form:

$$\partial_t \boldsymbol{Y} + \boldsymbol{C} \cdot \partial_x \boldsymbol{Y} = \boldsymbol{S} \, . \tag{6.1}$$

Where $\boldsymbol{Y}$ is the vector containing all the unknown variables of the system, for instance, n, $\boldsymbol{u}$, $\boldsymbol{B}$, and so on, and $\boldsymbol{S}$ contains the possible RHS terms. We will suppose $\boldsymbol{S} = 0$ in the following for the sake of simplicity (it is actually zero in an ideal medium without forcing). The tensor $\boldsymbol{C}$ contains all the coefficients of the differential system and can be, in general, function of $\boldsymbol{Y}$.

The method consists of deriving, from the original system, scalar characteristic equations which are of the form:

$$\boldsymbol{L} \cdot \left(\partial_t \boldsymbol{Y} + V_p \partial_x \boldsymbol{Y}\right) = 0. \tag{6.2}$$

Where $\boldsymbol{L}$ is a vector and V_p is a scalar, both depending possibly on the variable $\boldsymbol{Y}$, but not explicitly on x and t. The problem of finding such characteristic equations has generally several solutions. For a system of order N (that is involving N first order equations), the system is said to be hyperbolic when there are N characteristic equations with real coefficients. It can be seen that each equation has the form $\boldsymbol{L} \cdot D_t \boldsymbol{Y} = 0$ if one defines $D_t = \partial_t + V_p \partial_x$. The scalar V_p so appears as a local characteristic velocity, the field lines of this velocity field being called the "characteristic lines".

If $\boldsymbol{L}$ is independent of $\boldsymbol{Y}$ (linear problem), one can easily build a scalar variable $Y' = \boldsymbol{L} \cdot \boldsymbol{Y}$ and write the equation under the form:

$$D_t Y' = 0. \tag{6.3}$$

This means that, in this case, there is always a variable Y' which is invariant along the characteristic lines. It is called a "Riemann invariant". If $\boldsymbol{L}$ depends on Y (nonlinear general case), the existence of such a Riemann invariant is not guaranteed. When it can be defined, however, the result is the same as in the linear case (see next example). When it cannot be defined, it remains a differential equation anyway relating the derivatives $D_t(Y_i)$ along the characteristic lines of all the components of $\boldsymbol{Y}$.

In a hyperbolic system, there are N sets of characteristic lines. They can be used for solving completely any spatiotemporal problem, for instance, finding the spatial form $\boldsymbol{Y}$ of the perturbations at any time t when they are known at time $t = 0$ (initial condition problem) or finding the temporal evolution at any point x when they are known at one position $x = 0$ (boundary condition problem), or any hybrid problem when some variables are fixed at some points, others at other points, and others at an initial condition, and so on. The characteristic variables can also be used (in a nontrivial way) in numerical simulation, to build "open boundary conditions", canceling all the entering waves but letting free all the outgoing ones.

The case with a Riemann invariant is particularly simple and interesting since each characteristic line in the (x, t) plane then propagates the information about one single scalar variable from the boundary or initial condition. If the system admits Riemann invariants, but is not linear, the characteristic velocities are not constant: they depend nonlinearly on all the other variables of the system. However, when the system is excited so that all the other variables remain constant everywhere, this makes all the other characteristic sets without effect. The propagation V_p then becomes constant for the considered characteristic set, which is then simply made of straight lines. The solution is then called a "simple wave". It must be recalled that the propagation velocity is anyway variable in general from one line to another, so that the lines are not parallel to each other and this leads to the deformation of any initial profile. The nonlinear steepening comes from this effect.

This will be illustrated in the next example: all points, in a linear sound wave propagating in the positive direction, are supposed to propagate with exactly the same velocity $u_0 + c_{s0}$ characteristic of the zero order medium (with $c_{s0}^2 = \gamma P_0/\rho_0$). The complete nonlinear calculation will show that each point of the corresponding simple wave actually propagates at the velocity $u + c_s$ where u and c_s are the actual nonlinear values of the flow velocity and sound speed, as they are nonlinearly perturbed. This leads to a modification of the wave profile, some gradients being nonlinearly steepened and others flattened.

6.1.1.3 An Example: The Simple Sound Wave

In an ideal neutral medium, the density and velocity evolution can be described by only two equations, which are the transport equations for density and momentum. These two equations involve three variables: u, ρ and P. Adding a polytropic closure with $\gamma \neq 1$, they can be expressed as functions of two variables only. The most convenient is to keep u and c_s defined by $c_s^2 = \gamma P/\rho$. This leads to replacing the derivatives by: $\partial_x \rho = \alpha\rho\partial_x(c_s)/c_s$ and $\partial_x P = \alpha\rho c_s \partial_x c_s$, with $\alpha = 2/(\gamma - 1)$, and similar relations for the derivatives with respect to time.

With this notation, this initial two-equation system can be written under the simple form:

$$\text{Momentum:}\quad D_t u + \alpha c_s \partial_x c_s = 0 \tag{6.4}$$

$$\text{Continuity:}\quad \alpha D_t c_s + c_s \partial_x u = 0\,. \tag{6.5}$$

Which can be put under the tensorial form:

$$\begin{pmatrix} 1 & 0 \\ 0 & \alpha \end{pmatrix} \cdot \partial_t \begin{pmatrix} u \\ c_s \end{pmatrix} + \begin{pmatrix} u & \alpha c_s \\ c_s & \alpha u \end{pmatrix} \cdot \partial_x \begin{pmatrix} u \\ c_s \end{pmatrix} = 0\,. \tag{6.6}$$

Multiplying on the left by the inverse of the left tensor, this can finally be put under the form Eq. (6.1):

$$\partial_t \begin{pmatrix} u \\ c_s \end{pmatrix} + \begin{pmatrix} u & \alpha c_s \\ c_s/\alpha & u \end{pmatrix} \cdot \partial_x \begin{pmatrix} u \\ c_s \end{pmatrix} = 0 \,. \tag{6.7}$$

Looking for the characteristic form of the system means diagonalizing the tensor. This leads to determining its two eigenvalues, which are $\lambda_\pm = u \pm c_s$, and its two eigenvectors, which are:

$$\begin{pmatrix} \alpha \\ \pm 1 \end{pmatrix} . \tag{6.8}$$

This allows getting the second step Eq. (6.2) of the general method:

$$\begin{pmatrix} 1 & -\alpha \\ 1 & \alpha \end{pmatrix} \cdot \partial_t \begin{pmatrix} u \\ c_s \end{pmatrix} + \begin{pmatrix} u - c_s & 0 \\ 0 & u + c_s \end{pmatrix} \cdot \begin{pmatrix} 1 & -\alpha \\ 1 & \alpha \end{pmatrix} \cdot \partial_x \begin{pmatrix} u \\ c_s \end{pmatrix} = 0 \,. \tag{6.9}$$

The two lines of this tensorial equality have indeed the general characteristic form that we are looking for, with the vector $\boldsymbol{L}_- = (1, -\alpha)$ for the first line and $\boldsymbol{L}_+ = (1, \alpha)$ for the second one. Does a Riemann invariant exist in this case? The answer is "yes", since these two vectors are just constants, so that they can be put inside the derivatives as well as outside. This means that the first characteristic propagates the information about the variable $u - \alpha c_s$ at the velocity $u - c_s$ while the second one propagates the information about the variable $u + \alpha c_s$ at the velocity $u + c_s$.

A simple wave can, therefore, be built with this system. It can simply be done by just imposing, for instance, $u - \alpha c_s = \text{cst}$ everywhere in an initial condition and letting the system evolve. The quantity $u - \alpha c_s$, which propagates at $u - c_s$, cannot bring any change to this variable, since it is already constant everywhere. The only changes will come from the propagation of $u + \alpha c_s$ at $u + c_s$. Along each of these characteristic lines, $u + \alpha c_s$, by definition of these lines, will be constant, which means that u and c_s will both be constant separately all along the lines (since $u - \alpha c_s$ is constant everywhere). The propagation of u and c_s will, therefore, occur at the constant velocity $u + c_s$: the characteristic lines are straight lines. This allows us to calculate easily the deformation of any initial profile. In Figure 6.1, it is shown how, from an initial sinusoidal density profile, the negative slopes steepen and the positive flatten. The profiles reached at the end of the calculation, which present several values of the density at the same position, are obviously unphysical. The reason is that the calculation has been done with a purely ideal medium. If adding, for instance, some viscosity, even extremely small, the corresponding term will unavoidably become predominant when the width of the layer will tend toward zero: the effect of viscosity scales as $k^2 \nu$ if k is the inverse of the characteristic scale. This important remark shows the way a shock wave can form from an initial condition. Its stationary width will be smaller for smaller viscosity. In the next section, we will even see that, in a stationary shock, the kinetic energy jump is independent of the

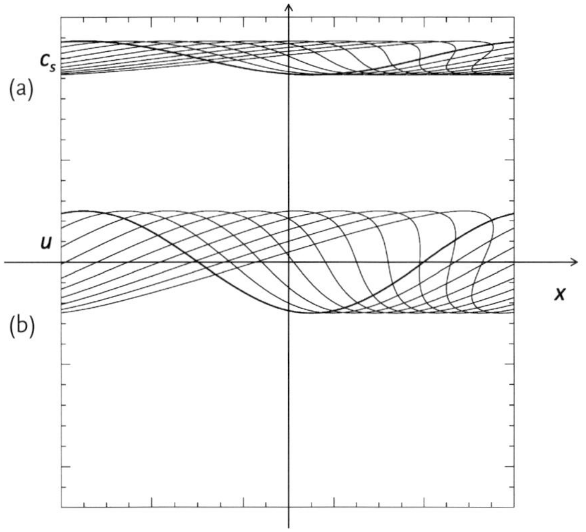

Figure 6.1 Propagation of a "simple" sound wave. The curves in (a) represent the sound speed c_s at different times, the ones in (b) the fluid velocity u. Each point of these curves just propagates at $u + c_s$, which leads to a steepening of a part of these profiles. The last curves obtained in this manner are unphysical (several values at the same x). When taking viscosity into account (which makes the ideal calculation of the simple wave invalid), the gradients stop steepening before reaching a vertical slope: the stationary solution obtained then is a shock wave.

viscosity value. This means that, this jump being fixed, the width has to decrease down to the point where the scale is sufficiently small to dissipate the corresponding energy with the given viscosity.

The same method of characteristics can be used to explain how shocks can be formed from spatial boundary conditions, for instance, when the velocity is imposed supersonic at one edge and null at the other one.

6.1.1.4 **Jumps**

Whatever the reason at the origin of a thin layer, nonlinear steepening or not, there are several general results that can be established for any such "discontinuity" (with quotes), at the condition that it has the two defining properties: 1D and stationary. These general results derive from the fact that some of the equations governing the system evolution from upstream to downstream are conservation laws. For each of these equations, the relation between upstream and downstream is, therefore independent of the profiles inside the layer. If some flux of particles enter the layer, the same flux must come out on the opposite side, and idem for the momentum and energy fluxes. In isotropic media without heat flux on both sides, the number of conservation laws is equal to the number of variables characterizing the downstream medium, so that this medium can be fully determined as a function of the upstream one, independently of the internal physics of the layer.

We know from MHD (although it is clearly much more general), that three vectorial conservation laws can be written, concerning the fluxes of density, momentum, and energy. Forgetting the ∂_t terms because of stationarity and replacing the ∇ terms by $\boldsymbol{n}\partial_x$ because of the 1D hypothesis ($\boldsymbol{n}$ being the unit vector along the normal), these equations can easily be rewritten from Chapter 2. Assuming an isotropic magnetized plasma and integrating them with respect to the normal coordinate x from upstream to downstream, they write:

$$\rho_2 u_{n2} = \rho_1 u_{n1} \tag{6.10}$$

$$\rho_2 u_{n2}^2 + P_2 + B_2^2/2\mu_0 = \rho_1 u_{n1}^2 + P_1 + B_1^2/2\mu_0 \tag{6.11}$$

$$\rho_2 u_{n2} \boldsymbol{u}_{t2} - B_n \boldsymbol{B}_{t2}/\mu_0 = \rho_1 u_{n1} \boldsymbol{u}_{t1} - B_n \boldsymbol{B}_{t1}/\mu_0 \tag{6.12}$$

$$\begin{aligned} &u_{n2}\left(5/2P_2 + \rho_2 u_2^2/2 + B_2^2/\mu_0\right) - B_n \boldsymbol{B}_2 \cdot \boldsymbol{u}_2/\mu_0 \\ &\quad = u_{n1}\left(5/2P_1 + \rho_1 u_1^2/2 + B_1^2/\mu_0\right) - B_n \boldsymbol{B}_1 \cdot \boldsymbol{u}_1/\mu_0 \,. \end{aligned} \tag{6.13}$$

It is useful to split the momentum equation, as done above, into two equations: one normal to the layer equation (6.11), and the other tangential equation (6.12). The divergence-free equation leads to $B_{n2} = B_{n1}$. It is the reason why no subscript is indicated for B_n. Note also that, in the last equation (energy flux), the Poynting flux $\boldsymbol{E} \times \boldsymbol{B}/\mu_0$ has been expressed as a function of $\boldsymbol{u}$ and $\boldsymbol{B}$ by using an ideal Ohm's law. It does not mean that this law is supposed to hold inside the layer, but only that it holds on both sides. Similarly, no heat flux has been taken into account, which also assumes that this heat flux is zero on both sides. No nonideal terms, such as viscosity in a collisional plasma or kinetic effects in a collisionless one, appear either for the same reason: even if they do allow momentum or energy exchanges inside the layer, they cannot change the net balance: in a stationary shock, all the energy/momentum flux entering the boundary must come out without loss. These terms are of course null on both sides where all gradients can be neglected.

The above system is still not complete since it contains only the evolution of the plasma as a function of the electromagnetic field, but not the reaction of the electromagnetic field as a function of the plasma. As in MHD, we, therefore, have to add the Ohm/Faraday law, which is written, with the same rules:

$$B_n \boldsymbol{u}_{t2} - u_{n2} \boldsymbol{B}_{t2} = B_n \boldsymbol{u}_{t1} - u_{n1} \boldsymbol{B}_{t1} \,. \tag{6.14}$$

An important remark then allows simplifying the problem. A change of reference frame in a tangential direction has no effect on the physics since the problem is completely invariant in these directions. Nevertheless, it changes the value of $\boldsymbol{u}_{t1}$ and $\boldsymbol{u}_{t2}$ in the above equation. Choosing the reference frame where $B_n \boldsymbol{u}_{t1} = u_{n1} \boldsymbol{B}_{t1}$ is, therefore, a clever choice since it makes $\boldsymbol{u}_{t2}$ and $\boldsymbol{B}_{t1}$ collinear as well. Noting that the coefficients between $\boldsymbol{u}_t$ and $\boldsymbol{B}_t$ (on both sides) are simply B_n and u_n, this means that the full 3D vectors $\boldsymbol{u}$ and $\boldsymbol{B}$ are collinear and that, in the chosen frame, the electric field is zero. This particular frame is called the "de Hoffmann–Teller" frame.

Considering the system in this de Hoffmann–Teller frame allows us to replace everywhere $\boldsymbol{u}_t$ by $\boldsymbol{B}_t$ so that it is written under the simpler form:

$$\rho_2 u_{n2} = \rho_1 u_{n1} \tag{6.15}$$

$$\rho_2 u_{n2}^2 + P_2 + \frac{B_2^2}{2\mu_0} = \rho_1 u_{n1}^2 + P_1 + \frac{B_1^2}{2\mu_0} \tag{6.16}$$

$$\left[\frac{u_{n2}^2}{B_n^2/(\mu_0\rho_2)} - 1\right] \boldsymbol{B}_{t2} = \left[\frac{u_{n1}^2}{B_n^2/(\mu_0\rho_1)} - 1\right] \boldsymbol{B}_{t1} \tag{6.17}$$

$$u_{n2}\left(\frac{5}{2}P_2 + \rho_2\frac{u_2^2}{2}\right) = u_{n1}\left(\frac{5}{2}P_1 + \rho_1\frac{u_1^2}{2}\right) . \tag{6.18}$$

It can be observed that the electromagnetic flux has disappeared from the energy flux equation in this frame, corresponding to the fact that the Poynting flux is zero in the frame where $\boldsymbol{E} = 0$.

The above system rules the jumps of all the variables across the boundary. It is called the "generalized Rankine–Hugoniot" system, the ordinary Rankine–Hugoniot system, commonly used in gas dynamics, being its reduction to the neutral systems or the unmagnetized plasmas (see, for instance, [28, 29]). It rules the jumps of all the variables across the discontinuities.

This system involves as many equations as unknowns (calling unknowns the downstream values when the upstream ones are known). One trivial solution is $X_2 = X_1$ for all variables X. The question is now: what are the other (nontrivial) solutions, if any? It is worth noting the similarity of this approach with the search of propagation modes in the linear case. We will indeed see that the nonlinear solutions of the discontinuity problem are strongly reminiscent of the linear solutions of the wave problem. One apparent difference just concerns the frame used for the calculation: linear waves are usually calculated in the frame where the structure moves with respect to the plasma, considered as steady when unperturbed, while the discontinuity calculation considers the structure as steady and calculates the motion of the plasma with respect to it. But this very superficial difference should not blur the numerous similarities.

6.1.2
Shocks and Other Discontinuities in a Magnetized Plasma

Considering first Eq. (6.17), it is clear that two kinds of discontinuities are possible:

1. The *rotational discontinuity* for which $u_n^2 = B_n^2/(\mu_0\rho)$ everywhere. Equation (6.17) does not bring then any information about $\boldsymbol{B}_{t2}$ when B_{t1} is given. This solution is nothing but an Alfvén wave: the theory of this wave indeed shows that the nonlinear solution is identical to the linear one. It means in particular that a rotational discontinuity cannot be formed, at least in this simple version, by any nonlinear steepening. From the very definition of this solution, it is clear that $u_n = V_{An}$ is constant since B_n and ρu_n are constant. From

the continuity equation, it can be concluded that, a rotational discontinuity does not involve any density jump. From the two remaining equations, it can checked that P and B^2 have also to be conserved separately since $P + B^2/(2\mu_0)$ and $5/2P + B^2/(2\mu_0)$ must be conserved at the same time. In summary, in a rotational discontinuity, as in any Alfvén wave, all the plasma parameters are conserved, except the direction of $\boldsymbol{B}_\mathrm{t}$, which rotates without changing its magnitude, and $\boldsymbol{u}$, which is proportional to $\boldsymbol{B}$ in the de Hoffman–Teller frame).

2. The *shocks* for which $u_n^2 \neq B_n^2/(\mu_0\rho)$. Equation (6.17) then immediately shows that B_{t2} and B_{t1} are collinear, which means that, contrary to the previous solution, the shocks show no rotation of $\boldsymbol{B}$ in the tangential plane. $\boldsymbol{B}_\mathrm{t}$ remains parallel to its upstream direction, just varying by its modulus. This is the so-called coplanarity property of the shock waves. There are several solutions of the shock type. We will see that they can correspond to an increase, a decrease, or a reversal of the tangential magnetic field.

It is important understanding that the Rankine–Hugoniot relations are necessary conditions for a discontinuity to exist. Whatever the internal physics of the layer, these conservation equations have to be verified (in the limits of the hypotheses done, in particular the isotropy). But they are not sufficient conditions: when two states have been found with the same fluxes, it is not guaranteed a priori that the internal physics, which involves nonideal effects, can always allow us to go from one state to the other. The first – obvious – restriction concerns the sense of the flow across the discontinuity: if an ordinary shock, for instance, allows the transition from a supersonic state to a subsonic one, the reverse solution cannot physically exist: the internal dissipation can oppose a steepening, but not a flattening of the gradient.

Since the above Rankine–Hugoniot relations provide as many equations as downstream variables, it can be solved. Figure 6.2 shows the result of this calculation for the shock solutions by displaying the downstream value u_{n2} as a function of the upstream one u_{n1}. It can be seen that, for given upstream conditions, there are generally two possible solutions respecting the same fluxes: u_{n1} (trivial) and another one (shock). There are even three solutions on a limited range of u_{n1}. It must be noticed that the different solutions are isolated: there is no other solution between them, meaning that one cannot go from one to the other by ideal solutions. This reinforces an important point already emphasized: the nonideal phenomena inside the layer are essential to get a stationary solution in spite of the nonlinear effects. The terms that have been ignored, for instance, the viscosity, make locally a significant departure from the conservation laws, although they cannot change the net balance when integrated over the whole layer.

The linear solutions stand in the vicinity of the crossings of the two curves $u_{n2} \neq u_{n1}$ and $u_{n2} = u_{n1}$ (weak shocks). The letters S, I, and F indicate the positions of the three MHD solutions: Slow magnetosonic, Shear Alfvén (or Intermediate), and Fast magnetosonic. When departing from these solutions, the shocks have been given the same names, depending on the value of u_{n1} with respect to these characteristic speeds: Slow Shock for $V_{\mathrm{F}}x < u_{n1} < V_\mathrm{I}$, Intermediate Shock

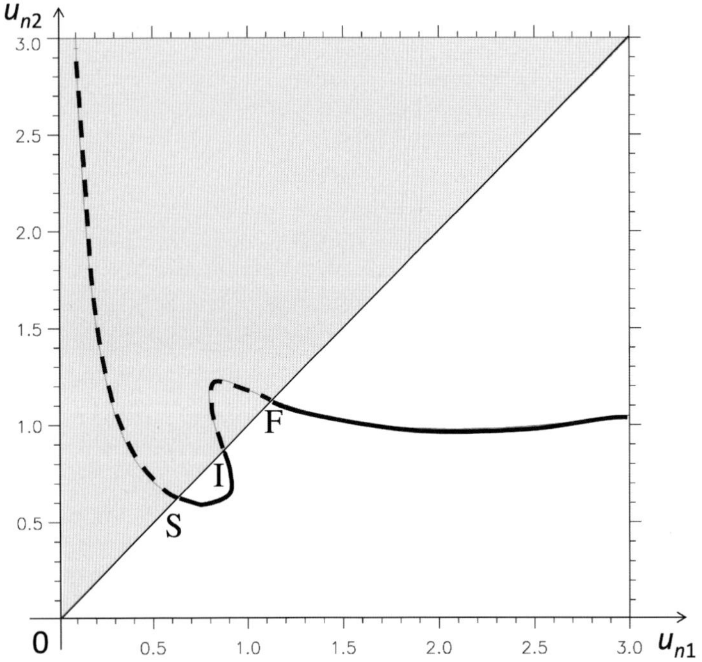

Figure 6.2 Solutions $u_{n2}(u_{n1})$ respecting the Rankine–Hugoniot conservation laws. The normal velocities are normalized to V_{A1}. The calculation has been done here for $\beta_1 = 0.8$ and a propagation angle between $\boldsymbol{B}_1$ and the normal direction equal to 30°. The dashed line indicates the solutions that are in the $u_{n2} > u_{n1}$ region: even if they respect the Rankine–Hugoniot relations, they are not realizable. The letters S, I, and F indicate the vicinity of the linear solutions: Slow, Intermediate (Alfvén), and Fast.

for $V_I < u_{n1} < V_F$, Fast Shock for $u_{n1} > V_F$. The presence of the Shear Alfvén solution in this shock calculation is surprising since it corresponds to a rotational discontinuity and not to a coplanar one. It is indeed a kind of singular point: even if $u_{n2} \approx u_{n1}$, this vicinity does not actually correspond to a linear solution: the magnitude B_t^2 does not change, but the vector $\boldsymbol{B}_t$ reverses. It is, therefore, a finite amplitude Alfvén solution, with a rotation of exactly π. The intermediate shocks close to this finite amplitude Alfvén wave are coplanar and they are, therefore, not classical Alfvén waves. The intermediate shocks are generally considered as unstable solutions. However, it must be noticed that two different jumps are possible in this region and that the two types of transitions have to be investigated separately: the largest jump appears similar to the slow type, the decrease just being larger than the incident value, so leading to a reversal. It must be noted also that, for increasing values of u_{n1}, one goes from slow to fast shocks, but that there is a gap between the two types of solutions: there is no possible shock for incident velocities in this limited range, the system necessarily introducing other types of transition layers (a priori nonstationary).

From Eq. (6.17), it is easy checking that the fast shocks correspond to an increase in the magnitude of B_t, the slow shocks to a decrease (without sign change), and the intermediate shocks to a sign change.

6.1.3
The Unmagnetized Shock Wave

The preceding results can of course be reduced to the usual case of a sound shock wave, as it is well known in neutral gas and which is valid also for unmagnetized plasmas. Let us just recall here one important result for this case, which will be useful in Chapter 8. The original Rankine–Hugoniot relations can easily be derived from the generalized ones, just by taking $\boldsymbol{B} = 0$ in Eqs. (6.10)–(6.13). In this case, the relation between u_{n2} and u_{n1} is much simpler and can be put under the following analytical form:

$$\frac{u_{n2}}{u_{n1}} = \frac{\rho_1}{\rho_2} = \frac{1}{4}\left(1 + 3/\mathcal{M}_1^2\right). \tag{6.19}$$

Where $\mathcal{M}_1 = u_1/c_{s1}$ is the upstream sonic Mach number. One can notice that the "compression ratio" $r = \rho_2/\rho_1$ goes from 1 for an infinitely weak shock ($\mathcal{M}_1 = 1$) to 4 for an infinitely strong shock ($\mathcal{M}$ infinite). For a polytropic closure with $\gamma \neq 5/3$ (gas with different forms of internal energies), the same results is written as:

$$r = \frac{\gamma + 1}{\gamma - 1 + 2/\mathcal{M}_1^2}\,. \tag{6.20}$$

6.1.4
A Particular Case: The Tangential Discontinuity

The particular case of perpendicular discontinuities ($B_n = 0$) is important and deserves a special comment. In this case, it can be checked that all discontinuities except the fast shock correspond to $u_{n1} = u_{n2} = 0$, which means that none of them involves plasma flow across the layer. These different discontinuities are, therefore, all immobile relative to each other so that they can be superposed or juxtaposed in any way without breaking the stationarity of the structure. This means that there is a degeneracy between the rotational discontinuity and the compressional ones, both kinds of variations, rotations and compressions, being allowed at the same time. It can be checked directly on the system of equations that all are verified when $B_n = 0$ and $u_n = 0$, at the only condition that $P + B^2/2\mu_0 = \text{cst}$. This degenerate solution is called a "tangential discontinuity". It is to be noted that this solution is based on the property that B_n is strictly zero; for an infinitesimal B_n the different kinds of solutions, rotational and compressional, will slowly separate from each other, so that the global set will not be an exactly stationary solution.

It must be recalled also that the remaining jump condition $P + B^2/2\mu_0 = \text{cst}$ is only a necessary condition: as in all the other discontinuities, the internal physics does not actually allow all the profiles verifying this condition. The internal model, for instance, can be a kinetic one, which is the case in particular when the layer is collisionless and when its width is of the order of the particle Larmor radii. Such models have been done for investigating the basic tearing mode instability (see Chapter 7). Modeling the internal physics brings other constraints on the possible profiles. There are presently few kinetic equilibria for tangential layers. The most famous (and the simplest) one is the "Harris sheet" [30], which is a quite particular one since it corresponds to a current layer surrounded by vacuum. This layer is supposed symmetrical, with a coplanar magnetic field reversal and no electric field. The temperatures T_{i} and T_{e} are supposed constant as well as the fluid velocities u_{i} and u_{e}. It is important to note that, in a Harris sheet, the localization of the current density $j_z(y) = n(y)e\delta u_z$ is intrinsically due to the localization of the density, since the difference $\delta u_z = \boldsymbol{u}_{z\mathrm{i}} - \boldsymbol{u}_{z\mathrm{e}}$ is supposed constant.

From the two-fluid point of view, the Harris sheet model is defined as follows (the discontinuity normal is supposed to be in the y direction, and the magnetic field in the x direction):

$$n = n_0/\cosh^2(y/L) \tag{6.21}$$

$$B_x = B_0 \tanh(y/L) \quad \text{with} \quad B_0^2/2\mu_0 = n_0(T_{\mathrm{i}} + T_{\mathrm{e}}) \tag{6.22}$$

$$j_z = -(1/\mu_0 L)\, B_0/\cosh^2(y/L) \tag{6.23}$$

$$A_z = L B_0 \ln\cosh(y/L) \tag{6.24}$$

$$P_{\mathrm{i}} = n_0 T_{\mathrm{i}}/\cosh^2(y/L) \tag{6.25}$$

$$P_{\mathrm{e}} = n_0 T_{\mathrm{e}}/\cosh^2(y/L) \tag{6.26}$$

$$\text{with} \quad u_{z\mathrm{i}} = -2V_{\mathrm{thi}}\, R_{\mathrm{i}}/L \quad \text{and} \quad u_{z\mathrm{e}} = 2V_{\mathrm{the}}\, R_{\mathrm{e}}/L. \tag{6.27}$$

R_{i} and R_{e} are the Larmor radii of the two populations. It is easy checking that this equilibrium does satisfy the force balance $P + B^2/2\mu_0 = \mathrm{cst}$, and that the constant fluid velocities agree with the current profile $j_z(y) = n(y)e\delta u_z$.

Let us now show that this equilibrium is also a kinetic one, valid in a collisionless plasma, if one assumes drifted Maxwellian distributions everywhere.

A kinetic equilibrium must be a stationary solution of the Vlasov equation for each population. Writing the Vlasov equation under the form $D_t(f) = 0$, where D_t is a convective derivative in phase space along a particle trajectory, there is a simple way of checking that the above solution is indeed a kinetic one: it is to notice that the distribution function can be written under the form $f(\boldsymbol{v}, y, t) = g(E, p_z)$ where E and p_z are two invariants of a particle trajectory, the energy $m(v_x^2 + v_y^2 + v_z^2)/2$ and the generalized momentum $p_z = mv_z + qA_z$. This means that f depends on y only implicitly through the dependence $A_z(y)$, and that it does not depend on t at all. It is, therefore, a kinetic equilibrium. This can be readily checked. For the ions, for instance:

$$\begin{aligned} f_i(\boldsymbol{v}) &= \frac{n}{\sqrt{2\pi}\,V_{\text{thi}}} e^{-\dfrac{v_x^2+v_y^2+(v_z-u_{zi})^2}{2V_{\text{thi}}^2}} \\ &= \frac{n}{\sqrt{2\pi}\,V_{\text{th}}} e^{-\dfrac{E+m_i u_{zi}^2/2-u_{zi}p_z+u_{zi}eA_z}{m_i V_{\text{thi}}^2}} \\ &= \frac{K}{\cosh^2(y/L)} e^{-\dfrac{E-u_{zi}p_z}{m_i V_{\text{thi}}^2}} e^{-\dfrac{u_{zi}eA_z}{m_i V_{\text{thi}}^2}} . \end{aligned} \tag{6.28}$$

The dependence in $1/\cosh^2(y/L)$ coming from the density dependence can be seen to be balanced by the dependence in $\cosh^2(y/L)$ coming from A_z (see expression (6.24)). This demonstrates the result.

Many attempts, more or less conclusive, have been made to generalize the Harris model and make it more realistic for modeling physical objects. These generalized models intend in particular to include a nonnull neighboring density, or a nonnull electric field, or a magnetic rotation, or a nonnull B_n, and so on. All the methods to find these new models have been based, as the original one, on the use of the trajectory invariants of the particles. Being able to put the distribution functions under the form of a single function of the invariants is – of course – a sufficient condition for obtaining a stationary solution, but it is to be noted that it is not necessary. The Vlasov equation does impose that the distribution function has to be constant along any particle trajectory, but not beyond. If two points are separated, perpendicularly to the magnetic field, by more than a Larmor radius, there is no particle trajectory joining them: they can, therefore, have the same invariants but correspond to different values of f. This is quite important, for instance, in the case of negligible Larmor radii, where "MHD-like" results are to be retrieved. This brings the necessity of considering multivalued functions, at least bivalued for this case of a discontinuity, corresponding to the two sides of the layer (see [31]). The following result deserves in particular to be kept in mind: remaining in the case of a symmetric current density profile with no electric field and no magnetic rotation, plasma boundaries layers with different densities and temperatures on both sides do exist. However, because of the symmetry of $A_z(y)$, it can be checked that none of them can be caught in the frame of single valued functions of the invariants.

The "contact discontinuity" is another particular solution of the initial system that has to be added to make the list exhaustive, although it is of much less practical importance. It is a solution in which $B_n \neq 0$ but $u_n = 0$. It theoretically allows a new solution where only the density ρ varies, all the other variables, $P = nT$, $\boldsymbol{u}$ and $\boldsymbol{B}$, being constant. Even if this is a solution of the system of conservation equations, it can hardly correspond to physical solutions because of the internal physics it would need: (i) no nonlinear steepening can create such a discontinuity, and (ii) if taking it as an initial condition, the kinetic effects as well as the collisional ones would rapidly diffuse the density gradient existing along the field lines.

6.1.5
Example: The Terrestrial Bow Shock, the Foreshocks

The solar wind impinges the terrestrial environment with a speed relative to it which is larger than all the MHD propagation speeds, up to 10 times the fast magnetosonic speed. It is, therefore, the cause of a "bow shock", in front of the terrestrial magnetic obstacle, which is a fast shock. This shock can be called collisionless since the mean free path in the solar wind is on the order of the Sun–Earth distance, that is, several orders of magnitude larger than the layer width. It stands at about 15 R_E (Earth radii) from the Earth, which is about 5 R_E upstream of the magnetopause obstacle. The solar wind region downstream of the bow shock and upstream of the magnetopause is called the magnetosheath (see the general presentation of the magnetospheres in Chapter 1). The magnetic field amplitude increases by a factor about 3 between solar wind and magnetosheath.

The great advantage of the terrestrial bow shock is that it is a permanent structure, within reach of spacecraft in situ measurements. It, therefore, has been explored in great detail since the beginning of the space physics era. Furthermore, due to its curvature and to the variations of the solar wind magnetic field, all the possible directions of the incident magnetic field with respect to the shock normal can be observed. This has allowed us to test and refine all the theories concerning the collisionless shocks.

We will present here only two of the properties that can be shown experimentally and which are important clues to understanding the collisionless physics involved: the difference between quasi-parallel and quasi-perpendicular shocks, and the existence of the upstream foreshocks.

The difference between *quasi-parallel and quasi-perpendicular* shocks is exemplary of the importance of the microphysics phenomena internal to the shock layer. In a collisional shock, nonideal phenomena such as viscosity are able to stationarize the nonlinear steepening, determining the internal profiles. In a collisionless one, the nonideal effects are kinetic. They are partly able to play the same role, but their efficiency is quite different in the directions perpendicular to the magnetic field (where the individual particles are naturally confined by their finite Larmor radius) and in the parallel direction where there is no such effect and where a collective electric field "stationary" at some scale cannot exist without microturbulence and smaller scales. This effect makes it much more difficult to maintain a stationary thin shock layer when the magnetic field is close to the normal direction than when it is almost perpendicular. This is illustrated in Figure 6.3. The quasi-perpendicular shocks appear as clean thin boundaries, close to stationary (the measurements of the four Cluster spacecraft can be compared), with a width of less than 100 km, while the quasi-parallel ones are extremely extended (on the order, typically, of 50 000 km), nonstationary, and involving large magnetic turbulence.

The terrestrial electron *foreshock* is a region starting where the shock is perpendicular (magnetic line tangential to the shock surface) and extending upstream of the bow shock, with a boundary slightly more inclined than the field lines. In this region, the electron distribution function shows back streaming electrons along the

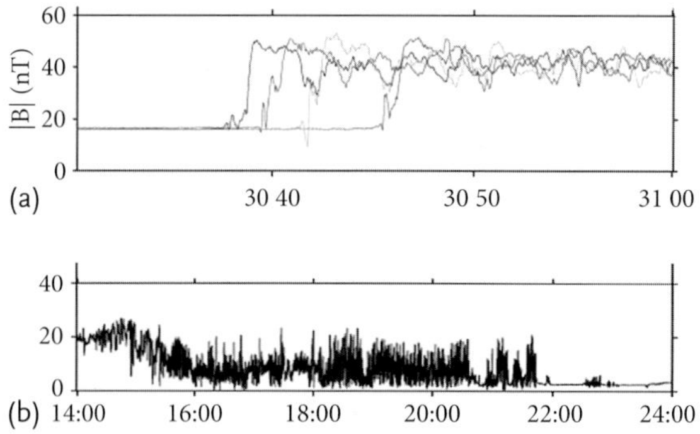

Figure 6.3 Two examples of shock crossings by the Cluster spacecraft. (a) shows the four measurements of the magnetic field magnitude at the four spacecraft in the case of a quasi-perpendicular shock (25 Dec. 2000). The total duration of the interval is 30 s, and the spacecraft goes from solar wind to magnetosheath. The ramp crossings last less than 1 s. (b) shows an example of crossing in a quasi-parallel region (2 Feb. 2001). The spacecraft goes from magnetosheath to solar wind. The total duration is 10 h and the magnetic decrease spreads over more than 30 mn. From [32, 33].

field lines, and an electrostatic turbulence is observed (mainly close to the boundary), due to this particular distribution function. A similar phenomenon exists for the ions and gives rise to the ion foreshock, with a boundary noticeably more inclined because of a smaller parallel velocity for the ion beams (the perpendicular velocity, fixed by the solar wind speed, being the same as the electron one). These effects are clear and geometrically localized thanks to the curvature of the bow shock and cannot be modeled by 1D shock simulations. They appear quite unusual in comparison with the usual collisional media since no perturbations can be observed in these media upstream of a shock.

The existence of these foreshocks is a consequence of the kinetic microphysics internal to the shock layer, but it is also quite informative about the necessity of going beyond the simplified fluid description for the incident medium. The internal microphysics makes possible the reflection/acceleration of some particles in the shock layer, for all the "connected" field lines (Figure 6.4). These reflected particles then stream back along the field lines as "beams". Combined with the field line velocity $\boldsymbol{E} \times \boldsymbol{B}/B^2$, this explains the directions of the foreshock boundaries. The existence of such particular distribution functions is of course a consequence of the collisionless character of the medium: they could not occur in a collisional one. However, when assuming that the media on both sides of the shock is isotropic and can be characterized by only three parameters n, $\boldsymbol{u}$, and P, all these phenomena have been ignored in the above description. In this context, which would be valid for collisional MHD (the distributions being Maxwellian), the fast mode is the fastest way of transporting information, and no perturbation can occur upstream of the shock. In the real collisionless case, on the contrary, the foreshocks show that some information is transported with the particles at speeds faster than the fast mode, so that perturbations can be observed upstream. It must be noted,

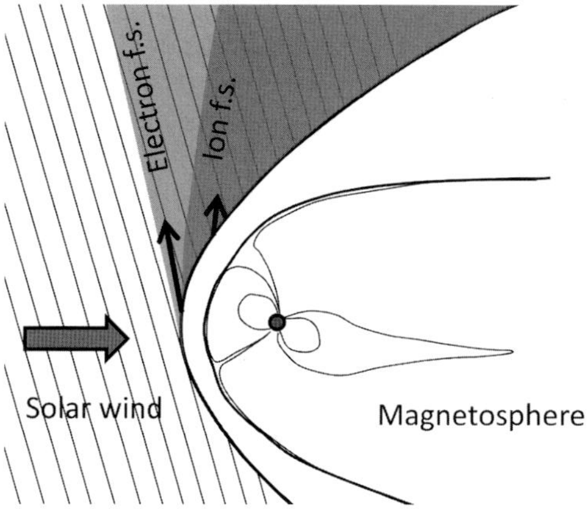

Figure 6.4 Electron and ion foreshocks. These regions (marked f.s.) are characterized by a turbulence likely caused by beam–plasma instabilities, for both species. The involved beams are believed to be due to particle reflection in the shock layer. The upstream boundaries of the two regions meet the shock at the first connected field line, that is, where the shock is "perpendicular" (shock normal perpendicular to the incident magnetic field).

however, that taking these kinetic differences would not change much the Rankine–Hugoniot relations, which are robust conservation laws, whatever the form of the distribution functions.

6.2 Turbulence (Mainly MHD)

In a turbulent flow such as found in a river, represented in drawings by the famous Renaissance artist Leonardo da Vinci, one sees a few large eddies mixed with numerous eddies of smaller sizes, the distribution of eddy sizes ranging continuously between large and very small. In such a hydrodynamic turbulence, small eddies form rapidly enough, starting with large-scale ones. A consequence of the coexistence of large and small eddies is twofold: (i) different parts of the flow become intimately mixed with one another ("spatial mixing"), (ii) the kinetic energy of the flow is rapidly lost, due to the rapid dissipation into heat of the gradients associated with the smallest eddies.

The development of small eddies starting from a few large eddies is commonly called a "turbulent cascade". It is triggered by the nonlinear couplings which progressively distribute energy to smaller and smaller scales, conserving the global kinetic energy, at least during the first phase of the process. In the standard case (called the "direct cascade"), energy is mostly redistributed from the largest to the smallest scales, in a process which resembles water falling down on successive steps of decreasing sizes, hence the name "cascade". The two mixing processes, in wavenumber and real space, respectively, are two main complementary aspects of turbulence.

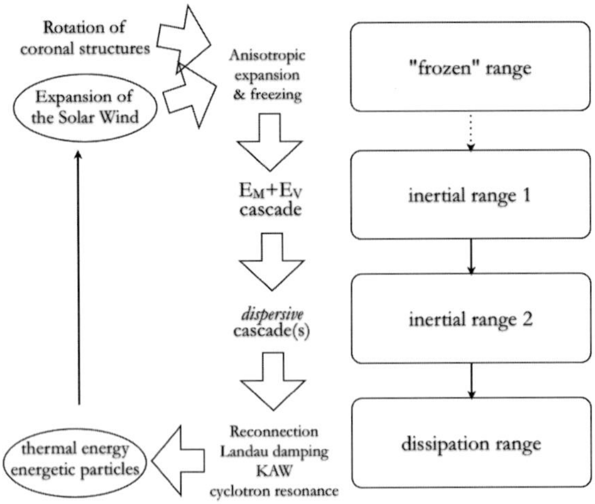

Figure 6.5 Schematics of turbulence in the solar wind case. The "inertial range 1" deals with a cascade of the kinetic + magnetic energies: it is the one considered here in detail. The following "inertial range 2" deals with cascade(s) in which the definition of the cascading quantities relies heavily on the increased importance at smaller scales of dispersive terms in the equations.

Although more difficult to visualize in turbulent plasmas than in neutral fluids, similar phenomena arise in plasmas. In principle, all these phenomena should be understood by using the Vlasov equation. However, due to their simpler structure, knowledge of turbulence based on hydrodynamic and MHD equations (and even more particularly, on incompressible versions of these equations) is more advanced than that of turbulence based on the Vlasov equations.

For these reasons, we will focus on hydrodynamic and incompressible MHD turbulence (the latter in more details). To make clearer what is gained and what is lost with such a choice, let us consider for a moment the example of the solar wind turbulence, of which Figure 6.5 gives a schematic representation. The different steps that appear from top to bottom of Figure 6.5 are meant to correspond to different parts of the total spectral range observed by spacecraft missions in the solar wind, shown in Figure 6.6. One may identify in the magnetic energy spectrum an "active" turbulent range, which spans about four or five decades of frequencies (or wavenumbers). This is actually made of two subranges, each showing a different power law scaling, the first one exhibiting the well known Kolmogorov scaling $k^{-5/3}$ law, and the second one exhibiting here a scaling close to $k^{-2.8}$. Note that this second scaling law is presently the object of some debate (see in particular [34, 35].

Apart from the second scaling law, which is most probably a consequence of dispersive terms becoming important at these smaller scales, two ranges deserve some comments: the "frozen" range and the dissipative range. The frozen range plays the role of an energy reservoir which is not directly accessible, as at very large scales, the damping terms due to the wind expansion dominate nonlinear coupling, which render them inefficient, and suppresses the cascade. The correspond-

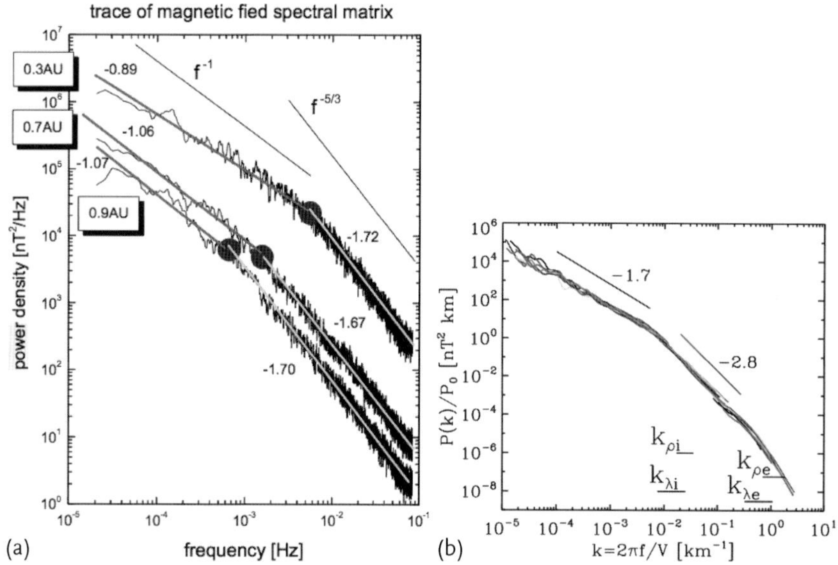

Figure 6.6 Magnetic energy spectra in the Solar Wind. (a) Large-scale (MHD) range, at 0.3, 0.7 and 0.9 astronomical units (AU), showing the break between the "frozen" f^{-1} range and the $f^{-5/3}$ inertial range (from [36]). (b) Higher frequencies at 1 AU, showing the break between the $k^{-5/3}$ and the $k^{-2.8}$ range. Abscissa is the radial wavenumber k, which is derived from frequency f and wind speed U, using the Taylor hypothesis $2\pi f \simeq kU$. The figure also shows wavenumbers associated with characteristic plasma length scales (inertial ion and electron lengths, ion and electron Larmor radii). From [34].

ing observed k^{-1} range is thus probably a remnant of coronal turbulence, with the spectral break with the $k^{-5/3}$ range changing with distance. The real consequence of this large-scale freezing on the smaller scale cascade is actually still unknown: indeed, the freezing effect is anisotropic, being mainly transverse to the radial. Finally, the cascade ends in the same way as it started: there, the nonlinear terms are dominated by damping terms, which transform energy into heat so rapidly that the cascade stops. Some of the effects that might be at work at this step are listed in Figure 6.5; among them, reconnection is considered in detail in Section 7.2, Landau damping in Section 5.6, and the cyclotron resonance (rather superficially) in Section 5.6.5. The term KAW denotes the kinetic Alfvén waves, which are a continuation of the Alfvén mode at large k for quasi-perpendicular propagation, where the electrostatic effects become important, and where some of the simplifying assumptions of MHD (negligible ion Larmor radius and/or negligible electron mass) have to be abandoned.

We adopt here (partially) the usual philosophy that, once there is a large enough inertial range where dissipation and forcing are unimportant compared to nonlinear coupling, then turbulence should develop in a universal way. Indeed, we do believe that the details of dissipation are unimportant, that is, have no effects on the development of turbulence. So, turbulence may be (and will be here) studied by using resistive and viscous dissipation, as is appropriated for collisional fluids.

However, we are convinced (and we hope to convince the reader) that even plain MHD turbulence is not universal, that is, may depend largely on large-scale configurations, and/or initial conditions of forcing.

In the following we shall introduce some basic concepts valid for the hydrodynamic case, and then focus on the MHD case, most of the time relevant for large-scale turbulence in astrophysical plasmas.

6.2.1
Hydrodynamics: Equations, Shocks

The equation of the movement is the Navier–Stokes equation:

$$\partial_t \boldsymbol{u} + \boldsymbol{u} \cdot \nabla \boldsymbol{u} + \nabla P/\rho = \nu \nabla^2 \boldsymbol{u} \tag{6.29}$$

where the viscosity ν does not need to be constant in the general compressible case. This equation has to be completed in general with two equations for the pressure P and density ρ. In the particular case of zero heat flux, these equations read

$$\partial_t P + \boldsymbol{u} \cdot \nabla P + \gamma P \nabla \cdot \boldsymbol{u} = 0 \tag{6.30}$$

$$\partial_t \rho + \boldsymbol{u} \cdot \nabla \rho + \rho \nabla \cdot \boldsymbol{u} = 0\,. \tag{6.31}$$

Shock formation The simplest example of a direct energy cascade is not the eddy cascade evoked above, but the shock formation in compressible fluids, already considered in Section 6.1.1.3, which we revisit now in a slightly broader context, including viscosity. Doing so is instructive because, although the shock formation process lacks the random and chaotic character of standard turbulence, it shares with it a simple form of direct transfer of kinetic energy from large to small scales.

We consider a planar shock formation, in which all quantities depend only on one coordinate x. We consider only the x component of the velocity. In such a flow, wave fronts are all parallel, hence there are no eddies (that would require at least curved wave fronts). Because of this we cannot speak of "small" or "large" eddies in the shock formation problem. We will speak instead of large and small wavenumbers k and associated Fourier modes $\exp ikx$. The "cascade" here consists of a transfer of energy between modes with small wavenumbers to modes with large wavenumbers. Let us thus begin with a single mode k_0, namely a velocity fluctuation of the type

$$u = U_0 \cos k_0 x\,. \tag{6.32}$$

We also have to specify the pressure and density. To simplify things, we consider $\delta\rho/\rho_0 = u/c$ and $\delta P/P_0 = \gamma\,\delta\rho/\rho_0$. Linear analysis shows that in this case the perturbation will progress in the rightward direction. We assume also that the Mach number is substantially smaller than unity.

The evolution of the initial velocity profile (Eq. (6.32)) is shown in Figure 6.7a from time $t = 0$ up to time $t = 2$. The profile is seen to propagate from left to right

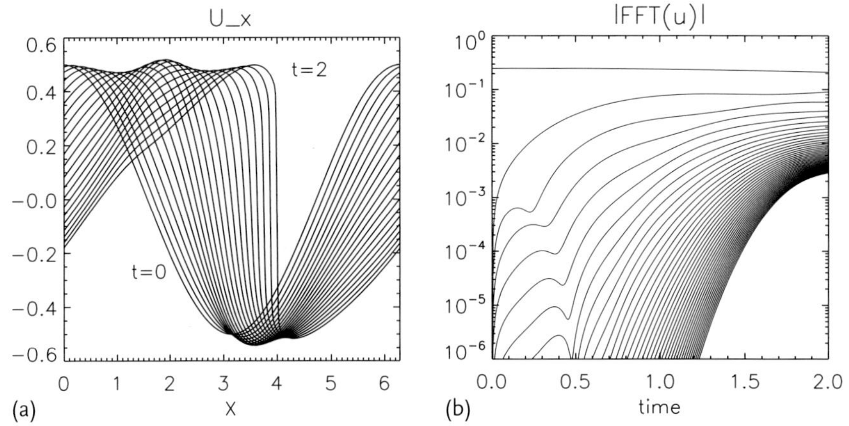

Figure 6.7 Evolution of a plane progressive (rightward propagating) sound wave during one nonlinear time. (a) Successive snapshots of velocity profiles up to one nonlinear time $t = t_{NL} = 2$. (b) The corresponding growth vs. time of successive harmonics of $k_0 = 1$. The quasi-horizontal line is harmonic $k_0 = 1$, and successive harmonics at a fixed time have decreasing amplitudes.

(due to the particular choice of initial condition) and to steepen as it propagates. The last profile, at $t = 2$, shows a quasi-discontinuity.

There are two ways to explain the steepening. The intuitive way amounts to saying that the peak of the wave (with higher velocity) catches the trough of the wave (with lower velocity), a simple effect of the advection term in the momentum equation. If this is true, the *steepening time* should be given by the distance between peak and trough ($l \simeq 1/k_0$), divided by the velocity difference U_0:

$$t_{NL} \simeq 1/k_0 U_0 . \tag{6.33}$$

This is also called the *nonlinear time*. This estimation is indeed correct: as seen in Figure 6.7a, the steepening time is $t \simeq 2$, which is equal to $t_{NL} = 2$ as predicted by Eq. (6.33).

Another way to understand steepening is to examine how the advection term adds harmonics to the initial wavenumber $k_0 = 1$. To do that, we neglect the pressure term in Eq. (6.29), thus considering the (zero pressure) Burgers' equation:

$$\partial_t u + u \partial_x u = \nu \partial_{xx} u . \tag{6.34}$$

At time $t = 0$, Eq. (6.34) becomes, after replacing in the advection term u by its initial value $U_0 \cos kx$ (neglecting viscosity):

$$\partial_t u = -k_0 U_0^2/2 \sin 2k_0 x . \tag{6.35}$$

The harmonic $2k_0$ thus grows linearly with time, at least for very short times. However, as the harmonic $2k_0$ grows, one can iterate the argument, introducing the harmonic $2k_0$ in the nonlinear term, thus showing that harmonics $4k_0$ grow as well, and so on.

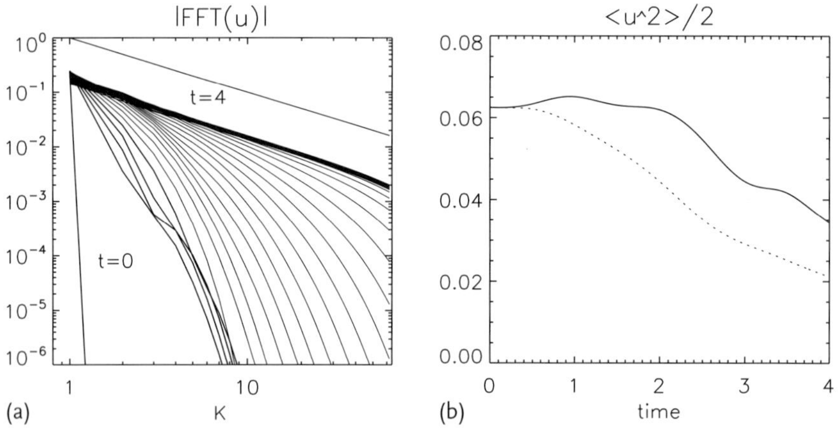

Figure 6.8 (a) Successive modal velocity spectra $|\hat{u}(k)|$ vs. wavenumber k; (b) time evolution of average kinetic energy $\overline{u^2}/2$ (solid thick line) and of energy in wavenumber $k = k_0 = 1$ (dotted line).

The full description in Fourier space is given by taking the Fourier transform of Eq. (6.34), which gives for the Fourier amplitudes $\hat{u}(k)$:

$$\partial_t \hat{u}_k + i\frac{k}{2}\sum_q \hat{u}_{k-q}\hat{u}_q = 0 \,. \tag{6.36}$$

We see that actually all modes $\hat{u}_p$ and $\hat{u}_q$ with $k = p + q$ contribute to the evolution of mode $\hat{u}_k$. The growth of the first 32 harmonics with time is shown in Figure 6.7b.

From Eq. (6.35) one can predict that the amplitude of harmonic $2k_0$ will be comparable to U_0 in a time $t \simeq 1/k_0 U_0$. But to infer anything for the rest of the spectrum, we have to use the full equation. Figure 6.7b shows that a quasi-equilibrium is reached for the whole spectrum after a time of the order of a nonlinear time.

The evolution of the modal spectrum $|\hat{u}(k)|$ vs. wavenumber k is shown in Figure 6.8a up to two nonlinear times $t = 4$. A $1/k$ range is seen to form at about $t = 2$ (corresponding to a $1/k^2$ *energy* spectrum), as expected for a quasi-discontinuous velocity profile. At later times, the spectrum decreases slowly in a quasi-self-similar way.

Shock dissipation The time evolution of the average kinetic energy per unit mass $\overline{u^2}/2$ is shown in Figure 6.8b (solid line). There are two phases. During the first phase, up to time $t = t_{\mathrm{NL}} = 2$, the average energy fluctuates around its initial value. The fluctuations of the total energy come from the fact that the kinetic energy is only an approximate invariant: reversible (quasi-linear) exchanges occur between the kinetic and the internal (thermal) energy.

During the second phase, the situation changes, the kinetic energy decreases steadily, because kinetic energy is systematically transferred to scales small enough for the viscous term to become important, so that kinetic energy is transformed continuously into thermal energy.

A more quantitative description comes as follows. During the first phase, energy is mostly concentrated on the largest scale $L \simeq 1/k_0$, so that the laplacian reads

$$\nu u'' \simeq -\nu u/L^2 \,. \tag{6.37}$$

and the energy dissipation rate thus reads

$$d\overline{u^2}/dt \simeq -2\nu\overline{u^2}/L^2 \tag{6.38}$$

where the average kinetic energy density per unit mass $\overline{u^2}/2$ is related to the energy spectrum $E(k)$, by $\overline{u^2}/2 = (1/L)\int E(k)dk$.

During the whole first phase, the progressive steepening of the velocity profile leads to a decrease of the characteristic scale l of the velocity profile, and thus to a growth of the advection term that scales as $uu' \simeq u^2/l$, with the viscous term growing more rapidly with $1/l$ as $\nu u'' \simeq -\nu u/l^2$. The viscous term thus grows more rapidly with $1/l$ than the advection term. The steepening stops when the viscous term balances (in the shock region) the advection term, which reads $u^2/l \simeq \nu u/l^2$, or

$$\nu = lu = lU_0 \,. \tag{6.39}$$

At the shock location, one thus has $\nu u'' \simeq -\nu u/l^2$ where l is given by Eq. (6.39). This leads to the energy decay at the location of the shock $du^2/dt \simeq -\nu u^2/l^2$. Hence, for the *average* energy:

$$d_t u^2 = -\nu u^2/l^2 \times l/L \tag{6.40}$$

$$= -u^3/L \,. \tag{6.41}$$

In the RHS of Eq. (6.40), the factor l/L is the relative width of the shock region.

The Figure 6.8b also shows (dotted line) the evolution of the energy in the first harmonic $k = k_0 = 1$. One sees that it actually decreases during both phases. This is because during both phases the energy of the first mode is transferred to the kinetic energy of the other modes.

To summarize, when the flow velocity varies smoothly, the averaged energy dissipation is controlled by viscosity and the characteristic length of the fluctuation L, its characteristic time scale being given by the large-scale viscous time L^2/ν. Once the shock is formed, the characteristic dissipation time becomes independent of viscosity and is controlled only by the velocity amplitude U and initial characteristic length L: it reduces to the nonlinear time L/U (Eq. (6.33)). Note that for the shock to form, the nonlinear time has to be much shorter than the large-scale viscous time, in other words the Reynolds number defined below has to be significantly larger than unity:

$$\mathrm{Re} = UL/\nu \,. \tag{6.42}$$

6.2.2
Hydrodynamics: 3D Incompressible Turbulence

In most natural flows, as in earth's atmosphere, rivers, oceans, and everyday life as well (wakes behind solid bodies), turbulence is commonplace, in fact it is almost a rule. As in the shock formation process, a main property of turbulence is to show a so-called developed spectrum, that is, a distribution of energy over a large range of scales or wavenumbers. Nevertheless, 3D turbulence differs from the shock formation process in several respects, closely related to the fact that it is three-dimensional. First, the flow is chaotic, that is, unpredictable. Second, the flow is (most of the time in nature) quasi-*incompressible*, that is, leads to no or negligible density variations, which is possible because the particle trajectories are turning, instead of being "compressive" as in the previous section.

Quantitatively, the incompressibility condition reads

$$\nabla \cdot \boldsymbol{u} = 0 \,. \tag{6.43}$$

In an incompressible flow, the growth of gradients is measured by the growth of "vorticity" $\boldsymbol{\omega} = \nabla \times \boldsymbol{u}$ which measures the local rotation rate of the fluid, the direction of the vorticity vector giving the rotation axis.

Figure 6.9 shows an example of a turbulent flow: there, turbulence above a bowl of hot tea is revealed by sunlight. Turbulence is driven by the temperature gradient between the hot tea surface and the cooler air layers above. Air in contact with the hot liquid gets heated, inflates, and so is pushed above by the buoyant force (Archimedes' principle). This generates the upward flow. During their upward movement, air particles generally cool and become denser than the surrounding unperturbed air layers, so they finally fall back, thus leading to downward flows. During this process, the small differences between trajectories of nearby particles are amplified, leading to chaos, unpredictability and amplification of gradients. We are interested here in the last effect: formation of the gradients and of the energy spectrum.

The Richardson–Kolmogorov cascade In the 1D case, the steepening of velocity profiles has been explained by using the simple zero pressure model of Burgers' equation (Eq. (6.34)) in which the crest of the wave catches the trough in a nonlinear time (Eq. (6.33)). In an incompressible flow, this is forbidden, because the velocity difference between two points is perpendicular to the direction of the wave vector, due to the incompressibility condition (Eq. (6.43)). Expressed in terms of wave vectors, one says that the straight 1D harmonics building by addition of wave vectors along a single direction (Figure 6.10a,b) is forbidden, as we will see in a moment. Basically, 1D harmonics building is transformed into the Richardson cascade in which eddies of smaller and smaller sizes are generated, with the interacting wave vectors exploring all directions, as shown in Figure 6.10c,d.

So the propagation of excitation to larger and larger wavenumbers in 3D is the result of coupling wave vectors in all directions. As a result, the energy spectrum should be isotropic, as there should be in general no preferred direction. Another

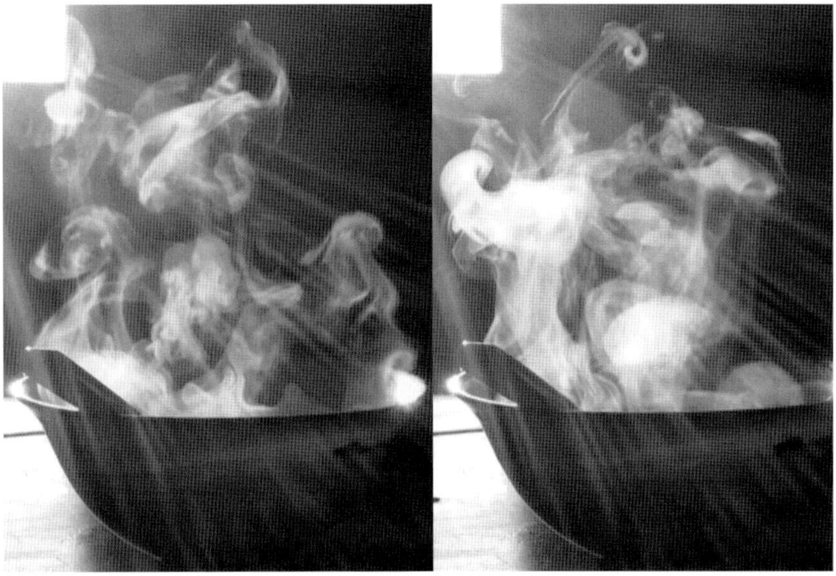

Figure 6.9 Two snapshots of the air above a bowl of tea. Sunlight makes visible turbulent vortices driven by the temperature gradient above the hot liquid surface.

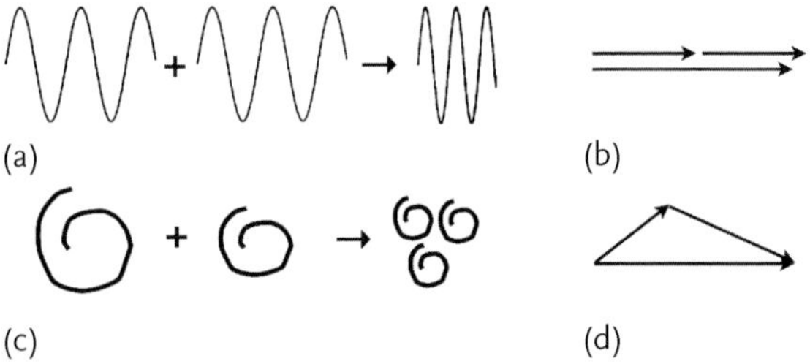

Figure 6.10 Schematics of the gradient formation in 1D (a,b) and 3D (c,d) flows. (a) Two (plane) waveforms with the same wavelength couple, leading to a signal with smaller wavelength. (b) The same coupling represented as an addition of three wave vectors: the daughter wave mode has a wave vector which is the algebraic sum of the parents wave vectors. (c) Two eddies of unequal sizes couple, leading to smaller eddies (d) a triad of wave vectors (among many others) representing this coupling, with the important constraint that the triad cannot be flat, due to the incompressibility condition (see text, Eq. (6.56)).

important point is that kinetic energy is strictly conserved by the nonlinear terms, so energy should be conserved during transfer, forgetting dissipative terms. (Recall that this was exactly true for the Burgers' equation with zero pressure, and almost true in the case of compressive 1D Navier–Stokes with small Mach number.)

Fourier space is represented in Figure 6.11, with a series of concentric spheres of (geometrically) growing radii, that is, wavenumbers $k_n = k_0 2^n$, $n = 1 \dots N$. We assume energy is continuously injected by a prescribed forcing in the first wavenumber shell $n = 0$. Denote by ϵ the (constant) injection rate, an energy per unit mass

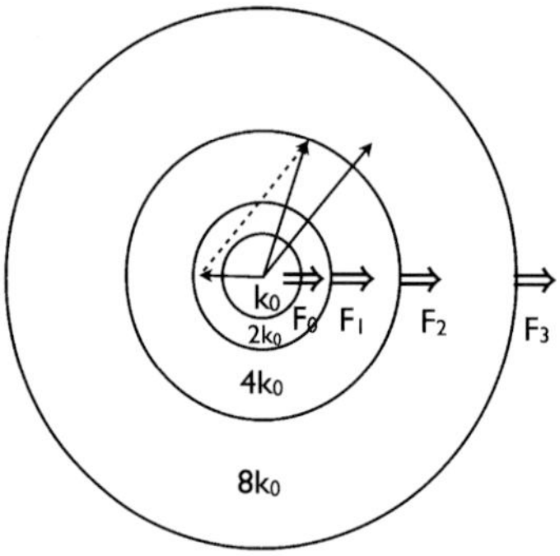

Figure 6.11 Sketch of isotropized energy flow in Fourier space for 3D turbulence. The transfer from one shell of wave vectors to another is represented, the shell radius k_n growing in a geometric progression. Also drawn is an example of a triad of wave vectors coupling the three concentric shells 0, 1 and 2.

divided by time. In the first shell, the energy equation is

$$d_t E_0 = \epsilon - F_0 \tag{6.44}$$

where F_0 is the energy loss rate (or energy flux) towards the shell $n = 1$ just below. In the other shells, the energy $E_n = u_n^2/2$ exchanges energy with both the shells above and below:

$$d_t E_n = F_{n-1} - F_n - D_n \tag{6.45}$$

where F_{n-1} is the energy flux entering the shell number n from the shell number $n - 1$ and F_n is the energy flux leaving shell n into shell $n + 1$. $D_n = -\nu k_n^2 u_n^2$ is the dissipation term, which is nonnegligible only at very small scales.

From dimensional considerations one can write the energy flux as

$$F_n = u_n^2/\tau_n = k_n u_n^3 = u_n^3/l_n \tag{6.46}$$

where $l_n = 1/k_n$ and the *nonlinear time* (or *turnover time*) is thus

$$\tau_n = l_n/u_n \, . \tag{6.47}$$

Equation (6.47) generalizes the nonlinear time defined previously in Eq. (6.33) for shock formation. Our description is both fully isotropic, and also local, in the sense that no direct energy exchange between distant shells (for instance, between shells n and $n+2$) is assumed. In Figure 6.11 is an isolated triad of wave vectors showing how adding wave vectors from shells $n - 1$ and n could contribute to the excitation of a wave vector in shell $n + 1$.

The stationary solution of the system is, in the so-called inertial range of scales $n = 1 \dots N$ where the dissipation D_n can be neglected:

$$\epsilon = F_0 = F_1 = \dots = F_N \, . \tag{6.48}$$

Replacing in $F_n = \epsilon$ the flux F_n by its expression in Eq. (6.46) we obtain for each shell n:

$$u_n = \epsilon^{1/3} l_n^{1/3} \tag{6.49}$$

or equivalently for the spectral energy density (per unit wavenumber) $E_k = (1/k) E_n$:

$$E_k = \epsilon^{2/3} k^{-5/3} \,. \tag{6.50}$$

This is the Kolmogorov spectrum, which is indeed observed in various natural turbulent flows. In the following, we will refer to K41 when dealing with the Kolmogorov scaling and phenomenology.

Intermittent flows We have found that 1D "turbulence" forms shocks, that is, quasi-discontinuities, which leads to energy spectra with a -2 spectral index, thus distinct from the $-5/3$ K41 index. This is a paradox, as using the K41 arguments should lead in both the 1D and 3D cases to the same flux estimate, hence the same spectral index. The solution of the paradox comes from remarking that the 1D and 3D cases differ by their spatial intermittency: in the 1D case, large gradients are localized in a small fraction of the accessible volume, much more than in 3D. For scale l_n, this fraction is l_n/l_0. The average energy E_n in shell n is thus smaller than $u_n^2/2$, being only $E_n = l_n/l_0 u_n^2/2$. Energy conservation leads again to the scale-invariance of the energy flux which reads $E_n/\tau_n = \epsilon$. From this, one deduces the scaling $u_n = \epsilon^{1/3} l_0^{1/3}$, hence $E_n = (l_n/l_0)^{1/3} (\epsilon l_n)^{2/3}$, thus for the spectral energy density $E_k = \epsilon^{2/3} k_0^{1/3} k^{-2}$, which is indeed what is expected from the localized discontinuities formed in shocks. Note that the same line of arguments applies in principle to other forms of intermittency, with the fraction of space occupied by active gradients varying as a power-law of the scale l: this is the so-called beta-model introduced by [37].

Dissipative scale Once turbulence has reached a quasi-stationary state, the injection rate, the energy flux, and the energy loss are all equal. To determine the scale at which energy is dissipated, one has to find the scale at which the (small) nonlinear time equals the viscous dissipative time. This leads to $u^2/l \simeq \nu u/l^2$ or $\nu \simeq ul$, which reads, adding the suffix D to indicate the dissipative scale l_{D}:

$$\nu \simeq u_{\mathrm{D}} l_{\mathrm{D}} \,. \tag{6.51}$$

The difference between the 1D expression (Eq. (6.39)) and Eq. (6.51) is that here $u_{\mathrm{D}} \ll U_0$ is the eddy velocity amplitude at the dissipative scale while in the 1D case one had $u_{\mathrm{D}} = U_0$. Using the Kolmogorov scaling of Eq. (6.49) to express u_{D} in terms of the large-scale parameters L and U, one finds $\nu = UL(l_{\mathrm{D}}/L)^{4/3}$, hence finally

$$l_{\mathrm{D}}/L = \mathrm{Re}^{3/4} \tag{6.52}$$

where $\mathrm{Re} = UL/\nu$ is the Reynolds number.

Navier–Stokes in Fourier space Our motivation to write down the 3D incompressible Navier–Stokes equations in Fourier space is threefold. First, it allows us to appreciate the distance between the primitive equations and the Kolmogorov model which has been used to derive the Kolmogorov scaling; second, it provides a guide to derive further model systems; third, it gives a hint of the complexity of 3D turbulence compared with 1D. We start from the Navier–Stokes equation (Eq. (6.29)) and take its Fourier transform. In order to eliminate the pressure, we apply the ∇ operator to Eq. (6.29), and impose $\nabla \cdot \boldsymbol{u} = 0$. We obtain

$$\nabla^2 P = \partial_i \partial_j (-u_i u_j) \tag{6.53}$$

or, as well, after Fourier transforming: $-k^2 \hat{P} = k_i k_j \widehat{u_l u_j}(k)$. Taking this into account to express the pressure in terms of the velocity, the spatial Fourier transform of Eq. (6.29) is:

$$\partial_t \hat{u}_i(k) + i P_{ijl}(k) \widehat{u_l u_j}(k) = 0 \tag{6.54}$$

where the operator P_{ijl} reads

$$P_{ijl}(k) = k_j \left(\delta_{il} - k_l k_i / k^2\right) . \tag{6.55}$$

We can now demonstrate our previous assertion that the harmonics building process cannot be as in 1D (Figure 6.10). To do so, we rewrite Eq. (6.54) after expanding the term $\widehat{u_l u_j}(k)$ in the form of a convolution integral:

$$\partial_t \hat{u}_i(k) = -i \sum_{k=p+q} k_j \hat{u}_j(q) \left(\delta_{il} - k_l k_i / k^2\right) \hat{u}_l(p) . \tag{6.56}$$

From this equation we see that the contribution of "flat" triads (k, p, q) is zero. Indeed, the scalar product $k_j \hat{u}_j(q)$ in the RHS of the equation is always zero as soon as $q \parallel k$, since then it is equal to $q \cdot \hat{u}(q) = 0$ which is zero whatever q, as $\nabla \cdot \boldsymbol{u} = 0$. The exclusion of flat triads demonstrates that a strict 1D cascade is impossible in 3D incompressible hydrodynamics.

Shell models To complete our introduction to the hydrodynamic cascade, we define now a model which we shall use again later with profit, namely a shell model, proposed first by [38, 39] (for a recent review see [40]). The starting assumptions are similar to those leading to the Kolmogorov scenario, but the arguments are closer to the equations, the aim being to derive equations simpler than the Navier–Stokes equations, but sharing the same main properties.

We assume again that the system is isotropic enough, so that we can map the whole Fourier space with shells of wave vectors, each being represented by its wavenumber $k_n = 1/l_n = 2^n k_0 = 2^n/L$. Mimicking Eq. (6.56), with the constraint that (i) total energy has to be conserved by the nonlinear terms (ii) we want to keep to a minimum (that is, here, just nearest neighbors) the number of interacting shells, we obtain equations for the velocity amplitude $u_n = u(l_n)$. One possible form is:

$$d_t u_n = k_n u_{n-1}^2 - k_{n+1} u_n u_{n+1} - \nu k^2 u_n + f_n \delta_{i1} \tag{6.57}$$

with $n = 1 \ldots N$ and $u_0 = 0$. That Eq. (6.57) indeed conserves energy (neglecting forcing and dissipation) can be checked directly (we have to assume energy in shell $n = 0$ is zero). The corresponding energy equation is $(1/2)du_n^2/dt = F_n - F_{n+1}$ with $F_n = k_n u_{n-1}^2 u_n$, which is not so different from Eqs. (6.45) and (6.46) in which $F_n = k_n u_n^3$. Both Eqs. (6.45) and (6.57) share the Kolmogorov scaling $u \propto l^{1/3}$ as a stationary solution.

Turbulent dissipation What happens when the energy transmitted from the largest scale towards smaller scales is not permanently refilled at the largest scale? We start from the shell energy equations. We consider only scales substantially larger than the dissipation scale, so that we can neglect dissipation. We thus have $d_t E_0 = -F_0$ and for $n > 0$, $d_t E_n = F_{n-1} - F_n$. In the decaying case (no forcing), the equation for shell $n = 0$ may be rewritten, adopting the simple form $F_n = u_n^3/l_n$:

$$d_t U^2/2 = -U^3/L \tag{6.58}$$

which is a familiar equation, already seen when we considered the shock formation problem. This will be compatible with Kolmogorov scaling if all scales have the same logarithmic decay, independently of the shell number n, that is, if

$$(1/E_n)d_t E_n = (1/E_{n+1})d_t E_{n+1} \; . \tag{6.59}$$

One finds that this is indeed satisfied for shells between the largest and the smallest dissipative shells. More precisely, integrating the shell models with no dissipation leads to $(F_{n-1} - F_n)/F_n \propto (l_n/L)^{2/3}$, which implies that the energy flux becomes more and more constant at small scales, converging to a finite value. This corresponds to the self-similar decay of the energy spectrum according to Eq. (6.58), the spectrum still following the Kolmogorov scaling as in the stationary case, with only a global amplitude decrease: this is verified experimentally. Finally, one should note that the large-scales do not evolve in a self-similar way; therefore, the size L of the energy-containing scale varies in general with time as well as the total energy. We will not insist on this point further here.

6.2.3 MHD Turbulence – Introduction

Two main issues arising in MHD turbulence are on the one hand, the dynamo problem, for which the primary goal is to understand how large-scale magnetic fields can be generated by the turbulent movements of the conducting fluids that can be found within stars and planets, and on the other hand, how turbulence is modified by a magnetic field, once it is present. Note that, contrary to a mean velocity field, which can be eliminated by a Galilean frame change, a mean magnetic field cannot: it is always there. We shall consider here the second problem, with some emphasis on the case of strong mean magnetic field. In fusion devices a strong mean magnetic field is commonplace, while in astrophysics both the zero mean field and the strong mean field cases are found.

We shall consider here the *incompressible limit*, because most theories and simulations have been elaborated in this framework. As we know from Chapter 3, the magnetic field contributes to the pressure tensor: it adds a magnetic pressure to the gas pressure and introduces a tension of the field lines. In the incompressible limit, the magnetic pressure plays the same role as the gas pressure: total pressure gradients balance nonlinear terms so as to constrain the flow to be divergenceless, that is, $\nabla \cdot \boldsymbol{u} = 0$. Magnetic tension has more important consequences: it leads to transverse waves, called Alfvén waves, which propagate along the magnetic field lines, and reduce the generation of small scales along this direction. Actually, in this limit, all magnetosonic modes either disappear (the fast ones, which propagate instantaneously), or degenerate into waves propagating with the same dispersion relation as the Alfvén waves. The latter are called pseudo-Alfvén waves, but we shall in the following call indifferently Alfvén waves the true shear Alfvén waves and pseudo-Alfvén waves (see for example [43, 44] for turbulent scenarios treating differently pseudo-Alfvén waves and shear Alfvén waves).

An example of Alfvén waves in nature is given in Figure 6.12a, where one can see thin, elongated magnetic field lines on the limb of the Sun, traced by density enhancements. Oscillations, at least partly in the form of Alfvén waves, propagate along the magnetic field lines. The turbulent dissipation of these oscillations could contribute to heat the corona, which is the rarefied hot atmosphere of the Sun. The signature of Alfvén waves emitted by the Sun is found as far as one AU (Figure 6.12)b. The high correlation observed between the velocity and magnetic field signals shows that waves are progressing in a given direction along the mean field, and most of the time, at least in fast winds, the direction is outward. These fluctuations show a well-developed spectrum which was interpreted early as an indication that the interplanetary space is a turbulent medium.

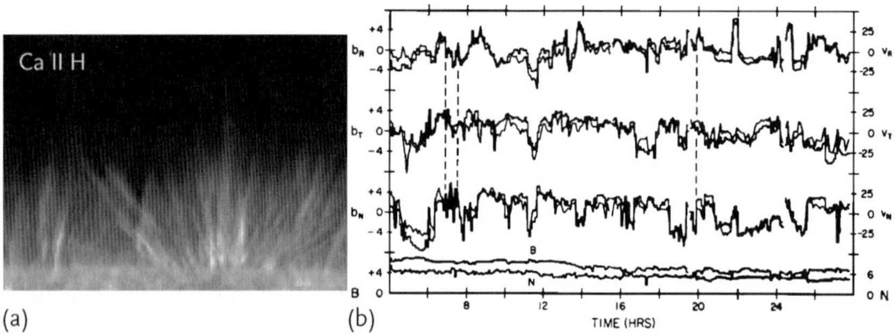

(a) (b)

Figure 6.12 MHD in the solar/solar wind context. (a) Snapshot of a solar limb region. Black features are density enhancements, called spicules, imprisoned in magnetic tubes. Spicules thus reveal the form of the magnetic field lines. These tubes contain ascending matter, but also oscillate transversely, triggered by surface convection, these oscillations probably revealing upward propagating Alfvén waves. From [41]. (b) Alfvén waves in the solar wind [42]. The top three waves each show a different orthogonal component of the solar wind magnetic field and the associated velocity component of the plasma. All show high correlation, thus revealing Alfvén waves. The bottom wave displays the magnitudes of the total magnetic field and density.

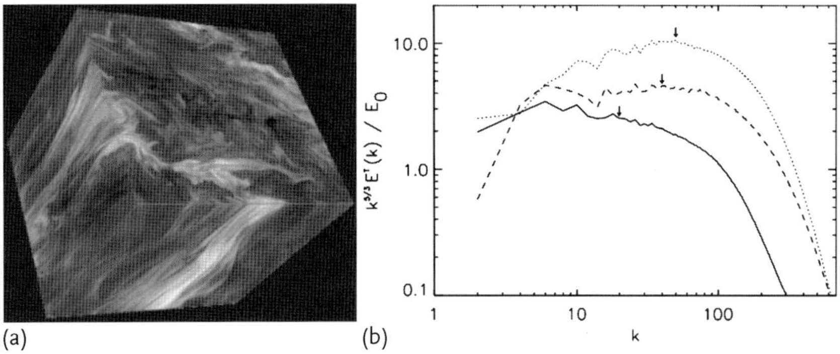

Figure 6.13 Direct numerical simulations of incompressible MHD turbulence. (a) Snapshot of the magnetic field modulus in a direct simulation of developed MHD turbulence with mean field [45]. (b) Energy spectra in three MHD regimes with zero mean field, decay of a Taylor–Green flow (from [46]). Energy spectra are compensated by $k^{5/3}$ and averaged over $\Delta t = 0.5$ (1.52 turnover times). The three arrows indicate the Taylor magnetic scales $\lambda_T = 2\pi \left(\int E_M(k)dk / \int k^2 E_M(k)dk\right)^{1/2}$.

Figure 6.13a gives a numerical example of (incompressible) MHD turbulence with mean magnetic field. The flow is clearly anisotropic, because the amplitude of the magnetic field fluctuation is smaller than the mean field B_0: $b_{\mathrm{rms}}/B_0 = 1/5$. The figure shows the modulus of the magnetic field on the faces of the cubic domain.

Contrary to homogeneous hydrodynamic turbulence, MHD turbulence offers various possible turbulent regimes, associated with different scaling laws. Figure 6.13b shows total (kinetic+magnetic) energy spectra for some of these regimes. The authors considered three MHD variants of the so-called Taylor–Green flow (meant to mimic a flow between two counterrotating disks), with no mean magnetic field, and equal magnetic and kinetic energies at start. The three runs evolve differently with time, due to different symmetries of the initial flow. The figure shows temporal averages of the total energy spectra $E(k)$, in the three cases, compensated by $k^{3/2}$. The spectral indices are found to be close to: -2, $-5/3$, $-3/2$. Each case is characterized by a different ratio $b_{\mathrm{rms}}/u_{\mathrm{rms}}$ which passes from very large to somewhat larger than unity, as the final energy spectrum goes from steep (-2 slope) to flatter ($-3/2$ slope).

Equations Incompressible MHD is described by only two fields: the velocity u and magnetic field B, with the constraint that $\nabla \cdot \boldsymbol{u} = \nabla \cdot \boldsymbol{B} = 0$. The density is set to $\rho = 1$. In the following, we will adopt the usual notations which consists of writing B instead of $B/(\mu_0 \rho)^{1/2}$. We define $b = B - B_0$ as the amplitude of the fluctuating field, $B_0 = \langle B \rangle$ being the mean field or, as well, the Alfvén speed. The two so-called Elsasser fields $z^{\pm}$

$$z^{\pm} = \boldsymbol{u} \pm \boldsymbol{b} \tag{6.60}$$

represent the amplitude of the modes propagating respectively to the left and right along the mean field (if B_0 points rightward). The incompressible MHD equations read in terms of the Elsasser fields

$$\partial_t z^{\pm} \mp \boldsymbol{B}_0 \cdot \boldsymbol{\nabla} z^{\pm} + z^{\mp} \cdot \boldsymbol{\nabla} z^{\pm} + \boldsymbol{\nabla} P = 0 \tag{6.61}$$

with $\boldsymbol{\nabla} \cdot z^{\pm} = 0$, and where we leave the diffusive terms aside.

The linear term $\mp \boldsymbol{B}_0 \cdot \boldsymbol{\nabla} z^{\pm}$ describes the linear propagation of Alfvén waves. The dispersion relation between the frequency ω and wave vector $\boldsymbol{k}$ is

$$\omega = \boldsymbol{k} \cdot \boldsymbol{B}_0 = k_{\|} B_0 \tag{6.62}$$

which indicates that the phase speed is maximal for wave vectors $\|$ to B_0, but goes to zero for perpendicular ones.

Using Elsasser variables $z^{\pm}$ instead of u and b makes apparent several important properties of the MHD system, in addition to showing the propagation in two directions of the two Alfvén modes. First, the nonlinear term is a function of the products $z_i^+ z_j^-$:

$$T_{\mathrm{NL}}{}_i^{\pm} = \partial_j \left(z_j^{\mp} z_i^{\pm} \right) + \partial_i P = \mathcal{F}\left(z_i^+ z_j^- \right). \tag{6.63}$$

To prove this, we need to show that the pressure term is indeed of this form. To do so, we apply the ∇ operator to Eq. (6.61), and use the incompressibility constraint $\boldsymbol{\nabla} \cdot z^{\pm} = 0$, which leads to the following relation between pressure and the Elsasser fields:

$$\nabla^2 P = \partial_i \partial_j \left(-z_i^- z_j^+ \right). \tag{6.64}$$

In other words, in *incompressible* MHD, a z^+ eddy interacts with z^- eddies, not with z^+ eddies.

Another property of the equations is that the energies of the two Elsasser species, $E_{\pm} = \langle z^{\pm 2}/2 \rangle$ are separately conserved by the nonlinear terms, neglecting the dissipative terms. To see this, it is sufficient to write down the equations for the energies $E^{\pm}$ in conservative forms (assuming the mean field is 0, but this is also true with a mean field):

$$\partial_t z^{\pm 2}/2 + \partial_i \left(z_i^{\mp} z^{\pm 2}/2 \right) + \partial_i \left(z_i^{\mp} P \right) = 0. \tag{6.65}$$

These properties have immediate consequences. Since each of the two fields is stirred only by the other one (and not by itself as in hydrodynamics), the nonlinear "turnover" times at the scale $l = 1/k$ will thus be

$$\tau_{\pm} = 1/k z^{\mp}. \tag{6.66}$$

Equation (6.66) generalizes the definition for the nonlinear time given previously by Eq. (6.47). Since the two energies $E^{\pm}$ are conserved separately, we expect, following the same line of arguments as for the hydrodynamical cascade, to arrive at two distinct energy flux, one for each Alfvén species $z^{\pm}$:

$$F_{\pm} = (z^{\pm})^2/\tau_{\pm} = k(z^{\pm})^2 z^{\mp}. \tag{6.67}$$

However, as we shall see, the propagation along the mean field may considerably alter the cascade, either leading to a strong anisotropy in Fourier space, and/or decreasing the energy flux at small scales, hence showing several possible regimes, with various spectral scaling.

6.2.3.1 Fourier Space Anisotropy

Because of the very structure of the nonlinear terms (Eq. (6.63)), the nonlinear coupling involves collisions between the two Alfvén species, z^+ and z^-, that is, between eddies propagating in opposite directions. The important points are that (i) when propagation is slow compared to the eddy turnover time $\tau_{\pm}$, then the energy exchange should proceed unchanged compared to usual hydrodynamics, while when propagation is fast, then the energy exchange might be altered; (ii) propagation is fast or slow for a given wavenumber k, depending on the angle between the wave vector and the mean field (see Eq. (6.62)). As an extreme example, fluctuations with purely perpendicular wave vectors will not propagate at all, or, more precisely, wave fronts will not move.

We now take the Fourier transform of Eqs. (6.61) and (6.64). After replacing the expression for the pressure field in the first equation, we obtain:

$$\partial_t \widehat{z_i^{\pm}}(k) \mp i k_{\parallel} B_0 \widehat{z_i^{\pm}}(k) + i P_{ijl}(k) \widehat{z_l^{\pm} z_j^{\mp}}(k) = 0 \tag{6.68}$$

where $k_{\parallel}$ denotes the projection of the wave vector on the mean field B_0 and $P_{ijl}(k)$ is defined in Eq. (6.55). We go on and rewrite the equations using the Heisenberg representation for the unknown amplitudes: $\widehat{z^{\pm}}(k) = Z^{\pm}(k) e^{\pm i k_{\parallel} B_0 t}$:

$$\partial_t Z_i^{\pm}(k) + i P_{ijl}(k) \int\limits_{k=p+q} d^3 p\, d^3 q\, Z_l^{\pm}(p) Z_j^{\mp}(q) e^{\mp i \omega t} = 0 \tag{6.69}$$

with

$$\omega = 2 q_{\parallel} B_0 \, . \tag{6.70}$$

The Navier–Stokes equation is recovered if we let $z^+ = z^-$ (which implies $b = 0$) *and* $B_0 = 0$ (which implies $\omega = 0$). In Eq. (6.69), for each pair of interacting modes $\boldsymbol{p}, \boldsymbol{q}$ contributing to the evolution of mode $\boldsymbol{k}$, there is an oscillating factor $e^{\mp i \omega t}$ in front of the nonlinear terms, with the frequency depending on the direction of the wave vector $\boldsymbol{q}$ with respect to the mean field. We thus expect a hydrodynamics-like contribution from triads with wave vectors $\boldsymbol{q}$ perpendicular to $\boldsymbol{B}_0$, while triads with a $\boldsymbol{q}$ such that the frequency is large will be attenuated. This difference between Navier–Stokes and MHD is illustrated schematically in Figure 6.14.

Forgetting for a moment the possible energy difference between the two Elsasser species, we summarize our findings by incorporating the Alfvén oscillating factor into the dimensional expression of the energy flux (assuming kinetic/magnetic equipartition and using u instead of $z^{\pm}$):

$$F = k \left\langle u^3 \cos \omega t \right\rangle \, . \tag{6.71}$$

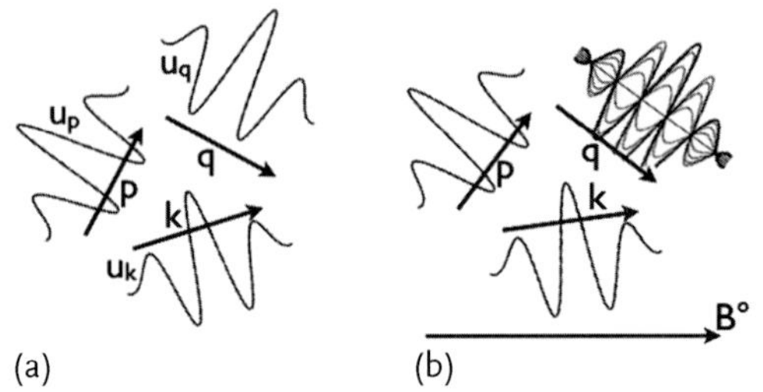

Figure 6.14 Triads of wave vectors $(\boldsymbol{k}, \boldsymbol{p}, \boldsymbol{q})$ illustrating the nonlinear couplings in hydrodynamics (a) and in MHD (b). Wave packets have been represented along with wave vectors. In (b) is the oscillating factor with frequency $\omega_q = q_{\parallel} B_0$ that modulates the mode Z_q in Eq. (6.69) by superposing the associated waveform with different phases. This oscillating factor prevents the coupling from being strong as in (a).

The brackets denote time-averaging over a time long compared to the Alfvén period $2\pi/\omega$. We will see that such a simple expression can possibly describe different regimes, depending on what is put into the frequency ω. At first sight:

1. Either one can neglect the Alfvén frequency ω compared to the nonlinear frequency ku (as, for instance, will be the case for the cascade in a direction perpendicular to B_0); in that case one expects to recover the Kolmogorov cascade.
2. Or this is not the case, and the nonlinear flux averages to zero, since u is almost constant during an Alfvén time $t_a = 1/\omega$, as it varies on the much larger nonlinear time scale $\tau = 1/ku$.

However, there exists an intermediate possibility as we will see below, namely that the velocity amplitude $u(k)$ varies on several time scales, including times much shorter than the nonlinear time $\tau = 1/(ku)$. This should lead to an energy flux weaker than the Kolmogorov flux ku^3, albeit nonzero. A reasonable guess is then that the adimensional weakening factor depends on the ratio of the two time scales $1/\omega$ and τ:

$$F = ku^3 A(\omega\tau) = ku^3 A(u/B_0) . \qquad (6.72)$$

The factor $A(u/B_0)$ should be smaller than unity and in general decrease with increasing wavenumber k.

6.2.4
Weak Isotropic (IK) Regime

A much studied regime is the Iroshnikov–Kraichnan regime (or IK), in which the weakening factor in Eq. (6.72) varies as $A \propto u/B_0$. We consider in turn the phenomenology, results of numerical simulations as well as extensions of the theory, and finally discuss the result of a simple shell model.

6.2.4.1 IK Theory

Let us assume that, in spite of the anisotropy in the dispersion relation, the cascade is globally isotropic enough so that, at a given scale $1/k$, we can consider an isotropized dispersion relation of the form

$$\omega \simeq k B_0 \,. \tag{6.73}$$

The analysis given below can in principle be applied to two different cases, either directly to the mean field configuration, or to the related problem with zero mean field. In the latter case, B_0 is meant for a *local* mean field, with amplitude of the order of b_{rms}. Actually, as we will see, the domain of validity of the original isotropic theory as exposed below is restricted to the 2D case with zero mean field. The application to the 3D case needs some modification to include anisotropy, as we will see later.

The theory [47, 48] (IK theory), assumes an isotropic inertial range, with an energy cascade that proceeds in all directions of Fourier space. One is interested in the small-scale behavior of turbulence. Since in general the turbulent amplitude decreases with wavenumber, at scales small enough one should have

$$u \simeq b \ll B_0 \,. \tag{6.74}$$

Hence, at small scales, the isotropized Alfvén time $t_{\text{a}} = 1/\omega$ is much smaller than the nonlinear time:

$$t_{\text{a}} = 1/\omega \ll \tau \tag{6.75}$$

where $\tau = 1/(ku) \simeq 1/(kb)$ is the nonlinear time (Eq. (6.47)).

The theory assumes that in this configuration the energy transfer does not proceed as in standard hydrodynamics in one nonlinear time. Instead, the transfer process is the sum of N uncorrelated successive independent uncorrelated energy transfers, each partial transfer event lasting about one Alfvén time t_{a}. The energy transfer is thus assumed to proceed as a random walk, with the energy transferred at time t being the sum of $N = t/t_{\text{a}}$ successive uncorrelated increments of order $\delta E \simeq \pm u^2 \times t_{\text{a}}/\tau$. Finally, the energy is completely transferred after $N = (u^2/\delta E)^2 = (\tau/t_{\text{a}})^2$ time steps, hence after a time

$$t_* = t_{\text{a}} \times (\tau/t_{\text{a}})^2 = \tau \times (\tau/t_{\text{a}}) \,. \tag{6.76}$$

In conclusion, this theory predicts that the weakening factor in Eq. (6.72) is $A = t_{\text{a}}/\tau = u/B_0$, so that the energy flux is

$$\langle F \rangle = k u^3 \times t_{\text{a}}/\tau = k u^3 \times (u/B_0). \tag{6.77}$$

The constraint of constant flux then leads to $k u^4 = \text{const}$, hence to $u \propto k^{-1/4}$ and finally, since $E(k) \simeq u^2/k$:

$$E(k) \propto k^{-3/2} \,. \tag{6.78}$$

6.2.4.2 IK Theory with Nonzero Cross-Helicity

The $-3/2$ spectral index has been found in numerical simulations: 2D MHD with no mean field [49, 50], 3D MHD with no mean field [46], and with mean field [51, 52]. The most convincing check of IK theory is actually to be found in systems with different energies in the two fields $z^{\pm}$. When this is the case, one says that there is nonzero *cross-helicity*, the cross-helicity being defined as the difference between the + and − energies.

The phenomenology has to be slightly generalized to apply to this case, that is, the flux reduction has to be applied separately to the energy of each Alfvén species.

Following the same line of arguments as previously for Eq. (6.77), one obtains for the energy flux of each Elsasser species [53, 54]:

$$F^{+} = A^{+}z^{+2}/\tau^{+} \times (t_a/\tau^{+}) = A^{+}kz^{+2}z^{-2}/B_0 \tag{6.79}$$

$$F^{-} = A^{-}z^{-2}/\tau^{-} \times (t_a/\tau^{-}) = A^{-}kz^{+2}z^{-2}/B_0 \tag{6.80}$$

accounting for the fact that the nonlinear times are built for each field on the amplitude of the other field (Eq. (6.66)): $\tau^{\pm} = 1/kz^{\mp}$. The factors $A^{\pm}$ are a priori of order unity [54], thus implying that the two energy fluxes are equal, which precludes the possibility to obtain a stationary state by forcing, for example, $z^{+} \gg z^{-}$. However, taking into account the nonlocality of couplings as in [53], one obtains that $A^{+} \neq A^{-}$, so allowing for $F^{+} \neq F^{-}$.

The conservation of the two energy fluxes $F^{\pm}$ then leads to the single relation between spectral indices ($E^{\pm}(k) \propto k^{-m^{\pm}}$):

$$m^{+} + m^{-} = 3 \tag{6.81}$$

thus leaving place for the possibility of different slopes for the two Elsasser fields.

As a rule, the cross-helicity $E^{+} - E^{-}$ is largest at large-scale, and vanishes when reaching the dissipative scales: Figure 6.15 shows one example of $E^{\pm}(k)$ spectra, obtained by solving numerically model spectral equations based on eddy-damped quasi-normal markovian spectral closure (EDQNM, cf. [49]) in the case of 88% normalized cross-helicity. Equation (6.81) has been shown to hold in direct simulations of decaying turbulent 2D MHD with no mean field (Figure 6.16). The figure reports several simulations starting with different values of the *normalized cross-helicity* $\sigma_c = (E^{+} - E^{-})/(E^{+} + E^{-})$ (denoted as ρ in the abscissa of the figure). During the decaying simulations, the cross-helicity increases because the dominant energy decreases at a slower rate than the dominated energy [54, 55], and the difference between the two spectral indices $m_{\pm}$ increases as well: these drifts during the simulations appear as rectangular boxes in the figure.

6.2.4.3 Shell Models Including the Alfvén Time

To gain some insight into the isotropic IK regime, we come back to the case with zero cross-helicity ($E^{+} = E^{-}$) and consider a simple shell model. We start with the hydrodynamic shell model of Eqs. (6.57). We want to include the basic physical ingredient which makes the difference between hydrodynamic and MHD coupling,

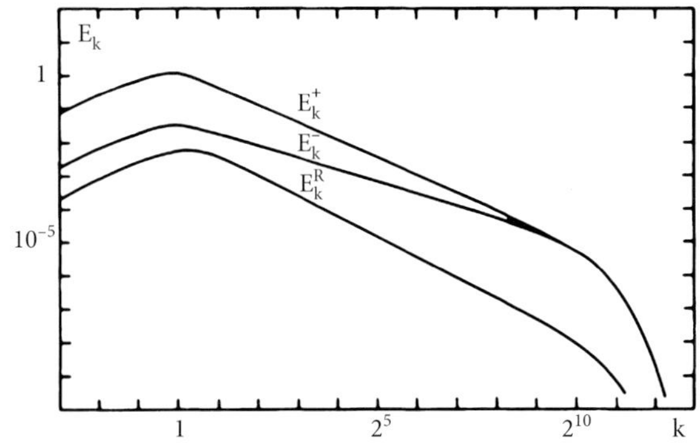

Figure 6.15 Energy spectra $E^{\pm}(k)$ with unequal amplitudes (cross-helicity is 88%), EDQNM spectral closure. Respective slopes are $m^{+} \simeq 1.8$, $m^{-} \simeq 1.2$. The residual energy $E_k^R = E_k^M - E_k^V$ is also represented, which adopts a k^{-2} slope (from [53]).

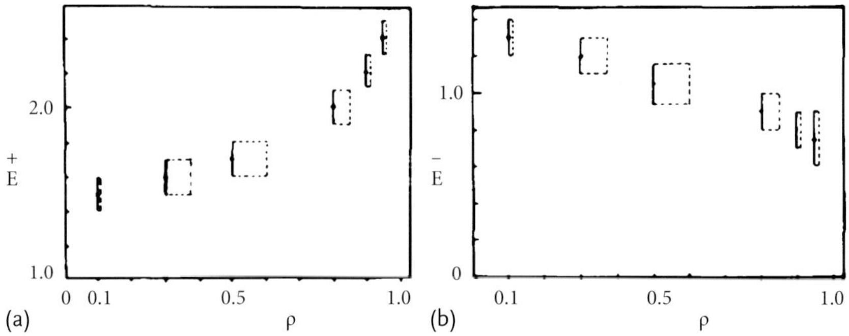

Figure 6.16 Decaying 2D incompressible MHD simulations with zero mean field: power-law indices m^{+} and m^{-} of the two Elsasser spectra. Abscissa: relative excess of z^{+} energy $\sigma_c = (E^{+} - E^{-})/(E^{+} + E^{-})$ (denoted as ρ in the figure). Each box represents a run, with its width representing the drift of slopes and cross-helicity during the calculation. From [49].

namely an isotropic version of the Alfvén dispersion relation $\omega = k B_0$, with k being the modulus of the wavenumber. There are two ways to do this. Either we use the complex Alfvén modulation $\exp(\mp i\omega t)$ of the original MHD equations, but in this case we have to consider the two (complex) Elsasser fields $z^{\pm}$, writing down additional quadratic coupling terms that conserve separately the two $E^{\pm}$ energies: this has been done by [56] based on the shell model by [57]. Or, as well, we can change a minimum in the Eqs. (6.57) that define the hydrodynamic shell model, keeping the single velocity field u_n, the only modification consisting in adding a factor $\cos \omega t$, which is the real value of the Alfvén term. We choose here the latter method which is simpler, thus rewriting Eq. (6.57) as follows:

$$d_t u_n = k_n u_{n-1}^2 \cos \omega_n t - k_{n+1} u_n u_{n+1} \cos \omega_{n+1} t - \nu k^2 u_n \quad (n = 2 \ldots N) \tag{6.82}$$

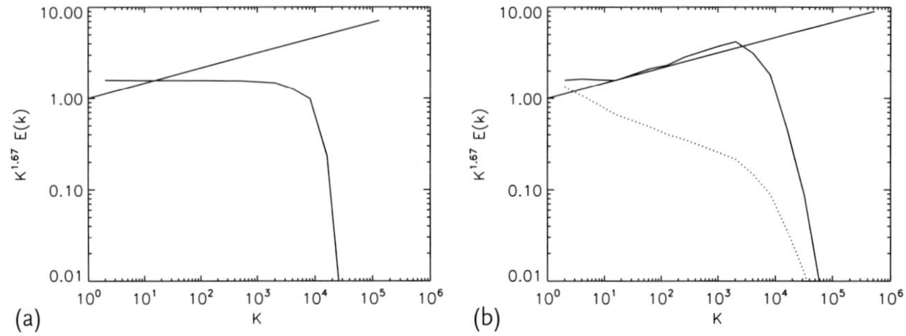

Figure 6.17 Shell model (Eq. (6.82)) with isotropized Alfvén term: compensated average spectra $k^{5/3}E(k)$ vs. wavenumber k. Abscissa: wavenumber $k_n = 2^n$. A plateau corresponds to the K41 scaling $E(k) \propto k^{-5/3}$ while the oblique line corresponds to the $k^{-3/2}$ scaling. (a) $B_0 = 0$, integration during 29 nonlinear times; (b) $B_0 = 0.75$ integration during 24 nonlinear times. The dotted line gives the ratio of Alfvén time over the nonlinear time.

with

$$\omega_n = k_n B_0 \,. \tag{6.83}$$

Continuous energy injection is achieved by fixing the shell amplitude $n = 1$ to unity: $u_1 = U_0 = 1$, which implies that the shell $n = 2$ is forced at the Alfvén frequency ω_2: $d_t u_2 = k_2 U_0^2 \cos\omega_2 t - k_3 u_2 u_3 \cos\omega_3 t$. As a result of this forcing, all shell amplitudes should show the local Alfvén frequency as well as the low frequency of shell number 2.

Figure 6.17 shows averaged compensated spectra $k^{5/3}E(k)$ (with $E(k_n) = u_n^2/k_n$) obtained by integrating the shell model (Eq. (6.82)) with $B_0 = 0$ (a) and $B_0 = 0.75$ (b). In both cases, the spectra are averaged during the second half of the calculation. In the case with $B_0 = 0$, the $-5/3$ scaling is obtained, while with $B_0 = 0.75$ the average spectrum follows closely the $k^{-3/2}$ IK scaling, albeit with a small large-scale $k^{-5/3}$ range. The dotted line gives the average ratio of Alfvén to nonlinear time u/B_0 vs wavenumber k: it is smaller than unity for $k \geq 4$, close to the observed spectral break.

The exact instantaneous energy flux $F_n(t)$ (cf. Eq. (6.45)) of the shell system (Eq. (6.82)) is

$$F_n(t) = k_n u_{n-1}^2 u_n \cos\omega_n t \tag{6.84}$$

which after time-averaging is close to the approximate expression given previously in Eq. (6.71). Several estimates of the energy flux are given in Figure 6.18a vs. wavenumber k, for the case $B_0 = 0.75$. The top panel of Figure 6.18a gives the average of the exact flux (Eq. (6.84)). It is reasonably flat, which proves that statistical equilibrium has been reached during the time interval considered. The second form in Figure 6.18a is the K41 form (Eq. (6.46)). One sees that it is approximately flat at large scales only. On the contrary, the IK form of the energy flux (Eq. (6.77)) is flat in the range $16 \leq k \leq 2000$ which is consistent with the respective scalings observed above for the energy spectrum for this case $B_0 = 0.75$ (bottom panel).

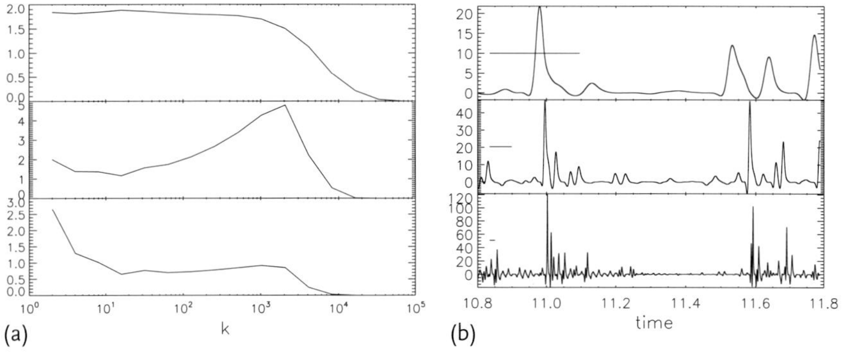

Figure 6.18 Shell model (Eq. (6.82)): energy flux in the case $B_0 = 0.75$. (a) Time-averaged energy flux versus wavenumber k. (Top) exact flux $F(k_n) = \langle F(n,t)\rangle$; (Middle) K41 flux $F_K(k_n) = kU_n^3$; (bottom) IK flux: $F_{IK}(k_n) = kU_n^4/B_0$ where $U_n = \langle |u_n(t)|\rangle$; (b) energy flux vs. time for shells $n = 4, 6, 8$, hence wavenumbers $k = 16, 64, 256$ (from top to bottom); in each shell, the length of an Alfvén period $t_a = 2\pi/\omega_n$ is indicated by a horizontal straight line.

Figure 6.18b shows the detailed time behavior of the (nonaveraged) energy flux during about two large-scale turnover times. The shells considered lie within the $k^{-3/2}$ range. One sees that the main events responsible for the energy transfer are bursts with time scale *shorter* than an Alfvén time and that all of them correspond to *direct* transfer events (in other words, transfer to smaller scale with a positive energy flux), for all three shells considered here. This requires the transfer event to be well correlated with the Alfvén oscillating term. Moreover, for the energy flux to be of the IK form (in other words, $\propto ku^4/B_0$ instead of ku^3), one expects the correlation with the Alfvén modulation $\cos \omega_n t$ to decrease progressively when wavenumber increases. A larger percentage of *inverse* transfer events (with negative flux) for shell 8 than for shell 4 is indeed apparent in the figure. When considering another chaotic shell solution with $B_0 = 0.1$, which leads to the $k^{-5/3}$ spectral scaling, such inverse transfer events are absent.

In conclusion, our shell model provides an interesting example of IK scaling in a chaotic system. However, the result appears difficult to generalize: increasing the B_0 amplitude does *not* systematically lead to the IK scaling. Similar results have been reported by [56], based on the complex MHD shell model of [57]. In this model, the $-3/2$ spectral index is obtained only if the linear coupling with the mean field is replaced by a coupling with the large-scale (time-dependent) magnetic field. Otherwise, a shallower spectrum with index -1.3 is obtained.

6.2.5 Anisotropic Regimes

What can be predicted when the mean field is large compared to the fluctuation amplitude b_{rms}, and when perpendicular and oblique/parallel wavenumbers are not coupled strongly enough so that most of the nonlinear coupling terms are nonresonant? A good approach for this situation is provided by the reduced MHD limit.

6.2.5.1 Reduced MHD and Shell-RMHD

We now consider the limit $\epsilon = b_{\rm rms}/B_0 \ll 1$. An important approximation can be derived from MHD in this limit. The approximation stems from the remark that the stiffness of the field lines has a tendency to suppress fluctuations with short wavelengths along the field lines, while gradients across the field lines are not constrained. As a result, typical eddies adopt an aspect ratio on the order of ϵ:

$$l_\perp/l_\parallel = O(b_{\rm rms}/B_0)\ . \tag{6.85}$$

This is compatible as we will see with the anisotropy observed in direct MHD simulations (for example, Figure 6.13b). Moreover, the linear dynamics lead to $u_\perp \simeq b_\perp$. We refer the interested reader to the complete derivation by [58, Chapter 8]. The main conclusion is that in this limit, called "reduced MHD", one can neglect the nonlinear terms involving parallel gradients, as well as parallel polarizations. The reduced MHD (RMHD) equations read:

$$\begin{aligned} \partial_t z^\pm \mp B_0 \partial_x z^\pm + z^\mp \cdot \nabla_\perp z^\pm + \nabla_\perp P &= 0 \\ \nabla_\perp \cdot z^\pm &= 0 \\ z_x^\pm &= 0\ . \end{aligned} \tag{6.86}$$

It is important to remark that, in spite of the suppression of the parallel components of the gradient in the nonlinear term, the resulting turbulence is not 2D, which would be the case only in the limit where the $z^\pm$ do *not* depend on x at all, but fully 3D. Moreover, the x variations are quite alive and active in the nonlinear coupling, in spite of the disappearance of the x component of the ∇ in the nonlinear term. To see this, remark that the different perpendicular planes develop their own (perpendicular) cascade, starting from initial (or forced) differences, that is, parallel gradients. As a consequence, the parallel gradients are permanently regenerated by the perpendicular turbulence. In other words, they are not simply frozen and/or "mixed" by the propagation along the mean field. We shall return to this point soon.

Shell RMHD models We present now shell models based on the RMHD equations, as these prove to be a very powerful tool to study anisotropic turbulence. By construction, shell models (as the ones considered here up to now) are isotropic. To keep the structure of the RMHD equations, one uses a shell model to describe the nonlinear coupling involving the (2D) perpendicular gradients, and, as in RMHD, we add the linear propagation terms along the mean field B_0. We will call this the "Shell RMHD" model. The equations read [59–62]:

$$d_t z_n^\pm \pm B_0 \partial_x z_n^\pm = T_n^\pm - \nu k_n^2 z_n^\pm + f_n^\pm \tag{6.87}$$

where $k_n = 2^n k_0 (n = 0 \ldots N)$ are the perpendicular wavenumbers, $f_n^+ = f_n^- = f_n(x,t)$ represent the large-scale (kinetic) forcing (nonzero for $n = 0, 1, 2$), and the

$T_n^{\pm}$ are the nonlinear terms. The nonlinear terms are akin to the RHS of Eq. (6.57) with the difference that they are here based on a larger set of quadratic terms allowing us to conserve separately a larger number of inviscid invariants, in particular E^+ and E^-. Formally they are of the form $T_n^{\pm} = Ak_m z_p^{\mp} z_q^{\pm}$ with m, p, q being close to n, and replace the convolution terms resulting from the Fourier transform with respect to the perpendicular coordinates of the RMHD equations Eq. (6.86).

6.2.5.2 Strong Regime – $k_\perp^{-5/3}$ Scaling – Critical Balance and Extensions

Consider a given point x on the mean field axis. The turbulent cascade in the perpendicular plane leads to excitation of a $k_\perp$ spectrum at this x coordinate. Consider the signal at scale $1/k_\perp$, namely $u(x, k_\perp)$; it varies with time in a chaotic manner. If all planes are identical (pure 2D case), then these fluctuations remain identical in all planes. But if it is not the case, then the gradients along the mean field should propagate as Alfvén waves along the mean field in both directions.

To quantify the anisotropy, we need to relate the parallel correlation length $L_\parallel$ of the signal with the perpendicular scale $1/k_\perp$. Let us thus return to the Alfvén waves propagating the parallel gradients. The random signal $u(x, k_\perp)$ should have a definite correlation time. If no smaller parallel scale is present from start (or forced into the system), then the Alfvén wave packets should reflect this correlation time by exhibiting a (parallel) correlation length, which should be equal to the Alfvén speed B_0 multiplied by the correlation time:

$$L_\parallel = B_0 t_{\text{cor}} . \tag{6.88}$$

Large eddies will in principle have a large correlation time and so the associated Alfvén perturbations (associated with the birth and death of these large eddies) will have time to travel a long parallel distance $L_\parallel$. On the contrary, a small eddy will show a much smaller parallel extension $L_\parallel$. This is schematically illustrated in Figure 6.19.

It is tempting to call "Alfvén time" the time $t_\text{a} = L_\parallel/B_0$. Equation (6.88) then just reads $t_\text{a} = t_{\text{cor}}$. This may be considered a mere expression of the causality principle, as expressed by a system in which there is just one kind of waves, the Alfvén waves.

One goes a step further if we assume that the perpendicular turbulent cascade proceeds according to the K41 phenomenology, in which case we have

$$t_{\text{cor}} = t_{\text{NL}} = 1/k_\perp u . \tag{6.89}$$

Then the causality relation (Eq. (6.88)) becomes

$$t_{\text{NL}} = t_\text{a} \tag{6.90}$$

which is generally called the "critical balance" relation. Using the K41 scaling $u = U(k_\perp/k_0)^{-1/3}$, Eq. (6.90) becomes

$$k_\parallel = (U/B_0) k_\perp^{2/3} k_0^{1/3} . \tag{6.91}$$

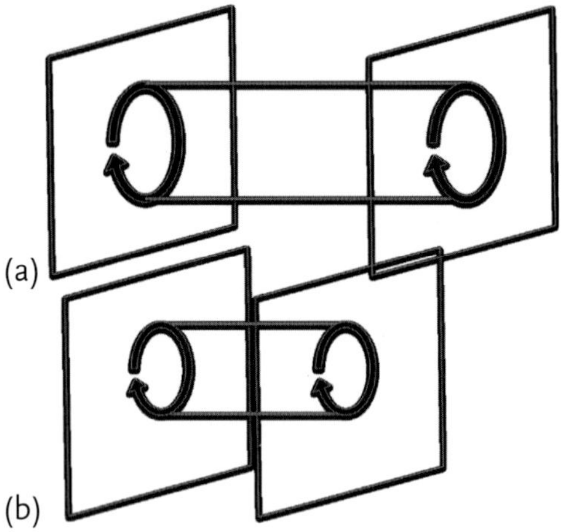

Figure 6.19 Reduced MHD: schematics illustrating the relation between eddy size, correlation time and parallel correlation length. Large eddies (a) have long correlation times (about a nonlinear or turnover time), which leads to a long parallel correlation length, here visualized by a long tube. The reverse is true for small eddies (b).

This relation should reveal in the $(k_\parallel, k_\perp)$ space, the boundary of the "strong" region where the nonlinear coupling remains resonant because it is relatively undisturbed by the Alfvén waves from the "weak" region where nonlinear couplings are nonresonant, hence either weak or vanishing. In the standard theory [63], the weak region is assumed to have vanishing nonlinear coupling, except if it is directly excited, as we will see below.

Numerical test of the critical balance boundary We now test the critical balance paradigm by integrating numerically the Shell RMHD equations. In order to allow the parallel extension of the spectrum to form as predicted by the critical balance, that is, to fill the "strong" part of the $(k_\parallel, k_\perp)$ space satisfying $t_{\mathrm{NL}} \le t_{\mathrm{a}}$, we must not fill from start the "weak" region $t_{\mathrm{NL}} \ge t_{\mathrm{a}}$, but instead the "strong" region:

$$t_{\mathrm{NL}} \le t_{\mathrm{a}} \, . \tag{6.92}$$

In Figure 6.20a, we show the result of such a numerical simulation. The figure shows a snapshot of the perpendicular energy spectrum, $E(x, k_\perp)$, at each point x along the mean field; the ordinate is $\log_2(k_\perp)$. The color indicates the amplitude of the spectrum compensated by $k_\perp^{5/3}$, so that a long vertical white "lane" represents the extent of a $k_\perp^{-5/3}$ inertial range along the vertical $k_\perp$ axis. If the activity would be uniform all along the mean field, one would get a uniform white strip with the same vertical extent. However, this is not the case: the turbulent activity, at the moment of the snapshot, is not uniform. This clearly reveals that the parallel "width" of an overactive region is decreasing when $k_\perp$ increases. Qualitatively, this picture is in agreement with the critical balance idea, as well as with the schematics given in Figure 6.19.

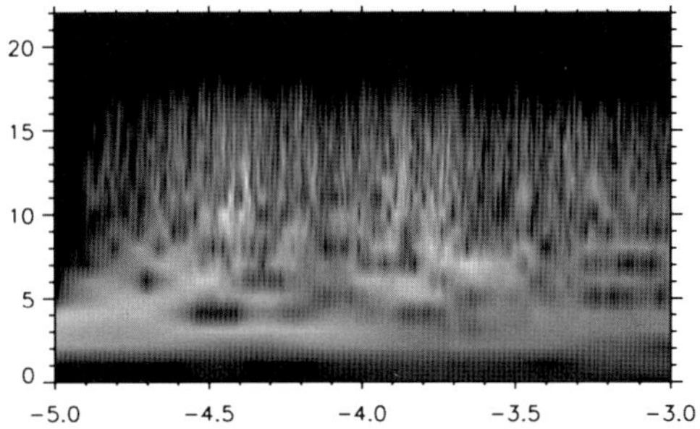

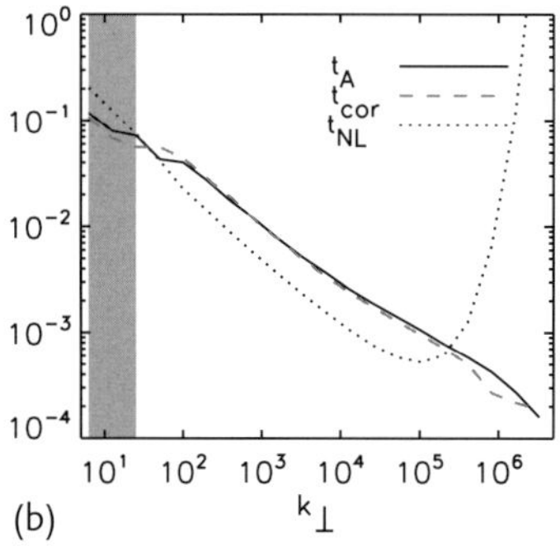

Figure 6.20 Shell-RMHD simulations with strong forcing ($t_{NL} \leq t_a$). (a) Snapshot of the energy spectrum $E^+(x, k_\perp)$; abscissa: coordinate x along the mean field B_0, ordinate: $\log_2(k_\perp)$. (b) Alfvén, correlation and nonlinear times vs $k_\perp$ for the signal $z^+(x, k_\perp, t)$. Forced scales are marked in gray. ((a) courtesy A. Verdini; (b) from [62]).

Figure 6.20b shows a more quantitative test. It gives the average scaling of the three characteristic times (Alfvén, correlation, and nonlinear times) as measured from the simulation data. One sees that the Alfvén time $t_a = L_\parallel / B_0$ and correlation times t_{cor} are equal on the whole range of $k_\perp$, while the nonlinear time $t_{NL} = 1/ku$ varies proportionally to the two other times in the inertial range $10^2 \leq k_\perp \leq 10^4$. The nonlinear time is systematically smaller by a factor of 2, which might be due to a small imbalance between the two Elsasser components (either $z^-_{rms}/z^+_{rms} \simeq 2$ or $1/2$ depending on time and x coordinate). Correspondingly, the energy spectrum $E(k_\perp)$ (averaged over the coordinate x) follows well the K41 scaling $k_\perp^{-5/3}$. Note that it is not expected that the spectrum follows the IK scaling $k_\perp^{-3/2}$, since (i) there is no mean field modeling in the perpendicular coupling (ii) there are no nonlocal coupling terms which could model the linear propagation along the local mean magnetic field.

The parallel width of the spectrum at a given wavenumber $k_\perp$ thus follows reasonably the predicted scaling $1/L_\parallel \propto k_\perp^{2/3}$. This should appear as a scaling of the "width" of the angular spectrum $E(k_\parallel, k_\perp)$ versus $k_\perp$. Figure 6.21 shows isocontours of the angular spectrum, in the $(\log k_\parallel, \log k_\perp)$ space. The critical balance line $t_a = t_{NL}$ is shown by two lines: the first (dotted) line corresponds to the ideal $k_\parallel \propto k_\perp^{2/3}$ law, while the dashed line corresponds to the effective $t_a = t_{NL}$ line. One sees that the isocontours of the spectrum are horizontal inside the region within the ideal critical balance boundary, showing that in this region the spectrum if a function of $k_\perp$ only. The ideal critical balance line coincides with a break in the isocontours; they become oblique outside the critical line. The oblique range is then followed by a region with quasi-vertical contours which denote a very steep spectral fall vs $k_\parallel$.

Figure 6.21b shows cuts of $E(k_\parallel, k_\perp)$ vs. $k_\parallel$ at fixed $k_\perp$. One can see that the oblique regions of Figure 6.21a correspond here to $k_\parallel^{-2}$ tails (the dot-dashed line indicating the $k_\parallel^{-2}$ scaling) which separate the strong coupling region from the region with negligible excitation. These parallel tails are most prominent for large-

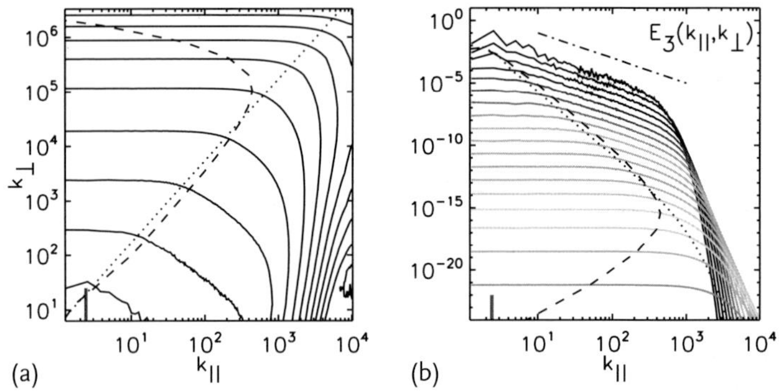

Figure 6.21 Shell-RMHD simulations with strong forcing: 3D and 1D spectra. (a) Contours of energy spectrum E_3 in the $(k_{||}, k_\perp)$ plane. (b) Horizontal cuts of E_3 vs. $k_{||}$ for $k_\perp = k_0\, 2^n$ with $n = 0$ to 17 from top to bottom. Dot-dashed line: $k_{||}^{-2}$ scaling. In both parts the dotted line is the theoretical boundary $t_a = t_{NL}$; the dashed line is the measured boundary $t_a = t_{NL}$. Forced scales are marked by a shaded area. (From [62]).

scale perpendicular eddies. They are actually generated by the transport by Alfvén waves of a *frequency* spectrum of perpendicular eddies that follows an f^{-2} scaling. The existence of this frequency spectrum is not contradictory with the fact that eddies of scale $1/k_\perp$ have a correlation time equal to $1/(k\,u)$; it simply reveals the coupling with the smaller perpendicular scales. The isotropic shell model (Eq. (6.82)) studied previously also showed high frequencies, which played an essential role in building a nonzero energy flux in this weak regime (see Figure 6.18b).

6.2.5.3 **Weak Regime – k^{-2} Scaling**

We just considered the case where the cascade is produced by forcing in a "strong" way, that is, in a region of Fourier space such as $t_{NL} \leq t_a$. In this way, the nonlinear coupling can proceed uninterrupted by the Alfvén waves, so leading to a standard Kolmogorov-like cascade along the perpendicular direction, with no special role played by the waves. A parallel extension of the spectrum $\Delta k_{||}$ thus grows in a such a way that the excited spectrum remains inside the region $t_{NL} \leq t_a$ at all perpendicular scales $k_\perp$, at least up to the dissipative range.

Now we consider what happens when we force in a "weak" region of the $(k_{||}, k_\perp)$ space, so that

$$t_{NL} \gg t_a\,. \tag{6.93}$$

In this case, the cascade is expected to be at least slowed down by the Alfvén waves. Let us thus try the Iroshnikov and Kraichnan idea again. Following the IK paradigm, the slow energy transfer time t_* is given by Eq. (6.76), that is, $t_* = t_a \times (\tau/t_a)^2 = \tau \times (\tau/t_a)$.

Now, the Alfvén time scale is not scaling here as in the previous isotropic case. In the isotropic IK regime, the Alfvén time is supposed to decrease as $1/(k\,B_0)$ as the cascade proceeds. Here, on the contrary, the Alfvén time actually is expected to keep its original large-scale value as cascade proceeds towards larger $k_\perp$.

$$t_a = t_a^0 = 1/(k_0 B_0) . \tag{6.94}$$

The reason why the Alfvén time remains fixed is that, as the cascade proceeds (presumably mainly along the $k_\perp$ axis), critical balance is able to generate Alfvén times larger, not smaller, than the already existing time (Eq. (6.94)). However, as the perpendicular cascade proceeds at larger $k_\perp$, the nonlinear time decreases, so that progressively the ratio t_a/t_{NL} should increase, and thus should reach unity at some wavenumber $k_\perp^*$.

Let us consider what happens before this critical wavenumber is reached, that is, in the range between the starting wavenumber k_0 and $k_\perp^*$. The characteristic parallel wavenumber, and the associated Alfvén time remaining fixed: $t_a = t_a^0 = 1/k_\parallel^0 B_0$, the effective transfer time t_* (Eq. (6.76)) becomes

$$t_* = t_a^0 \times \left(\tau/t_a^0\right)^2 = \tau \times \left(\tau/t_a^0\right) . \tag{6.95}$$

Hence, the energy flux reads

$$\langle F \rangle = k u^3 \times t_a^0/\tau = k u^3 \times \left(k u/k_\parallel^0 B_0\right) . \tag{6.96}$$

This leads to the scaling $u \propto k^{-1/2}$ and hence to a steep spectrum

$$E(k) \propto k^{-2} . \tag{6.97}$$

This weak $k_\perp^{-2}$ regime can be derived in a rigorous way [64].

One might be tempted to call this regime the anisotropic IK regime. However, as we shall see in a moment, this name is better given to a family of regimes including the 2D IK regime as a limiting value.

6.2.5.4 Transition from Weak to Strong Cascade

As mentioned above, as the weak perpendicular cascade proceeds to smaller and smaller perpendicular scales, the associated nonlinear time τ, initially much larger than the Alfvén time, should decrease continuously while the parallel Alfén time remains constant. So, at some step, the two time scales become equal, and then the cascade should switch to the strong cascade with its parallel extent dictated by the critical balance boundary.

The strong and weak anisotropic cascades have been looked for by several authors, in various situations. The control parameter considered is the large-scale ratio χ_0 between Alfvén and nonlinear times:

$$\chi_0 = t_{a0}/\tau_0 = k_{\perp 0} u_0 / k_{\parallel 0} B_0 \tag{6.98}$$

the index 0 indicating the forced scales. One expects to obtain a break in the spectrum at some critical scale $1/k_*$, with the break shifting to smaller and smaller scales as the parameter χ_0 is decreased. For reasonable values of χ_0, one should be able to make the transition visible at a given k_* within the available spectral range.

However, direct numerical simulations of the RMHD equations have failed to show this break. On the contrary, the simulations show a continuous variation of the spectral index as the parameter χ_0 is varied [65–67].

This negative result may be interpreted as being because the Reynolds number is too low in these simulations to allow a clear detection of two distinct slopes in a numerical simulation with moderate resolution. The issue has thus been reconsidered using the shell RMHD model which is able to reach a Reynolds number of order 10^6.

Shell RMHD results The control parameter is again the ratio $\chi_0 = t_a/\tau = k_\perp u/k_\| B_0$ associated with the forcing term. In Figure 6.22 we show the average *perpendicular* total energy spectrum for the strong ($\chi_0 = 1$) and weak ($\chi_0 = 1/64$) forcing cases, compensated by $k_\perp^{5/3}$. With strong forcing (Figure 6.22a), the scaling is close to $k_\perp^{-5/3}$ in the interval $100 \leq k_\perp \leq 10^4$. With weak forcing (Figure 6.22b) one sees two power-laws, the scaling being $k_\perp^{-2}$ at large scales and $k_\perp^{-5/3}$ at small scales. Long integration times ($25 t_{NL}^0$) at $Re = 10^6$ are needed to reveal this composite spectrum [62]. One finds equipartition between magnetic and kinetic energies in the weak $k_\perp^{-2}$ range, while magnetic energy dominates by a uniform factor $\simeq 2$ in the strong range (not shown). Also, the characteristic times are found to follow all the phenomenological rules mentioned above, namely, (i) in the weak range with k^{-2} scaling, the correlation time of the signal is found to be constant, with the nonlinear time decreasing (ii) at the scale where the two times become comparable, then we enter the range with $k^{-5/3}$ scaling, with the correlation time and the nonlinear time decreasing together at comparable levels. In both inertial ranges, the correlation time and the Alfvén time ($t_a = L_\|/B_0$) are equal, which is satisfying, and shows that the elementary arguments used previously are correct.

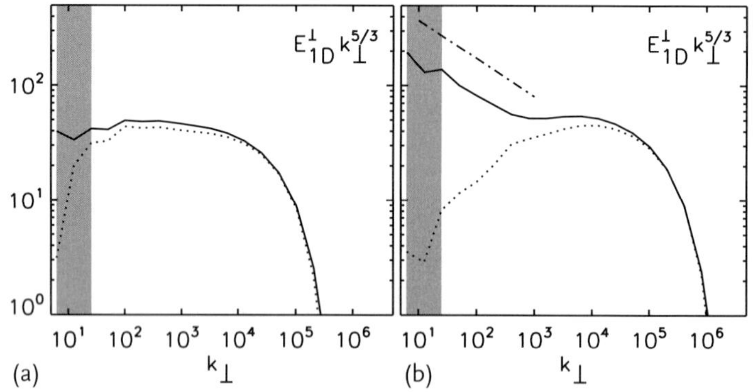

Figure 6.22 Shell RMHD simulations with (a) strong forcing: $\chi_0 = 1$. (b): weak forcing $\chi_0 = 1/32$. Average *perpendicular* (total) energy spectra compensated by $k^{-5/3}$ (solid lines). Perpendicular forcing scales are indicated as shaded areas in each part. Dotted lines: reduced spectra built from the 3D spectrum with $\chi < 1/2$ excitation suppressed. The dot-dashed line is the $k_\perp^{-2}$ scaling. (From [62]).

6.2.5.5 "Strong" Regime with Small-Scale Cross-Helicity – $k_\perp^{-3/2}$ Scaling

We have seen previously that in the case of the isotropic IK regime (see above the Section 6.2.4.2) a large-scale crosshelicity ($E^+ \gg E^-$) leads to differing spectral slopes for the two spectra E^+ and E^-. Here we consider the role of small-scale crosshelicity, which has been proposed to explain the existence of a $k_\perp^{-3/2}$ spectrum in the anisotropic strong regime, that is, excluding any cascade in the nonresonant region of $(k_\parallel, k_\perp)$ space. By small-scale helicity we mean that globally the two species $z^\pm$ have equal energies, but that locally one energy dominates the other, with the normalized crosshelicity growing with wavenumber.

This is illustrated in Figure 6.23, which shows the crosshelicity normalized as $\sigma_C = \boldsymbol{u} \cdot \boldsymbol{b}/(|u||b|)$, while our usual definition is $\sigma_c = (z^{+2} - z^{-2})/(z^{+2} + z^{-2})$. Simple algebra shows that these two definitions coincide if equipartition between kinetic and magnetic energies holds. Note that the definition of σ_C coincides with the cosine of the angle between velocity and magnetic field. From this viewpoint, regions with highly aligned fields correspond to regions with $E^+ \gg E^-$, while the reverse is true for antialigned regions. The figure shows that alignment and antialigned regions indeed coexist inside a region with zero global crosshelicity, with possible increase of the alignment when filtering the fields at smaller scales.

The important point is that, if a precise scaling exists for the local alignment then the nonlinear coupling will be slowed down (in the perpendicular plane) in a scale dependent way, that will lead to departures from the usual strong $-5/3$ index. Let us write the strong expression of the two energy flux, starting from Eq. (6.67) (we omit the $\perp$ index for wavenumbers):

$$F_\pm = k(z^\perp)^2 z^\mp . \tag{6.99}$$

We assume that, although globally both energies $(z^\pm)^2$ are globally comparable, they are not when looking at subdomains as in Figure 6.23. In other words either $R^- = z^-/z^+ < 1$ or $R^+ = z^+/z^- < 1$. Let us denote for a while z^- the local minor field and by z^+ the local dominant field. Then we assume that

$$z^-/z^+ \propto k^{-\alpha} . \tag{6.100}$$

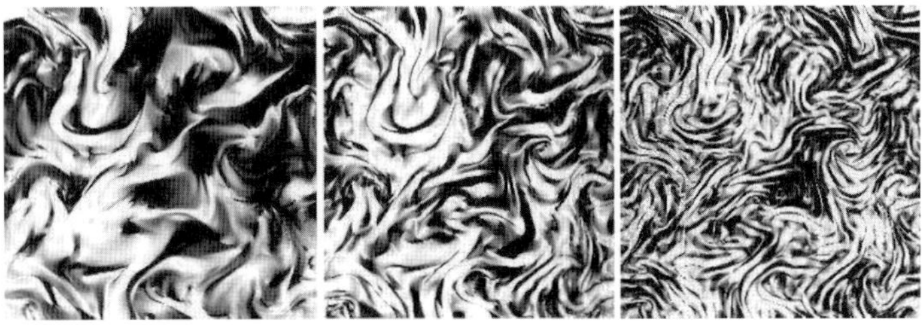

(a) (b) (c)

Figure 6.23 RMHD simulations, alignment regions at different scales. Plotted is $\cos\theta = \tilde{u}.\tilde{b}/(|\tilde{u}||\tilde{b}|)$ in a plane perpendicular to the mean field B_0, where $\tilde{u}$ and $\tilde{b}$ are filtered at the scale $1/k_f$. (a–b) $k_f = 1, 4, 10$. (From [68]).

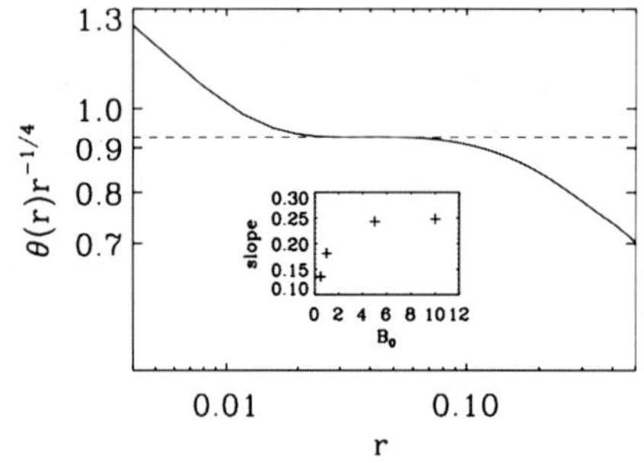

Figure 6.24 RMHD simulations, anisotropic turbulence with $k_\perp^{-3/2}$ energy spectrum: velocity–magnetic field alignment angle θ (compensated by $r^{1/4}$) vs. scale $r \simeq 1/k_\perp$. The theoretical scaling $\theta \propto r^{1/4}$ is reached in the center of the domain. The insert shows how the scaling index of θ reaches the value 1/4 as the mean field ratio B_0/b_{rms} increases. From [69].

We will thus have in a subdomain where the dominant species is z^+:

$$F_+ = k(z^+)^2 z^- \simeq k(z^+)^3 k^{-\alpha} . \tag{6.101}$$

Hence, the resulting scaling law will be $u \propto k^{(\alpha-1)/3}$, and the corresponding energy spectrum $E(k) \propto u^2/k \propto k^{(2\alpha-5)/3}$. Now, if the scaling index α of the imbalance factor R is precisely $\alpha = 1/4$, then the solution of Eq. (6.101) is $z^+ = k^{-1/4}$ for the locally dominant field, which means that the energy spectrum is $E(k) = k^{-3/2}$.

The z^-/z^+ scaling has been measured in direct simulations by [69] and is found to follow precisely the index $\alpha = 1/4$, while the energy spectrum scales as $k_\perp^{-3/2}$. This proves that, at least in the case of these simulations, the slope is indeed a direct consequence of the small-scale crosshelicity scaling. Figure 6.24 shows the 1/4 scaling index of the related angle θ between the velocity and the magnetic fields, which shows the same scaling as z^-/z^+ in the limit of high alignment.

Let us see how we may apply critical balance in this context. In this regime, contrary to the isotropic IK regime in which the correlation time of the signal is the shortest one, namely the Alfvén time based on b_{rms}, the correlation time is long, namely, it is (assuming z^+ is the dominant field) $t_+ = 1/kz^- = 1/kz^+ \times k^{1/4} \propto k^{-1/2}$. Applying the critical balance thus leads to the following anisotropic scaling [70]:

$$k_\parallel \propto k_\perp^{1/2} . \tag{6.102}$$

The anisotropy is thus increasing faster with perpendicular wavenumber than for the strong anisotropic $k_\perp^{-5/3}$ regime with strict balance between Elsasser species.

6.2.5.6 **3D IK Regime**

None of the previous 3D regimes with mean field considered up to now has the IK cascade as the perpendicular cascade. However, there is no reason why this should be so. Below, we first attempt to define theoretically what could be the 3D version of the IK regime, then we turn to numerical simulations to show what it really is, and finally we explain why it is so.

Perpendicular IK cascade with critical balance The 2D IK regime is isotropic and purely in the 2D plane: in a sense, it is an extreme example of the 3D anisotropy models. Indeed, in this regime, fields do not depend on the x coordinate along the mean field: the parallel extension "width" of the spectrum in Fourier space is zero. Can we extend the 2D IK regime to a 3D version with finite anisotropy?

A natural idea is to see what comes up by applying the critical balance scenario to the perpendicular 2D IK regime. Let us first remark that assuming balance between the long characteristic times of the cascade (either the transfer time or the nonlinear time) and the parallel Alfvén time would be wrong. Indeed, as seen in Section 6.2.5.2, the basic relation underlying the usual critical balance relation is the causal relation (Eq. (6.88)) which relates directly the correlation time of the perpendicular cascade and the Alfvén time.

In the strong regime, this relation is equivalent to the usual critical balance statement that relates the nonlinear time and the Alfvén time, because the nonlinear time *is* the correlation time of the signal at the scale $1/k_\perp$. However, in the present case of a weak perpendicular cascade, the correlation time of the signal $u_\perp(k_\perp)$ is by definition shorter than the nonlinear time: the correlation time is actually the Alfvén time *in the perpendicular plane*, that is the one based on the local mean field within the perpendicular plane, that is, b_{rms}:

$$t_{\mathrm{a}\perp} = 1/(k_\perp b_{\mathrm{rms}}) . \tag{6.103}$$

So the basic anisotropy relation reads

$$t_{\mathrm{a}\perp} = t_{\mathrm{a}} \tag{6.104}$$

or as well

$$k_\perp b_{\mathrm{rms}} = k_\parallel B_0 . \tag{6.105}$$

In spite of its clear difference with standard usage, we will still call in the following "critical balance" the relations above.

The anisotropy we have just found has two main properties: (i) it is scale-independent, and (ii) it leads to a full 2D regime in the limit of infinite B_0. Does a turbulent regime with such an anisotropy exist? Several authors have reported linear (or quasi-linear) relations between the spectral anisotropy and the ratio b_{rms}/B_0 (for example, [71]). We consider now in some detail a turbulent regime which seems to be the true extension to 3D of the 2D IK regime.

The 3D IK regime: simulations We just described a possible extension of a 2D IK regime to a 3D case with strong mean field. Its main property is that it remains a perpendicular cascade, with a scale-independent anisotropy resulting from the critical balance between the perpendicular and parallel Alfvén times. No published simulation with this property has been described. We now describe simulations [72] which instead of showing a mere perpendicular scaling, show isotropic scaling, with the anisotropy being thus concentrated on the amplitude, not the scaling. As we will explain in a moment, this regime with isotropic scaling is actually built on the previous one, which in a sense is an intermediary stage. It may be considered the true 3D extension of the 2D IK regime.

A snapshot of the angular spectrum $E(k_\parallel, k_\perp)$ is shown in Figure 6.25c, obtained via a simulation of incompressible MHD in the case with $b_{\rm rms}/B_0 = 1/5$, with isotropic large-scale forcing within the range $1 \leq k \leq 2$. Though the forcing is isotropic, the conditions for our "critical balance" condition (Eq. (6.105)) are expected to hold for $k \geq 5{-}10$.

For comparison, we show in Figure 6.25a the theoretical perpendicular IK regime with its critical balance extension (IK$_{\rm CB}$ in the following), namely with the $k_\perp^{-3/2}$ scaling, filling the space $k_\parallel B_0 \leq k_\perp b_{\rm rms}$ (Eq. (6.105)). Figure 6.25a,c are not directly comparable as the former shows the spectrum expected when measured in the local mean field frame, while the latter shows the spectrum as measured (for example, [73]) in the absolute frame of reference defined with respect to the global mean field B_0. To allow a direct comparison, we show in Figure 6.25b the spectrum expected from the IK$_{\rm CB}$ regime when measured in the absolute frame. This is obtained from the spectrum in Figure 6.25a by rotating the spectrum randomly around the perpendicular axis with an random angle having a gaussian distribution of rms amplitude $(2/\sqrt{3})(b_{\rm rms}/B_0)$. In drawing the spectrum with critical balance in Figure 6.25a, we assumed a rapid decrease of the 3D energy spectrum outside the critical balance domain $\chi = t_{{\rm a}\parallel}/t_{{\rm a}\perp} = k_\perp b_{\rm rms}/(k_\parallel B_0) \geq 1$ proportional to χ^{15}. This choice is arbitrary, but adopting a much shallower slope as χ^5, while resulting in a strong smoothing of the spectrum boundary in Figure 6.25a, actually does not change substantially the spectrum in Figure 6.25b.

This change of frame results in a substantial increase of the parallel width of the IK$_{\rm CB}$ spectrum. In spite of this transformation, Figure 6.25b,c show substantial differences. In Figure 6.25c, the angular spectrum shows a region around the mean field direction that is almost devoid of excitation, which is not the case in Figure 6.25c. Moreover, in Figure 6.25b the energy contours are not homothetic, while in Figure 6.25c they are. Actually, the scaling with k in Figure 6.25c has been found to be isotropic, that is, the 3D spectrum actually follows the same $k^{-3/2}$ scaling along all radial directions, including the radial one. In this sense, this regime deserves the name of 3D IK regime.

The self-similarity of the spectral isocontours of the simulation actually extends into the dissipative range: all contours are homothetic. The dashed line in Figure 6.25c actually marks the critical balance boundary: it does not bound in any way the energy containing part of Fourier space, even approximately, contrary to the case in Figure 6.25b.

Actually, Eq. (6.105) happens to describe one of the properties of the turbulent regime shown in Figure 6.25c, if we interpret it in a way different from usual, namely as a measure of the respective inertial range in the parallel direction $k_{d\parallel}$ versus the inertial range in the perpendicular one $k_{d\perp}$. In other words, it gives the *aspect ratio* of the angular spectrum, which indeed scales as:

$$k_{d\parallel}/k_{d\perp} = b_{\rm rms}/B_0 \,. \tag{6.106}$$

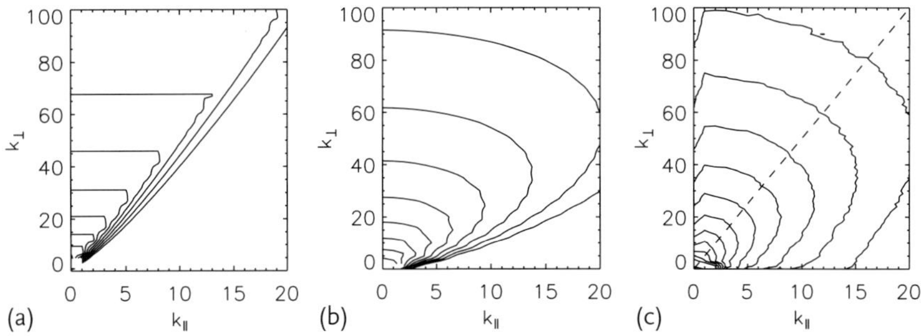

Figure 6.25 Perpendicular IK regime with critical balance vs. 3D IK regime – angular spectrum $E(k_\parallel, k_\perp)$ with $b_{rms}/B_0 = 1/5$. (a,b) The theoretical 2D IK regime + critical balance; (a) angular spectrum viewed in the local frame of the local mean field. (b) The same spectrum viewed in the absolute frame, taking into account a fluctuation of the direction of the local frame $\simeq (2/\sqrt{3})b_{rms}/B_0$. (c) The numerical regime found by [72], with isotropic forcing on $1 \le k \le 2$. Dashed line: line $k_\perp = k_\parallel/5$ marking the critical balance (Eq. (6.105)). From [74].

More generally, the form of the spectrum is

$$E(k_\parallel, k_\perp) = f(k)A(\theta) . \tag{6.107}$$

This relation holds as well in the inertial range where $f(k) \propto k^{-3/2}$ and in the dissipative range. We will call this turbulence "3D IK regime".

The 3D IK cascade: scenario A scenario for the previous 3D quasi-isotropic cascade has been proposed in [74]. It has three steps. The first is the 2D IK cascade in the perpendicular plane. This cascade is based on the perpendicular Alfvén time $t_{a\perp} = (kb)^{-1}_{rms}$, *not* on the mean field B_0. Note that, indeed, numerical simulations have shown that the inertial range in the perpendicular plane does *not* vary with B_0: only the global aspect ratio varies, that is, the extension of the parallel range along the parallel axis. In a second step, we allow for a parallel extension of the spectrum satisfying the critical balance between the parallel and the perpendicular Alfvén times (Eq. (6.105)), which generates the reservoir of quasi-perpendicular wave vectors that will be used for the third, final step. In this final step, a quasi-stationary energy flux feeds the rest of the Fourier plane, that is, oblique and parallel directions outside of the critical balance domain.

The energy flux associated with the latter oblique cascade is based on resonant (or quasi-resonant), that is, strong interactions. This is to be contrasted with the perpendicular IK cascade which is weak, that is, driven by nonlinear coupling permanently interrupted by the Alfvén oscillating factor, with the weakness increasing with wavenumber. The constraint of quasi-resonance imposed on the oblique cascade restricts the number of available interacting triads in such a way that the energy flux, although based on strong interactions, is reduced with growing wavenumber in exactly the same way as it is in the standard weak 2D IK cascade. The origin of the triad reduction in the oblique cascade, as well as its nature is explained now.

Starting with the angular spectrum sketched in Figure 6.25a, we want to arrive at the figure shown in Figure 6.25c, that is, we want to fill the void region outside the critical balance region.

The cascade uses triads of wave vectors that allow the excitation (and hopefully energy) to progress along pairs of oblique radial axes. The Figure 6.26 shows a series of such triads. The excitation is transmitted alternatively from one ray to the other (deserving the name of *ricochet* process) via an intermediary mode with wave vector q, which is to be taken within the critical balance region:

$$q_{\parallel} B_0 \geq q_{\perp} b \,. \tag{6.108}$$

The reason why we impose that the third mode of each triad satisfies this condition is simple: we want the coupling to be quasi-resonant. Examining Eqs. (6.69)–(6.70) shows that Eq. (6.108) is a sufficient condition, as it implies:

$$\omega t_{\mathrm{NL}} \leq 1 \,. \tag{6.109}$$

It is important to remark that, although the constraint of Eq. (6.108) resembles the usual critical balance condition, it is actually a much weaker condition, as it is applied only on *one* of the three vectors of each triad, instead to been applied to all the three wave vectors. This is why our condition allows energy to find a path that fills the whole $(k_{\parallel}, k_{\perp})$ plane instead of remaining confined to the usual critical balance domain around the perpendicular axis. We will call Eq. (6.108) the *detailed quasi-resonance* condition.

The second consequence of selecting the third side of the triads in this way is that it forces us to eliminate a large number of triads, keeping a percentage of triad $\simeq q_{\parallel}/q \simeq b/B_0$. The energy flux is thus reduced in this proportion, compared to the usual strong expression of the flux $F = k u^3$, that is, the resulting flux is $F = k u^3 b/B_0 \simeq k u^4/B_0$, which is exactly the expression of the IK flux that leads to $u \propto k^{-1/4}$ hence to the IK spectrum $E(k) \propto k^{-3/2}$. Thus, we proved that the spectral scaling in the oblique and parallel directions should be the same as the perpendicular one.

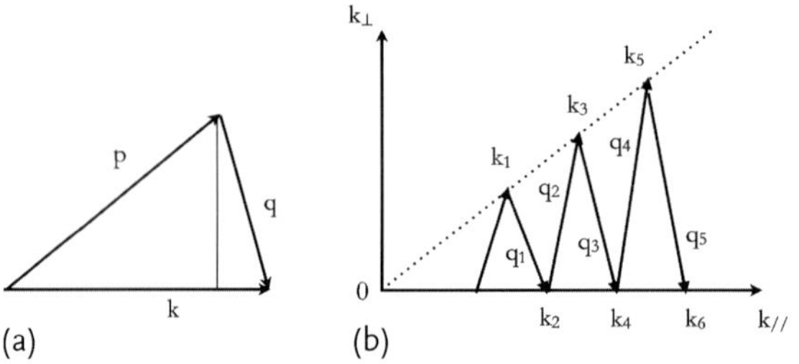

Figure 6.26 Schematics of the ricochet process for the 3D IK cascade. (a) One elementary triad of wave vectors satisfying detailed critical balance, that is, with the third wave vector of the triad (q) being quasi-vertical, thus leading to quasi-resonance (see text). (b) Ricochet process along two oblique rays (particular case with one ray being along parallel axis) showing the path followed by the excitation towards increasing wavenumbers.

Finally, balancing the dissipation and nonlinear flux in the perpendicular direction (where the flux is the IK expression based on the local frame, $k u^3/b_{\rm rms}$) and in the parallel/oblique directions (where it is $k u^3/B_0$) leads to the observed relation between spectral anisotropy and $b_{\rm rms}/B_0$ (Eq. (6.106)).

Residual and total energy spectra An important property of the 3D IK regime, which might be a general property of MHD, is that the so-called residual energy spectrum, that is, the difference between the magnetic and kinetic energy spectra, follows its own scaling law. A common idea is that in MHD the residual spectrum is zero, since Alfvén waves have on average equal kinetic and magnetic energies. However, in the 3D IK regime, there is a systematic magnetic excess, which goes to zero in the dissipative range, so that the scaling law of the residual energy spectrum is (see Figure 6.27a):

$$E_{\rm R}(k) = E_B(k) - E_V(k) \propto k^{-2} \,. \tag{6.110}$$

This scaling actually results from a more general relation which relates the total energy spectrum and the residual energy spectrum, that has been derived in the framework of closure equations (eddy-damped quasi-normal approximation or EDQNM). We give here a simplified and short derivation. The basic idea is that in a turbulent state the Alfvén linear equipartition effect is balanced by a local nonlinear dynamo effect. The linear Alfvén effect drives towards equipartition the kinetic and magnetic fields in a time of order $t_{\rm a} = 1/k b_{\rm rms}$, while the *local* dynamo effect drives on the contrary the system away from equipartition in a time of order $t_* = t_{\rm NL}^2/t_{\rm a}$ (Eq. (6.76)). So the spectrum is solution of the equation

$$E_{\rm R}(k)/t_{\rm a} = E_{\rm tot}(k)/t_* \tag{6.111}$$

or as well

$$E_{\rm R}(k) = (t_{\rm a}/t_{\rm NL})^2 E_{\rm tot}(k) = k E_{\rm tot}(k)^2 \tag{6.112}$$

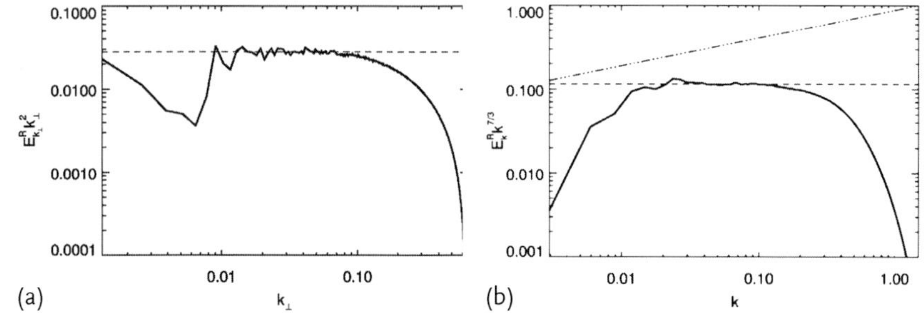

Figure 6.27 Residual energy spectrum in 3D MHD turbulence. (a) 3D IK regime, $B_0 = 5b_{\rm rms}$, residual spectrum compensated by $k_\perp^2$; (b) isotropic turbulence, $B_0 = 0$, residual spectrum compensated by $k_\perp^{7/3}$ (dotted dashed line: k^{-2} scaling). From [52].

which indeed leads to Eq. (6.110) when replacing $E_{tot}(k)$ by $k^{-3/2}$. The −2 index for the residual spectrum E_R has been found successively using the EDQNM spectral closure (Figure 6.15) and direct numerical simulations (Figure 6.27a).

The validity of Eq. (6.112) is actually not restricted to the 3D IK regime. It also works when the mean field is zero. In this case the derivation based on the EDQNM equations also leads to the long time scale t_* for the local dynamo effect [52]. In the isotropic case with no mean field, the authors [52] have found a $k^{-5/3}$ scaling for the total energy. When this scaling is injected in Eq. (6.112), one finds a residual spectrum scaling as $k^{-7/3}$. This is precisely what is obtained numerically (Figure 6.27b).It would be interesting to investigate whether the residual spectral index again becomes −2 again in the zero mean field regime with −3/2 index for the total energy found by [46].

6.2.6 Discussion

We have considered several different approaches to the problem of incompressible MHD turbulence with some emphasis on the case with strong mean field. Note that, while the results (numerical or theoretical) cannot be transposed immediately to other domains than MHD turbulence, like, for example, the electrostatic turbulence in tokamaks, one can hope to transpose with profit the approach, models, and methods. In the following we summarize our main points and discuss some of the unsolved issues.

6.2.6.1 Ubiquity of the −3/2 Scaling – Weak to Strong Transition

The case of the −3/2 spectral slope is particularly interesting. It can be obtained in many different flavors. First, it comes up in 3D MHD with no mean field [46]. In the 2D case with no mean field, the −3/2 scaling is clearly due to the phase decorrelation due to Alfvén waves (weak turbulence, for the “2D IK” perpendicular cascade). Note that in these first two cases, no mean field comes in: this role is taken by a *local* mean field of order b_{rms}. When a true global mean field B_0 is present, two ways to the −3/2 scaling exist: (i) the strong perpendicular $k_\perp^{-3/2}$ cascade with flux depletion due to a reduced coupling at small scale (in other words, the small-scale crosshelicity scaling); (ii) the isotropic $k^{-3/2}$ scaling resulting from a weak perpendicular cascade associated with a strong oblique/parallel cascade, but with flux depletion imposed by the quasi-resonance condition (“3D IK regime”). In both cases, B_0 has no effect on the (dominant) perpendicular cascade, which is controlled directly by the local mean field b_{rms} again. This remark is important to evaluate the dissipation rate, as we see below.

The case of the weak (k^{-2}) to strong ($k^{-5/3}$) transition is also interesting. We have seen that it has been well reproduced by numerical simulations of the Shell RMHD system (Figure 6.22). However, this transition is not working well in the several published works using RMHD simulations: in this case, a whole family of slopes (varying continuously with the control parameter) is instead obtained.

This negative result might be due to the limited Reynolds number reached in direct simulations. Another explanation would be that the simplified structure of the Shell RMHD system allows it to follow the predictions of the weak-strong anisotropic theory, which would be incomplete and not adapted to the real RMHD/MHD system. In fact, in two of the works dealing with the weak-strong transition [65, 66], the "weak" -2 slope is not a bounding value: indeed, the spectrum further steepens as the parallel Alfvén time decreases when the parallel forcing wavenumber is increased. To explain this further steepening, [66] have generalized the weak anisotropic regime, with the energy transfer time scaling as (cf. Eq. (6.95))

$$t_* = \tau \times (\tau/t_a^0)^\alpha \quad (\alpha \geq 1)\,. \tag{6.113}$$

The parameter α would grow from $\alpha = 1$ to larger and larger values when the large-scale forcing becomes weaker, resulting in a spectrum $E(k) = k^{-m}$ with $m = (5+3\alpha)/(3+\alpha)$, so allowing the further steepening from the -2 spectral index to -3.

6.2.6.2 Dissipation/Heating: B_0 vs. b_{rms}

Among the various regimes which have been considered here, several do adopt different forms for their energy flux. Does the turbulent heating change in this case? The answer depends on what becomes the flux at the injection scale, since this is precisely the turbulent heating, when turbulence is in a quasi-stationary state.

In the case of an ordinary hydrodynamic direct cascade, the energy flux reads $F = ku^3$, which means that the injection rate at scale $1/k_0$ (and dissipation rate as well) reads

$$F = k_0 u_{rms}^3\,. \tag{6.114}$$

This is also true for the standard anisotropic strong regime (with $k^{-5/3}$ scaling) with critical balance.

In the large-scale weak regime which is on "top" of the strong critical balance regime (Section 6.2.5.3), the flux reads as (Eq. (6.96)) $F = k_\perp^2 u^4/(k_\parallel B_0)$ which gives an injection flux at the largest scale $1/k_0$ which is much smaller than the strong form:

$$F = k_0 u_{rms}^3 (k_0 b_{rms}/k_\parallel B_0) \ll k_0 u_{rms}^3\,. \tag{6.115}$$

The inequality in the RHS holds because, by definition of the weak regime, one has at the injection scale $1/k_0$: $t_{a\parallel}/t_{NL} = k_0 b_{rms}/k_\parallel B_0 \ll 1$.

In the isotropic 2D IK regime, the flux reads $F = ku^3 u/b_{rms}$, which leads at the injection scale $1/k_0$ to the heating/injection rate $F = k_0 u_{rms}^3$, as $u_{rms} \simeq b_{rms}$. This is also true in the 3D IK regime, as the main cascade remains the perpendicular one: the dissipation rate does not depend on the mean field B_0, which has been checked numerically. The regime with strong perpendicular cascade and small-scale crosshelicity (with $-3/2$ spectral index), still has a zero global crosshelicity, hence in this case also the expression of the injection rate is the strong expression.

So, changing the regime will not automatically change the flux entering the system (that is, equal to the flux leaving the system, which is the dissipation rate). However, it will generally change the extent of the inertial range (for a given form of the dissipation terms). And, of course, the extent of the dissipation range, as well as the associated anisotropy will determine the kind of process able to dissipate the energy (or whatever invariant that holds at these small scales).

So the detailed process of dissipation is drastically dependent on the kind of phenomenology that holds for turbulence, but *the dissipation rate is most of the time not.*

This becomes wrong only if we consider the very anisotropic weak cascade (with k^{-2} spectral scaling), in which it is truly the mean field B_0 that gives the clock of the cascade, and not the largest scale eddies. One may also note that in real cases like the solar wind, the energy injection scale may be not so easy to determine.

6.2.6.3 The Magnetic Excess

We have seen that the simulations by [52] show a definite relation between the total energy spectral slope and the residual spectral slope. The same relation holds for two regimes studied in this paper: one with no mean field and $E_{\text{tot}} \propto k^{-5/3}$, one with strong mean field and $E_{\text{tot}} \propto k^{-3/2}$. In the former case, the residual spectrum is relatively flat ($E_R \propto k^{-2}$), in the latter case it is steeper ($E_R \propto k^{-7/3}$). Because of this slope difference, the magnetic excess at large scale is much larger in the latter case with large residual energy slope. This is to be compared with the zero mean field simulations by Lee *et al.* [46] in which different scalings appear, depending on the magnetic excess at large scale: in these simulations also, the $-5/3$ slope is associated with a larger magnetic excess than the $-3/2$ spectral slope.

6.2.6.4 Application to the Solar Wind Turbulence: A Short Summary

The solar wind turbulence provides the only observable set of turbulent data to which one can attempt to apply results/theories of MHD turbulence. The solar wind plasma is (i) quasi-collisionless, (ii) not incompressible, (iii) unsteady, and (iv) inhomogeneous. All these properties make a direct application of the theory a bit perilous. Nevertheless, this has been done from the very beginning of observations.

Spectral slopes and magnetic excess As has been shown in Figure 6.6, the magnetic field spectrum in the solar wind shows a nice $k^{-5/3}$ scaling on average. However, the *kinetic spectrum* shows an extremely clear scaling in $k^{-3/2}$ [75, 76]. Superficially, this resembles the situation encountered in the simulations already mentioned by [52]: there is an excess of magnetic energy which is largest at the largest scale. However, in the solar wind, the average kinetic spectral slope measured in the wind has a definite slope close to $-3/2$, while the residual energy scaling is not so well defined.

An additional important point is that the solar wind is actually not homogeneous: its average properties (wind speed, Alfvén speed, proton temperature) are systematically varying with the distance to the heliospheric current sheet (HCS) where the mean magnetic polarity changes sign. As spacecraft samples the plasma parameters at different distances to the HCS, turbulent properties also change: the *magnetic excess* varies, and at the same time the spectral magnetic and kinetic slopes do vary as well, in such a way that the magnetic excess is larger when the spectra are flatter [77]. So, qualitatively, the correlation between spectral slopes and magnetic excess is the same as that found in the simulations [52]. However, a basic difference between the solar wind turbulence and the latter simulations remains: the scaling quantities in the solar wind are the pair (kinetic, magnetic), while in the simulations they are the pair (total, residual). This is not equivalent.

Spectral anisotropy The anisotropy with respect to the mean magnetic field in the solar wind was studied early. There are basically two kinds of results: (i) global spectral anisotropy and (ii) local anisotropy.

The method for measuring the global anisotropy of a given field (mostly the magnetic field) is as follows: the fluctuations are sampled along the flow (which is basically radial), and the Taylor hypothesis applied, which allows us to deduce the spatial fluctuations from the temporal fluctuations. One then notes the relative direction between the mean magnetic field and the radial direction, and, adopting a symmetry hypothesis around the magnetic field axis, one builds progressively the autocorrelation function of the various fields vs. the distance between two points along different directions with respect to the mean field. The resulting autocorrelation figure depends strongly on whether the spacecraft is far or close to the HCS: when it is far from it, turbulence develops small scales preferentially along the mean field, while when close to the HCS, small scales develop more in the perpendicular directions [78].

The method for measuring the anisotropy with respect to the *local* mean field relies on structure functions and/or wavelets. The published results indicate an anisotropy which seems akin to the one predicted by the critical balance [79], but the variation of the anisotropy with the distance to the HCS remains to be studied.

6.3 Nonlinear Kinetic Physics

We have seen in Section 6.1 that thin interface layers (shocks and discontinuities) can be interpreted as the result of nonlinear propagation of MHD waves. We have seen in Section 6.2 how the turbulent interaction of a large number of wave modes is the cause of a cascade of energy from the larger to smaller spatial scales. These phenomena and others treated in the above section are compatible with an approach based on the plasma fluid equations. But waves can also interact with small subsets of plasma particles in a way that would not be compatible with a predictable closure equation. This requires the direct use of the Vlasov–Maxwell equations.

In Section 5.6, we dealt with the linear theory of wave–particle interaction. As we did, we base our approach on the simplified case of electrostatic plane waves in a nonmagnetized plasma. This context, in which Langmuir waves are the dominant wave mode, offers a good frame to show, with a limited amount of algebra, the most important concepts arising in the nonlinear physics of wave–particle interaction.

In this section, we insist on two important concepts: particle trapping and quasi-linear diffusion. They both, in different circumstances, control the process of saturation of wave growth and of wave damping.

6.3.1
Nonlinear Electrostatic Waves

We consider the simple case of electrostatic plane waves in a uniform nonmagnetized plasma. The ions are considered a neutralizing background. The oscillation are longitudinal (here along the x axis) and the Vlasov–Maxwell–Gauss equations for the electrons simply read:

$$\partial_t f + v\partial_x f + q/mE\partial_v f = 0 \tag{6.116}$$

$$\partial_x^2 \Phi = -\partial_x E = -q/\epsilon_0 \int f dv \,. \tag{6.117}$$

These equations describe the isotropic nature of the electrostatic waves in a nonmagnetized plasma. They are also appropriate for the longitudinal electrostatic oscillations in a magnetized plasma.

The linear theory of these equations was developed in Section 5.6, where the distribution function was considered the sum of an equilibrium function $f_0(v)$ and a perturbation $f_1(x, v, t)$ caused by the waves. It was shown that the wave modes for a given wavelength k are (for regular initial conditions) Langmuir waves. The amplitude of these waves was allowed to vary following an exponential law (exp $\gamma_k t$ where γ_k is the wave growth rate). The linear theory showed that according to the shape of the unperturbed distribution function $f_0(v)$, the waves can be damped ($\gamma_k < 0$) or unstable ($\gamma_k > 0$).

Because the electric field in Eq. (6.116) depends on the distribution function in Eq. (6.117), there is a quadratic dependence on the distribution function in the last term of the Vlasov equation. This is the origin of the nonlinear effects, neglected in Section 5.6, that are discussed below.

We focus our analysis on two cases that illustrate the importance of particle trapping and quasi-linear diffusion.

Particle trapping (Section 6.3.2) can be shown when there is only one sinusoidal wave, whose slowly varying amplitude is either large or small.

When there are several waves of small amplitude, the slow evolution of these amplitudes can be computed on the basis of a polynomial expansion. These waves can have peculiar characteristics that allow coherent interactions (6.3.3), or they can be the cause of diffusion processes acting on the particle distributions (6.3.4).

6.3.2 Particle Trapping

Let us start with a numerical experiment. A bump-in-tail instability is simulated with the δf code (Section 2.3), in a quasi-1D domain of 128×4 cells of size λ_D. The ions are considered a neutralizing background, and there are 1,000 macroelectrons per cell. The time step is $\Delta t = 0.05\omega_{pe}$. The boundary conditions are biperiodic. Figure 6.28 shows the space averaged distribution functions. The initial condition consists of a Maxwellian distribution of thermal velocity $v_{te} = 0.075c$, and a bump of mean velocity $v_b = 2.5v_{te}$ and of thermal velocity $v_{tb} = 0.00375c$. On the right-hand side, the time evolution of the electric energy is displayed in logarithmic scale. We can see that until $t \sim 150$, the growth of energy is exponential, as expected with linear theory. In this phase (for instance at time $t = 102$) the bump is still present.

After $t \sim 150$, the electric energy growth is no longer exponential, and after a few larger oscillations with a period $T_{NL} \sim 50$, it saturates with a finite level.

What happened?

Let us consider a single wave characterized by a sinusoidal electric field $E_0 \sin(kx)$, where E_0 is a slowly varying function of time. It is possible to find an exact solution to Eq. (6.116) and (6.117). Because there is a single wave, there is a single resonance velocity $v_{\phi k} = \omega/k$ and in the reference frame that moves at this velocity, the wave electric field is independent of time. In this frame, Eq. (6.116) is

$$\partial_t f + v\partial_x f + (q/m)E_0 \sin(kx)\partial_v f = 0 . \tag{6.118}$$

This equation can be solved through the method of characteristics, where, from the Vlasov equation (often called the "Liouville theorem" in this context), the characteristics are just the particle trajectories $[x(\tau), v(\tau)]$ and the invariant is $f[x(\tau), v(\tau), \tau]$. The particle trajectories are given by:

$$d_\tau^2 x = \frac{qE_0}{m}\sin(kx) \quad \text{and} \quad d_\tau x = v . \tag{6.119}$$

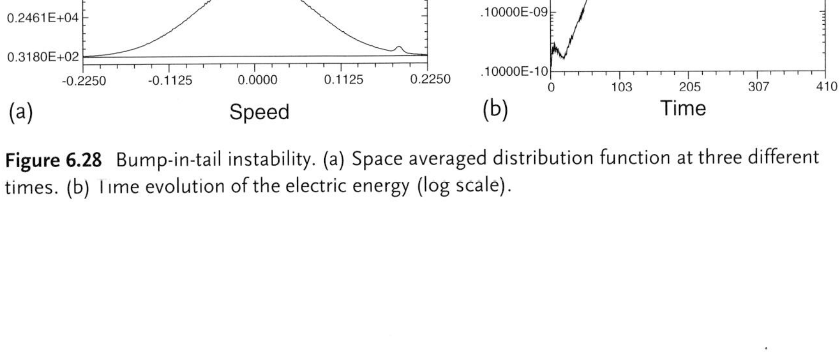

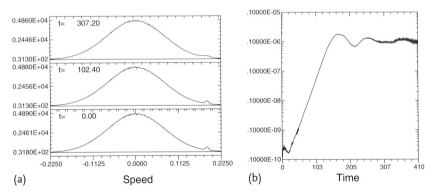

Figure 6.28 Bump-in-tail instability. (a) Space averaged distribution function at three different times. (b) Time evolution of the electric energy (log scale).

This equation has been solved in Section 1.4.3.2. From the Vlasov equation

$$d_\tau f\left[x(\tau), v(\tau), \tau\right] = 0 . \tag{6.120}$$

or equivalently, the evolution from time 0 to t is provided by

$$f\left[x(\tau), v(\tau), \tau\right] = f\left[x(0), v(0), 0\right] . \tag{6.121}$$

Then, from the motion of the untrapped and the trapped particles, and from the Vlasov equation, the evolution of the distribution function can be computed. The computation is tedious. Without deriving it formally, some features of the x, v space can appear. The trapped particles area is delimited by the two curves of equation

$$v \pm \frac{\omega_B}{k}(1 - \cos(kx))^{1/2} = 0 , \tag{6.122}$$

in the frame of the wave, or

$$\left(v - \frac{\omega}{k}\right) \pm \frac{\omega_B}{k}(1 - \cos(kx))^{1/2} = 0 , \tag{6.123}$$

in the frame of the observer. The bounce frequency ω_B is defined by Eq. (1.142).

When the wave is a result of an instability of linear growth rate γ_L, at $t\omega_B < 1$, the trapped particles do not yet influence the wave evolution and $\gamma \sim \gamma_L$. Later, the electric field is influenced by the trapped particles. Inside the area of the x, v space of the trapped particles, delimited by Eq. (6.123), the particles describe a loop. From Eq. (1.148), the particles with a larger motion amplitude β (closer to the separatrix) turn more slowly than the less energetic particles (closer to the phase velocity); therefore, there is a kind of phase mixing in the loop formed by the trapped particles. The consequence on the space averaged distribution function is the formation of a plateau, between the two corresponding bounding velocities $v \sim (\omega \pm \omega_B)/k$ given by Eq. (6.123). The time necessary to form that plateau depends on the dispersion of bouncing periods as a function of the energies of the trapped particles. It is of the order of a few ω_B^{-1}. Outside the trapping area, the particle motion is slightly perturbed, the perturbation being mostly sinusoidal. The averaged distribution function is deformed only in the vicinity of the trapping area.

The formation of a plateau in the averaged distribution function caused by the trapped particles quenches the instability; therefore, these slow oscillations tend to be damped, and the waves reach a more or less constant level on the time scale of a few ω_B^{-1}. O'Neil has made a precise analytical derivation of the evolution of the electric field.

Let us go back to the numerical experiment. Figure 6.29 shows the time evolution of the electric field and of the electron density at an arbitrary position (grid number 3, 2). Figure 6.30 shows the perturbation of the electron distribution function in the (x, v) phase space. More precisely, this is the distribution function from which the initial main Maxwellian (without the beam) has been subtracted. At time $t = 0$, only the unperturbed beam appears. At time $t = 102$, in the linear phase, the beam

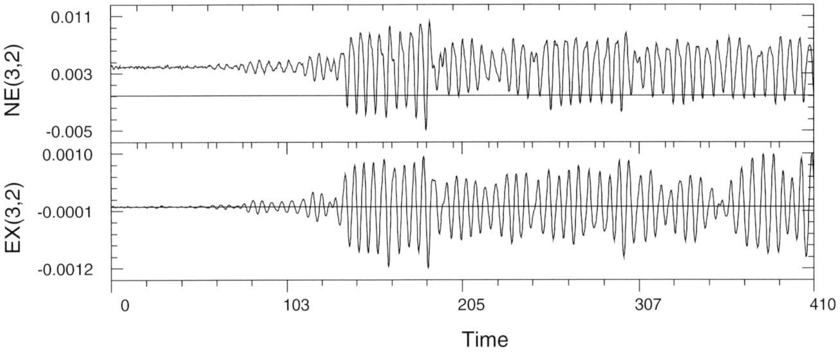

Figure 6.29 Same simulation as in Figure 6.28. Time evolution of the electric field E_x and of the electron density n_e measured at a fixed arbitrary position.

is perturbed as well as the distribution of velocities far from the beam velocity. At time $t = 307$, the beam distribution is more perturbed than for velocities far from the beam velocity. The beam is actually reshaped by the trapped particles that form loops. The shape of these phase space loops is delineated by a curve that is like the separatrix given by Eq. (6.123). We count nine loops, and this means that there are nine wavelengths along the simulation domain; this corresponds to $kc/\omega_{pe} = 5.85$. The high frequency oscillations of the electric field correspond to the wave frequency $\omega = 1.13\omega_{pe}$. The envelope of this signal starts ($t < 150$) with an exponential growth rate (and, therefore, a very weak level of excitation at the beginning). Then, the envelope is pulsed with the period $T_{NL} \sim 50$. We can compare this period with the bounce period $T_B = 2\pi/\omega_B = 2\pi(m/keE)^{1/2}$. From Figure 6.29, at time $t > 200$, we have $qE/mc \sim 10^{-3}$. With $kc/\omega_{pe} = 5.85$, we find $T_B = 43$, that is, very close to the roughly estimated nonlinear time $T_{NL} \sim 50$. This confirms that the bounce period of the trapped particles defines the characteristic time of the nonlinear evolution. From Eq. (6.123), the loops formed by the trapped particles are centered on the phase velocity $v_B = \omega/k$ of the wave, and their range of velocities is $\Delta v_B = 2\omega_B/k$. On the figure, we can read $v_B = 0.19 \sim v_b$ and $\Delta v_B = 0.045$. Numerically, we find $\omega/k = 1.13/5.85 = 0.19$ and $\omega_B/k = 0.126/5.85 = 0.021$. The figures correspond quite well, proving that simple estimates derived in the theoretical approach characterize effectively the nonlinear evolution of a quasi-monochromatic wave.

6.3.2.1 **Vortices of Trapped Particles in a Three-Dimensional Magnetized Plasma**

So far, we have not considered the velocity components v_y and v_z, but only v_x (noted v in the above section). In a real space (6D phase space) it is also possible to form vortices of trapped particles. They can have a 1D structure, being invariants along the directions y and z. But if the vortices are related to the x and v_x components, they can become very sensitive to small perturbations in the transverse directions y and z. Consequently, in a real multidimensional phase space, trapping can lose its efficiency on a time scale that correspond to the particle escape. This time depends

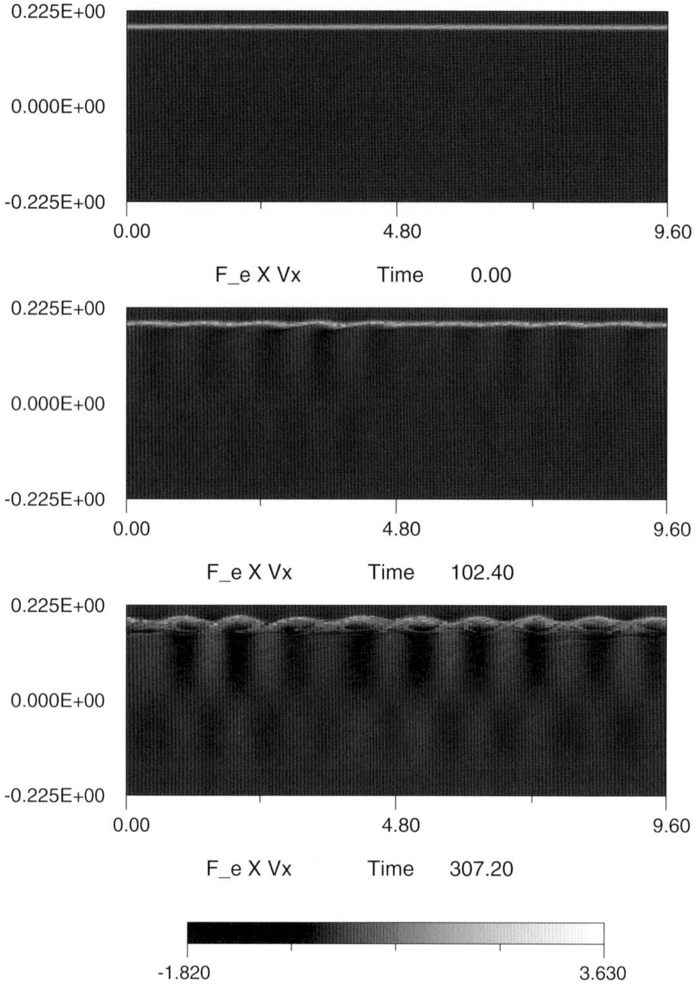

Figure 6.30 Same simulation as in Figure 6.28. Distribution function in the x, v_x space. In order to show the effect of the instability, the initial main Maxwellian distribution function has been subtracted. Therefore, this plot represents the initial bump-in-tail (time 0) together with the perturbations caused by the wave (time > 0).

on the transverse size of the vortices, and on the transverse thermal velocity. The trapping structures can also be destabilized by transverse plasma instabilities.

Nevertheless, if there is a magnetic field B_x in a 3D space, the instability of 1D trapping structures can be compensated by the cyclotron motion of the electron in the magnetic field. When the cyclotron frequency ω_c exceeds the trapping frequency ω_B, a transverse phase mixing combining these two components of the electron motion stabilizes the electron holes [80].

In other circumstances, the simple 1D vortices of trapped electrons can be destabilized. This does not imply that they disappear. They can also break out in a series of 2D or 3D structures, still populated with trapped particles. This is why these vortices, also called "electron holes", "solitary waves", and "weak double layers" are so often observed in space plasmas, as we will see in the next section.

6.3.2.2 Application: Nonlinear Waves in Space Plasmas

Lasting a few ω_B^{-1}, the phase of linear growth is rather short. In that sense, most of the unstable plasmas found in nature are only marginally unstable, unless a strong source of free energy can supply renewed flux particles with an unstable distribution function. In space plasmas, zones with unstable growing waves are seldom crossed. But distribution functions with a plateau are like fossils of past instabilities, and they are very commonly met. It is also common to measure coherent electrostatic structures, called weak double layers, that are the consequence of the evolution of phase space loops of trapped particles. Actually, the particle detectors require many seconds to measure the dependence of the distribution on the particle velocities, and it is not possible to make direct measurements of f in the (x, v) space. Instead, measurement of the electric field spikes and of unperturbed magnetic fields are used as a proxy for electrostatic coherent structures. These structures have been observed in space almost everywhere, in particular, they are common in the auroral acceleration zone [81, 82], in the magnetotail [83, 84], and in the solar wind. Figure 6.31 shows various wave forms captured with electric antennas in the solar wind [4].

The first three figures represent quasi-monochromatic waves whose envelope is modulated. The case (a) corresponds to Langmuir waves. Their envelope is modulated and it can be compared to those of Figure 6.29. It has been noted that these very localized wave packets are often associated to plasma density depletion. This can be interpreted in terms of ponderomotive force (see Eq. (1.140)), where the modulated envelope of a high frequency wave tends to push the particles away from the regions of high wave intensity. This is also interpreted in terms of wave coupling (with a low frequency compressional wave), and again, these models of wave coupling include the ponderomotive force.

The waves in (b) and (c) have a lower frequency and they are still modulated. In particular, case (c) looks like an isolated wave packet. Some of these structures are interpreted as waves trapped in the plasma density depletion [85] or in plasma magnetic depletions [86]. The case (d) corresponds more to what is seen in the above simulations: a series of regularly spaced nonsinusoidal spikes. Concerning the solar wind, it has been interpreted as vortices of trapped particles.

6.3.2.3 Solitary Waves

We have seen that a wave instability can be a source of vortices (in the phase space) of trapped particles. The growth of these structures is connected to a wave. The vortices are consequently as numerous as the number of wavelengths in the unstable domain. Figure 6.31 shows isolated structures, and it is suggested by its authors that they are phase space vortices too. From a theoretical point of view, a way of solving the Vlasov–Poisson equation allows one to build a very large set of models

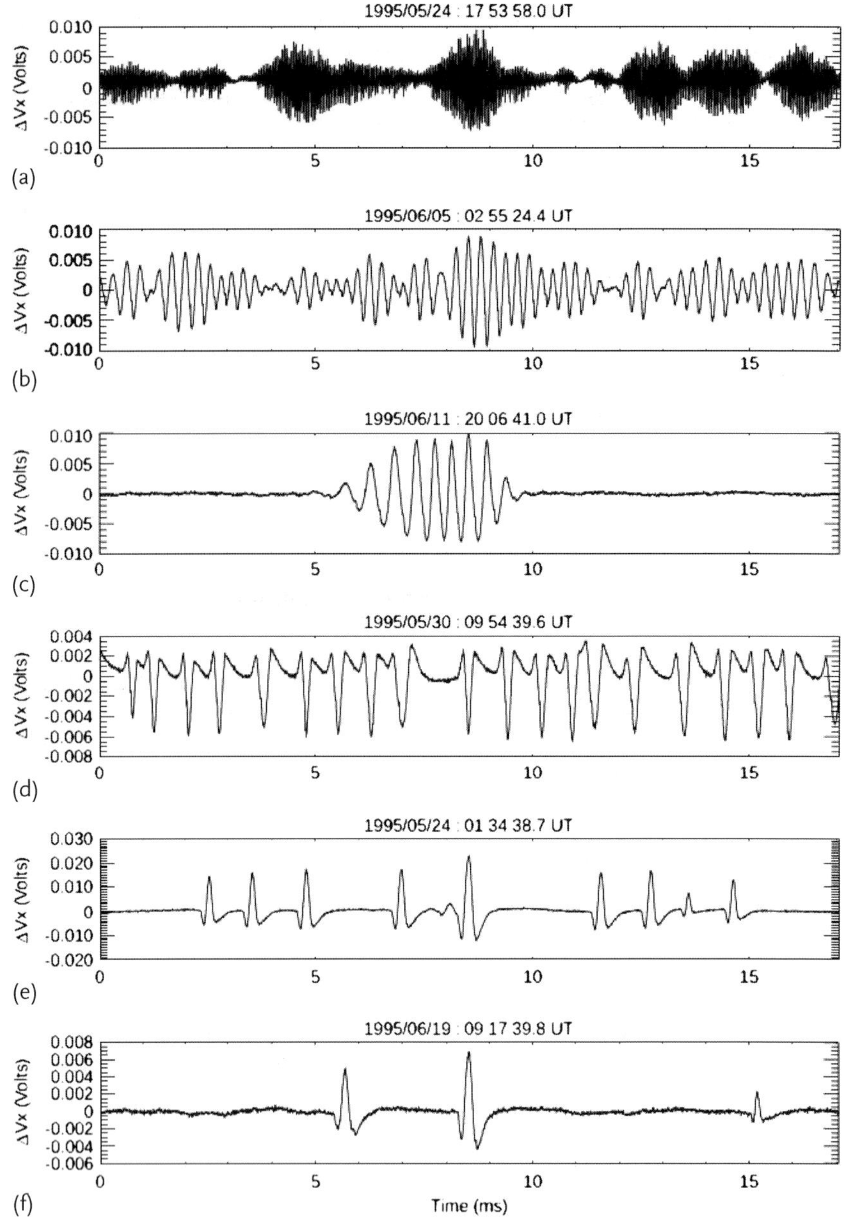

Figure 6.31 Electric potential (in volts) of six typical wave forms observed in the solar wind. (a) Langmuir waves; (b) and (c) low frequency quasi-sinusoidal wave packets; (d,e,f) nonsinusoidal wave packets and isolated electrostatic structures (IES). From [4].

of vortices called Bernstein–Greene–Kruskal (BGK) modes [87]. The first stage in building a BGK mode is to set a wave profile. It does not have to be sinusoidal.

A Gaussian pulse can be established, as well as an isolated square function, periodic saw-tooth wave, and so on. We look for electrostatic waves that are stationary in the wave frame. The equations are the Vlasov equation (6.116) without the partial time derivative and Eq. (6.117). The position x and the velocity are related through the energy equation

$$W = \frac{1}{2} m v^2 + q \Phi(x) , \tag{6.124}$$

where $\Phi(x)$ is the electrostatic potential associated to the chosen wave profile. The energy is a constant of motion, therefore, any distribution function $f(W)$ that depends only on W is a solution of the Vlasov equation. The Poisson equation reads

$$\partial_x^2 \Phi = \sum_s \frac{q_s}{\epsilon_0} \frac{f(W) d W}{\sqrt{2 m_s (W - q\Phi)}} . \tag{6.125}$$

After a multiplication by $\partial_x \Phi$, a first integral is

$$\frac{1}{2} (\partial_x \Phi)^2 + V(\Phi) = \text{constant} , \tag{6.126}$$

where

$$V(\Phi) = \frac{1}{\epsilon_0} \sum_s \int d W \, f_s(W) \sqrt{\frac{2}{m_s} (W - q_s \Phi)} . \tag{6.127}$$

This may be solved as

$$x - x_0 = \pm \frac{1}{2} \int_{\Phi_0}^{\Phi} \frac{d\Phi}{\sqrt{V(\Phi) - V(\Phi_0)}} . \tag{6.128}$$

Quite often, the shape of the electric potential is imposed as well as the distribution function of the passing particles. Then, the above equation is used to compute the population of trapped particles, provided that the obtained distribution (passing plus trapped particles) is positive. (This is not guaranteed.) The computation of the trapped particle distribution can be fully analytic. For instance, for a Gaussian potential and a Maxwellian distribution of passing electrons, with ions considered a uniform neutralizing background,

$$\Phi = \Phi_0 \exp\left(-x^2/2\delta^2\right) \quad \text{and} \quad f_p = 2\pi^{-1/2} \exp - W . \tag{6.129}$$

Turikov [88] has shown that the trapped electrons distribution is

$$f_t = \frac{4\sqrt{-W}}{\pi \delta^2} \left[1 - 2 \ln \left(\frac{-4 W}{\Phi_0} \right) \right] + 2\pi^{-1/2} \exp(-W)[1 - \operatorname{erf}\sqrt{-W}] . \tag{6.130}$$

Then we see that vortices of trapped electrons can be solitary coherent structures. They are often called solitary waves.

6.3.2.4 Interaction between Solitary Waves

The vortices of trapped electrons seen in Figures 6.30 and 6.34 correspond in the phase space (x, v) to regions of lower plasma density. The same property is generally observed with solitary waves. For this reason, they are often called "electron holes", or "electron phase space holes". Then, as noted in [89], because of their coherence and their negative density, electron holes can be considered large particles having the electric charge $-q$, but with a negative mass $-m_e$. As other particles in an electrostatic system, they interact through the electric force qE, and because of the negative mass, they attract each other. This attractive force causes collisions between electron holes. This is illustrated with Figure 6.32, that is issued from a PIC

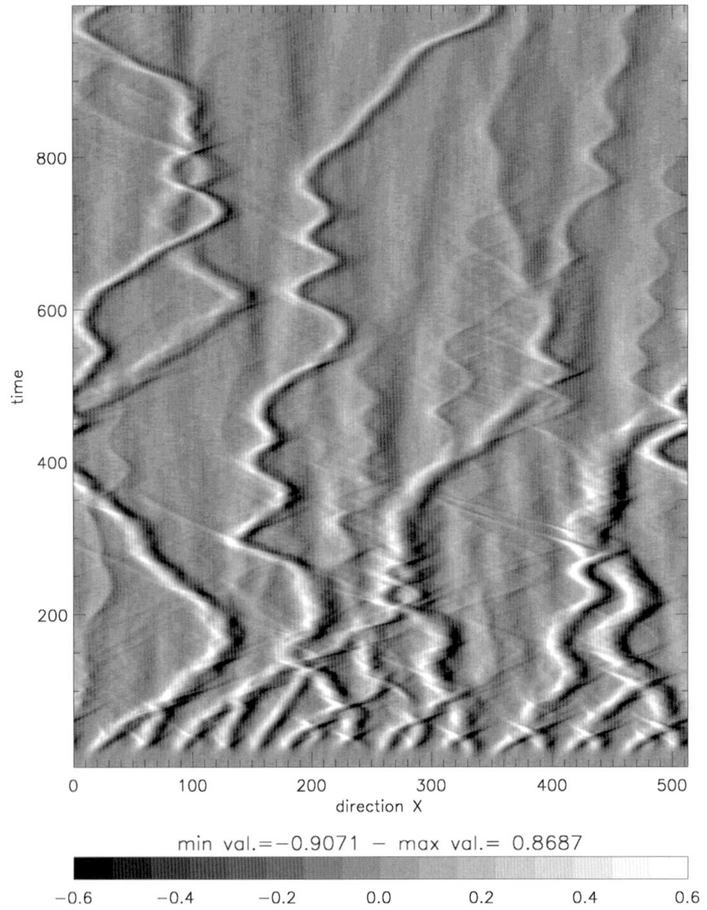

Figure 6.32 Instability of two opposite streams of electrons, in a 1D numerical simulation (PIC code). A uniform external magnetic field is set along the x axis of the simulation domain. The plot shows the amplitude of the electric field $Ex(x, t)$ as a function of the position x and time t. The electric field fine structures are solitary waves, that is, markers of electron holes. We can see that the electron holes have a tendency to attract each other and to merge. From [90].

simulation of an instability triggered by two Maxwellian distributions of electrons (temperatures 400 and 100 eV) of the same density and opposite mean velocities (2.5 and −2 times the thermal velocity). We can see that initially, the electron holes were disposed regularly, separated by one wavelength. Then, the electron holes start to interact with one of their neighbors, merging and also forming two small faster structures (these collisions between coherent structures are not elastic). And the process repeats. In the end, we are faced with a reduced number of isolated solitary structures evolving almost independently from each other. This spatial repartition of the electric spikes resembles those corresponding to the cases (d,e) and (f) of Figure 6.31, as if the case (d) corresponded to early times, and (f) to late times.

6.3.3 The Nonlinear Interaction of Many Electrostatic Waves of Low Amplitude

The nonlinear behavior is different when there are many waves. The case of many waves is very complex, and it is quite challenging to derive a general theory of their interactions with the plasma particles. But, under the assumptions that these waves have a low amplitude (in a sense that we will define below), useful analytical results can be derived.

We write the distribution function as the sum of the equilibrium distribution f_0, and the nonlinear contribution f_1. The nonlinear contribution is Fourier transformed,

$$f = f_0 + f_1 = f_0 + \sum_k f_k = f_0 + \sum_k \hat{f}_k \exp i(\omega_k t - kx) . \qquad (6.131)$$

More precisely, f_1 is Fourier transformed, but only in the frequency domain of the waves of interest. Slower evolutions are kept in the coefficients of the exponential as well as in the spatially uniform function f_0. Therefore, $\hat{f}_k$ and $\hat{E}_k$ are considered slowly varying functions of time (and spatially uniform).

The hypothesis of low amplitude waves consists of the following assumption: for any wave vector k, the distribution f_k is the solution of the linearized Vlasov–Maxwell–Gauss equations, and consequently, ω_k is the solution of the linear dispersion equation.

Let us note that $\hat{f}_k$ is a complex number. Because f is real, the sum

$$\hat{f}_k \exp i(\omega_k t - kx) + \hat{f}_{-k} \exp i(\omega_{-k} t + kx)$$

is real. This requires that $\hat{f}_{-k}$ is the complex conjugate of $\hat{f}_k$, and $\omega_{-k} = -\omega_k$. Therefore, the indices k and $-k$ represent two conjugate Fourier components of the same wave mode.

Equation (6.116) becomes

$$\partial_t f_0 + \sum_k \left(\partial_t f_k + v \partial_z f_k + \frac{q}{m} E_k \partial_v f_0 \right) = - \sum_k \sum_{l \neq 0} \frac{q}{m} E_{k-l} \partial_v f_l . \quad (6.132)$$

Writing the exponentials explicitly, and taking into account the fact that f_k is the solution of the linearized equations,

$$\partial_t f_0 + \sum_k \partial_t \hat{f}_k \exp i(\omega_k t - kx)$$
$$= -\sum_k \sum_{l \neq 0} \frac{q}{m} \hat{E}_{k-l} \partial_v \hat{f}_l \exp i[(\omega_{k-l} + \omega_l)t - kx] \,. \qquad (6.133)$$

According to this equation, different wave coupling scenarios can happen.

Three-wave Interaction Provided that there exist waves of numbers k, l and $k - l$ that fit the relation $\omega_{k-l} + \omega_l = \omega_k$, the right-hand term acts on the slow time-variation of f_k on the left-hand side of the equation. This represents nonlinear three-wave interactions, or wave coupling.

Four-wave Interaction Even in the case where this relation does not happen, it is possible that there exist two values l, l' with $\omega_{k-l} + \omega_l = \omega_{k-l'} + \omega_{l'}$. Then, two terms on the right-hand side of the equation interfere, causing four-wave interactions.

Wave and Particle Coupling Here, the three- and four-wave nonlinear interaction appears in the context of the Vlasov–Maxwell equations, but it is not specific to the plasma kinetic theory. It also happens with the somewhat simpler plasma fluid theories. Wave interaction is specifically important in the theory of *weakly turbulent plasmas,* and more precisely in what we call *wave-turbulence.* But the terms in Eq. (6.133) also reveal a kind of wave interaction that is specific to the kinetic theory. This is the coupling of two waves with particles whose velocity v_r (the index r is for resonant velocity) is defined by the relation

$$k v_r = [(k - l) + l]v = (\omega_{k-l} + \omega_l) \,. \qquad (6.134)$$

For these particles, the right-hand side of Eq. (6.133) couples with the slow time evolution of f_0 (first term on the left-hand side). In the present case of 1D electrostatic waves in a nonmagnetized plasma, the waves are Langmuir waves, and their dispersion relation $\omega^2 = \omega_{pe}^2(1 + 3/2k^2\lambda_D^2)$ does not allow for three-wave interactions, because it is not possible to find wavenumbers $k, l, k - l$ for which $\omega_{k-l} + \omega_l = \omega_k$. All these phenomenon are called *coherent,* because they are based on a specific relation connecting some properties of a small number of linear modes. We don't develop further the theory of coherent wave interactions. Instead, in the next section, we consider what happens in the opposite case of unrelated linear modes, but with close phase velocities. Useful reading on coherent waves interaction in plasmas can be found in [91].

6.3.4 Quasi-Linear Theory

The quasi-linear theory is based on the assumption that three- and four-wave interactions, as well as wave–particle–wave interactions do not occur. We define a time average operator (noted $\langle\rangle$) over a duration that is long comparatively to the periods $2\pi\omega_l^{-1}$, but short comparatively to the time of evolution of the $\hat{f}_k(t)$ coefficients. (A multitime scale expansion could be done, but the simplicity of the present problem allows us to skip it.) When applied to Eq. (6.133), many terms are eliminated. Then Eq. (6.133) is decoupled into two independent parts: the linear equation governing k_k; and

$$\partial_t f_0 = -\sum_k \sum_{l\neq 0} \frac{q}{m} \hat{E}_{k-l} \partial_v \hat{f}_l \langle \exp i[(\omega_{k-l} + \omega_l)t - kx] \rangle \ . \tag{6.135}$$

The terms coupling the different modes disappear for all values of $k \neq 0$. The terms with $k = 0$ are not associated with wave coupling, but with the terms $-l$ and l that represent the same mode; therefore, they are kept. Considering that $\omega_{-l} + \omega_l = 0$,

$$\partial_t f_0 = -\sum_{l\neq 0} \frac{q}{m} \hat{E}_{-l} \partial_v \hat{f}_l \ , \tag{6.136}$$

From the linearized Vlasov equation, for which f_l is the solution, a simple expression of $\hat{f}_l$ would be

$$\hat{f}_l = -\frac{iq}{m} \frac{1}{\omega_l - lv} \hat{E}_l \partial_v f_0 \ . \tag{6.137}$$

Eliminating $\hat{f}_l$ from Eq. (6.136) we find a classical expression [19],

$$\partial_t f_0 = \partial_v (D \partial_v f_0)$$
$$D = i\frac{q^2}{m^2} \sum_k \frac{\hat{E}_k \hat{E}_{-k}}{\omega_k - kv} = \frac{iq^2}{m^2} \sum_k \frac{\hat{E}_k^2}{\omega_k - kv} \ . \tag{6.138}$$

(We have replaced the dummy variable l with k.) Diffusive phenomena are generally known to happen in the space of configurations. Equation (6.138) characterizes diffusion in the velocity space.

Because of the importance of the linear equation that controls the evolution of $\hat{f}_k$, we speak of *quasi-linear diffusion* (QL diffusion). Because this concerns the velocity space, proving it with fluid equations like in Chapter 2 would be extremely challenging. Arbitrarily high order moment equations would be required. This is why QL diffusion is considered a typical kinetic effect.

Actually, the diffusion coefficient D found in Eq. (6.138) presents a singularity. This results from the choice of Eq. (6.137) that does not correspond to the analysis of the linear eigenmodes developed in Section 5.6.4.3. We have seen that the condition of no singularity in the initial value of the distribution leads, instead of Eq. (6.137), to the more general Eq. (5.106), that in terms of $\hat{E}_k$ reads:

$$\hat{f}_k = i n_0 \frac{q}{m} \frac{\hat{E}_k}{k} \left[D_{pk} \frac{k^2}{\omega_{\text{pe}}} (v_{\phi k}) \delta(v - v_{\phi k}) + \frac{1}{n_0} \text{p.v.} \left\{ \frac{f_0'(v)}{v - v_{\phi k}} \right\} \right] , \tag{6.139}$$

where D_{pk} is the dispersion term that contains a principal value (noted p.v.),

$$D_{pk} = 1 - \frac{\omega_{\text{p}}^2}{\hat{n}_0 k^2} \text{p.v.} \int \frac{f_0'(v)}{v - v_{\phi k}} dv . \tag{6.140}$$

In the above equations, the phase velocity $v_{\phi k} = \omega_k / k$ is a real number (ω_k is a real frequency, as expected in a Fourier transform). The diffusion coefficient D is given by

$$D f_0'(v) = -i n_0 \frac{q^2}{m^2} \sum_k \frac{\hat{E}_k^2}{k} \left[D_{pk} \frac{k^2}{\omega_{\text{pe}}} (v) \delta(v - v_{\phi k}) + \frac{1}{n_0} \text{p.v.} \left\{ \frac{f_0'(v)}{v - v_{\phi k}} \right\} \right] . \tag{6.141}$$

Let us introduce the solution, called the kinetic or Landau mode, that corresponds to a regular distribution (see Eq. (5.107)),

$$D_{pk} \frac{k^2}{\omega_{\text{pe}}} (\tilde{v}_{\phi k}) - i \frac{\pi}{n_0} f_0'(\tilde{v}_{\phi k}) = 0 . \tag{6.142}$$

Here, $\tilde{v}_{\phi k}$ is the complex resonant velocity $v_{\phi k} + i\gamma_k / k = (\omega_k + i\gamma_k)/k$. In the approximation $\gamma_k \ll \omega$, a Taylor expansion shows that the correction of this expression in the vicinity of $v_{\phi k}$ brings finite terms proportional to γ_k that we neglect. Then, we can write

$$D_{pk} \frac{k^2}{\omega_{\text{pe}}} (v_\phi) - i \frac{\pi}{n_0} f_0'(v_\phi) = 0 , \tag{6.143}$$

where $v_{\phi k}$ is real. The diffusion coefficient becomes

$$D f_0'(v) = -i \frac{q^2}{m^2} \sum_k \frac{\hat{E}_k^2}{k} \left[i\pi f_0'(v) \delta(v - v_{\phi k}) + \text{p.v.} \left\{ \frac{f_0'(v)}{v - v_{\phi k}} \right\} \right] . \tag{6.144}$$

This Eq. (6.144) generalizes the Eq. (6.138). It can be obtained as well through the classical Landau approach, with its analytical continuation and its integration contour deformation, or by direct application of the Plemelj relation:

$$\frac{1}{v - v_{\phi k}} = \text{p.v.} \left(\frac{1}{v - v_{\phi k}} \right) + i\pi \delta(v - v_{\phi k}) . \tag{6.145}$$

This relation is actually a convenient short-cut for the complete Landau theory. It is written here in the notation of the theory of distributions.

When the opposite terms k and $-k$ are gathered in this sum, considering that $\omega_{-k} = -\omega_k$, the principal value just cancels (it would not with a complex resonant velocity $\tilde{v}_{\varphi k}$ instead of the real $v_{\varphi k}$). One can write

$$\begin{aligned} D f_0'(v) &= 2\frac{q^2}{m^2} \sum_{k>0} \frac{\hat{E}_k^2}{k} \pi \delta(v - v_{\varphi k}) f_0'(v) \\ &= 2\frac{q^2}{m^2} \sum_{k>0} \hat{E}_k^2 \pi \delta(kv - \omega_k) f_0'(v) \,. \end{aligned} \tag{6.146}$$

This result can be found in [92], where it is derived from the classical contour integrals method inspired by the Landau calculation. The resonant character of the QL diffusion appears clearly in this relation. Let us note that before Eq. (6.143), we have made the approximation $\gamma_k \ll \omega_k$. Then we have neglected the terms of an asymptotic development that are proportional to γ_k, and that would bring a supplement in the diffusion coefficient D in the form of weak and nonresonant terms.

The diffusion equation is coupled to the equation that governs the wave intensity on the slow time scale, that is, still in the linear approximation

$$d_t E_k^2 = 2\gamma_k E_k^2 \,. \tag{6.147}$$

Let us consider a gentle bump-in-tail instability. Then $0 < \gamma_k \ll \omega_k$, and the wave intensity grows. The resonant velocities are in the vicinity of the velocity domain where $f_0'(v) > 0$. Then, the diffusion tends to increase the value of f_0 where $f_0'(v) > 0$ and to reduce it at larger velocities where $f_0'(v) < 0$. Therefore, the bump tends to be replaced by a plateau. Progressively, the bump get smoother, and the growth rate γ_k (that depends on $f_0'(v)$) tends to diminish. Then, Eq. (6.147) shows that the intensity of the waves tends to reduce its growth, until the bump in $f_0(v)$ is replaced by a flat plateau bordered by areas where $f_0'(v) < 0$. Then, the instability stops and the waves reach a saturation level (no matter what they were initially, damped or growing). In conclusion, the QL diffusion in velocity space tends to smooth the distribution function in the set of resonant velocities, and to freeze the evolution of the wave spectrum.

The QL theory has also been developed in the more general context of photon and particle distribution coupling through a so called semiclassical mechanism [93]. The kinetic equations describing the distribution functions evolution are characterized by emission and absorption of the waves by the particles, managed in terms of probabilities of transition. The quasi-linear equations are relatively easily found, and they are equivalent to Eqs. (6.136) and (6.147), when applied to plasma waves. But the estimate of the diffusion coefficient, related to the transition probabilities requires some care.

No matter what the formalism, the QL theory can be developed in a 3D velocity space, where a magnetic field makes the plasma anisotropic. In that case, the dispersion can be computed in terms of particle pitch-angle α and of modulus p of the momentum,

$$\partial_t f_0 = \frac{1}{\sin\alpha}\partial_\alpha(D_{\alpha\alpha}\sin\alpha\partial_\alpha f_0) + \frac{1}{\sin\alpha}\partial_\alpha(D_{\alpha p}\sin\alpha\partial_p f_0) + \frac{1}{p^2}\partial_p(p^2 D_{p\alpha}\partial_\alpha f_0) + \frac{1}{p^2}\partial_p(p^2 D_{pp}\partial_p f_0)\,. \tag{6.148}$$

The phase angle around the direction of the magnetic field would be ϕ. Because of the symmetry of the unperturbed plasma relatively to the magnetic field direction, and because of the phase mixing, the angle ϕ is excluded from the above diffusion equation. Equation (6.148) shows that in a 3D magnetized plasma, the equilibrium distribution of the particles is subject to diffusion relatively to the momentum (or velocity, or energy), but also relatively to the pitch angle. The pitch angle diffusion has important consequences in space plasmas.

6.3.4.1 Application: Pitch Angle Diffusion in Space Plasmas

In the Sun's coronal loops and in the closed field lines of the magnetosphere of magnetized planets, particles can be trapped in regions of low magnetic field (see Section 1.4.2.1). Along the particle trajectory (of abscissa s), the pitch angle $\alpha(s)$ increases with the magnetic field. When $\alpha(s)$ reaches $\pi/2$ the particle is mirrored back in the direction of decreasing magnetic field. If a particle reaches a region $s > s_c$ filled with a collisional plasma before being mirrored, it is captured in the collisional plasma. Therefore, only the particles with a pitch angle $\alpha(s) > \alpha_m(s)$ where $\alpha_m(s)$ corresponds to $\alpha_m(s_c) = \pi/2$ are trapped by the magnetic field. The distribution functions of trapped particles have a loss cone that corresponds to the particles lost in the collisional plasma. The loss cone is characterized by the angle $\alpha_m(s)$.

If there was no source of trapped plasma, the loss cone would be set in a time that is necessary for particles to bounce between the mirror points. Then it would be stable. But it happens that, even when no source of trapped plasma is identified, the flux of precipitated particles into the collisional plasma is finite. This means that in the trapping region there exists something that populates the loss cone.

Quite generally, the trapped plasma is excited with various wave modes. These waves, through pitch angle quasi-linear diffusion, are proposed as an explanation for the refilling of the loss cone. Particles initially with $\alpha(s) > \alpha_m(s)$ would be diffused, as shown in Eq. (6.148), and reach the zone $\alpha(s) < \alpha_m(s)$ of the particles that will stop to be mirrored and that will be precipitated into the collisional plasma.

Diffuse auroras are low particle energy precipitation, hardly visible to the naked eye, but much more frequent than the bright auroral arcs. Most of the energy associated with the auroral precipitation is brought by diffuse auroras. These auroras can be explained as electrons precipitated from the Earth's plasma sheet, as a result of QL pitch angle diffusion by waves. These waves are related to the electron–cyclotron mode and to the whistler mode. The dominant effect is caused by a kind of whistler radiation called a "chorus" [94].

According to [95], decimetric short-lived radio-emissions of the Sun's corona, known as "spikes", can be explained as the result of wave instabilities called an "electron–cyclotron maser". These waves are excited by a loss cone instability, that is, by the gradient in velocity space associated with the loss cone. The saturation level of these waves is reached through QL diffusion of the loss cone, and the consequent reduction of the gradient in velocity space.

The QL theory can also explain the diffusion of ions in the velocity space, through their interaction with Alfvén waves [96]. Then, as is usual for MHD waves, the wave amplitudes is expressed through its magnetic field B_1, and not the electric field, and

$$\partial_t f = \mathrm{Re}\frac{iq^2}{m^2c^2}\sum_{\sigma=-1,1}\int_{-\infty}^{+\infty} dk_{\parallel}\times \left[(v_A - v_{\parallel})\frac{1}{v_{\perp}}\partial_{v_{\perp}} + \partial_{v_{\parallel}}\right]\langle B_1^2(k_{\parallel})\rangle \times \frac{v_{\perp}^2}{4(\omega + \sigma\omega_{ci} - k v_{\parallel})}\left[(v_A - v_{\parallel})\frac{1}{v_{\perp}}\partial_{v_{\perp}} + \partial_{v_{\parallel}}\right] f\,, \tag{6.149}$$

where all the particle data is related to the ions, and v_A is the Alfvén velocity. The interaction of ion distribution functions with Alfvén waves is expected in the environment of comets, where the Alfvén waves are triggered by pick-up ions. The ions distribution can then become isotropic under the influence of the QL diffusion by these waves.

An equation analogous to Eq. (6.148) exists for relativistic plasma. It has practical applications in models of radiation belts. Wave–particle quasi-linear diffusion, as well as other forms of diffusion, are taken into account in numerical simulation codes of the radiation belts. Because of the regularity of the magnetic field on a large scale, the diffusion equation is expressed in terms of the adiabatic invariants of the particles [97].

6.3.5 Trapping versus Quasi-Linear Diffusion

In the quasi-linear diffusion, the size of the plateau $\Delta v \sim \Delta v_{\varphi k}$ is proportional to the interval of phase velocities associated with the most intense waves present in the system. Nevertheless, each of these waves is presumed to be of negligible amplitude (quasi-linear hypothesis). The characteristic time is proportional to the square E_k^2 of the wave amplitudes, and as the waves have a low intensity, this is a rather long time scale.

With particle trapping, the size of the plateau $\Delta v \sim \omega_B/k = (qE_0/(km))^{1/2}$ depends on the amplitude of the wave. The characteristic time is also proportional to $E_0^{1/2}$.

Therefore, the quasi-linear effect and particle trapping are of a totally different nature. Nevertheless, in 1D plasmas, they both imprint the same trend on the wave evolution, with the creation of a plateau in the space averaged distribution function, or at least the reduction of the slope $\partial_v f$ of the distribution function. In the case of an unstable wave, this has the consequence to quench the instability, and to set the wave amplitudes at a finite saturation level.

In the real world, wave spectra are never monochromatic, and their amplitudes are never negligible. Therefore, both the spread of the phase velocities and the wave amplitudes contribute to the formation of a plateau, and possibly, at least with our experience of measurements of space plasmas, the distinction between quasi-linear processes and trapping will be impossible, if not a bit too academic. The main point, that is common to these two idealized processes is that a plateau is formed, and that this plateau sets a limit to the linear growth of unstable waves. Nevertheless, the nonlinear evolution of an instability is currently dominated by either quasi-linear processes, or by particle trapping. Actually, for most of the unstable waves, the linear growth rate taken as a function of the wavenumber generally has a clear maximum. Because of the exponential growth, the corresponding wave quickly dominates the others, and most often, this results in a narrow spectrum (quasi-monochromatic) and particle trapping.

A typical example is shown in Figure 6.33. In this simulation, the electron beam is hotter, slower and denser. The growth rate is strong and large loops of trapped particles appear in the (x, v) phase space, as can be seen in Figure 6.34. In this simulation, the simulation box size is 1024×4, and it contains a larger number of wavelengths; this is why more vortices are visible. Their amplitudes (in term of velocity range) are higher because the wave's electric field reaches also a much higher value.

With a bump-in-tail instability, trapping is clearly the easiest nonlinear regime to catch. Nevertheless, under very peculiar circumstances, it is possible to reach a regime where a plateau is formed without the formation of loops of trapped par-

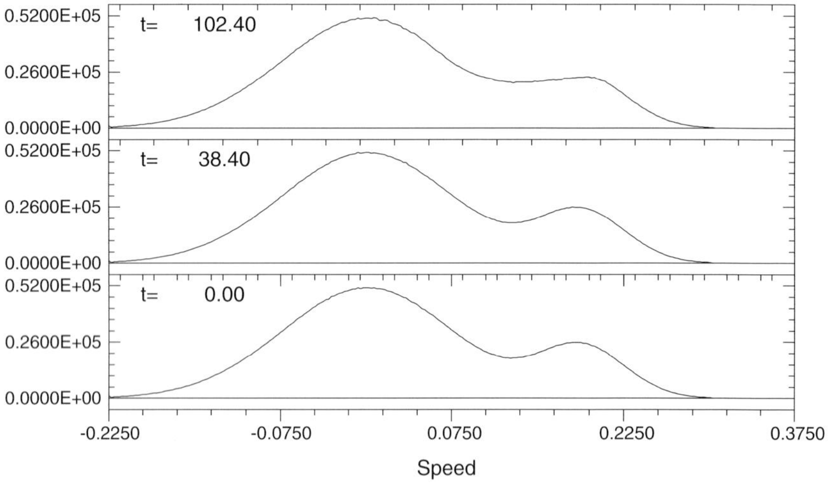

Figure 6.33 Instability caused by a dense and hot electron beam, simulated with a 1D perturbative code. The plots represent the spatially averaged distribution function at three different times in the simulation.

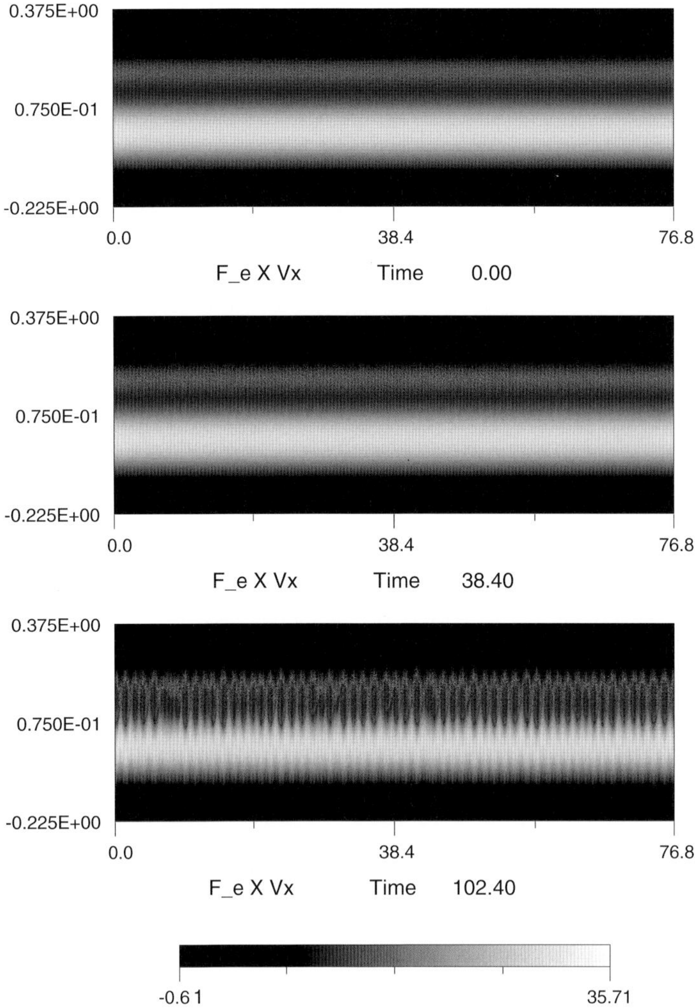

Figure 6.34 Same simulation as in Figure 6.33. Distribution function in the x, v_x space. The initial main Maxwellian distribution function has been subtracted.

ticles. An example is given in Figure 6.34 that represents the perturbation of the distribution function in the (x, v) phase space. As in Figure 6.34, the simulation is performed over a long domain (1024 cells). We can see that a wave brings perturbations of the distribution, even for nonresonant particles. But the beam is not "modulated" with vortices of trapped particles. A view of the suprathermal part of the space averaged distribution function is given in Figure 6.36. At time $t = 0$, we can see a fast and hot beam (mean velocity $v_b = 3.5v_{te}$, thermal velocity $v_{tb} = 0.5v_{te}$) of low density. The bump is actually hardly visible, the distribution seems more like having a long suprathermal component. We can see that a plateau is formed on

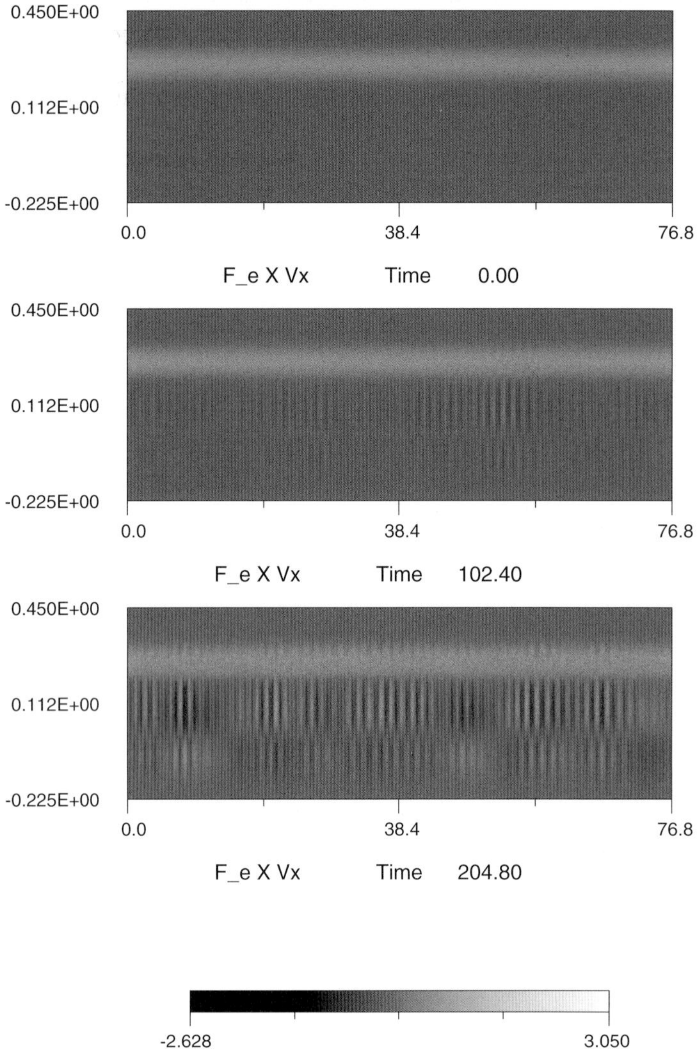

Figure 6.35 Instability of a fast, hot, and weak electron beam. Distribution function in the x, v_x phase space, the direction x being the direction of the static magnetic field. The initial main Maxwellian distribution function has been subtracted.

this initially very soft bump; therefore, the evolution is "subtle". To catch a quasi-linear diffusion and not trapping, it was necessary to set a hot bump, in order to have a larger and flatter spectrum of unstable waves, weak bump in order to have a weak growth rate and stay in the linear regime during a time exceeding the phase mixing time, and a fast beam in order to have a positive derivative $\partial_v f$ at $t = 0$, in spite of a low density beam (for instability). The analogous simulations with a higher density beam display vortices of trapped particles.

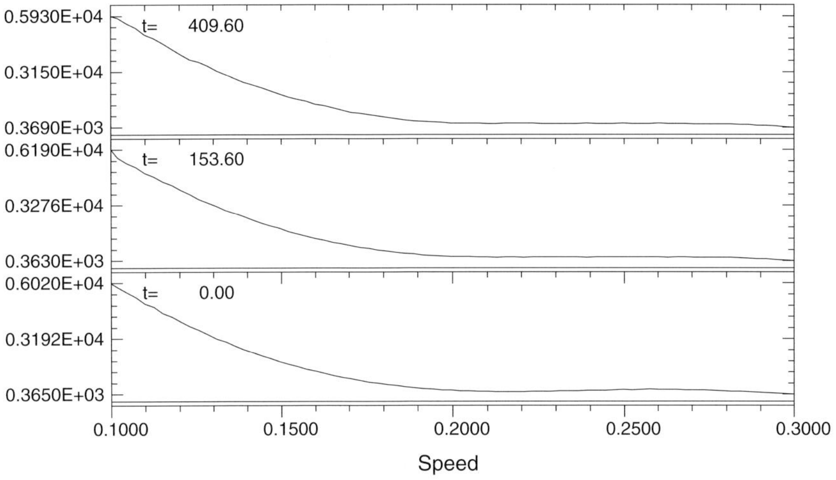

Figure 6.36 Same simulation as in Figure 6.35. Tail of the space-averaged distribution function. At time $t = 0$, the tail is slightly bumpy (with a positive slope for $0.21 < v < 0.26$). At times $t = 153$ and $t = 409$, it is flat.

Concerning 3D plasmas, the situation is more complex. The longitudinal electric field of a wave with $k = k_{\|}$ can favor trapping and the formation of a plateau in a way similar to 1D plasmas. But more efficient mechanisms can go into action. Most often, they involve waves with a transverse polarization. These transverse waves are not necessarily the ones initially in the system. They may have been generated by wave–wave coupling processes such as the parametric instability. These transverse waves can act, through particle trapping simultaneously on the particle energy and pitch angle distributions. Nevertheless, pitch angle diffusion is not easy to characterize with 3D monochromatic wave trapping as it is for QL wave–particle interaction. If the formation of a plateau seems to be favored by vortices of trapped particles, QL linear diffusion is mostly invoked for the interaction of waves and the pitch angle of the particle distribution function.

7
Flow and Particle Acceleration Processes

7.1
Flow Acceleration and Heating in a Collisional Fluid

7.1.1
Basic Equations

There are many possible ways to accelerate a a plasma, depending on the dominant force acting on it. We shall consider the case of a nonrelativistic, neutral, nonresistive, and inviscid plasma subject to the plasma's own pressure force, to magnetic forces (Laplace force), and to a conservative force expressible as the gradient of a time independent potential ψ (for example, the gravitational potential of a static mass distribution). As already discussed in Chapter 2, the equation of motion can be written as

$$\partial_t \boldsymbol{u} + \boldsymbol{u} \cdot \nabla \boldsymbol{u} = -\frac{\nabla P}{\rho} + \frac{\boldsymbol{j} \times \boldsymbol{B}}{\rho} - \nabla \psi \tag{7.1}$$

where $\boldsymbol{u}$ is the fluid motion, P the thermal pressure, ρ the mass density, $\boldsymbol{j} = \nabla \times \boldsymbol{B}/\mu_0$ the electric current, $\boldsymbol{B}$ the magnetic flux, and ψ the potential for the relevant force. In the case where all terms on the right-hand side of Eq. (7.1) are zero, every fluid parcel of the plasma moves at constant velocity. In many particular physical systems where one or more forces act on the fluid it is more convenient to replace Eq. (7.1) by a conservative equation for the per volume energy content: ∂_t(Energy per volume) $= -\nabla\cdot$(Energy flux). A conservative equation does explicitly pin down a physical quantity $\boldsymbol{X}$ such that, under stationary conditions, $\nabla \cdot \boldsymbol{X} = 0$. For example, the familiar continuity equation

$$\partial_t \rho = -\nabla \cdot (\rho \boldsymbol{u}) \tag{7.2}$$

which expresses the fact that mass is neither created nor destroyed, predicts that in a stationary system $\nabla \cdot (\rho \boldsymbol{u}) = 0$. Thus, in a cartesian, one-dimensional, and stationary system where all quantities depend on one spatial coordinate x, Eq. (7.2) implies that $\rho(x)u(x) =$ constant from where one deduces that $\rho(x)$ and velocity $u(x)$ do spatially vary in antiphase.

Collisionless Plasmas in Astrophysics, First Edition. Gérard Belmont, Roland Grappin, Fabrice Mottez, Filippo Pantellini, and Guy Pelletier.

A few more equations and assumptions are needed to complement Eq. (7.1) in order to write down a conservation equation for the energy. The zero resistivity assumption is one of these assumptions which allows us to use the ideal Ohm law $\boldsymbol{E} = -\boldsymbol{u} \times \boldsymbol{B}$ (see Chapter 2) to replace the electric field in Faraday's equation viz

$$\partial_t \boldsymbol{B} = -\nabla \times \boldsymbol{E} = \nabla \times (\boldsymbol{u} \times \boldsymbol{B}) . \tag{7.3}$$

Equation (7.3), best known as the ideal MHD equation, establishes a direct relationship between magnetic field and fluid velocity.

The last essential equation which we do need in order to be able to write down the aspired energy equation is a relation linking the internal energy per unit mass U in terms of two other thermodynamic state variables. The first law of thermodynamics states that in a reversible process the internal energy per molecule changes as a function of the volume $V = m/\rho$ it occupies as

$$dU = -P dV = \frac{mP}{\rho} \frac{d\rho}{\rho} . \tag{7.4}$$

After some juggling around with Eqs. (7.1)–(7.4) we obtain the energy equation

$$\partial_t \left(\rho \frac{u^2}{2} + nU + \frac{B^2}{2\mu_0} + \rho\psi \right) = -\nabla \cdot \boldsymbol{F} \tag{7.5}$$

where the adiabatic energy flux vector $\boldsymbol{F}$, is given by

$$\boldsymbol{F} = \rho \boldsymbol{u} \left(\frac{u^2}{2} + \frac{w}{m} + \psi \right) + \boldsymbol{S} . \tag{7.6}$$

and where $\boldsymbol{S} = \boldsymbol{E} \times \boldsymbol{B}/\mu_0$ is the Poynting flux vector. In Eq. (7.6) w is the enthalpy per unit mass defined as

$$w = U + PV = U + \frac{Pm}{\rho} . \tag{7.7}$$

Equation (7.5) is obviously incomplete as it does not include the possibility of nonreversible modifications of the energy content of a portion of fluid. One possibility is the nonreversible heating (or cooling) through addition (subtraction) of energy from outside the fluid itself. Examples are the heating of the terrestrial atmosphere by the Sun's radiation and the cooling of water inside a bottle placed in a refrigerator. These nonreversible contributions to the internal energy can be taken into account by adding a term of the form $\rho\delta W/\delta t$ on the right-hand side of Eq. (7.5), where $\delta W/\delta t$ is the energy per unit mass and time added to the fluid from external sources. Nonreversible energy exchanges can also occur within the fluid itself. Let $\boldsymbol{Q}$ designate the nonreversible energy flux vector in units of energy per surface and time, not to be confused with the third order tensor of Chapter 2 obtained by computing the third order moment of the velocity distribution function $f(\boldsymbol{v})$. Let us consider a fixed volume V. The time variation of the energy content in the volume V can than be obtained by computing the volume integral of Eq. (7.5) and by adding the nonreversible energy flux $\mathbf{Q}$ through the volume surface ∂V and the external heating $\rho\delta W/\delta t$:

$$\int_V \partial_t\left(\rho\frac{u^2}{2} + \rho U + \frac{B^2}{2\mu_0}\right) dV = -\int_V \nabla\cdot\mathbf{F}\,dV - \int_{\partial V} \mathbf{Q}\cdot d\mathbf{A} + \int_V \rho\frac{\delta W}{\delta t}\,dV \tag{7.8}$$

where $d\mathbf{A}$ is vector surface element pointing outwards with respect to the volume V. Using the divergence theorem to transform the surface integral into a volume integral and noting that Eq. (7.8) must be valid for any volume we obtain an improved energy equation including a nonadiabatic internal energy transfer term and an external heating term

$$\partial_t\left(\rho\frac{u^2}{2} + nU + \frac{B^2}{2\mu_0}\right) = -\nabla\cdot(\mathbf{F}+\mathbf{Q}) + \rho\frac{\delta W}{\delta t}. \tag{7.9}$$

A few words on the nonreversible energy flux $\mathbf{Q}$ are in order here. In a collisional fluid the nonreversible energy flux is generally associated with thermal conduction, viscous transfer, or both. Thermal conduction occurs in the presence of temperature gradients. If the latter are sufficiently weak so that the collisional mean free path is small compared to the scale of variation of the temperature, the conductive heat flux can be Taylor expanded in terms of powers of the temperature gradient. To the lowest order, the conductive flux, is then given by the already widely mentioned "Fourier law"

$$\mathbf{Q}_\kappa = \mathbf{q} = -\kappa\nabla T \tag{7.10}$$

where κ is a positive thermal conduction coefficient, which generally depends on the temperature, and where the minus sign indicates that energy flows against the temperature gradient towards colder regions of the fluid. In Figure 7.1a heat flows

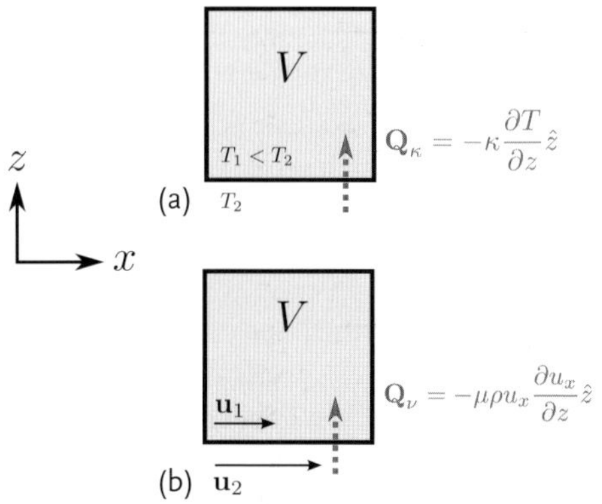

Figure 7.1 Nonreversible energy transport into the volume V through thermal conduction (a) and viscous transport (b).

into the volume V as a consequence of the temperature being higher outside of V. The conductive flow is an irreversible process as it does not change sign under reversal of the fluid velocities. This is rather trivial in the particular case of heat conduction as it occurs even in the absence of any macroscopic fluid motion. The actual energy flowing into the volume V through thermal conduction across the bottom face of area A in the example of Figure 7.1 is just $A\kappa\nabla T$.

Nonreversible energy transfer into V can be a consequence of the viscous dissipation of a velocity gradient across the volume boundary as illustrated in Figure 7.1b. In a collisional fluid the viscous energy flux is the product of the stress tensor $\boldsymbol{\sigma}$ and the velocity $\boldsymbol{u}$:

$$Q_v = \boldsymbol{\sigma} \cdot \boldsymbol{u} \, . \tag{7.11}$$

If the collisional mean free path is small with respect to the characteristic spatial variations of the fluid velocity $\boldsymbol{u}$, one may be tempted to define the stress tensor as $\nu \partial_i u_k$. As already pointed out in Chapter 4 this form of the stress tensor is far too general as it does not take into account some obvious symmetry conditions which can be deduced from macroscopic considerations not discussed in Chapter 4. For example, in the case of a rigid rotation where adjacent fluid parcels do not move with respect to each other and friction should be zero. The fluid velocity field in the case of a rigid rotation with angular velocity $\boldsymbol{\Omega}$ is given by $\boldsymbol{u} = \boldsymbol{\Omega} \times \boldsymbol{r}$ where $\boldsymbol{r}$ is the vector connecting an arbitrary point on the rotation axis and the observation point with fluid velocity $\boldsymbol{u}(\boldsymbol{r})$. In cartesian coordinates the three components of the fluid velocity are thus given by $v_1 = \Omega_2 r_3 - \Omega_3 r_2$, $v_2 = \Omega_3 r_1 - \Omega_1 r_3$, and $v_3 = \Omega_1 r_2 - \Omega_2 r_1$. There are only two possible ways to construct a stress tensor which vanishes in the case of rigid rotation independently of the coordinate systems. These are $\delta_{ik}\nabla \cdot \boldsymbol{u}$ and $\partial_i u_k + \partial_k u_i$. The general form of the stress tensor could, therefore, be written as a linear combination of these two forms. However, as already shown in Chapter 4, it is common practice to write the stress tensor as the sum of a trace-free and a diagonal tensor respectively:

$$\sigma_{ik} = \mu \left(\partial_k u_i + \partial_i u_k - \frac{2}{3} \delta_{ik} \nabla \cdot \boldsymbol{u} \right) + \zeta \delta_{ik} \nabla \cdot \boldsymbol{u} \, . \tag{7.12}$$

For isotropic fluids the shear viscosity μ and the volume viscosity ζ must be velocity independent scalars which can also be shown to be strictly positive if viscosity is supposed to transform mechanical energy into internal energy. A simple example of sheared flow causing a nonreversible energy transfer into the volume V is shown in Figure 7.1b. As in the conductive case we can compute the viscous energy flux across the bottom face of the volume V as $A\mu\rho\partial_z(u_x^2/2)$. Like the conductive flux, this flux is not reversible as it does not change sign under reversal of the fluid velocities $u_x \rightarrow -u_x$.

7.1.2
Expressions for the Polytropic Fluids

According to the first principle of the thermodynamics, the variation of the internal energy of a fluid dU is the sum of the adiabatic term $-PdV$ and a nonadiabatic contribution $\delta\mathcal{E}$ according to

$$dU = \delta\mathcal{E} - PdV = \delta\mathcal{E} + \frac{mP}{\rho}\frac{d\rho}{\rho} . \tag{7.13}$$

The first term on the right-hand side of Eq. (7.13) includes the heating or cooling of a fluid parcel through nonreversible processes like heat as the heat conduction and the viscous energy transfer discussed in the previous section and even the energy input from external sources. In classical thermodynamics the irreversible alteration of the thermodynamic state of a fluid parcel is measured by the variation ds of its entropy s defined through

$$\delta\mathcal{E} = Tds . \tag{7.14}$$

It is common practice to define the adiabatic exponents γ_1, γ_2 and γ_3 as

$$\gamma_1 = \frac{\rho}{P}\left(\frac{\partial P}{\partial \rho}\right)_s , \quad \frac{\gamma_2 - 1}{\gamma_2} = \frac{P}{T}\left(\frac{\partial T}{\partial P}\right)_s , \quad \gamma_3 - 1 = \frac{\rho}{T}\left(\frac{\partial T}{\partial \rho}\right)_s . \tag{7.15}$$

In the general case the three adiabatic exponents are all different from each other, but are identical for an ideal gas (see Section 7.1.2.1). In situations where these exponents remain approximately constant it is possible to establish simple and useful relations between the variations of the thermodynamic variables describing the fluid and the energy $\delta\mathcal{E}$ added (or subtracted) from the fluid parcel. From Eqs. (7.13) and (7.14) we find that $(\partial U/\partial s)_\rho = T$ and $(\partial U/\partial \rho)_s = mP/\rho^2$. Noting that $\partial^2 U/\partial\rho\partial s = \partial^2 U/\partial s\partial\rho$ it follows that

$$\left(\frac{\partial T}{\partial \rho}\right)_s = \frac{m}{\rho^2}\left(\frac{\partial P}{\partial s}\right)_\rho . \tag{7.16}$$

On the other hand, from $dP = (\partial P/\partial\rho)_s d\rho + (\partial P/\partial s)_\rho ds$ it also follows that

$$\left(\frac{\partial s}{\partial \rho}\right)_P = -\left(\frac{\partial P}{\partial \rho}\right)_s \left(\frac{\partial P}{\partial s}\right)_\rho^{-1} . \tag{7.17}$$

Combining the above two expressions and choosing P and ρ as thermodynamic variables we obtain

$$\delta\mathcal{E} = Tds = T\left\{\left(\frac{\partial P}{\partial s}\right)_\rho^{-1} dP + \left(\frac{\partial s}{\partial \rho}\right)_P d\rho\right\} \tag{7.18}$$

$$= \frac{m}{\rho(\gamma_3 - 1)}\left(dP - \gamma_1\frac{P}{\rho}d\rho\right) . \tag{7.19}$$

This equation immediately shows that a fluid motion is adiabatic $Ds/Dt = 0$ provided pressure and density evolve in space and time according to $D/Dt(P/\rho^{\gamma_3}) = 0$. It is sometimes useful to write $\delta\mathcal{E}$ as a function of the adiabatic exponents and either the heat capacity at constant volume $c_V = (\partial U/\partial T)_V$ or the heat capacity at constant pressure $c_P = (\partial w/\partial T)_P$:

$$\delta\mathcal{E} = c_P\left(dT - \frac{\gamma_2 - 1}{\gamma_2}\frac{T}{P}dP\right) \tag{7.20}$$

$$\delta\mathcal{E} = c_V\left(dT - (\gamma_3 - 1)\frac{T}{\rho}d\rho\right). \tag{7.21}$$

7.1.2.1 Adiabatic Exponents for Ideal Gases

For an ideal gas the following equation of state holds

$$P = \rho T/m. \tag{7.22}$$

The differential form of Eq. (7.22) $dP/P = d\rho/\rho + dT/T$ can be used to show that for an ideal gas the three adiabatic exponents in Eq. (7.15) are identical $\gamma = \gamma_1 = \gamma_2 = \gamma_3$. In addition, from Eqs. (7.13) and (7.14) one deduces that the entropy of an ideal gas varies according to

$$ds = \frac{dU}{T} - \frac{d\rho}{\rho} \tag{7.23}$$

and thus, using Eq. (7.15), one finds that during an adiabatic process (for examples, during a slow compression or expansion of the volume occupied by the gas) the internal energy varies as a function of temperature

$$\left(\frac{\partial U}{\partial T}\right)_s = \frac{T}{\rho}\left(\frac{\partial\rho}{\partial T}\right)_s = \frac{1}{\gamma - 1}. \tag{7.24}$$

As an example, statistical mechanics teaches us that the internal energy for a monatomic gas at thermal equilibrium is $U = 3T/2$ per particle (in other words, $T/2$ per degree of freedom) corresponding to a unique adiabatic exponent $\gamma = 5/3$.

Summarizing the above results we can write the following expressions for an ideal gas:

$$U = \frac{T}{\gamma - 1} \tag{7.25}$$

$$w = U + \frac{mP}{\rho} = \frac{\gamma}{\gamma - 1}T \tag{7.26}$$

$$ds = \frac{1}{\gamma - 1}\left(\frac{dP}{P} - \gamma\frac{d\rho}{\rho}\right) \tag{7.27}$$

$$= \frac{1}{\gamma - 1}\left(\frac{dT}{T} + (1 - \gamma)\frac{d\rho}{\rho}\right) \tag{7.28}$$

$$c_V = \frac{1}{\gamma - 1} \qquad (7.29)$$

$$c_P = c_V + 1 = \frac{\gamma}{\gamma - 1} \,. \qquad (7.30)$$

7.1.3
Bernoulli's Principle

In this and the remaining subsections of Section 7.1 we will use simplified versions of Eq. (7.9) to address different mechanisms of flow acceleration in stationary situations.

The simplest, stationary, and nontrivial form of Eq. (7.9) is obtained by retaining only the kinetic energy flux and the enthalpy flux on the right-hand side of Eq. (7.9) thus neglecting the explicit contribution from conservative forces (for example, gravitational and electrostatic forces) or an external magnetostatic field. We do also assume there is no external heating or cooling, that is, $\rho\delta W/\delta t$. The resulting energy equation then reads

$$\nabla \cdot \left[n\boldsymbol{u} \left(\frac{m u^2}{2} + w \right) \right] = 0 \qquad (7.31)$$

which, according to the continuity equation (7.2), reduces to the even simpler expression:

$$n\boldsymbol{u} \cdot \nabla \left(\frac{m u^2}{2} + w \right) = 0 \,. \qquad (7.32)$$

Equation (7.31) expresses the fact that the total energy flux $\int [\ldots] \cdot d\boldsymbol{A}$ going through any two surfaces A_1 and A_2 connected by a bundle of streamlines must be equal, as in the example of Figure 7.2 showing a flow through an hourglass shaped pipe. Equivalently, Eq. (7.32) expresses that $u^2/2 + w$ is an invariant of the flow, that is,

$$\frac{m u^2}{2} + w = \text{const} \qquad (7.33)$$

along streamlines. Equation (7.33) generally known as Bernoulli's principle where, in the adiabatic limit with $ds = 0$ and following Eqs. (7.13) and (7.7), the enthalpy is given by

$$w = V \int dP = \frac{1}{n} \int dP \,. \qquad (7.34)$$

7.1.4
Venturi Effect

In the preceding section it has been established that in a stationary adiabatic flow the quantity $u^2/2 + w$ is invariant along streamlines. This invariant is a function

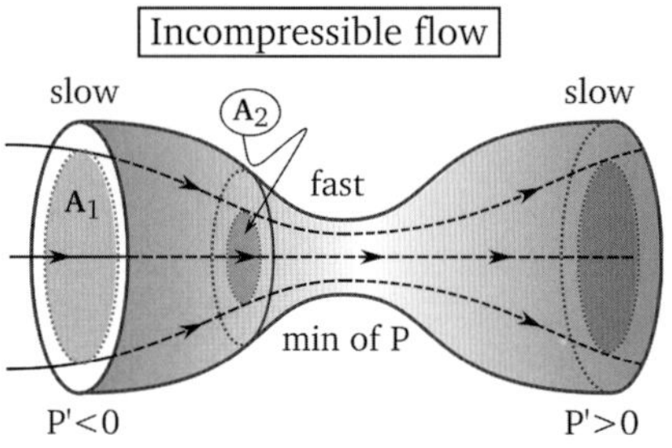

Figure 7.2 Stationary incompressible flow through an hourglass shaped pipe. The fluid pressure P is weakest where the pipe cross section A is minimum and the fluid velocity is highest.

of the enthalpy given by (7.34) and, therefore, on both the pressure and the density for which no particular relation has been specified yet. An equation of state pertinent to the problem at hand must be invoked. We shall see indeed, that the flow characteristics may strongly change when changing the equation of state. Let us come back to the particular problem of the flow through an hourglass shaped pipe (Figure 7.2) and let us suppose that the fluid is incompressible so that the pertinent equation of state can be set to $\rho = \rho_0 = \text{const}$. The compressible case will be discussed later in Section 7.1.5.

After setting $\rho = \rho_0$ in the expression for the enthalpy Eq. (7.34) and after plugging the latter into Bernoulli's Eq. (7.32) we deduce that pressure P and kinetic energy $\rho_0 u^2/2$ must compensate exactly along streamlines so that we can write

$$\rho_0 \frac{u^2}{2} + P = \text{const}\,, \tag{7.35}$$

that is, the pressure in the fluid increase/decreases as $\rho_0 u^2/2$ decreases/increases. As to determining the velocity profile within the system we make use of the continuity equation which tells us that in a stationary system the mass per time unit flowing through any two surfaces A_1 and A_2 delimited by the same bundle of streamlines must be equal, that is $d(A\rho u) = 0$ along streamlines. Given the axial symmetry of the system shown in Figure 7.2 we chose the surface A to be a disk oriented perpendicularly to the symmetry axis so that A is a function of the coordinate x along the cylinder axis only. In this case the two surfaces A_1 and A_2 could be like showed in Figure 7.2 and mass conservation could be written as

$$\partial_x (A\rho_0 \bar{u}) = 0 \tag{7.36}$$

where $\overline{u_x(x)}$ represents the flow velocity along x, averaged over the surface $A(x)$. This equation shows that the flow velocity $\overline{u_x}$ within the pipe increases/decreases as the tube section $A(x)$ decreases/increases so that $\overline{u_x}$ reaches a maximum where $A(x)$ goes through a minimum. Assuming that the flow velocity is directed predominantly along axial (in the case of an extremely elongated hourglass or case of a small flux tube near the pipe axis), we can assume $u^2 \approx u_x^2$ and deduce that $P(x)$ and A vary in phase so that the lowest pressure in the system occurs at the narrowest point in the pipe.

7.1.5
De Laval Nozzle

Things are quite different when the fluid cannot be assumed to be incompressible. Let us again concentrate on the fluid motion along the axis of the flux tube of Figure 7.3. We can then use Eq. (7.33) in its differential form $udu + dw = 0$ (differential operators must be taken along streamlines) together with the adiabatic expression for the enthalpy $dw = dP/\rho = c^2 d\rho/\rho$, where $c^2 = (\partial P/\partial\rho)_s$ is the adiabatic sound speed squared, to establish a relation between density and fluid velocity differentials along streamlines:

$$\frac{d\rho}{\rho} = -\frac{udu}{c^2} . \tag{7.37}$$

On the other hand, as already discussed in the previous section, the continuity equation for a stationary system can be written as $d(A\rho u) = 0$ along streamlines. Thus,

$$\frac{d(A\rho u)}{A\rho u} = \frac{dA}{A} + \frac{d\rho}{\rho} + \frac{du}{u} = 0 \tag{7.38}$$

where contrary to the incompressible case discussed in the previous section ρ is a function of position. Combining Eqs. (7.37) and (7.38) we obtain the instructive relation

$$\frac{dA}{A} = (\mathcal{M}^2 - 1)\frac{du}{u} \tag{7.39}$$

$$= (1 - \mathcal{M}^2)\frac{d\rho}{\rho\mathcal{M}^2} \tag{7.40}$$

$$= (1 - \mathcal{M}^2)\frac{dP}{\rho u^2} \tag{7.41}$$

where $\mathcal{M} = |u|/c$ is the sonic Mach number. We shall remember that expressions (7.39)–(7.41) must be taken along streamlines. Thus, to facilitate the analysis we concentrate on the central streamline along the pipe axis, where $u = u_x$ and consider a small surface A. Equation (7.39) shows that u and A vary in opposition when the Mach number of the flow is $\mathcal{M} < 1$ (the subsonic case) as in Section 7.1.4 while u and A vary together when $\mathcal{M} > 1$. The variation of A being given by the boundary conditions, the only possible steady state solution for the flow is to turn supersonic exactly at the position of the narrowest point of the tube (see Figure 7.3).

Thus, instead of going through a maximum the velocity profile has a positive first derivative. Assuming, for simplicity, a constant sound speed c (corresponding to $\gamma = 1$ for a polytropic gas) and Taylor expanding the tube section $A(x)$ and the flow velocity $u_x(x)$ around the throttle position $x = x_0$ where A reaches its minimum, that is, $A = A_0 + A''\delta x^2$ and $u_x = c + u'_x\delta x$, we do easily get from Eq. (7.39)

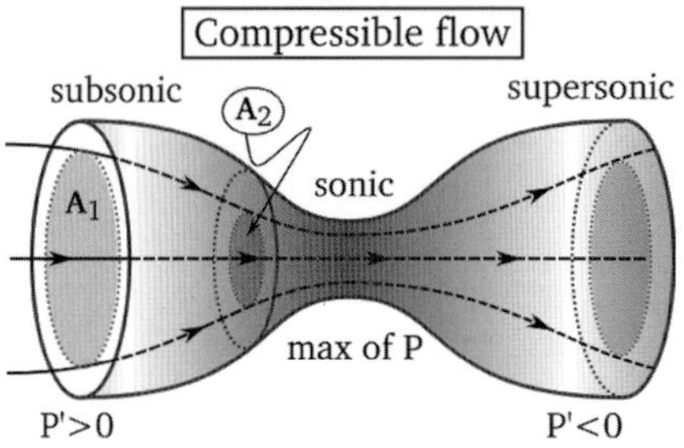

Figure 7.3 Stationary compressible flow through an hourglass shaped pipe. The fluid pressure P is strongest where the pipe cross section A is minimum corresponding to the place where the flow becomes supersonic.

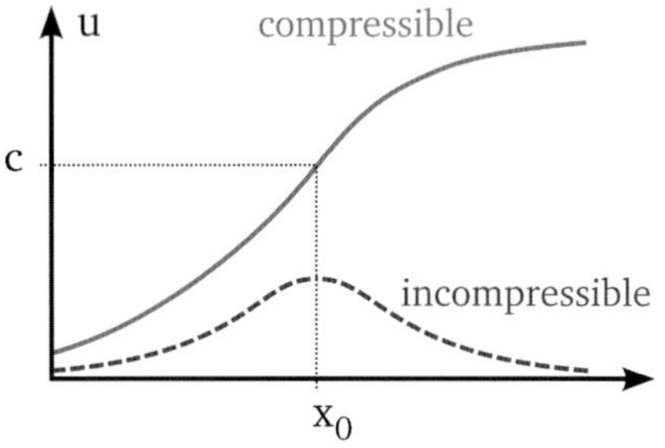

Figure 7.4 Time stationary velocity profiles for the flow through a hourglass shaped pipe for both compressible and the incompressible fluids. For the incompressible fluid the velocity peaks at the throttle position in x_0.

that $2u'_x/c = (A''/A)^{1/2} > 0$. The qualitative difference between the velocity profiles for compressible and incompressible flows is illustrated in Figure 7.4. In the compressible case treated here, the density is obviously not constant as in the (incompressible) Venturi effect. The density, as well as the pressure, do both rise in the subsonic region when approaching the throttle and decrease thereafter.

As we shall see in the next section a similar constraint exists for a compressible fluid in a spherically symmetric gravitational field such as the extended atmosphere of a star.

7.1.6 Stellar Winds

The kinetic energy required to escape from the surface of a star is finite. Assuming a close to Maxwellian distribution of the particles' velocity distribution in the Sun's atmosphere where the transition from collisional to collisionless occurs, some particles have a high enough energy to escape to infinity. For a given star mass $\mathcal{M}_*$ and a distance r_0 from its center, the flux of escaping particles of average mass $\bar{m}$ depends essentially on the ratio between the square of the escape velocity $2G\mathcal{M}_*/r_0$ (G is the gravitational constant) and the square of the thermal velocity $2T/\bar{m}$, that is $\gamma_G^2 \equiv G\mathcal{M}_* \bar{m}/T$. The smaller γ_G the stronger the flux. For the solar corona where the temperature is on the order of 10^6 K, $r_0 \approx 7 \times 10^8$ m we have $\gamma_G \approx 3.6$ (assuming a typical solar composition with 92% protons, 8% alphas, and the corresponding population of charge neutralizing electrons leading to an average particle mass of $0.57 m_p$). This value of γ_G is larger, but not horrendously larger, than unity so that a flow from the solar corona into the interstellar space may be expected. The existence of a supersonic plasma flow escaping from the Sun has effectively been confirmed by in situ observations starting in the 1960s. The flow is actually so tenuous and so hot that it turns nearly collisionless only a few thousands km above the photosphere. Despite the weak collisionality of the solar wind in particular and stellar winds in general, simple hydrodynamic models have been shown to provide a convenient understanding of its macroscopic behavior. In this section we shall discuss in some details these hydrodynamic wind models and postpone the discussion on the stellar wind kinetics to Chapter 9.

Let us assume a spherically symmetric atmosphere above the surface of a star of radius r_0 in an otherwise empty universe so that no fluid, other than the star's atmosphere itself must be treated. We neglect the magnetic field and state that the atmosphere total mass is small with respect to the star mass $\mathcal{M}_*$ so that the gravitational potential for $r \geq r_0$ is the Newtonian potential for a point mass $\psi(r) = -G\mathcal{M}_*/r$, where G is the gravitational constant. Let us further assume that the atmosphere is in a stationary state so that all time derivatives vanish $\partial_t = 0$. Given the spherical symmetry the continuity Eq. (7.2) becomes $\partial_r(r^2 u\rho)$ the right-hand side of Eq. (7.9) reduces to

$$\partial_r\left(\frac{u^2}{2} + \frac{w}{m} + \psi\right) = 0 \tag{7.42}$$

where we have assumed that the external heating $\delta W/\delta t = 0$ and the heat flux $\boldsymbol{q} = 0$ are both zero thus implying a strict adiabatic expansion. We shall immediately relax this constraint by implicitly including the possibility of a nonadiabatic flow by assuming a general polytropic relation $P = \rho T/m \propto \rho^\Gamma$ where the index Γ must be interpreted as an effective index rather than the ordinary adiabatic index $\gamma \equiv c_P/c_V$. With the differential form of the enthalpy being inspired by the adiabatic formulation, that is $dw = dP/n = \Gamma dT/(\Gamma - 1)$ Eq. (7.42) can be written as (assuming $\Gamma \neq 1$ throughout this section)

$$\partial_r\left(\frac{u^2}{2} + \frac{\Gamma}{\Gamma - 1}\frac{T}{m} + \psi\right) = 0 \tag{7.43}$$

or, defining the integration constant K:

$$K = \frac{u^2}{2} + \frac{\Gamma}{\Gamma - 1}\frac{T}{m} + \psi = \text{const} . \tag{7.44}$$

We note that for the normal case $\Gamma > 1$ (see Section 7.1.6.1) and because of the vanishing potential $\psi \propto -1/r$ for $r \to \infty$, $K > 0$ is required for the flow to reach infinity. In the following we shall, therefore, only consider the case $K > 0$. If a radial heat flux q_r is present in the system and the enthalpy given by its adiabatic form the above equation should be written extensively as

$$\partial_r \left(\frac{u^2}{2} + \frac{\gamma}{\gamma - 1} \frac{T}{m} + \psi + \frac{q_r}{mnu} \right) = 0 \, . \tag{7.45}$$

Thus the implicit conductive, nonadiabatic, energy flux in Eq. (7.43) is

$$q_r = \left(\frac{\Gamma}{\Gamma - 1} - \frac{\gamma}{\gamma - 1} \right) mnuT \tag{7.46}$$

a form for the heat flux sometimes used as a model for the collisionless conductive flux in the solar wind. After rearrangement of Eq. (7.43) and defining the polytropic sound speed $c_\Gamma^2 \equiv \Gamma T/m$ we obtain the equation for the Mach number $\mathcal{M} = u/c_\Gamma$

$$\left(\mathcal{M} - \frac{1}{\mathcal{M}} \right) \left(\partial_r \mathcal{M} + \frac{1}{\mathcal{M}} \frac{1}{c_\Gamma} \partial_r c_\Gamma \right) + \left(\frac{G\mathcal{M}_*}{c_\Gamma^2 r^2} - \frac{2}{r} \right) = 0 \tag{7.47}$$

or, using the integration constant K from Eq. (7.44), its equivalent form

$$\frac{\mathcal{M}^2 - 1}{2\mathcal{M}^2} \partial_r \mathcal{M}^2 = \frac{1}{n} \partial_r n \left(\frac{\Gamma - 1}{2} \right) (1 - \mathcal{M}^2) + \frac{2}{r} - \frac{G\mathcal{M}_*}{c_\Gamma^2 r^2} \, . \tag{7.48}$$

In the general case, both the Mach number and the sound speed are functions of r, so that Eq. (7.47) must be supplemented by the continuity equation which may be written as a function of $\mathcal{M}$ and n

$$\frac{2}{r} + \frac{\mathcal{M}'}{\mathcal{M}} + \frac{\Gamma + 1}{2} \frac{n'}{n} = 0 \tag{7.49}$$

or as a function of $\mathcal{M}$ and c_Γ

$$\partial_r \left[r^2 \mathcal{M} c_\Gamma^A \right] = 0 \, , \quad \text{with} \quad A \equiv \frac{\Gamma + 1}{\Gamma - 1} \tag{7.50}$$

for which, the solution at an arbitrary distance r is related to the Mach number and sound speed at a reference level r_0 via $r^2 \mathcal{M} c_\Gamma^A = r_0^2 \mathcal{M}_0 c_{\Gamma 0}^A$. From Eq. (7.47) it follows that a transonic solution with finite values of the derivatives for the Mach number and the sound speed, requires the sonic point r_s, where $\mathcal{M}(r_s) = 1$ to be given by the implicit solution of

$$r_s = \frac{G\mathcal{M}_*}{2c_\Gamma^2} \, . \tag{7.51}$$

7.1.6.1 Considerations on the Effective Polytropic Index Γ

Equation (7.50) shows that for $\Gamma < 1$ the spatial derivative of the Mach number $\mathcal{M}$ and the temperature $T = c_\Gamma^2 m/\Gamma$ have the same sign at the sonic point. Thus, if we consider the normal case where the Mach number is smaller than unity below the sonic point and larger than unity above the sonic point, the temperature must increase through the sonic point. We shall exclude this case as it implies that the heat flux is directed towards the star and coherently assume $\Gamma > 1$, with the particular case $\Gamma = 1$ being discussed in some details in the next section.

A further restriction on the value of the effective polytropic index Γ comes from the requirement of the integration constant K being positive, in order for the fluid to escape from the star. From Eqs. (7.44) and (7.51) we deduce that $K \geq 0$ provided

$$\frac{\mathcal{M}^2}{2} + \frac{1}{\Gamma - 1} - 2\frac{r_s}{r} \geq 0 \tag{7.52}$$

evaluation of this equation at the sonic point where $r = r_s$ and $\mathcal{M} = 1$ implies $\Gamma < 5/3$. An even stronger restriction on the possible values for the effective polytropic index Γ can be established if one requires the fluid velocity u to be a growing function of r for $r \to \infty$. From Eq. (7.44), the continuity equation $nur^2 = \text{const}$ and the polytropic relation $T \propto n^{\Gamma-1}$ we obtain

$$\frac{u^2}{2} = K - Au^{1-\Gamma} r^{2(1-\Gamma)} + \frac{G\mathcal{M}_*}{r} . \tag{7.53}$$

If we require the asymptotic velocity u to grow with distance as $r \to \infty$, that is, $\partial_r u^2 > 0$ we set a constraint on Γ. Taking the spatial derivative of (7.53) for the limiting case of a constant asymptotic velocity $u = u_\infty$, so that $u' \to 0$, then implies a balance between the enthalpy term and the gravitational term, viz.

$$2A(\Gamma - 1)u_\infty^{1-\Gamma} r^{2(\Gamma-1)-1} - \frac{G\mathcal{M}_*}{r^2} = 0$$

which has a solution only if $\Gamma = 3/2$. Thus, $\Gamma < 3/2$ is required for the velocities being a growing function of r at large distances. The same constraint can be obtained by requiring the derivative of the velocity being positive as $r \to 0$ the limiting case $\Gamma = 3/2$ corresponding to a vanishing derivative $u' \to 0$ for $r \to 0$. From now on, we will thus assume that $\Gamma < 1.5$. Since a plasma made of protons has $\Gamma = 5/3$, the polytropic index considered here can only be an effective index representing the effects of non-adiabatic energy input, to be found either in the heat flux, or in the transformation of coherent fluctuations into heat during the transport by the wind.

7.1.6.2 **Isothermal Wind Approximation**

Despite the effective Γ in the solar wind being closer to the upper limit 1.5 with a constant asymptotic wind speed other than 1 we shall discuss in more details the isothermal case $\Gamma = 1$ as it doesn't differ qualitatively from the real case while being more easily treatable as the characteristic speed c_Γ does not depend on r. In this particular case Eq. (7.47) reduces to

$$\left(\mathcal{M} - \frac{1}{\mathcal{M}}\right) \partial_r \mathcal{M} = -\left(\frac{G\mathcal{M}_*}{c_1^2 r^2} - \frac{2}{r}\right) . \tag{7.54}$$

Possible solutions of Eq. (7.54) describing the outflow from a star are shown in Figure 7.5.

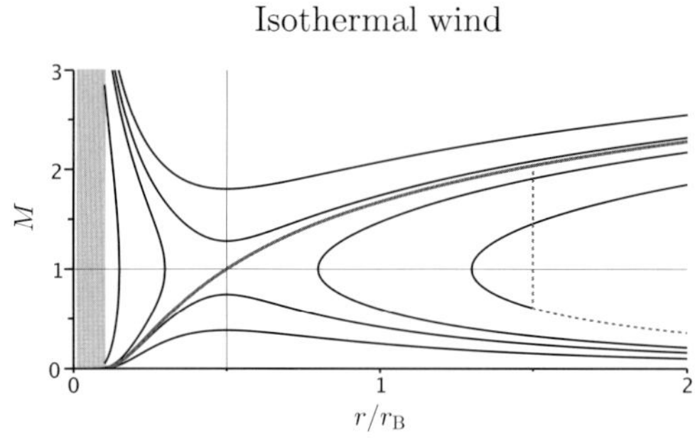

Figure 7.5 Possible solutions of Eq. (7.54) for the case of an isothermal outflow from a stellar surface located at then outer edge of shaded region of the plot. The characteristic length $r_B \equiv G\mathcal{M}_*/c_1^2$ has been used to normalize distances.

Let us suppose that surface gravity at $r = r_0$ is sufficiently strong for a thermal particle with characteristic velocity $\sim c_1$ to be bound to the star, that is $G\mathcal{M}_*/c_1^2 r_0 \gg 2$ so that the right-hand side of Eq. (7.54) is negative. For such a strong gravity one expects the fluid motion near the surface to be subsonic $\mathcal{M} \ll 1$ with a positive acceleration $\partial_r \mathcal{M} > 0$ as most of the particles have order c_1 velocity and are not energetic enough to escape to infinity. As the Mach number grows with distance it may eventually reach the value $\mathcal{M} = 1$ for which the left hand side of Eq. (7.54) necessarily vanishes. Equation (7.54) implies that this happens at the sonic point

$$r_s \equiv G\mathcal{M}_*/2c_1^2 = r_B/2 \,. \tag{7.55}$$

If $\partial_r \mathcal{M} > 0$ at the sonic point, then the flow velocity must be supersonic for $r > r_s$, exactly as in the case of the compressible flow going through a nozzle (see Figure 7.4). We note that the just mentioned transonic solution is not the only possible one for a subsonic surface condition. Indeed, setting $\partial_r \mathcal{M} = 0$ at the sonic point r_s does also allow to satisfy Eq. (7.54). In this case the Mach number reaches a maximum at the sonic point, but without exceeding $\mathcal{M} = 1$ similar to the compressible profile in Figure 7.4. The effective solution chosen by the system obviously depends on the conditions imposed at its boundaries, for example at the surface r_0. However, only the transonic solution can be shown to be stable in general [98, 99] unless the flow is already supersonic near the stellar surface, in which case the system adjusts to one of the top curves in Figure 7.5 for which the fluid velocity is supersonic everywhere. Obviously, at infinity the fluid velocity must go to zero in order to merge with the interstellar medium. Under such circumstances, the transonic solution which has a monotonically growing velocity profile, must break down at some distance and "jump" to a subsonic solution with decreasing velocity profile at a position which is compatible with the given interstellar medium density and the isothermal shock jump conditions $\rho_- \mathcal{M}_- = \rho_+ \mathcal{M}_+$ and $\mathcal{M}_- \mathcal{M}_+ = 1$.

The transonic solution is the one with largest mass loss since it is the solution which connects the stellar surface to infinity with the largest fluid velocity u_0 at the base and, therefore, the largest mass flux $\dot{\mathcal{M}} = \rho_0 u_0 4\pi r_0^2$ for a given surface density ρ_0. Making use of the isothermal wind Eq. (7.54) and the continuity Eq. (7.49) with $\Gamma = 1$ we find the useful relation

$$\frac{\mathcal{M}^2 - \mathcal{M}_0^2}{2} + \ln\frac{\rho}{\rho_0} + \frac{2r_s}{r_0}\left(1 - \frac{r_0}{r}\right) = 0 \tag{7.56}$$

where ρ_0 and $\mathcal{M}_0$ are the density and the Mach number at the coronal radius r_0. With Eq. (7.56) we can express the density at the sonic point as a function of the base density ρ_0 in the normal case $\mathcal{M}_0 \ll 1$

$$\rho_s = \rho_0 \exp\left[-\frac{\mathcal{M}_s^2}{2} - \frac{2r_s}{r_0}\left(1 - \frac{r_0}{r_s}\right)\right] \tag{7.57}$$

where $\mathcal{M}_s = 1$ for the transonic solution and $\mathcal{M}_s < 1$ for breezes. This expression allows us to compute the mass flux $\dot{\mathcal{M}} = \rho_s c_1 4\pi r_s^2$ provided the coronal values for the density and the temperature are known. In 1958 Parker [100] suggested that the solar wind, and more generally, winds from main group stars, are driven to supersonic velocities by a thermal pressure gradient in the solar corona according to Eq. (7.43). Assuming a coronal temperature of $T_0 = 10^6$ K a coronal density and a an average mass per particle $m = 1.15m_p$ we obtain an isothermal sound speed $c_1 = 85$ km/ s. Setting $r_0 = 7 \times 10^8$ m and $\mathcal{M}_* = 2 \times 10^{30}$ kg we find that the sonic point is located at $r_s \approx 13.3r_0$ which is in excess of present estimates by at least a factor two mainly because of the unrealistic assumption of isothermal stratification. With a coronal density $n_0 = 10^8\,\text{cm}^{-3}$, Eq. (7.57) predicts $n_s = 1.5 \times 10^{-3}$ cm for the transonic solution which is already below the density of the wind at earth's orbit by a factor 10^4. The mass flux $\dot{\mathcal{M}} = \rho_s c_1 4\pi r_s^2 \approx 240\,\text{kg}\,\text{s}^{-1}$ is also a clear underestimate with respect to the observed flux of 8×10^8 kg/ s. A better agreement between theory and observation can be obtained by solving the more general Eq. (7.47) with a polytropic index Γ compatible with the observed cooling of the wind with distance [101]. However, the applicability of the hydrodynamic equations to such an extremely rarefied plasma where collisions are rare, has soon be questioned leading to the appearance of the so-called exospheric models [102, 103] which will be discussed in Section 9.2.2.

7.1.6.3 Lowering the Sonic Point by the Effect of the Alfvén Wave Pressure

We have seen previously numerical examples of Alfvén waves increasing the wind speed. We now give analytical estimates of this acceleration in the isothermal, spherically symmetric case. Taking into account the wave pressure $\delta B^2/(2\mu_0)$, we write for the radial velocity u:

$$\partial_t u + uu' + c^2\rho'/\rho + (\delta B^2)'/(2\mu_0\rho) = -G\mathcal{M}/r^2. \tag{7.58}$$

Time averaging over one wave period, and neglecting fluctuations of the radial velocity and density (thus assuming circularly polarized Alfvén waves), this reduces to:

$$uu' + c^2\rho'/\rho + \frac{\langle\delta B^2\rangle'}{2\mu_0\rho} = -G\mathcal{M}/r^2 \tag{7.59}$$

u, ρ now denoting now time-averaged quantities. Using as previously the constancy of mass flux, we arrive at (cf. Eq. (7.54)):

$$(u'/u)(u^2 - c^2) = 2c^2/r^2\,(r - r_s^0) - \frac{\langle \delta B^2/2 \rangle'}{2\mu_0 \rho} \tag{7.60}$$

where r_s^0 is the usual sonic point (with no Alfvén wave pressure, Eq. (7.55)).

To go on, we must have a law for the radial variation of the Alfvén wave pressure δB^2. In the WKB limit (high frequency), without damping, and with radial mean magnetic field, the invariance of the wave action flux $\epsilon = \rho r^2 \langle \delta B^2/(2\rho\omega) \rangle\,(u + v_a)$ implies:

$$\frac{\delta B^2}{2\mu_0 \rho} = \alpha u v_a/(u + v_a)^2 \propto X/(1 + X)^2 \tag{7.61}$$

where $\alpha = \epsilon\omega_0/F$ is a constant combining action flux ϵ, wave frequency ω_0 and mass flux F. $X = u/v_a$ is the Alfvén Mach number. Relying on the invariance of the magnetic and mass flux, one finds:

$$\rho'/\rho = -2X'/X. \tag{7.62}$$

Using this relation, we arrive at the following expression for the wave pressure term in Eq. (7.60):

$$\frac{\langle \delta B^2/2 \rangle'}{2\mu_0 \rho} \propto -X'(1 + 3X)/(1 + X)^3. \tag{7.63}$$

This expression can be integrated and used to compute the additional speed gained, due to the Alfvén wave pressure. We merely examine here the resulting displacement of the sonic point. From Eq. (7.60), we obtain that, to first order in the Alfvén wave pressure term, the new sonic point r_s is given by

$$r_s = r_s^0 + \frac{(r_s^0)^2}{2c^2} \left. \frac{\langle \delta B^2 \rangle'}{2\mu_0 \rho} \right|_{r=r_s^0}. \tag{7.64}$$

But from Eq. (7.63), using Eq. (7.62), $(u'/u)|_{r=r_s^0} = 1/r_s^0$ and mass the conservation condition $(\rho u r^2)' = 0$ we get with $X_0 = X|_{r=r_s^0}$:

$$\frac{\langle \delta B^2 \rangle'}{2\mu_0 \rho} \propto -\frac{3}{2r_s^0} \frac{X_0(1 + 3X_0)}{(1 + X_0)^3}. \tag{7.65}$$

Finally the new sonic point is

$$r_s \simeq r_s^0 \left[1 - \frac{3}{4} \frac{\alpha}{c^2} \frac{X_0(1 + 3X_0)}{(1 + X_0)^3} \right]. \tag{7.66}$$

The constant α is the constant appearing in Eq. (7.61) giving the Alfvén wave pressure. This shows that the sonic point is indeed lower due to the Alfvén wave pressure. Note again that this expression holds only if we neglect wave dissipation, and assume pure spherical symmetry. These conditions are generally not satisfied; see, for example, the numerical simulation in Figure 7.6.

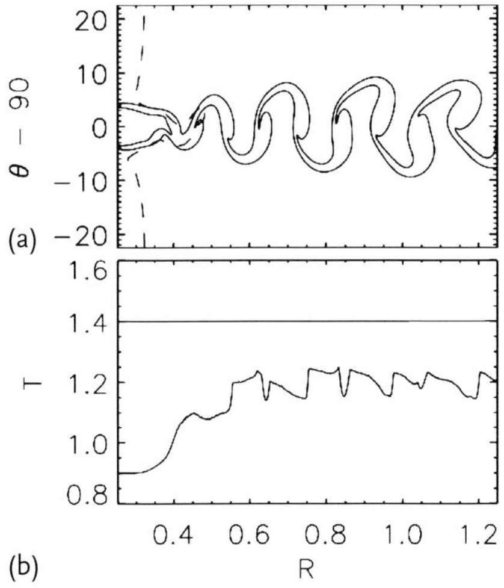

Figure 7.6 Unstable wake above low temperature coronal region. (a) Temperature contours in the (R-latitude) plane (dashed line: sonic surface); (b) temperature extrema versus R. A diffusive coefficient (conductive-like) is responsible for the progressive increase of the minimum temperature in the unstable cold zone [104].

7.1.7
Possible Routes to Turbulence in Stellar Winds

An illustration of the importance of the concept of control of the wind speed by the coronal temperature is shown by considering multidimensional stellar winds, that is, departing from spherical symmetry. We consider again simple isothermal ($\Gamma = 1$) case, which allows explicit expressions, for example, for the sonic point.

Let us consider a multidimensional situation and assume several distinct subsets of the corona with differing temperatures; each will have its own attractor forming a local quasi-radial flow, and the sonic surface (the set of points where $\mathcal{M} = 1$) will not be the surface of a perfect sphere as in the case of a uniform coronal base. In the case of a nonuniform coronal base, the sonic surface will be a deformed spherical surface as the wind is expected to be fast/slow above the hot/cold regions of the corona. Hence, spatial temperature variations at the coronal base can generate velocity shears. A velocity shear can lead to turbulent von Karman-like streets in the case of nonmagnetic winds as shown in the hydrodynamical simulation of Figure 7.6. As seen in the figure, the shear flow is destabilized at the level of the sonic point in the cold region around the equator. The rapid growth of the fluctuations as a function of distance is due to transverse pressure gradients generated by differing scale heights in the cold and hot regions. The generation of turbulence through mixing of hot and cold streams could well proceed in this way in nonmagnetized stellar winds.

A magnetic field (say, parallel to the flow) has a stabilizing effect on shear flows and the mixing instability is easily suppressed. However, if the velocity shear occurs in regions of magnetic field reversal as is the case of the so-called coronal helmet streamers, other instabilities implying the reconfiguration of the magnetic topology (for example, formation of magnetic islands) can occur leading to turbulence similar to the unmagnetized case of Figure 7.6.

Such configurations have been much studied and have led to scenarios for opening closed magnetic regions in the solar corona (coronal streamers). Such a work including the expansion of the solar wind is illustrated in Figure 7.7. In this simulation, one models a planar section of a solar streamer belt, in the region beyond the cusp of a helmet streamer, that is, above the closed magnetic region around the solar equator, as a magnetic current sheet embedded in the center of a broader wake flow. As already seen in Section when describing the Alfvén pressure effect in a solar wind model, the plasma flows at the speed of the fast solar wind at the edges and at a much lower velocity at the sheet. This is taken into account in the initial conditions. The development of the instability forms a magnetic blob which is accelerated by the reconfiguration of the magnetic topology when taking into account the effects of the expansion of the wind above the closed region (coronal streamer).

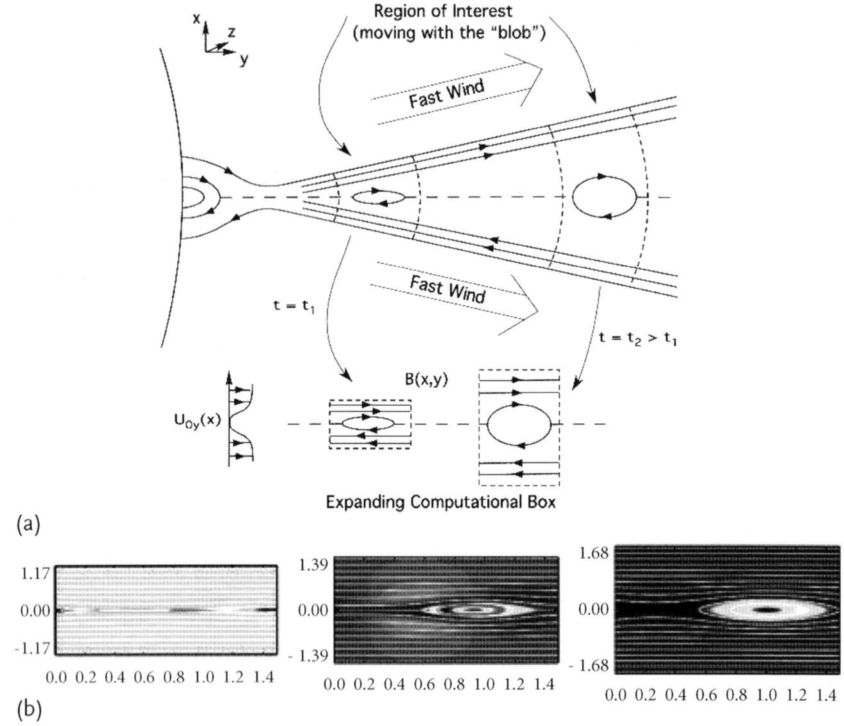

Figure 7.7 Acceleration of the slow solar wind in the heliospheric current sheet above a closed magnetic region. (a) Schematics of the numerical setup, that allows to follow the evolution of a plasma parcel embedded in a current sheet and a given mean radially expanding flow (with low speed within the current sheet). (b) Evolution of the plasma parcel density showing the formation of a reconnecting blob with larger density and fast acceleration. function of cross-stream distance to the current sheet at selected times. (From [105]).

This should change significantly the global velocity of the flow. Note however that this scenario has not been tested up to now via a full global simulation of the wind.

7.1.8 Accretion

Accretion of material onto compact objects, like stars or black holes, is another rather common feature in astrophysics. In the general case the infalling material carries an angular momentum which prevents the material from falling radially on the stellar surface eventually leading to the formation of accretion discs. Here we limit ourselves to the case of a radial infall of material onto a central, spherically symmetric object. Under such circumstances the relevant equations are those of the stellar wind from Section 7.1.6 with reversed flow velocity $u \to -u$, a transformation which leaves both Eqs. (7.47) and (7.54) unchanged.

As in the case of an outflow, the stable solution is the transonic one unless the stellar radius is larger than $r_B/2$ and as in the outflow case, the transonic solution corresponds to the one with the fastest accretion rate compatible with a subsonic velocity at infinity. Once again, it is instructive to write a conveniently normalized version of Eq. (7.43) for isothermal radial accretion in a central potential

$$\partial_r \left(\frac{\mathcal{M}^2}{2} + \ln \frac{\rho}{\rho_\infty} - \frac{r_B}{r} \right) = 0 \tag{7.67}$$

where ρ_∞ is the density at large distances for the solutions with vanishing Mach number $\mathcal{M}$ at $r \to \infty$ for which the expression enclosed by the brackets is zero. We note that every solution corresponds to a different value for ρ_∞. The typical Bondi radius for an interstellar medium at $T = 100$ K and an average per particle mass $m = 1.15 m_p$ in the field of a solar mass star is $r_B \approx 1.8 \times 10^{14}$ m which is roughly 2.5×10^5 times the solar radius. The Bondi radius being much larger than a typical stellar radius implies that the subsonic solutions, that is, the solutions for which the Mach number $\mathcal{M}$ is everywhere smaller than unity (cf. Figure 7.8), are unrealistic. Indeed, for the subsonic solution one has $\mathcal{M} \to 0$ for $r \to 0$. Under such circumstances Eq. (7.56) implies that the density profile must approach the hydrostatic equilibrium profile $\rho(r) = \rho_\infty \exp(r_B/r)$ meaning that near the

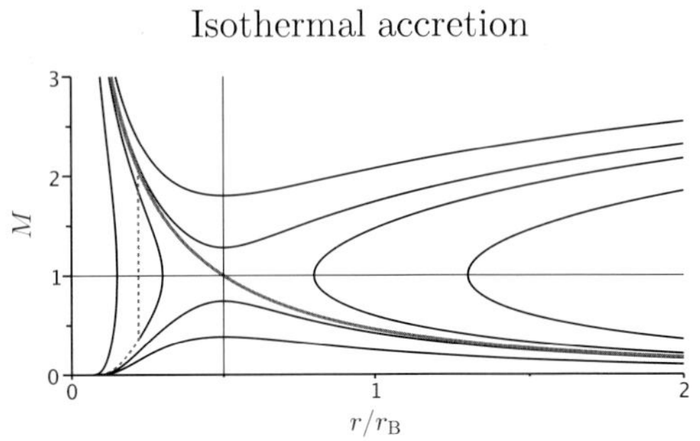

Figure 7.8 Possible solutions of Eq. (7.54) for the case of an isothermal radial inflow.

stellar surface, at $r = r_0$, the number density n should grow to a horrendously large value $n(r_0) = n_\infty \exp(2.5 \times 10^5) \approx n_\infty 10^{10^5}$. Given an interstellar density n_∞ of 1 particle per cm^{-3} the number of particles per cm^{-3} should then exceed the estimated number of particles in the whole universe! The subsonic solutions being unrealistic, the only acceptable solution is the transonic one for which we can compute the accretion rate $\dot{\mathcal{M}}$. Indeed, from Eq. (7.56) we get that at the sonic point $r_B/2$ where $\mathcal{M} = 1$, the density is $\rho_s = \rho_\infty \exp(3/2) \approx 4.48\rho_\infty$. The mass flux at the sonic point, and through any other spherical shell outside r_0, is then given by

$$\dot{\mathcal{M}} = \rho_s u_s 4\pi r_s^2 = 14.1\rho_\infty c_1 r_B^2 .$$

This is the largest possible accretion rate compatible with a vanishing inflow velocity at infinity. It is even larger than the flux of the unrealistic subsonic solutions. At some point, on its way to the surface of the star the fluid will have to slow down through a standing shock wave.

7.2 Magnetic Reconnection

7.2.1 Conservation of Connections vs. Reconnection

As explained also in Chapter 1, a magnetic field line motion, at the magnetic *velocity* $\boldsymbol{v}_m = \boldsymbol{E} \times \boldsymbol{B}/B^2$, can be defined whenever $E_\parallel = 0$, that is, when the electric field does not have a component parallel to the magnetic field. In these conditions, two points of a magnetic field line, if moved at $\boldsymbol{v}_m$, always remain on the "same" (moving) field line. The points of the same field line are said to be "magnetically connected" and there is, therefore, "conservation of magnetic connections" during the system evolution if $E_\parallel = 0$. The exact condition on $E_\parallel$, as shown in this chapter, is even a little bit wider, but this extension has little practical consequences. The conservation of the magnetic connections is guaranteed in ideal MHD and in any model, fluid or kinetic, where an ideal Ohm's law is verified since $\boldsymbol{E} = -\boldsymbol{u} \times \boldsymbol{B}$ does imply the nullity of $E_\parallel$. The inclusion of a Hall electric field, which amounts to replacing the ideal Ohm's law by $\boldsymbol{E} = -\boldsymbol{u}_e \times \boldsymbol{B}$ ($\boldsymbol{u}_e$ being the electron velocity), also does not break the conservation of connections since it does not more allow the existence of a parallel electric field.

As explained in Chapter 3, the conservation of magnetic connections brings an important constraint on the plasma flows when it is imposed. It forces in particular the creation of thin impervious interfaces when two magnetized plasmas are pushed toward one another. Reconnection occurs when this constraint is broken somewhere on the surface and when this breaking allows important changes in the connectivity (it is the case in particular when the magnetic field directions are strongly different on both sides of the surface). The most obvious consequences of

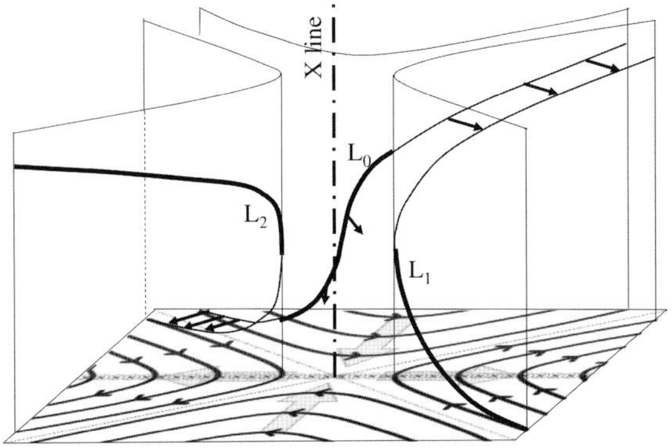

Figure 7.9 Loss of connectivity in a X line geometry with a nonnull guide field. In this geometry, the vertical B_z component, even if small, becomes always predominant when close enough to the X line. Three particular field lines are drawn: L_0, L_1, and L_2. Assuming the geometry is stationary and moving each point of L_0 with the local magnetic velocity $\boldsymbol{v}_m = \boldsymbol{E} \times \boldsymbol{B}/B^2$ transports all the lower left points onto L_2, while all the top right points are transported onto L_1. Between these two extremities, the points reach different field lines. There is, therefore, partial conservation of the connectivity on each side separately, far enough from the X line, but no global conservation for the full line. No field line is "broken" in the process, however, although the horizontal projection of this motion does appear discontinuous.

reconnection are the different magnetic topologies that become accessible, allowing, for instance, the penetration of magnetic flux (together with particle fluxes) across surfaces that would have remained impervious otherwise. But there are also other consequences which can be as important, and even sometimes more important: the transfer of magnetic energy into kinetic energy in the process. This leads to acceleration and heating of the plasma in the exhaust region, driven by the relaxation of the magnetic tension existing in this region.

The most typical geometry considered for reconnection is a 2D geometry with an "X line" (see Figure 7.9). The component B_z of the magnetic field in the direction perpendicular to the plane is constant in this case and is called a "guide field". The absence of a guide field, that is, $B_z = 0$, is the situation most often investigated: this situation actually allows for the simplest analytical calculations and it corresponds to the most unstable situation with regard to the appearance of spontaneous reconnection (tearing instability). Nevertheless, the apparent "breaking" of the field lines, which is characteristic of this absence of guide field, is purely due to the presence in this case of a "null line", where B is strictly zero. It is the only case where field lines may seem to "cross" each other. In all other cases, no such breaking can be involved, as shown on the figure. The breaking is, therefore, rather an anecdotal feature, only associated to the particular case of a null line, and it should not be taken as a definition of reconnection.

It is important to distinguish clearly the field line geometry and the notion of reconnection. For the same geometry presented in Figure 7.9, there is reconnection if there is an electric field, corresponding to an ideal motion of the field lines across the given structure, but there is no reconnection if there is no electric field, that is, if everything is steady.

7.2.2
Departure from the Ideal Ohm's Law: Microscopic Mechanisms and Macroscopic Consequences

Reconnection is due to a departure of the field line motion from the frozen-in motion at $\boldsymbol{v}_{\mathrm{m}}$. The above X line example has furthermore shown that this departure can be localized in a very small region (close to the X line); however, it changes all the connectivity properties of the plasma at large scales. As the particles can circulate almost freely along the field lines, two regions magnetically connected will obviously have a behavior quite different from two nonconnected regions with respect to the plasma properties. As the conservation of connections is linked to the nullity of $E_{\parallel}$, the question arises to know where this condition can be broken and why it is generally localized at small scales. Whatever the plasma model, fluid or kinetic, the answer lies entirely in the generalized Ohm's law, or the electron momentum equation:

$$\boldsymbol{E} = -\boldsymbol{u}_{\mathrm{e}} \times \boldsymbol{B} - \frac{1}{n_{\mathrm{e}} e} \left[\rho_{\mathrm{e}} d_t \boldsymbol{u}_{\mathrm{e}} + \nabla \cdot \boldsymbol{P}_{\mathrm{e}}\right] + \eta \boldsymbol{j} . \tag{7.68}$$

In this expression, the electron fluid velocity $\boldsymbol{u}_{\mathrm{e}}$ can be expressed as a function of the global one $\boldsymbol{u}$ and the current density $\boldsymbol{j}$. If neglecting the ratio $m_{\mathrm{e}}/m_{\mathrm{i}}$ with respect to unity, it is simply: $\boldsymbol{u}_{\mathrm{e}} = \boldsymbol{u} - \boldsymbol{j}/ne$. The first term, therefore, contains both the ideal term $-\boldsymbol{u} \times \boldsymbol{B}$ and the Hall term $\boldsymbol{j} \times \boldsymbol{B}/ne$. As already emphasized, both are unable to allow parallel electric field.

The three remaining terms involve derivatives, which means that they can give noticeable changes with respect to the ideal motion only at small scales. The characteristic scales at which they become important can be calculated by the general methods of dimensional analysis (see Chapter 3), but in the most difficult case where an accurate description of the polarization is needed, distinguishing the directions parallel and perpendicular to the magnetic field, and the directions parallel and perpendicular to the gradients (longitudinal and transversal). The calculation then greatly resembles the linear calculation of the propagation modes, in its full tensorial form. Let us consider, for the sake of simplification, the different additional terms of the Ohm's law as perturbations with respect to the ideal shear Alfvén mode (incompressible). The main results, deriving from a crude dimensional analysis (see Chapter 3), can then be summarized as follows:

- The term in $\rho_{\mathrm{e}} d_t \boldsymbol{u}_{\mathrm{e}}$ is maximum in the parallel direction and scales as $(\omega/\omega_{\mathrm{ce}})(u_{\mathrm{e}1\parallel}/u_{\mathrm{e}1\perp})$. Assuming $u_{\mathrm{e}1\perp} \approx u_{1\perp}$ and $u_{\mathrm{e}1\parallel} \approx j_1/n_0 q$, the ratio $u_{\mathrm{e}1\parallel}/u_{\mathrm{e}1\perp}$ can be estimated as $k^2 d_{\mathrm{e}}^2 \omega_{\mathrm{ce}}/(\omega)$. The resultant scaling for this first term is, therefore, $k^2 d_{\mathrm{e}}^2$: the electron inertia term becomes important for scales less than the electron inertial length $d_{\mathrm{e}} = c/\omega_{\mathrm{pe}}$.
- The pressure term scales as: $k^2 R_{\mathrm{e}}^2 (\omega_{\mathrm{ce}}/\omega)(u_{\mathrm{e}1\mathrm{l}}/u_{\mathrm{e}1\mathrm{t}})$, where R_{e} is the electron Larmor radius, and $u_{\mathrm{e}1\mathrm{l}}$, $u_{\mathrm{e}1\mathrm{t}}$, the components of the electron velocity variation,

longitudinal and transversal. Assuming $u_{e1l}/u_{e1t} \approx u_{1l}/u_{1t}$, this means that the electron pressure term does not bring any contribution for purely Alfvénic variations, that is when $u_{1l} \ll u_{1t}$. However, the ion inertia makes the ratio $u_{1l}u_{1t}$ vary as ω/ω_{ci}. This makes the resultant scaling of the electron pressure term on the order of $k^2 R_e^2 m_i/m_e$, bringing the small scale $R_e^* = R_e\sqrt{m_i/m_e}$.

- The resistive term ηj scales as $k_\perp \eta/\mu_0 V_A$, bringing the small scale $\lambda_\eta = \eta/\mu_0 V_A$, called the "resistive length".

The last term is the only nonideal term used in resistive MHD (collisional case), but the two previous ones are obviously more relevant for collisionless reconnection.

7.2.3 Flow Acceleration by Reconnection

When the reconnection is triggered by an instability, it can be characterized by two numbers: (i) the characteristic time for going from an initial state without reconnection to a final one with reconnection (one can take the inverse growth rate of the instability γ^{-1}, see next section), and (ii) the stationary reconnection rate which characterizes the final state if the system does evolve nonlinearly toward such a stationary state. If the system is not unstable, but the reconnection is driven by some external forcing, the triggering problem is not relevant, but the question of a stationary reconnection rate is similar, as long as the forcing does lead also to such a stationary state.

The reconnection rate is defined as $\tau_r = u_{y1}/V_{A1}$ in the simplest case of a symmetric 2D reconnection with an X line geometry without guide field, x being the direction of the reversing magnetic field, y the direction normal to the reconnecting layer, and z the invariant direction. The reconnection rate is, therefore, a dimensionless number, which is the ratio between the incident velocity u_{y1} on the axis $x = 0$, far enough from the layer, and the incident Alfvén speed at the same place. This Alfvén speed chosen for normalizing the incident velocity can be viewed as some "maximum" value for it, at least whenever the compressible effects can be neglected (the fast magnetosonic speed could be taken otherwise). Actually, the fastest reconnection rates are usually about $\tau_r \approx 0.1$. Considering that far enough from the layer an ideal Ohm's law is satisfied, the motion of the field lines – and, therefore, of the flux tubes – occurs at the velocity u_{y1}. The reconnection rate can thus be defined as the ratio between the reconnected flux during any time δt and the maximum value of this reconnected flux. In the simple case of a 2D stationary geometry, it is easy checking that the plasma motion is related to a constant electric field E_z perpendicular to the (x, y) plane of the gradients (see Faraday equation). This electric field is also a proxy frequently used to measure the reconnection rate.

For estimating the reconnection rate, it is necessary to distinguish an "internal" region, generally called a "diffusion region", close to the X line and where the nonideal effects make the reconnection possible, and an external one, where the flow motion is considered ideal. The variables in the input and output external regions are related by two kinds of relations: (i) relations involving the internal physics and (ii) fundamental conservation laws, independent of the internal nonideal physics (depending only on the global geometry). The separation between the two kinds of relations is reminiscent of the calculations of the thin layers ("discontinuities", see Chapter 6): the fundamental conservation laws give there the Rankine–Hugoniot laws, and, therefore, the jumps, and the internal physics determines the possibility of the discontinuity, its width, and the different variable profiles.

The first estimation of the reconnection rate in the 2D symmetric case (without guide field) was made by Sweet and Parker in 1957 and 1958 [106, 107]. This work was done for an elongated plane sheet, in the context of resistive MHD, but the basic method that they used can easily be generalized for other 2D geometries and other nonideal effects. These generalizations are still used today, in particular for all the more recent collisionless models. The considered 2D stationary hypothesis imposes that the electric field is in the z direction, and that this field E_z is constant (Faraday's equation). Far enough from the X line, it imposes the ideal motion in the (x, y) plane.

Here are the main arguments, assuming that the ratio $\varepsilon = \delta y/\delta x$ of the width of the layer (see for instance Figure 7.10) to its length is small.

1. The mass conservation imposes $u_{x2} = u_{y1}/\varepsilon$ if u_{y1} and u_{x2} are the characteristic speeds, respectively, in input and in output, on the symmetry axes, far enough from the X line so that the motions can be considered ideal. This relation derives from an assumption of incompressibility, which is correctly satisfied as long as the flow speeds remain sufficiently sub-Alfvénic in the region under study. Because of the hypothesis of an ideal motion out of the inner region, the same relation also draws $B_{y2} = \varepsilon B_{x1}$ and, therefore, $V_{A2} = \varepsilon V_{A1}$. It can be derived directly as well from the conservation of the magnetic flux of a tube when it crosses a separatrix far from the X line.
2. The energy conservation imposes that: $\rho u_{y1}^2/2 + 5/2 P_1 + B_{x1}^2/\mu_0 = \rho u_{x2}^2/2 + 5/2 P_2 + B_{y2}^2/\mu_0$. The third term corresponds to the Poynting flux in the same hypotheses of ideal motions in input and output. This relation can be written in the form: $u_{y1}^2 + 5V_{th1}^2 + 2V_{A1}^2 = u_{x2}^2 + 5V_{th2}^2 + 2V_{A2}^2$ (thanks to the hypotheses of incompressibility and symmetry).
3. Injecting the relations $u_{x2} = u_{y1}/\varepsilon$ and $V_{A2} = \varepsilon V_{A1}$ in the energy equation, and assuming that $\varepsilon \ll 1$, one gets: $5V_{th1}^2 + 2V_{A1}^2 = u_{x2}^2 + 5V_{th2}^2$. In low beta plasmas, this means that $u_{x2} \approx V_{A1}$, that is, due to energy conservation, the flow speed in the exhaust region of reconnection is fixed by the Alfvén velocity in the incident region. The incident speed flow is smaller in the ratio ε so that $u_{y1} \approx V_{A2}$, and, therefore:

$$\tau_r \approx \varepsilon . \tag{7.69}$$

This "cold" case is the typical one, the most generally treated in the literature. We will restrict ourselves to this case hereafter although the energy conversion of the input magnetic energy into output thermal energy is generally on the same order, and even often larger, than its conversion into bulk flow energy

Nevertheless, this introduces numerical factors that can be ignored at the level of the crude estimations of order of magnitude that we are presenting here.

4. The above "external" physics fixes the reconnection rate τ_r as a function of one single geometrical factor ε. The internal physics can be expected to give another relation between the two same parameters, so allowing us to determine both. The internal physics must be such that the shortest dimension is sufficiently small to "defreeze" the field line motion from the ideal velocity field at $\boldsymbol{v}_m = \boldsymbol{E} \times \boldsymbol{B}/B^2$. One can generally assume that this shortest scale concerns the y direction (normal to the layer), so that one can write $\delta y' \approx d_{ni}$, where d_{ni} is the characteristic scale associated with the effective nonideal effect. This scale is taken as λ_η in resistive MHD, but it is rather d_e or R_e^* in a collisionless plasma. For a crude estimation, it is sufficient assuming that the nonideal term, which is negligible in the external region is dominant in the internal one. This leads to:

$$\tau_r \approx \frac{d_{ni}}{\delta y'} = \frac{1}{\varepsilon} \frac{\delta y}{\delta y'} \frac{d_{ni}}{\delta x} . \tag{7.70}$$

5. Putting together the external result $\tau_r \approx \varepsilon$ and the internal result $\tau_r \approx (1/\varepsilon)(\delta y/\delta y')(d_{ni}/\delta x)$, we get the general conditions:

$$\tau_r \approx \varepsilon \approx \sqrt{\frac{\delta y}{\delta y'} \frac{d_{ni}}{\delta x}} . \tag{7.71}$$

We can see that the maximization of τ_r must satisfy two contradictory requirements: on one hand, the aspect ratio ε must be as large as possible to allow the best reconnection rate compatible with the external conservation equations (when the ratio is small, the exhaust region is small, allowing for little possibility for flux ejection there); on the other hand, the same aspect ratio must be as small as possible to allow defreezing in the internal region (with a too large aspect ratio, there is little possibility of departing from the frozen in motions, which opposes reconnection). The above result gives the only possible compromise between the two contradictory requirements. In addition to the small value of the reconnection rate, it provides also a relation between the two characteristic widths: $\delta y \delta y' = d_{ni} \delta x$.

The above points are common to all the models of 2D stationary reconnection presently existing, which makes the result (7.71) quite general and robust. The differences come afterward, in the assumptions done about the internal geometry and the nature of the nonideal effects. The original calculation by Sweet and Parker assumes that the internal physics can be described by ideal MHD, so that $d_{ni} = \lambda_\eta$ and that there is one single characteristic scale, the defreezing scale being equal to the width of the layer, so that $\delta y' = \delta y$. The result then reads $\tau_r \approx \varepsilon \approx \sqrt{\lambda_\eta/\delta x}$. This corresponds to extremely small values for all the known stationary layers: the lengths of the layers, fixed by large-scale phenomena, are always extremely larger than the resistive scale, not only in the collisionless or weakly collisional plasmas, where λ_η is almost zero, but even in the more collisional ones. It also corresponds to very thin layer with a width called the "Sweet–Parker length" given by: $\delta_{SP} = [\lambda_\eta \delta x]^{1/2}$.

This discouraging result seems to indicate that the reconnection phenomenon is so slow that it is a negligible one. Actually, it is not the case for two reasons: (i) the hypotheses of the Sweet–Parker model (resistive MHD and a geometry with a single structure imposing $\delta y' = \delta y$) have to be questioned, in particular in collisionless or weakly collisional plasmas; (ii) the imposed condition of stationarity may be too strict: the layer can be unstable to the creation of small magnetic islands and plasmoids with much smaller aspect ratios (see next section about the tearing instability). The combination of all these nonstationary structures can then give rise to an effective reconnection rate which is statistically stationary and much larger than the strictly stationary one [108, 109].

Figure 7.10 sketches the most important geometries that have been investigated in the literature. Figure 7.10a corresponds to the original Sweet–Parker configuration, where the scale of the defreezing layer, in black, is identical to the scale of the variation of $\boldsymbol{B}$. Figure 7.10b corresponds to the Petschek model (1964) [110], where the defreezing layers are two slow shocks, much thinner than the global layer. If such a geometry really could exist in a stationary state, it would correspond to a much faster reconnection thanks to the factor $\delta y'/\delta y$, but if remaining in the resistive MHD frame, it is not the case. In this case, one single characteristic scale is present in the equations, so that one single characteristic scale can be present in a stationary solution (if not directly imposed in the boundary conditions). Starting from Petschek's geometry as an initial condition, this one will naturally evolve: the large scale δy will decrease toward the small one $\delta y'$, so coming back to the Sweet–Parker geometry, the reconnection rate rapidly decreasing from its initial fast value to a slow (quasi-null) one. Figure 7.10c shows the geometry currently accepted nowadays for the collisionless plasmas (see for instance [111]). The resistive MHD model is replaced by a two-species description, which can be Hall-MHD, two-fluid, hybrid, fully-kinetic, and so on. In any case, two different characteristic scales are embedded: δy is fixed by the minimum ion scale, d_i or R_i, but the defreezing is associated to electron scales where the electrons are "demagnetized" (d_e or R_e^*), which are smaller. This allows faster reconnection in a stationary state via the ratio $\delta y'/\delta y$. The ion region, in which the major part of the magnetic field reversal occurs, is dominated by the Hall electric field, due to the difference of perpendicular velocity between ions (demagnetized) and electrons (magnetized). Its aspect ratio determines the reconnection rate $\tau_r = \varepsilon_i$, while the electron ensure the defreezing at a smaller scale without limiting the ejected flux.

Whatever the physical mechanism in the internal region, all the above results show that reconnection can allow transferring a flux of magnetic energy $(\rho V_{A1}^2/2)$ in input to a flux of kinetic energy $\rho u_2^2/2$ in output (and of thermal energy, see [112]). This energetic consequence, even if it cannot be taken as a definition of reconnection, is so important that it is often the main reason why reconnection is invoked in astrophysical conditions, independently of the defining notions of magnetic connections. It is the case, for instance, in solar physics where reconnection is invoked

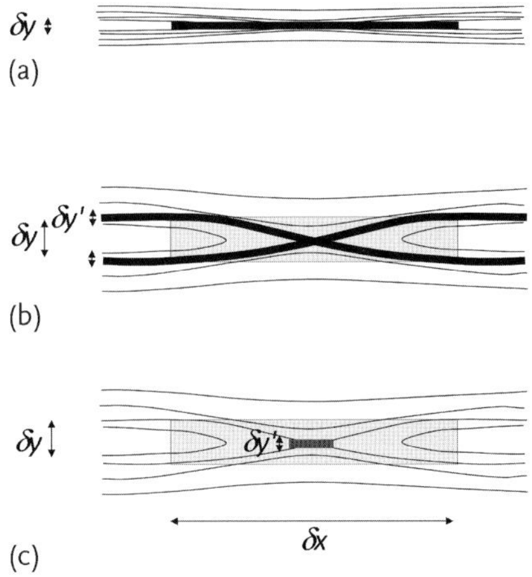

Figure 7.10 Classical model geometries for stationary reconnection. (a) Sweet–Parker model: one single characteristic scale λ_η in the model (resistive MHD), which determines the width of the layer. (b) Petschek model: still in resistive MHD, the defreezing is localized in thin substructures (slow shocks) of width λ_η, but the full layer can be much wider, allowing a much faster reconnection rate. This demands that shocks do exist in a stationary state and that the large scale of the layer does not collapse to the small one. (c) Collisionless model: two different characteristic scales (therefore, beyond resistive MHD). The defreezing is localized in a substructure at the electron scale while the global layer width is fixed by the ion one.

for explaining the heating of the corona. The acceleration of the flow can be understood simply: the field lines are strongly curved in the exhaust region, so that the tension force accelerates the plasma. In spite of this simple argument, however, it must be kept in mind that at least one-half of the output energy is actually in the form of thermal energy and does not go into bulk flow acceleration.

Let us finally briefly mention that reconnection is also often invoked in the domain of nonthermal energetic particles. It is particularly the case, for instance, in solar physics for explaining some of the solar cosmic rays emitted by the Sun. It can be the case also in magnetospheric physics where the energetic electrons that precipitate in the polar auroras may be related to reconnection events in the magnetotail during magnetic substorms. In all these contexts, a few particles are supposed to be accelerated up to very high energies by the mechanism. It must be emphasized, however, that such effects cannot exist in the simple frame of a stationary single X line geometry that we have considered hitherto. To explain these energetic events, it is necessary to take into account, not only nonstationary and 3D effects, but also the presence of multiple magnetic islands and/or the coupling with the surrounding physics, as mentioned in Section 8.2.

7.2.4
Tearing Instability

Reconnection, as shown in the previous section, can be stationary. The nonstationarities are, therefore, clearly not essential to its definition. When one has $\partial_t = 0$ in some proper frame, all variations are purely spatial in this frame. However, if the reconnection is presented as a temporal effect in this case, as it is often done, ("change" of magnetic connections, energy "change", and so on), it must be understood that it is not in the fixed frame, but in a local frame moving at $\boldsymbol{v}_\mathrm{m} = \boldsymbol{E}\times\boldsymbol{B}/B^2$. If, in a given X line geometry, one has $\boldsymbol{E} = 0$, that is, no velocity $\boldsymbol{v}_\mathrm{m}$ across the spatial variations, there is no reconnection. If $\boldsymbol{E} \neq 0$, there is reconnection and the spatial variations are viewed as temporal changes when following the ideal motion of a field line.

In this sense, sentences such as "reconnection is a mechanism which allows a system to reach a lower energy state" or "reconnection is always accompanied by a violent release of magnetically-stored energy" are misleading: in stationary reconnection, the total magnetic energy does not change at all. These changes are not involved in the very definition of magnetic reconnection, but only in its possible nonstationarities. The most typical problem of this kind consists of considering a tangential layer in equilibrium (see Chapter 6), and looking for the conditions making this layer unstable to the creation of magnetic islands. This instability is called "tearing instability" and it necessarily implies reconnection since the ideal field line motion prevents any topology change. When such an instability arises, its consequence is of course to decrease the system energy, as for any other instability, with or without reconnection.

The linear calculation of the tearing instability, as the calculation of a stationary reconnection rate, needs to distinguish the external MHD physics, far from the X lines, from the internal physics, close to them. The general method of calculation was settled in the 1960s [113]. The instability gives rise to a surface wave: the external perturbation is assumed to tend to zero at infinity; the internal perturbation is such that it matches the two external ones. The result is of course sensitive to the nature of the internal description, kinetic or resistive MHD, the kinetic one being generally more unstable than the MHD one. In the case of resistive MHD, the main results are:

1. The instability growth rate depends on the relative values of four lengths, the total length of the layer δx, its width δy, the x wavenumber k, and the resistive length λ_η, with: $\lambda_\eta \ll \delta y \ll k^{-1} \ll \delta x$. The result can be expressed as a function of two main dimensionless parameters: the Lundquist number $S_y = \delta y/\lambda_\eta$ and $k\delta y$ (the resistive length is calculated with the asymptotic magnetic field in the external region). The Lundquist number S_y is a very large number. The global one, $S_x = \delta x/\lambda_\eta = \delta x/\delta y\, S_y$, is of course still larger.
2. The layer is unstable for a limited range of wavenumbers, with a maximum growth rate at: $k\delta y \approx S_y^{-1/4}$.

3. The corresponding maximum growth rate is given approximately by: $\gamma_{\max} = \tau_A^{-1} S_\gamma^{-1/2}$, which is much smaller than the inverse Alfvén transit time $\tau_A^{-1} = V_A/\delta y$ (V_A is also based on the asymptotic magnetic field in the external region).

Coming back to the dimensioned variables, the maximum growth rate can be written:

$$\gamma_{\max} = V_A \lambda_\eta^{1/2} \delta y^{-3/2}. \tag{7.72}$$

This result shows that for a fixed V_A, the maximum growth rate of the tearing instability depends on two parameters: the resistivity, via the resistive length λ_η, and the layer width δy. If these two parameters are considered independent, it is clear that the growth rate tends to zero when the resistivity tends to zero, if the width is kept arbitrarily fixed. On the other hand, choosing a width small enough may always make a significant growth rate, for any given resistivity. To solve this problem more consistently, one can assume that the width of the layer, as it seems reasonable, is determined by the above Sweet–Parker estimation $\delta y = \delta_{SP} = [\lambda_\eta \delta x]^{1/2}$. The width so becomes also a function of the resistivity. The corresponding result is:

$$\gamma_{\max} = V_A \lambda_\eta^{-1/4} \delta x^{-3/4}. \tag{7.73}$$

The resistive length now appears at a negative power, implying that the layer is more and more unstable when the resistivity decreases: the effect of the width decrease is stronger than the direct effect of the resistivity on the growth rate. This can allow reconnecting thin layers in very large Lundquist media, not by a stationary process, but via the formation of multiple plasmoids. It is worth noting that this effect has been shown only recently because it could not been observed in numerical simulations before the computation facilities allow resolving very thin layers with very low resistivity.

The relevant theory, in a collisionless plasma, is not resistive MHD. One has then to take into account the different characteristic lengths and the different geometries, but the same principle can be used. In this collisionless case, which is ubiquitous in space plasmas, one has to take care also that studying the stability of a tangential layer demands to be able to describe the initial equilibrium in a kinetic manner, which is far from easy in general (see Chapter 6).

Let us mention in addition that many extensions of the above treatments have been done, in particular for studying the possibility of a tearing mode in the case of a layer with a nonnull B_n, that is, a layer which is not tangential. This is of interest, for instance, in the Earth's magnetotail for understanding the question of magnetic substorms at the origin of auroras. The kinetic results show less instability, in this case, than the MHD ones.

7.2.5
3D Reconnection

The principles of reconnection have been presented in the previous section in the simplest geometry, which is 2D, with an X line and a coplanar reversal of the magnetic field. This is of course quite restrictive. Even if the definitions have been given in a general form, applicable in any case, the calculations presented have been restricted, for the sake of simplicity, to this simple case. They can certainly not be used without generalization in solar physics for instance, where coronal magnetic loops can meet during their motion, create thin layers with complex shapes and reconnect.

We will not present in this book specific calculations for the case of 3D reconnection, but only give a short example of a typical geometry, just to give a taste of it. Most of the work in this field must be numerical (generally in MHD up to now). In Figure 7.11, one can see the kind of geometry that can arise in solar events as mentioned above. It involves a surface separating inside and outside field lines. At the top of this structure, a null point is present, but without null line. The singular line going through this point is called a spine (it can be an axis of symmetry in the simplest modeling). The nonideal region where the reconnection can occur is close to the null point, and the concerned field lines are, therefore, close to the structure surface and to the spine. As in 2D, the reconnection allows the field line to slip with respect to their ideal motion and penetrate from outside to inside the structure. In addition, however, the third dimension allows slipping of these lines also azimuthally, leading to fast changes of connectivity for lines that remain outside the structure.

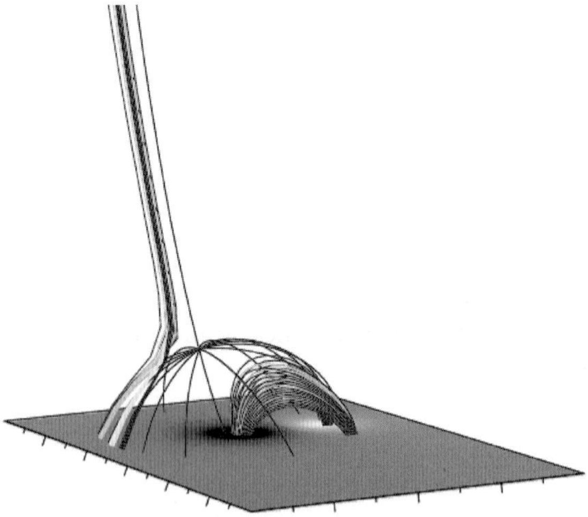

Figure 7.11 Example of 3D field line geometry with a null point. This geometry is typical of those assumed to lead to reconnection in the solar corona (from [114]).

7.3
Kinetic Acceleration Processes in Magnetospheres

Nature has developed a great variety of processes to accelerate charged particles, even when, as is mostly the case in this book, only nonrelativistic processes are considered. Therefore, it is not possible in this chapter to develop a full theory of acceleration. Here, we give an overview of the processes that do not concern the bulk of the plasma, but only a part of it, and sometimes, a minority of particles. The expression "kinetic acceleration processes" is used here in opposition to "plasma flow acceleration process". Such processes are expected in astrophysical shocks, a topic that is presented in Section 6.1. In the present section, we emphasize the kinetic acceleration occurring in the magnetospheres. Because the Earth's magnetosphere has been abundantly explored in situ, it is the basis of the explanations given in this section.

Many references are made in the following section to the various regions of the magnetosphere; for memory, they are briefly defined in Section 1.2.4.

7.3.1
Substorms and Auroras in the Earth's Magnetosphere

Polar auroras are the first evidence of energetic electrons precipitating into the ionosphere. Their striated structures with almost vertical lines of enhanced luminosity are the signature of a series of atoms excited (then de-excited) by a series of collisions with electrons having a large velocity along the magnetic field. In the late 1950s, rocket flights in bright auroral arcs revealed their energy is in the range 1–10 keV, and that they constitute a minority of the ambient plasma electrons. The ambient plasma is much denser, less energetic, and without noticeable bulk velocity.

An event implying the development of bright auroral arcs over a large portion of the auroral zone (the area where auroras generally occur) is called a *substorm*.

The first explanation for auroral arcs was that the solar wind electrons directly come into the auroral zone, flowing along the terrestrial magnetic field lines delineating the polar cap. This explanation was believed by many scientists for a few years, and is still widely spread among non specialists, but it is wrong.

Let us suppose that the electrons arrive above the ionosphere with the same energy as in the solar wind. The equation of motion of the electron guiding center Eq. (1.122) can be simplified into

$$d_t \boldsymbol{U}_{\parallel 0}(t) = \frac{q}{m} \boldsymbol{E}_{\parallel} - \frac{\mu}{m} \nabla_{\parallel} B \,. \tag{7.74}$$

(We have neglected the last term.) Let us consider an electron from the solar wind with a typical 100 eV energy. If it arrives with the same energy into the ionosphere, then, on average $d_t \boldsymbol{U}_{\parallel 0} = 0$. Then, a parallel electric force $q E_{\parallel}$ must be set in order to compensate the magnetic field gradient force $\mu \nabla B$. Considering typical figures[1], the average electric field is $E_{\|} \sim 2 \times 10^{-2}\,\mathrm{V\,m^{-1}}$. Considering the length of the magnetic field line, this implies a potential drop of 640 kV! Therefore, the simple image of electrons flowing along magnetic field lines does not hold. A strong acceleration is required, all the more since the measured energy of the auroral electrons is not 100 eV but several keV.

There is a second objection to the direct precipitation of electrons along the polar cap field lines. If this was true, the brightest auroras would happen at the feet of the magnetic field lines where the solar wind electrons arrive. This would correspond to the polar cap, that is, at high latitudes on the dayside's ionsophere. On the contrary, the brightest auroras are observed around midnight (local time) at magnetic latitudes typically between 60° and 80°.

Actually, the parallel electron acceleration that causes the bright polar auroras does not happen along the polar cap field lines. Instead, it involves the dynamic behavior of the whole magnetosphere, especially in the tail. To be convinced by this idea, let us evaluate the electric potential drop across the whole magnetosphere induced by the solar wind convection $\boldsymbol{E}_{\text{convection}} = \boldsymbol{v}_{\text{SW}} \times \boldsymbol{B}_{\text{SW}}$. With the currently used figures for the Earth's magnetosphere[2] the potential drop is $\Delta\Phi \sim 1200\,\mathrm{kV}$. This is more than the previously computed potential drop, and this indicates that the solar wind energy captured by the whole magnetosphere could be the source of the auroral acceleration. Because this capture of the solar wind energy concerns the whole magnetospheric volume, the acceleration processes can be at work in various regions of the magnetosphere.

A substorm can be observed simultaneously in the auroral zone and in the magnetotail. It starts in the magnetotail, with a growth phase characterized by an amplification of the cross tail current. The auroral counterpart is a quiet aurora, with a very simple geometrical structure. The quiet phase is followed by a break-up that consists of a local reconfiguration of the magnetic field in the magnetotail. From a tail-like structure ($\|B_z\| \ll \|B_x\|$ where x points southward and z points northward), it becomes more dipole-like (smaller B_x) with a smaller east-west tail current intensity. Then, during the expansion phase, the magnetic reconfiguration propagates in the whole magnetotail. In the ionosphere, it is characterized by very dynamical and bright auroral patterns expanding over a large range of longitudes and latitudes. The final phase of a substorm is the recovery, during which the auroral pattern decreases. It also happens that local magnetic reconfigurations do not propagate all over the tail. The event is then called a pseudobreak-up. It can have an auroral counterpart with locally bright arcs.

1) Let us consider a field line length $l \sim 5R_{\text{E}} \sim 32 \times 10^6$ m, a solar wind magnetic field $B_{\text{sw}} \sim 5$ nT, and an ionospheric magnetic field $B_{\text{i}} \sim 50\,\mu$T. The typical electron energy K in the solar wind is 100 eV, both in the perpendicular and parallel directions, and we can, therefore, expect $\mu = K_{\perp}/B \sim 0.3 \times 10^{-8}\,\mathrm{J\,T^{-1}}$. The average magnetic force (see Eq. (1.122)) is $\mu \Delta B / l \sim 4 \times 10^{-21}$ N.

2) The solar wind velocity is $v_{\text{sw}} \sim 4 \times 10^5\,\mathrm{m\,s^{-1}}$. We consider a southward magnetic field component of about 5 nT, and a transverse magnetospheric size of $10\,R_{\text{E}} \sim 6.4 \times 10^7$ m.

In the following sections, various acceleration processes are presented. Most of them play a role in the origin of the polar auroras. Most of these processes can be effective in other astrophysical contexts.

7.3.2
Fermi Acceleration in the Magnetosphere

The Fermi acceleration in the Earth's magnetosphere is based on the periodic bounce motion of particles trapped in a magnetic field. The conservation of the second adiabatic invariant J defined in Eq. (1.132) implies that on average, $p_{\|}$ increases when the distance between the mirror points reduces. This can be seen when we write $J = \langle v_{\|} \rangle d_{\mathrm{mp}}$ where d_{mp} is the distance between the mirror points. The Fermi acceleration process is possible only when the evolution of the magnetic field is slow, in regards to the bounce period of the trapped particle, otherwise, the adiabatic invariant J would not be conserved.

For a gently evolving magnetic field, and from Eq. (1.126), the pitch angle increases with the magnetic field amplitude B; therefore, the mirror points (reached when α goes up to $\pi/2$) become closer. Because the position of the mirror points depends only on a reference pitch angle (for instance, the equatorial pitch angle of the particle) and not on the particle energy, the Fermi acceleration is not energy dependent. The low energy plasma is accelerated as well as the high energy plasma. The selection criteria for acceleration is rather based on the initial pitch angle of the particles.

A compression of a magnetosphere, where consequently the magnetic field is amplified, favors the acceleration of the particles with a large pitch angle. For instance, this is what happens during *magnetic storms*, when the Earth's magnetosphere is exposed to the increase of solar wind pressure following a coronal mass ejection, and/or a change of interplanetary magnetic field direction. Then, the ring current region, populated with trapped ions and electrons, tends to be accelerated. The current in this region of dipole magnetic field is carried by the charge dependent sum of the gradient and the curvature drifts given in Eq. (1.130). Because this drift is proportional to v^2, because v increases with the Fermi acceleration, this results in an increase of the ring current during magnetic storms.

Fermi acceleration in the magnetotail during magnetospheric *substorms* seems more marginal. It is sometimes observed in the magnetic tail of the Earth during events when the strength of $\boldsymbol{B}$ increases, in regions of closed magnetic field lines.

The Fermi acceleration is invoked in other regions of the universe, in particular in shocks, for the acceleration of cosmic rays. We will see that the conservation of the adiabatic invariants is not necessarily required (Chapter 8).

7.3.3
Acceleration by a Forced Current Forced along Convergent Magnetic Field Lines

If Fermi acceleration and magnetic reconnection are often invoked by astrophysicists, other processes can play an important role, in particular in the highly magnetized regions where the magnetic field configuration cannot be transformed easily. Among these processes are those resulting from a forced current into a plasma.

An important feature of the magnetospheric substorms is the current interruption in the current sheet of the tail. Some observations suggest that the dawn to dusk current that goes across the whole plasma sheet is interrupted during the break-up phase of substorms. This interruption is local. Because of the charge conservation, this current must find a path to connect the two sides of the current sheet, that contours the current sheet. As soon as the current leaves the plasma sheet, a magnetic field is seen. Because the conductivity along the magnetic field lines is higher than across, the current then flows in the form of a parallel current density $J_{\parallel}$. The magnetic field lines connect the current sheet to the ionosphere (downward parallel current). In the ionosphere, which is a collisional plasma, the transverse conductivity, governed by the so-called Perdersen and Hall integrated conductivities σ_{P} and σ_{H} is high (these are finite conductivities in the horizontal direction, integrated over the vertical axis). Then the electric circuit can be closed, and after crossing a portion of the ionosphere, the current flows up to the current sheet (return upward parallel current).

The forcing of this current is caused by something happening in the plasma sheet. Without considering the nature of what happens in the plasma sheet, it is possible to draw a model of plasma acceleration resulting from the parallel current forcing. The simplest system was described by Knight ([115]). Because of the repulsive effect of the magnetic field, most particles tend to be reflected. Therefore, only the electrons with a low reference pitch angle (taken in the current sheet) reach the ionosphere. Because these electrons are few and must carry the same parallel current (roughly $n v_{\parallel}$ with n lower at low altitude) as at high altitude, they must get a higher parallel velocity. To sustain their acceleration, an accelerating electric potential is set. Practically, the Knight theory [115] provides a current–potential relation.

By hypothesis, the parallel current is carried by the electrons

$$J_{\parallel} = -e \int_{\Sigma} d^3 v\, v_{\parallel} (f_{\mathrm{S}} - f_{\mathrm{I}}) \,, \qquad (7.75)$$

where f_{S} and f_{I} are the plasma sheet and the ionosphere distribution functions. This integral is computed over the set σ of electrons that pass through the ionosphere ($v_{\parallel}^2 > 0$ is required) and that come from the plasma sheet. The conservation of energy and the adiabatic invariant give

$$m(v_{\perp}^2 + v_{\parallel}^2) - 2e\Phi = \text{const} \quad \text{and} \quad \frac{2}{m}\mu = \frac{v_{\perp}^2}{B} = \text{const}' \,. \qquad (7.76)$$

The condition for a particle between the ionosphere and the plasma sheet to have passed through a region R is

$$2e(\Phi_R - \Phi) + m v_{\perp}^2 \left(1 - \frac{B_R}{B}\right) + v_{\parallel}^2 > 0 \,. \qquad (7.77)$$

Let us consider that R is either the ionosphere I or the plasma sheet S. In the $(v_{\parallel}, v_{\perp})$ space, particles passing through the ionosphere lie between the hyperbola

$(B_I/B - 1)v\perp^2 - v_\parallel^2 = (2e/m)(\Phi_I - \Phi)$ and the $v_\perp = 0$ axis, and the particles coming from the plasma sheet lie in the ellipse $(1 - B_S/B)v\perp^2 = v_\parallel^2 = (2e/m)(\Phi - \Phi_S)^2$. This defines the set σ of particles that contribute to the current $J_\parallel$. After a rather tedious computation, it is found that

$$J_\parallel = eB\left[n_E\left(\frac{T_I}{2\pi m}\right)^{1/2} g(\Delta, T_E) - n_S\left(\frac{T_S}{2\pi m}\right)^{1/2} g(\Delta, T_S)e^{\Delta/T_S}\right], \quad (7.78)$$

where $\Delta = e(\Phi_I - \phi_S) > 0$ is the net electric potential drop and

$$g(\Delta, T) = \frac{1}{B_S}\left[e^{-\Delta/T} - \frac{B_I - B_S}{B_I}e^{-(\Delta/T)(B_I/(B_I - B_S))}\right]. \quad (7.79)$$

This effect is not symmetric. It associates a potential drop to the current carried by downward electrons (upward current). It does not act upon the downward current carried by upward electrons, because the motion of electrons coming from the ionosphere is not decelerated by the magnetic field variation. Therefore, according to this model, the polar auroras should happen in regions of upward current. This is indeed what is mostly observed.

Let us notice that for the Earth's auroral zone, we can write a simplified version of the above relation,

$$\text{for} \quad 1 \ll \frac{e\Delta\Phi}{T_I} \quad \text{and} \quad \frac{e\Delta\Phi}{T_M} \ll \frac{B_I}{B_M}, \; j_\parallel \sim -eN_M\left(\frac{T_M}{2\pi m_e}\right)^{1/2}\frac{e\Delta\Phi}{T_M}. \quad (7.80)$$

Numerical evaluations of the field aligned current and their comparison with the potential drop estimated from the distribution of the electron energy shows that the effect of the convergence of the magnetic field lines on a forced electric current cannot explain alone the auroral acceleration. Nevertheless, many of the properties of the accelerated electrons, and of the regions of acceleration show that this phenomenon plays an important part in the acceleration process.

From a general point of view, it is interesting to note that the mirroring of the electrons plays, from an electrodynamical point of view, the role of resistivity. In a collisional medium, resistivity conducts to dissipation of energy and heating, and certainly not to acceleration with the creation of free energy in the form of a beam of particles. It could seem paradoxical that a cause of dissipation triggers, in the present case, particle acceleration. Because the auroral plasma is not collisional, because there is no trend toward a local thermal equilibrium, we can see that the resistivity acts on the majority of the particles that are mirrored, but accelerates a minority of particles, which at low altitude become the main current carriers.

Actually, Knight theory does not deal with the back-reaction of the plasma to the electric field. Experimental [117] and theoretical [87, 118–120] works show that when an electric potential drop appears in a plasma, it tends to concentrate over a region of small extent (a few Debye lengths). It forms a coherent electrostatic structure, called *double layer* (DL) which is said to be *strong* when the potential drop exceeds the plasma thermal energy, and *weak* otherwise.

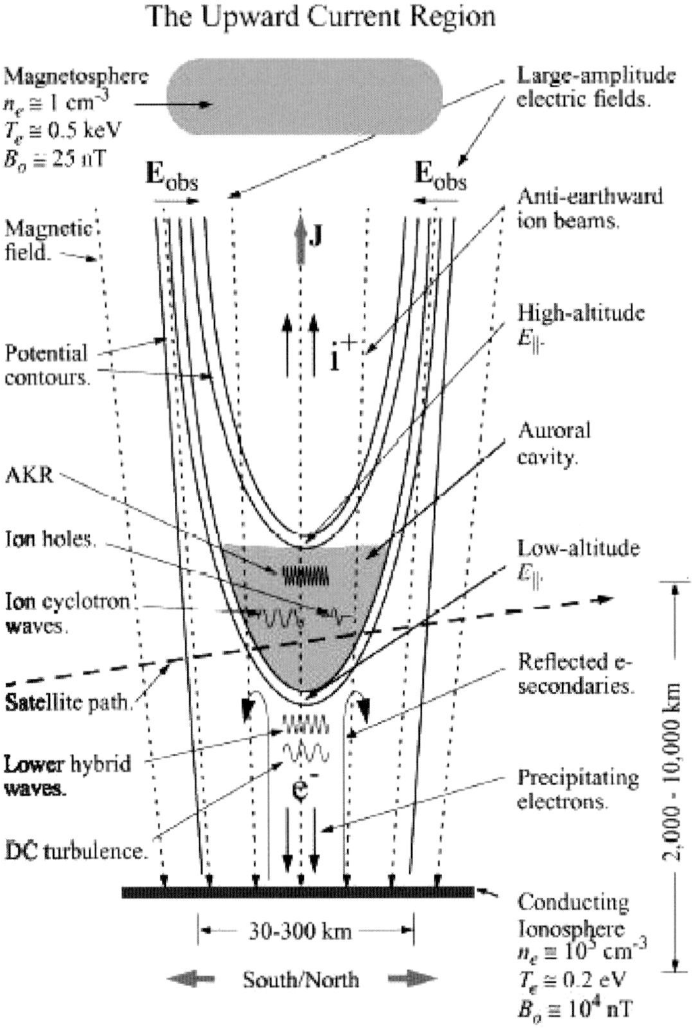

Figure 7.12 A strong double layer region: sharp potential drops and V structures. These features are observed above many auroral arcs. The acceleration occurs in the double layer. Below it, downward accelerated electrons are observed, while upward accelerated ions are observed above. From [116].

A sketch of a strong DL in an upward FAC is shown in Figure 7.12. The continuous lines are electric equipotentials and the dotted lines are the magnetic field lines. In the DL, their gradient, and, therefore, the electric field are aligned with the magnetic field. This is the region of acceleration. Other details on the figure show various phenomenon that are triggered by the accelerated particles, like plasma turbulence, streams of solitary waves (Section 6.3.2.2) auroral kilometric radiation (AKR) and other waves, and plasma cavities.

When one restricts the study of the double layers in one dimension (two in phase space), the double layers can be considered BGK structure (see Section 6.3.2.3). The only difference with solitary waves is that they carry a net potential drop. Three categories of particles trajectories must be considered: trapped, reflected, and passing. Then, the BGK theory can be applied quite straightforwardly [121]. It is shown that the beam of accelerated electrons have a narrow range of energy. It is said to be quasi-monoenergetic.

A few theories take into account the difference of magnetospheric and ionospheric temperatures [122, 123], emphasize the role of mass transfer [124], take into account the vertical density gradients [125] or the convergence of the magnetic field lines and the effect of various populations of particles [116, 126]. Many of these models and many numerical simulations impose an electric potential drop across the box, instead of a forced current. They do not really explain the cause of the accelerating electric field, but show how it evolves under the reaction of the accelerated plasma.

7.3.4 Acceleration by an Electric Current Forced by a Wave

The Knight theory [115] considered that the forced current is not time dependent. But the auroral phenomenology presents a lot of time dependent acceleration events, in the form of bursts, or quasi-periodic events of various periods and amplitudes. Therefore, it is natural to consider that the parallel current might be time dependent too. And unless a high degree of turbulence is reached, a time dependent current in a plasma is expected to be carried by waves.

The effectiveness of magnetic field lines convergence for the particles energy amplification implies that the wavelength is comparable to or larger than the gradient scale of the magnetic field. The right family of waves to fulfil this condition is those of the MHD waves.

Time varying field aligned currents were inferred well before the space age, thanks to the observations of the magnetic micropulsations. These are magnetic fluctuations registered from the ground. They have been classified, relatively to their period and to the shape of the wave envelop. Pi1 (> 100 s) and Pi2 (< 40 s) are irregular micropulsations [127] that are correlated with magnetic fluctuations in space [128], with substorm onsets (start of bright and active auroras) [129, 130], and with current disruption in the magnetotail [131]. From a theoretical point of view, the irregular Pi1 are seen as Alfvén wave *field line resonance*, that is an Alfvénic oscillation of a whole field line. Their frequencies correspond quite precisely to the values expected from the theory. The irregular Pi2 are seen as field line resonances with nodes at low altitude [132, 133]. The bursts of pulsations PiB (0.2–1 Hz) correspond to Alfvén wave trapped at low altitude [134–136] as can be seen in Figure 7.13. The other pulsations (PC1, pearls, etc.) are associated with other kinds of waves or instabilities, without direct link to Alfvén wave.

From Table 5.1 we can see that, in the linear MHD approximation, an Alfvén wave carries an electric current with a parallel component. In ideal MHD there is

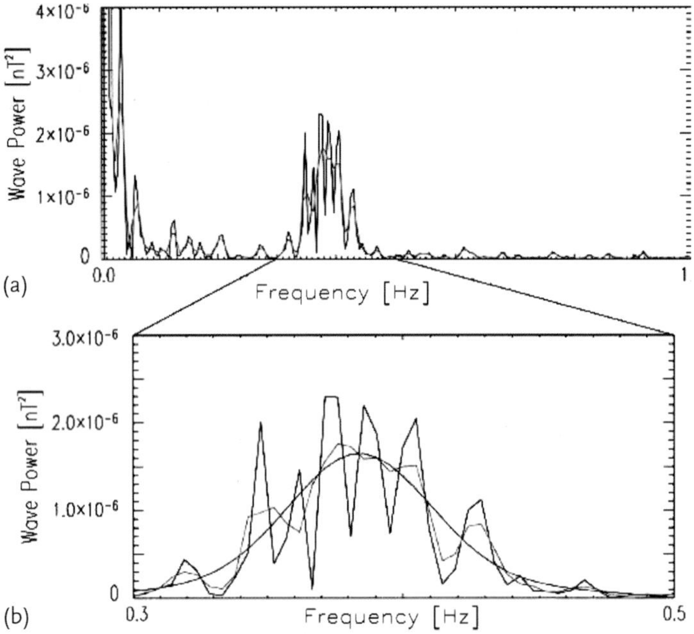

Figure 7.13 Spectrum of a PiB pulsation. PiB pulsations are observed during a substorm, associated to auroral arc activity. PiB have a typical frequency 0.2 Hz $< f <$ 1 Hz. This event was recorded on 4 June 2008 with a magnetometer located at Longyearbyen in Svalbard, Norway. It lasted about 4 min. This measurement was completed with data from other instruments and the event is interpreted as the ground magnetic field counterpart of an Alfvén wave trapped in the ionospheric resonator. From [137].

no corresponding parallel electric field. This field can be induced from a variation of the ambient magnetic field intensity, as in the Knight effect. The treatment of an Alfvén wave in these circumstances involves the effect of trapped, passing, and reflected particles, and it cannot be considered fully with the MHD theory. Quite often, a fluid treatment of the MHD wave in the inhomogeneous plasma is completed by a study of the particle dynamics in its electromagnetic field. The theory of waves in inhomogeneous plasma can be found in [19]. In the particular case of Alfvén waves trapped near the ionosphere, a theory has been developed in [138]. The study of the particle dynamics in conjunction with the imposed electric current of the wave involves the Knight effect. A parallel electric field is then established, that accelerate electrons of low pitch angle [116, 139].

7.3.5
Acceleration by an Alfvén Wave (NonMHD) Parallel Electric Field

Alfvén waves can accelerate particles without the requirement of a large parallel wavelength $2\pi k_{\parallel}^{-1}$. In that case, the magnetic field variation cannot be invoked, and $E_{\parallel}$ does not come from the Knight effect. A parallel electric field $E_{\parallel}$ also occurs for small perpendicular wavelengths, when the MHD approximation validity breaks. In the low altitude of the auroral zone, $\beta \ll m_e/m_i$ (β is the ratio of the plasma pressure to the magnetic pressure.) The motion of the particles in the strong magnetic field is given by the guiding center approximation,

$$\frac{d\boldsymbol{v}_\parallel}{dt} = +\frac{e}{m}\boldsymbol{E}_\parallel \quad \text{and} \quad \boldsymbol{u} = \boldsymbol{E} \times \frac{\boldsymbol{B}}{B^2} + \frac{m}{qB^2}\frac{d\boldsymbol{E}_\perp}{dt} . \tag{7.81}$$

Because of their low mass, the electrons have a high mobility along the magnetic field lines, while the ions are sensitive to the polarization drift (last term of the second equation). When the ions see the perpendicular electric field of the Alfvén wave, they move accordingly across the magnetic field lines. The much lighter electrons cannot do so, and a charge density establishes, that causes a parallel electric field $E_\parallel$ which favors the resetting of charge neutrality. Therefore, the electrons are accelerated along the magnetic field direction. Then [141], the dispersion relation is not exactly those provided by the MHD equations, and the parallel electric field becomes important as soon as $k_\perp^2 c^2/\omega_p^2 \geq\sim 1$

$$\omega^2 = \frac{k_\parallel^2 V_A^2}{1 + k_\perp^2 c^2/\omega_p^2} , \quad \frac{E_\parallel}{E_\perp} = \frac{k_\parallel}{k_\perp}\frac{k_\perp^2 c^2/\omega_p^2}{1 + k_\perp^2 c^2/\omega_p^2} . \tag{7.82}$$

When, $1 \gg \beta \gg m_e/m_i$, the inertial effect are dominated by kinetic effects linked to finite temperature [142, 143],

$$\omega^2 = k_\parallel^2 V_A^2 \left(1 + k_\perp^2 \rho_i^2 \left(\frac{3}{4} + \frac{T_e}{T_i}\right)\right) \sim k_\parallel^2 V_A^2 \left(\frac{k_\perp^2 \rho_i^2}{1 - \Gamma} + \frac{k_\perp^2 \rho_i^2 (T_e/T_i)}{\Gamma[1 + \zeta Z]}\right) , \tag{7.83}$$

where $Z = Z(\omega/k_\parallel a_e)$ is the Fried and Conte function, $a_e = (2T_e/m_e)^{1/2}$, and $\Gamma = \Gamma(k_\perp^2 \rho_e^2)$ is the gamma function. The parallel electric field is

$$\frac{E_\parallel}{E_\perp} = \frac{k_\parallel}{k_\perp}\frac{T_e}{T_i} k_\perp^2 \rho_i^2 . \tag{7.84}$$

In all cases, a parallel electric field is set as soon as $\beta < m_e/m_i$ and $k_\perp^2 c^2/\omega_p^2 \sim 1$ or $m_e/m_i < \beta \ll 1$ and $k_\perp^2 \rho_i^2 \sim 1$. Figure 7.14 sketches the regions of the auroral zones where these conditions can be met. The inertial region corresponds to inertial Aflvén waves and $\beta < m_e/m_i$, the kinetic region correspond to the regime where $m_e/m_i < \beta \ll 1$. The plasma $\beta = v_{thi}^2/V_A^2$ is estimated after the profile of the Alfvén velocity and the electron (or ion) thermal velocity along an auroral magnetic field lines, such as Figure 7.15.

Knowing that a parallel electric field exists is important, but not sufficient to understand how it can accelerate the plasma. This point is discussed in the following section.

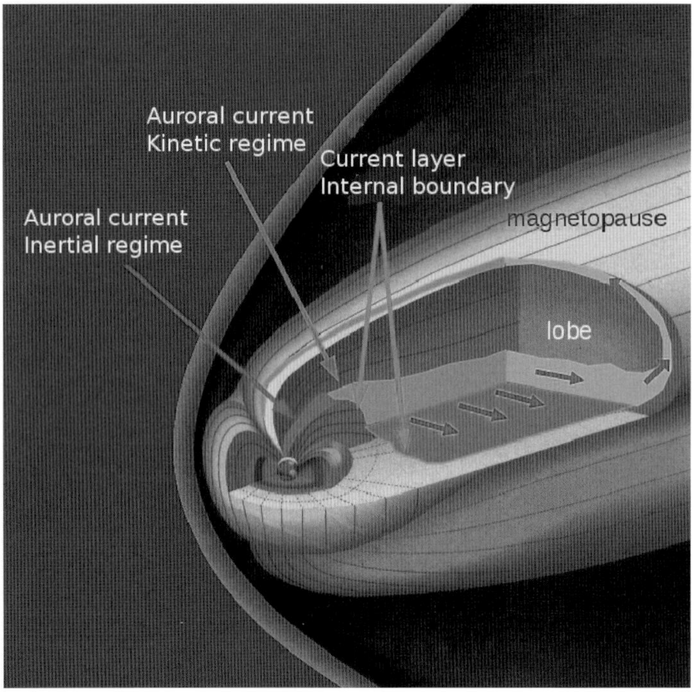

Figure 7.14 Schematic view of the magnetosphere of the main currents and of the auroral zone, shown in this figure through the auroral currents. Two areas important for Alfvénic acceleration are: the kinetic region, characterized by $m_e/m_i < \beta \ll 1$ and the inertial region, corresponding to $\beta < m_e/m_i$.

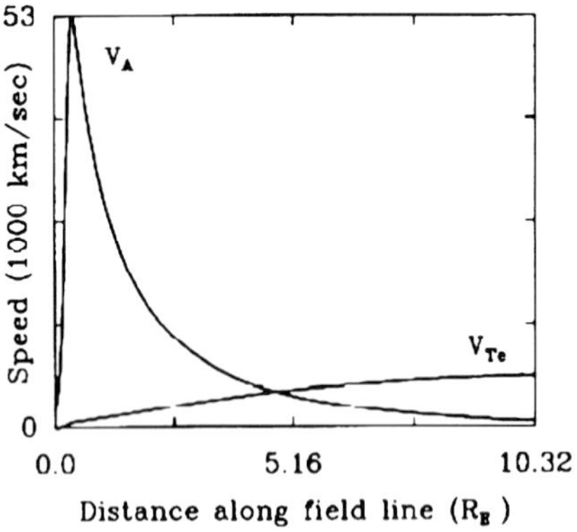

Figure 7.15 V_A and V_{the} as functions of the altitude along a magnetic field line whose footprint (intersection with the ionosphere) corresponds to auroral latitudes. From [136, 140]. The region of high Alfvén velocity is favorable to wave trapping; it is called the ionospheric Alfvénic resonator. The waves trapped in this region typically have a frequency in the range 0.1–1 Hz.

7.3.6 Resonant Acceleration by a Wave

Let us consider the polarization of a linear wave in terms of transverse and longitudinal electric fields. Equation (1.74) shows that the transverse fields $\boldsymbol{E}_{1t}$ and $\boldsymbol{B}_1$ are always perpendicular to each other. Furthermore, the ratio of their modules is always equal to the phase velocity of the wave:

$$\frac{E_{1t}}{B_1} = \frac{\omega}{k} = v_\varphi . \tag{7.85}$$

This property has an interesting consequence. The electric field is, as recalled in Chapter 1, frame dependent. For a monochromatic wave, let us calculate it in the frame moving at the phase velocity:

$$\boldsymbol{E}'_1 = \boldsymbol{E}_1 + \boldsymbol{v}_\varphi \times \boldsymbol{B}_1 = \boldsymbol{E}_1 + v_\varphi \boldsymbol{l} \times \left(\boldsymbol{l} \times \boldsymbol{E}_1 / v_\varphi\right) = \boldsymbol{E}_1 - \boldsymbol{E}_{1t} = \boldsymbol{E}_{1l} . \tag{7.86}$$

The vector $\boldsymbol{l}$ is the longitudinal unit vector: $\boldsymbol{l} = \boldsymbol{k}/k$. The same relation can be derived from the relativistic equations. This result shows that the longitudinal electric field is independent of the frame but that the transverse one is zero in the reference frame of the wave.

This is important when investigating the particle acceleration by waves.

Waves with a longitudinal electric component are seen by the particles with a velocity close to the phase velocity as a slowly varying electric field. Then, as described in Section 6.3.2, the particle can be trapped by the wave. The trapped particles then oscillate, keeping an average longitudinal velocity equal to the phase velocity. Let us now suppose that along the direction of propagation of the wave, the phase velocity progressively increases. Then the particles are accelerated.

This process is based on trapped particles, it is a resonant process that cannot concern the bulk of the plasma distribution (otherwise the wave would be too efficiently damped). From Eqs. (6.123) and (1.142) the particles that can be trapped, and, therefore, accelerated, are located in the phase space inside the area delimited by the curves of equation

$$\left(v - \frac{\omega}{k}\right) \pm \left(\frac{q E_0}{k m}\right)^{1/2} (1 - \cos(k x))^{1/2} = 0 . \tag{7.87}$$

We can see that the interval of trapped particles, centered on the phase velocity, increases when the wavelength is large, and when the particles are light. Physically, this can be explained by the fact that a single wavelength is seen by the particles during a time that can exceed their time of inertia. The velocity jump $\Delta v = (q E_0 / k m)^{1/2}$ appearing in Eq. (7.87) provides at the same time the range of particles that can be accelerated, and their gain of velocity by acceleration in a uniform plasma.

We have seen that MHD Alfvén waves are purely transverse. Therefore, they cannot change the energy of the particles in this particular frame: they can just make their velocity rotate with a constant modulus. On the contrary, kinetic and inertial

Alfvén waves have a longitudinal electric component, and they can, therefore, accelerate the particles. From Eq. (7.87), we can see that the long Alfvén wavelengths are favorable to acceleration, and that the electrons, lighter than the ions, are preferentially accelerated, in terms of velocity. Ions and electrons are equally favored in terms of kinetic energy.

Because Δv is large, and because it is the gain of velocity expected in a uniform plasma, we can see that Alfvénic acceleration can be efficient without invoking an increase of the phase velocity along the wave trajectory.

The accelerated particles by an Alfvén wave have a large range of velocities, and, therefore, of energy. Actually, in the auroral zone, the broadness of the spectrum of the accelerated particles is considered as a proxy of Alfvénic acceleration.

On the contrary, waves of short length can only accelerate particles in a narrow range of velocities Δv. Then, in a uniform plasma, the acceleration is weak. A large acceleration rate can only be obtained if the phase velocity increases along the wave trajectory.

7.3.7
Acceleration by a Wave of Short Length

Nevertheless, a wave of short length can also accelerate a plasma, even if its phase velocity is constant, provided that its envelope in not uniform. In order to avoid the possible cumulative and complex effect of particle trapping, we consider a wave with a phase velocity that is much larger than the particle velocities. Then, we are not in the context of resonant acceleration by the wave. In these circumstances, the particles are in a situation described in Section 1.4.3.1, subject to the ponderomotive force exerted by the slowly varying variations of the wave amplitude $E_0(\boldsymbol{R}, t)$. This force is given by Eq. (1.140), and repeated here for a discussion:

$$d_t U = -\frac{q^2}{2m^2\omega}\nabla E_0^2 \,. \tag{7.88}$$

In terms of gain of energy, this force acts equally on particles of various masses, provided that they have the same electric charge. This force pushes both positively and negatively charged particles in regions of lower wave amplitude and it is independent of the phase velocity, provided that it is much larger than the particle velocities.

The ponderomotive force is invoked in laboratory, for plasma acceleration by a modulated laser beam. In space plasmas, the ponderomotive force is more often considered to explain how the plasma is pushed away, creating the density depletions that develop in place of localized wave packets. Then, the ponderomotive force in space plasma is more related to questions of plasma density than of kinetic energy.

7.3.8
Application: Acceleration in the Earth's Magnetosphere

Most acceleration processes are enhanced in the Earth's magnetosphere during the substorms, or during pseudobreak-ups (a pseudobreak-up is a local magnetic reconfiguration in the magnetotail that stops before affecting the global structure of the magnetosphere).

Brief flows of plasma are observed in the tail of the magnetosphere. They are called *bursty bulk flows*; they invoke acceleration of the whole plasma.

The tearing instability is a bulk acceleration process invoked mostly in the tail of the magnetosphere, as the first acceleration process that characterizes the break-up of some substorms. The plasma in the plasma sheet, at distances of typically 20 R_E, can then flow both earthward and antiearthward, with energies $\sim$ 0.1–1 keV.

Other instabilities occurring at a closer distance to the Earth are at work with other substorms. The ballooning instability may be one of these, occurring on the inner boundary of the plasma sheet (distance $\sim$ 8 R_E). Whatever the substorm is initially driven by, a tearing instability or ballooning, the important fact, apart from plasma acceleration, is the change of magnetic configuration of the tail.

When the part of the tail plasma accelerated earthward penetrates into the auroral zone, a supplement of acceleration pushes a part of the plasma up to energies of a few keV. This is not bulk acceleration, but kinetic acceleration, with the energization of a small component of the plasma, with the possible deceleration of the rest, as seen, for instance, with the Knight effect (Section 7.3.3).

Many acceleration processes have been identified. The acceleration by strong double layers is stable enough for phases of acceleration lasting a few tens of minutes. The signature of these processes consists of a reversal of the electric field (it is perpendicular to the isopotential lines shown in Figure 7.12) and a rather monoenergetic flux of accelerated electrons and ions (not necessarily observed simultaneously) of opposite directions.

Additionally, auroral acceleration by Alfvén waves of various scales (trapped waves in the ionospheric resonator, as well as more or less isolated wave packets of short wavelengths) is identified. Alfvénic acceleration is characterized by a large spectrum of energies of the accelerated electrons.

In some phases of the development of auroral arcs, such as the substorm break-up, the Alfvénic acceleration seems to be the dominant process, while double layers seem to dominate during the expansion phase. Reciprocally, other observations show that the plasma accelerated by double layers triggers Alfvén wave instabilities that cause in turn Alfvénic acceleration.

In the magnetosphere of the giant planets Jupiter and Saturn [144, 145], the auroral oval is continuously excited. It seems that the acceleration processes happen every time, even if some phases of enhanced activity are observed.

Some peculiar regions of acceleration can be probed from Earth with radio-telescopes. This is the case of the magnetic flux tube connecting Jupiter and its closest satellite Io [146, 147]. The analysis of the radio emissions of the Io–Jupiter flux tube associated indicate that Alfvénic acceleration and acceleration by strong double layers happen simultaneously, if not at the same altitude.

8
Transport and Acceleration of Cosmic Rays

This chapter is made up of two parts. The first deals with the transport of suprathermal particles due to magnetic disturbances, especially in collisionless plasmas. The second part deals with the main aspects, in the nonrelativistic regime, of the Fermi acceleration process which is at the origin of the cosmic rays.

The transport of suprathermal particles is important for both fusion experiments and astrophysical phenomena. In collisional plasmas, diffusion of matter, temperature, resistivity, and viscosity are transport phenomena that are governed by Coulomb collisions and are described by transport coefficients that tend either to produce a relaxation towards the thermodynamic equilibrium, or to maintain the distribution functions close to it. In this first part, we will see that magnetic field irregularities in a collisionless plasma can generate diffusion of matter, energy, etc., but do not produce a relaxation towards thermal equilibrium. In some conditions it can even maintain the distribution functions out of thermal equilibrium, such as in power law distributions of suprathermal particles. Actually, this first part can be viewed as a step towards the next one devoted to Fermi acceleration that accounts for the generation of suprathermal distribution in astrophysics.

Despite its simple and well understood basis proposed by Enrico Fermi in the 1940s, the topic of Fermi acceleration viewed in the second part of the chapter is still intensively investigated by high-energy astrophysicists. This is the key to the physics of nonthermal processes in astrophysics, which are revealed by continuum radiation spectra that have been out of equilibrium for billions of years. This is a special development of the so-called Quasi-Linear Theory of plasma physics that describes the evolution of distribution functions for a plasma in a turbulent state. The first stage of its development was essentially focused on the so-called second order Fermi process until the end of the 1970s, where its association with a collisionless shock in the form of the "first order Fermi process" opened the way for understanding of the "power law" spectra and their quantitative estimates. In this chapter, the elementary Fermi process will be presented and its nonrelativistic development with its two versions of second order in a turbulent magnetic field without shock and first order at shocks.

Collisionless Plasmas in Astrophysics, First Edition. Gérard Belmont, Roland Grappin, Fabrice Mottez, Filippo Pantellini, and Guy Pelletier.

8.1
The Problem of Transport

Why do we observe distributions of energetic particles that seem to be maintained out of equilibrium forever, as, for instance, in the form of power law distribution, as revealed by the spectrum of cosmic rays received on Earth or by the synchrotron spectra of various energetic astrophysical sources? The collisional processes are inefficient at high energy. Indeed, Coulomb interactions have a typical impact parameter b_0 such that, for an energy ϵ in the center of mass, $b_0 = Ze^2/\epsilon$. The cross section is thus $\sigma \sim \pi b_0^2 \ln \Lambda$, where Λ is the Coulomb parameter that stems from the cut off of the long range interaction limited by the "Debye screening" (see Section (4.7)). Then the mean free paths increase proportionally to ϵ^2 and let non-Coulomb interactions take place. In this way, relaxation towards thermal equilibrium is inhibited.

Especially magnetic turbulence of astrophysical media acts on particle distributions at scales shorter than their collisional mean free path. This is the reason why those distributions do not undergo relaxation towards thermal equilibrium and are not subject to collisional transport phenomena (diffusion, thermal conduction, viscosity, and so on) and are maintained out of thermal equilibrium.

How is the transport of suprathermal particles realized? Those particles are essentially sensitive to the magnetic field carried by the thermal plasma of astrophysical media such as the interstellar medium, the intergalactic medium, the environment of galactic or extragalactic compact sources. The magnetic field carries out the coupling between suprathermal particles or cosmic rays with the thermal plasma of low energy.

The Fermi processes, which are thought to be the phenomena that generate power law distribution of suprathermal particles, is tightly linked with the magnetic transport processes and their efficiency directly depends on the efficiency of particle momentum scattering. In particular, we will see that the characteristic acceleration times are of the form $\tau_{acc} \sim \tau_s/\beta_c^2$, where τ_s is the characteristic time of particle momentum scattering by magnetic field irregularities and β_c is the velocity of an accelerating front.

8.1.1
The Magnetic Field: Obstruction to Transport

When the field lines are slightly curved (in other words, the curvature radius is much larger than the Larmor radius), particles keep a helicoidal motion that follows approximately the field lines, except for some slow drift motion and the possibility of reflection, as shown in Section 1.4.1. Then, at first glance, the magnetic field is not at all a transport agent; it is the contrary. Indeed, it tends to channel, to inhibit particle motions, to lock phase space.

8.1.1.1 Magnetic Barrier

Looking at how a charged particle of high energy interacts with a magnetic barrier is a simple and instructive exercise. Consider a magnetic field of the form $\boldsymbol{B} = B_0\phi(x)\boldsymbol{e}_z$, where $\phi(x)$ is a function of bell shape with a characteristic width δ_0. The particle motions are governed by three invariants: energy ϵ (or the impulsion norm p), p_z, and the conjugate momentum of y, namely $\pi_y \equiv p_y + (Ze/c)A_y(x)$, the vector potential being $\boldsymbol{A} = B_0\Phi(x)\boldsymbol{e}_y$ with $\Phi(\infty) \equiv \delta \sim \delta_0$. In this problem, it is convenient to define a Larmor radius $r_L \equiv p_\perp c/(ZeB_0)$, where $p_\perp$ is the norm of the transverse component of the particle momentum, which is a conserved quantity, and a pitch angle α such that $p_x = p_\perp \cos\alpha$, $p_y = p_\perp \sin\alpha$. The interaction only produces a change of the pitch angle α between the momentum and the direction normal to the magnetic sheet ($\boldsymbol{e}_x$). Writing the conservation of p^2 (or $p_\perp^2$), we obtain an equation that governs the evolution of p_x:

$$m^2\gamma^2\dot{x}^2 = p^2 - \left(\pi_y - \frac{Ze}{c}B_0\phi(x)\right)^2 - p_z^2 = p_\perp^2\left[1 - \left(\sin\alpha_i - \frac{\Phi(x)}{r_L}\right)^2\right]. \tag{8.1}$$

Because ϵ is conserved, $\gamma = \epsilon/mc^2$ is an invariant. The particle motion reduces to studying a differential equation of the form $\dot{x}^2 = f(x)$, which leads to the following conclusions, bearing in mind that the function $\Phi(x)$ is monotonic and takes values between 0 and δ:

- There is crossing of the magnetic barrier if $f(x) > 0 \forall x$, which is realized if and only if $(1 + \sin\alpha_i)r_L > \delta$, and since the previous equation can simply be rewritten as $\sin\alpha = \sin\alpha_i - \Phi(x)/r_L$ (which is nothing but the invariance of π_y), the momentum deflection is such that $\Delta \sin\alpha = \pm\delta/r_L$;
- There is a reflection on the magnetic barrier at the point where $f(x) = 0$, which exists if $(1 + \sin\alpha_i)r_L < \delta$.

This simple example illustrates an essential aspect of the interaction of a cosmic ray with a magnetic perturbation: it is sensitive only for Larmor radii not larger than the perturbation scale and leads to a mere deflection. As an example, if the cosmic magnetic field has been sufficiently amplified during the collapses leading to large cosmic structure formations of scale $\sim$ 10 Mpc so that it reaches intensity on the order of a few 0.1 μG in superclusters, then ultrahigh-energy cosmic rays ($\epsilon \sim 10^{20}$ eV) are subjected to that type of deflection.

8.1.1.2 Adiabatic Invariant

Another example of magnetic barrier is that of the motion of a particle along a curved field line, when its Larmor radius is much smaller that the curvature radius. We have seen that the conservation of the first adiabatic invariant (Eq. (1.125)) causes particle trapping in regions of low magnetic field, provided that the particle pitch angle is not too small (Section 1.4.2.1). As we have seen in Section 7.3.2, acceleration of the trapped particles occurs when the second adiabatic invariant is conserved, and when mirror points have convergent motions toward each other. This is an example of Fermi acceleration.

8.1.2
Magnetic Irregularities: Transport Agent

Irregularities of the magnetic field cause erratic motions of suprathermal particles, especially when they encounter disturbances of size comparable to their Larmor radius, which is widely occurring when the magnetic field displays a turbulent state characterized by a wide spectrum of Fourier modes, as will be make precise further on. Processes similar to diffusion processes can then take place and allow the elaboration of a complete description of suprathermal particle transport.

8.1.2.1 Breaking the Adiabatic Invariant by Landau-Synchrotron Resonances

In order to obtain a diffusive type of particle transport, it is necessary that the adiabatic invariant be broken. We illustrate the approach to realize that purpose in a simplified situation where the mean field is homogeneous so that the invariant to be broken is nothing but the pitch angle α. The Fermi acceleration process works efficiently when the particle momentum can be frequently reversed like in a Brownian motion; the situation must be such that the pitch angle evolves randomly with a high frequency of reversal.

The pitch angle variations are governed by a simple stochastic equation, resulting from the projection of the dynamic equation along the mean field while taking account of energy conservation (and thus also of p and of v).

$$\dot{\alpha} = f(t) \equiv \omega_{\mathrm{L}}[\cos\phi(t) b_2(t) - \sin\phi(t) b_1(t)] , \tag{8.2}$$

where $\omega_{\mathrm{L}} \equiv Ze\bar{B}/m\gamma c$ and $\boldsymbol{b} \equiv \delta\boldsymbol{B}/\bar{B}$ is the irregular part of the magnetic field experienced by the particle along its trajectory. $\phi(t)$ is the gyro-phase, that is, $\phi(t) = \omega_{\mathrm{L}} t + \phi_0$ when perturbations are neglected.

First, let us investigate a toy model (see [148]). Let us consider perturbations transverse to the mean field $B_0 \boldsymbol{e}_x$ and depending on the coordinate x only. It is suitable to introduce the reduced vector potential $\boldsymbol{a}$ such that $\boldsymbol{b} = \nabla \times \boldsymbol{a}$. The motions are described by a simple Hamiltonian system for the two conjugate variables (α, x), knowing that $\dot{x} = v\cos\alpha$:

$$H(\alpha, x) = v\sin\alpha - \omega_{\mathrm{L}}[\cos\phi(t) a_1(x) + \sin\phi(t) a_2(x)] , \tag{8.3}$$

where we suppose that the perturbation is weak enough for assuming $\phi = \omega_{\mathrm{L}} t + \phi_0$.

Let us consider a discrete ensemble of Fourier modes of the following form:

$$\boldsymbol{a}(x) = \sum_n a_n (\boldsymbol{e}_1 \cos(k_n x) + \varepsilon_{\mathrm{c}} \boldsymbol{e}_2 \sin(k_n x)) , \tag{8.4}$$

where $\varepsilon_{\mathrm{c}} = +1$ for modes of right polarization and -1 for the modes of left polarization. The Hamiltonian is then:

$$H(\alpha, x) = v \sin \alpha - \omega_{\mathrm{L}} \sum_{n} a_n \cos(k_n x - \varepsilon \omega_{\mathrm{L}} t + \phi_0) , \quad (8.5)$$

where $\varepsilon = \varepsilon_{\mathrm{c}} \operatorname{sgn}(q)$. A resonance occurs for various values α_n of the pitch angle α so that $k_n \dot{x} = \varepsilon \omega_{\mathrm{L}}$, or $k_n \mu_n = \varepsilon \omega_{\mathrm{L}}$ with $\mu_n = \cos \alpha_n$. A negative charge moving forwards undergoes a resonant interaction with a right mode ($\epsilon = 1$), whereas it undergoes resonance with a left mode when moving backwards and vice versa for a positive charge. The conclusions are the opposite ones for receding modes. Those resonances are synchrotron resonances. When these discrete resonances are isolated by separatrixes, the Hamiltonian can be approximated by a pendulum Hamiltonian in the vicinity of each resonance by using the following canonical transformation:

- $\theta = k_n x - \varepsilon \omega_{\mathrm{L}} t + \phi_0$,
- $J = (\alpha - \alpha_n)/k_n$,
- $H' = H - \varepsilon \omega_{\mathrm{L}} \alpha / k_n + \text{const}$,

the approximate Hamiltonian is

$$H'(J, \theta) = -k_n^2 \sin \alpha_n \left(\frac{J^2}{2} - \Omega_n^2 \cos \theta \right) , \quad (8.6)$$

where the nonlinear pulsation Ω_n is such that

$$\Omega_n^2 = \frac{\bar{\omega} a_n}{k_n^2 \sin \alpha_n} . \quad (8.7)$$

The pendulum approximation differs from the exact Hamiltonian by oscillating terms. The half-width of the nth nonlinear resonance in phase space (J, θ) is $\Delta J = 2\Omega_n$ and the resonances overlap when that half-width is larger than the half-separation between resonances $\Delta\alpha_n / k_n$; which leads to the Chirikov criterium for intrinsic stochasticity (Hamiltonian chaos) (see [149, 150]):

$$\bar{\omega} a_n \sin \alpha_n > (\Delta \mu_n)^2 / 4 . \quad (8.8)$$

In fact, it is known that chaos occurs even for a lower threshold. As mentioned previously, the dynamical description with H' differs from the exact dynamics by oscillating contributions, among which there are the mode propagating in the opposite direction. Particles cannot resonate simultaneously with both progressive and regressive modes. The smallest value of μ_n controls the jump of the pitch angle around 90°. A particle can jump from the resonance with the right mode ($\varepsilon = 1$) to the resonance with the left mode ($\varepsilon = -1$) if $k_n^2 a_n > \bar{\omega}/4$. This is the nonlinear solution to the problem of momentum reversal.

When the amplitude of the modes is sufficiently above the stochasticity threshold, the chaotic jumps of the pitch angle behave like a diffusion process. Only the momentum reversal could be slowed down by a "sticky" regime around 90°, leading to a subdiffusion process.

8.1.2.2 Theorem on Symmetries and Transport

In the previous model, although parallel diffusion has been made possible (see also Section 8.1.2.3) by the chaotic behavior of trajectories due to perturbations depending on the variable x, transverse diffusion is impossible, because phase space is still locked by invariants. Indeed the following theorem is important to understand the conditions for transport; it was established by Jones, Jokipii, and Baring [151]. When the Lagrangian is invariant under translations along some direction, in other words, the coordinate associated with that direction (straight or circular) is "ignorable", then

- if the mean field $\boldsymbol{B}_0$ points in that direction, there is no restriction on the trajectory wandering in that direction;
- otherwise, the trajectory is confined in a layer or a tube whose thickness is of few Larmor radii.

For example, if $\boldsymbol{A}(x)$ depends on x only, then a trajectory is confined in the intersection of a layer C_y, of thickness of a few Larmor radii in the y-direction with a layer C_z of thickness of few Larmor radii also in the z-direction; in other words, any trajectory is confined in a tube $C_y \bigcap C_z$ of few Larmor radii size along the x-direction.

Let us consider an oblique shock and perturbations depending on the curvilinear abscissa along the mean field. Any cosmic ray trajectory is then confined in a bent tube of a few Larmor radii size. Cosmic rays can move back and forth in the tube and even cross the shock several times, but cannot diffuse out of the tube.

Another example: if $\boldsymbol{A}(r, z)$, any trajectory is confined in a layer defined by $A_\phi(r, z) = C_0 \pm \text{few} r_\text{L}$. But, if $\boldsymbol{B}(r, \phi) = B_z(r, \phi)\boldsymbol{e}_z$, there is no constrain.

8.1.2.3 Diffusion along the Mean Field and Angular Diffusion

Spatial diffusion along the mean field stems from random jumps of the momentum direction: $\dot{x} = v\mu$, where $\mu = \cos\alpha$, the pitch angle cosine with respect to the mean field. As long as the self-correlation function of the random process $\mu(t)$ displays a relaxation time τ_s,

$$\tau_\text{s} \equiv 3 \int_0^\infty \langle \mu(t)\mu(t-\tau)\rangle d\tau \,, \tag{8.9}$$

then a coefficient of spatial diffusion can be estimated. Indeed, for a location jump Δx during $\Delta t \gg \tau_\text{s}$,

$$\langle \Delta x^2 \rangle = v^2 \int_0^{\Delta t} dt_1 \int_0^{\Delta t} dt_2 \langle \mu(t_1)\mu(t_2)\rangle = 2\Delta t \int_0^{\Delta t} \langle \mu(\tau)\mu(0)\rangle d\tau \,, \tag{8.10}$$

$$\langle \Delta x^2 \rangle \simeq 2D_\parallel \Delta t \,, \tag{8.11}$$

with $D_\parallel = (1/3)v^2\tau_\text{s}$. The main question for the transport is to know the diffusion time of the pitch angle, which is nothing but the time τ_s for a given spectrum of turbulence.

In weak turbulence [152, 153] described by a continuous spectrum of Fourier modes, one can calculate a frequency of angular diffusion, defined as follows:

$$\nu_{\mathrm{s}} \equiv \frac{\langle \Delta \alpha^2 \rangle}{\Delta t} , \tag{8.12}$$

calculated from Eq. (8.2) using the self-correlation function of the perturbation $f(t)$ at the lowest order, which amounts to insert the unperturbed trajectory in $\boldsymbol{b}(t)$. Thus,

$$\nu_{\mathrm{s}} = \omega_{\mathrm{L}}^2 \int_0^{\infty} \langle \boldsymbol{b}(\tau) \cdot \boldsymbol{b}(0) \rangle \cos \omega_{\mathrm{L}} \tau \, d\tau . \tag{8.13}$$

When $\boldsymbol{b}$ is expanded in Fourier modes and that the unperturbed motion in inserted in the phases, the previous integral takes the following form:

$$\nu_{\mathrm{s}} = \omega_{\mathrm{L}} \int_{\mathcal{R}^3} \frac{d\boldsymbol{k}}{(2\pi)^3} F(\boldsymbol{k}) g(\boldsymbol{k}, \boldsymbol{p}) \tag{8.14}$$

where $F(\boldsymbol{k})$ is the 3D-correlation spectrum of the magnetic irregularities and the function $g(\boldsymbol{k}, \boldsymbol{p})$ describes the resonant interaction between particles and modes:

$$g(\boldsymbol{k}, \boldsymbol{p}) \equiv \omega_{\mathrm{L}} \int_0^{\infty} e^{i\boldsymbol{k}\cdot\Delta \boldsymbol{x}(\tau) - i\omega(k)\tau} \cos \omega_{\mathrm{L}} \tau \, d\tau , \tag{8.15}$$

where $\Delta \boldsymbol{x}(\tau)$ is the jump of the unperturbed trajectory during a time τ and $\omega(k)$ the mode pulsation. It displays resonances, for it is of the form:

$$g(\boldsymbol{k}, \boldsymbol{p}) \propto \delta(\omega(k) - k_{\parallel} v_{\parallel} \pm n\omega_{\mathrm{L}}) . \tag{8.16}$$

These are the Landau-synchrotron resonances that we partially saw in the example of the toy model.

For an isotropic spectrum, we define $S(k)$ such that $F(k) 4\pi k^2 dk = S(k) dk$, with a degree of irregularity

$$\eta \equiv \frac{\langle \delta \boldsymbol{B}^2 \rangle}{\langle \boldsymbol{B}^2 \rangle} = \int S(k) dk , \tag{8.17}$$

and a coherence length ℓ_{c} can be defined as the spatial range of the magnetic field correlation:

$$\ell_{\mathrm{c}} \equiv \int_0^{\infty} C(r) dr = \frac{\pi}{2} \int_0^{\infty} \frac{S(k)}{k} dk . \tag{8.18}$$

For the particles interacting with turbulence, it is convenient to define a "rigidity" parameter ρ such that $\rho \equiv r_{\mathrm{L}}/\ell_{\mathrm{c}}$.

When this spectrum is a power law $S(k) \propto \eta k^{-\beta}$ between $k_{\rm min}$ and $k_{\rm max}$, for the resonant range where

$$\frac{k_{\rm min}}{k_{\rm max}} < \rho < 1\,, \tag{8.19}$$

one obtains

$$\nu_{\rm s} \sim \omega_{\rm L} \eta \rho^{\beta-1}\,. \tag{8.20}$$

It is easy to check that $\tau_{\rm s} \sim \nu_{\rm s}^{-1}$. The scattering time $\tau_{\rm s}$ can also be defined as the time required for a particle momentum to be deflected of an angle $\pi/2$, so that

$$\tau_{\rm s} = \frac{\pi^2}{4\nu_{\rm s}}\,.$$

However, this definition can be slightly modified for formal convenience.

The transport theory generally deals with two different times that characterize the random process, the one is the correlation time $\tau_{\rm c}$ associated with the force experienced by particles, the other is the correlation time associated with the momentum, in particular, the pitch angle scattering time $\tau_{\rm s}$. In weak turbulence these two times are very different, whereas they are comparable in strong turbulence.

In strong turbulence, the gyro-resonances broaden; however, the scaling with the rigidity ρ and the degree of irregularity (properly defined!) η can be extrapolated [154]. However, the correlation time and the scattering time become comparable and the memory of the initial value of μ is lost; thus the variations of μ are the main cause of resonance broadening.

8.1.2.4 **Scattering in Strong Large-Scale Turbulence**

When the plasma is in a state of fully developed turbulence with an isotropic spectrum of magnetic disturbances $S(k)$ (normalized to 1), we can derive a general formula for the scattering frequency of relativistic particles within a pre-factor of order unity. Indeed we have shown that the deflection angle of a particle crossing a magnetic disturbance of size λ with a Larmor radius $r_{\rm L} > \lambda$ is $\Delta\alpha \simeq \lambda/r_{\rm L}$. The particle crosses it on a time $\Delta t \sim \lambda/c$. Therefore, the scattering frequency

$$\nu_{\rm s} = \frac{\langle\Delta\alpha^2\rangle}{\Delta t} \sim \left\langle \frac{\lambda^2}{r_{\rm L}^2}\frac{c}{\lambda}\right\rangle. \tag{8.21}$$

This average can be calculated by introducing a size distribution $\rho(\lambda)$ such that $\rho(\lambda)d\lambda = S(k)dk$. And thus

$$\nu_{\rm s} \sim \frac{c}{r_{\rm L}^2}\int_{\lambda<r_{\rm L}} \lambda\rho(\lambda)d\lambda \sim \frac{c}{r_{\rm L}^2}\int_{kr_{\rm L}>1} \frac{S(k)}{k}dk\,. \tag{8.22}$$

For a power law spectrum, $S(k) \propto k^{-\beta}$, for $k\ell_{\rm c} > 1$ and $1 < \beta < 2$ (like Kolmogorov and Kraichnan spectra) and declining rapidly on scales larger than the coherence length (that is, $k\ell_c > 1$), the coherence length then characterizes a turbulence state concentrated on large scales and when $r_L < \ell_c$ (see Figure 8.3),

$$\nu_s \sim \frac{c\ell_c}{r_L^2}\left(\frac{r_L}{\ell_c}\right)^{\beta} \sim \frac{c}{\ell_c}\left(\frac{r_L}{\ell_c}\right)^{\beta-2} \propto \epsilon^{\beta-2} . \tag{8.23}$$

This estimate is very similar to the one previously obtained with a perturbative theory. Beyond the coherence length,

$$\nu_s \sim \frac{c\ell_c}{r_L^2} \propto \epsilon^{-2} . \tag{8.24}$$

There are numerical simulations and a firm theory that support this intuitive derivation [154, 155]. Note that we can also define a rate of reflection for particles having $r_L < \lambda$. If $r_L > \ell_c$, that rate obviously vanishes. For $r_L < \ell_c$, the reflection rate can be estimated as

$$\nu_r \sim \left\langle\frac{c}{\lambda}\right\rangle \sim \frac{c}{\ell_c}\int_{k r_L<1} k\ell_c S(k)dk \sim \frac{c}{\ell_c}\left(\frac{r_L}{\ell_c}\right)^{\beta-2} . \tag{8.25}$$

This reflection rate is thus similar to the scattering rate, except for Larmor radii larger than the coherence length where the reflection rate vanishes immediately.

8.1.3
Other Diffusion Coefficients

We have introduced two types of diffusion coefficient, one is a spatial coefficient due to a random motion that implies position jumps Δx during a time Δt longer than the scattering time τ_s, the other is an angular diffusion, the scattering frequency ν_s, that implies random deflection angle $\Delta\alpha$ of a particle momentum due to a random force during a time Δt, that is, longer than another time scale, the correlation time τ_c of the force experienced by a particle. This correlation time τ_c is very often much shorter than the scattering time τ_s. So we deal with two levels of random processes, the shortest time scale one, characterized by the time τ_c, describes the momentum variation due to the random force, and the longest one, characterized by the scattering time τ_s, describes the spatial variations due to the random motions. Several types of diffusion coefficients are introduced related to both random processes.

8.1.3.1 Diffusion Across the Mean Field

Particle diffusion across the mean field is an important issue, since it controls the confinement loss of particles in a galaxy, an astrophysical jet, or a tokamak. Let us assume that a plasma is invariant under rotation about the direction of the mean field. Let $\Delta \boldsymbol{x}_\perp$ be a transverse random jump during $\Delta t \gg \tau_s$; the transverse diffusion coefficient is then defined by

$$\left\langle \Delta \boldsymbol{x}_\perp^2 \right\rangle = 4D_\perp \Delta t . \tag{8.26}$$

In weak turbulence, the coefficient can be estimated in the following way.

$$\langle \Delta x_\perp^2 \rangle \simeq v^2 \int_0^{\Delta t} dt_1 \int_0^{\Delta t} dt_2 \langle \sin\alpha_1 \sin\alpha_2 \rangle \cos(\phi_1 - \phi_2) \,. \tag{8.27}$$

where the gyro-phase $\phi(t)$ is supposed to rotate at Larmor pulsation without significant nonlinear perturbation and thus is not correlated with the pitch angle, for example, $\phi_1 - \phi_2 \simeq \omega_{\rm L}(t_1 - t_2)$. Moreover, let us assume that the self-correlation function of $\sin\alpha$ exponentially decreases with a characteristic time $\tau_{\rm s}$ (a more detailed derivation can be done, see [155]), we obtain

$$D_\perp = \frac{D_\|}{1 + \omega_{\rm L}^2 \tau_{\rm s}^2} \,. \tag{8.28}$$

In the framework of weak turbulence theory, $\omega_{\rm L}^2 \tau_{\rm s}^2 \gg 1$, which makes $D_\perp \ll D_\|$ and we may expect a good confinement of particles. We will see in the next subsection that this law is deeply modified even for a moderate level of turbulence by effects of "magnetic chaos".

8.1.3.2 Cross Diffusion and Magnetic Chaos

Large-scale irregularities of the magnetic field result from chaotic behavior of the field lines when a 3D power law spectrum is prescribed. Indeed, the integration of the field line system displays a generic chaotic behavior essentially characterized by two quantities. The first one, the Kolmogorov or Lyapounov length $\ell_{\rm K}$, measures the rate of exponential divergence of two field lines initially close together, divergence revealed in 3D-simulations, but not in 1D nor 2D simulations. The second one, $D_{\rm m}$, measures a kind of diffusion of the separation of two field lines when they are distant by more than a coherence length: $\langle \Delta r^2 \rangle = 2 D_{\rm m} \Delta s$. The transport of cosmic rays is very evident by this chaotic behavior of field lines, as is clearly shown in the study by [153]. In particular, the transverse diffusion is completely under the control of magnetic chaos and depends on the two characteristic quantities previously mentioned. The average rate of divergence between two field lines depends on the irregularity parameter η and the coherence length $\ell_{\rm c}$. Indeed, the Kolmogorov length $\ell_{\rm K} \propto l_{\rm c}/\eta^{1+\varepsilon'}$ and the magnetic diffusion coefficient $D_{\rm m} \propto \ell_{\rm c}\eta^{1+\varepsilon}$. An analysis of the consequence of chaos on the particle diffusion leads to the following result:

$$D_\perp = \eta^{2+\varepsilon} D_\| \,. \tag{8.29}$$

Moreover, numerical simulations show that $\nu_{\rm s}$ and $D_\|$ keep the same dependence as a function of ρ and η as in quasi-linear theory (weak turbulence), except that $\nu_{\rm s}$ decreases in ρ^{-2} for $\rho > 1$. The result in Eq. (8.29) about transverse diffusion, in agreement with numerical simulations, rules out the prediction of quasi-linear theory and also the Bohm conjecture that states that $D \propto v r_{\rm L}$ and emphasizes the importance of chaos in the understanding of transport phenomenon. The simulations previously mentioned have made the exploration of both the moderate and strong regimes of turbulence, provided reliable laws for ν_s and $D_{\|}$ in both regimes and proposed a fairly accurate law of transverse diffusion due to chaos.

8.1.3.3 Diffusion in Energy Space

From a general point of view, a random force, characterized by a short correlation time τ_c, leads to a diffusion tensor in momentum space:

$$\Gamma_{ij} = \frac{\langle \Delta p_i \Delta p_j \rangle}{2\Delta t}, \tag{8.30}$$

where Δp_i is the random variation of the i-component of the momentum during a time Δt larger than the correlation time τ_c. Often we use coefficients that are valid when the turbulent state that produces the random force is isotropic. Thus, when one considers a particle of momentum $\boldsymbol{p}$ and intends to calculate the momentum diffusion tensor, one can simplify the derivation by assuming invariance under rotations in momentum space around the considered momentum before the jump. The diffusion tensor is, therefore, characterized by two eigenvalues, Γ_ℓ and Γ_t, Γ_ℓ describes diffusion of the momentum along the direction defined by $\boldsymbol{p}$ and thus describes an energy change during the momentum jump, whereas Γ_t describes a momentum deflection without energy change.

When particles experience a magnetic force, they undergo transverse diffusion only when the magnetic field is static; which is often an approximation when the time variation of the field is due to propagation at a velocity much smaller than the particle velocity. In that case only Γ_t is important and is, in fact, a scattering process described by the scattering frequency ν_s related to Γ_t by $\Gamma_t = \nu_s p^2$. The previous estimates of diffusion coefficient, namely ν_s and $D_{\|}$, can be done in the approximation where the magnetic perturbations are perceived as static by suprathermal particles (for instance, $\omega(k) \ll \omega_L$ or $V_A \ll \mu c$). Taking into account that the magnetic disturbances (or Alfvén waves) have some motion needs to introduce the induction electric field that modifies the particle energy. Therefore, particles undergo an energy diffusion, which is the modern view of the so-called second order Fermi process. In the case of Alfvén waves, for instance, the diffusion coefficient Γ_ℓ is of second order in V_A/c and depends also essentially on the scattering frequency. Indeed, let us show this property for each Fourier mode. The magnetic field and the electric field are derived from the vector potential that can be expanded in Fourier modes of the form

$$\boldsymbol{A} = \sum_k \mathcal{A}(\boldsymbol{k}) \exp[i\boldsymbol{k} \cdot (\boldsymbol{x} - \boldsymbol{V}t)], \tag{8.31}$$

where V can be the Alfvén speed of any slow speed of a magnetic perturbation. The longitudinal projection of the force is, for each Fourier mode:

$$F_\ell = q\frac{\boldsymbol{v}}{v} \cdot \boldsymbol{E} = -qvv \cdot \partial_t \boldsymbol{A} = q(\boldsymbol{k} \cdot \boldsymbol{V})\left(\frac{\boldsymbol{v}}{v} \cdot \boldsymbol{A}\right) \tag{8.32}$$

and for the magnetic part of the force

$$\boldsymbol{F}_t = q\boldsymbol{v} \times \boldsymbol{B} = iq[(\boldsymbol{v} \cdot \boldsymbol{A})\boldsymbol{k} - (\boldsymbol{k} \cdot \boldsymbol{v})\boldsymbol{A}]. \tag{8.33}$$

We remark that, because div $\boldsymbol{A} = 0$, $\boldsymbol{k}$ and $\boldsymbol{A}$ are orthogonal. Then the ratio

$$\frac{|F_\ell|^2}{|F_\mathrm{t}|^2} = \frac{(\boldsymbol{k}\cdot\boldsymbol{V})^2(\boldsymbol{v}\cdot\boldsymbol{A})^2/v^2}{(\boldsymbol{v}\cdot\boldsymbol{A})^2k^2 + (\boldsymbol{k}\cdot\boldsymbol{v})^2|\boldsymbol{A}|^2} \sim \frac{V^2}{v^2} . \tag{8.34}$$

Being an estimate for each Fourier mode, the ratio is generally the same for any superposition of independent modes. As a consequence we obtain an estimate of the energy diffusion coefficient:

$$\Gamma_\ell \equiv \frac{\langle \Delta p^2 \rangle}{2\Delta t} \sim \nu_\mathrm{s}\frac{V^2}{v^2}p^2 . \tag{8.35}$$

In particular, for $V = V_\mathrm{A} \ll c$ and relativistic particles having $v \simeq c$, that second order energy diffusion coefficient is much smaller than the transverse diffusion coefficient that describes pitch angle diffusion (scattering). The pre-factor of order unity that accompanies the ratio $(V/v)^2$ can be obtained by calculating the integrals over angles, with specific choices of their distribution. That energy diffusion coefficient is the essential ingredient of the theory of second order Fermi acceleration as will be seen in the next chapter. In the resonant range of a power law spectrum of magnetic turbulence, we saw that $\nu_\mathrm{s} \propto \epsilon^{\beta-2}$, thus the energy diffusion coefficient scale as ϵ^β. The acceleration time scale associated with that energy diffusion can be defined as $\tau_\mathrm{acc} \equiv p^2/\Gamma_\ell$ and thus it becomes longer and longer with increasing particle energy for $1 \le \beta \le 2$:

$$\tau_\mathrm{acc} \sim \frac{\ell_\mathrm{c}}{c}\left(\frac{c}{V}\right)^2 \frac{\rho^{2-\beta}}{\eta} . \tag{8.36}$$

It is worth noting that the knowledge of the scattering frequency ν_s allows us to describe all the transport phenomena of cosmic rays. And it has been calculated for all possible regimes of turbulence.

8.1.4
Transport Equation of Cosmic Rays

We will introduce the Fokker–Planck equation that governs the evolution of the distribution function of the suprathermal particles simply by a generalization of the diffusion equation, for which the reader is probably familiar. This is the simplest approach to this topic. There are more sophisticated derivations of this transport equation of cosmic rays with all the details, as for instance in [156].

Spatial Diffusion

Let us consider a random coordinate $x(t)$, that locates the position of a particle at time t, the particle diffusing in a fluid of Eulerian velocity u. During a time Δt, very short compared with the diffusion time (which will be determined a posteriori), the position of the particle has varied of an amount $\Delta x = u\Delta t + \delta x$; the first contribution is due to the bulk motion of the fluid and the second δx is due to the purely random diffusion process. Its variance is proportional to Δt:

$$\langle \delta x^2 \rangle = 2D\Delta t ; \tag{8.37}$$

the average of δx vanishes if the process is rigorously homogeneous. This can be derived in the following way. One considers the velocity $\delta\dot{x}(t) \equiv \zeta(t)$ as a purely random process with a short correlation time, stationary, and homogeneous for the moment. Then

$$\langle\delta x(t)^2\rangle = \int_0^{\Delta t} dt_1 \int_0^{\Delta t} dt_2 \langle\zeta(t_1)\zeta(t_2)\rangle = 2\Delta t \int_0^{\Delta t} \langle\zeta(\tau)\zeta(0)\rangle d\tau \,. \tag{8.38}$$

For Δt much longer than the correlation time τ_c that characterizes the decay of the correlation function, then one can approximate

$$\langle\delta x(t)^2\rangle \simeq 2\Delta t D \,, \tag{8.39}$$

where

$$D \equiv \int_0^{\infty} \langle\zeta(\tau)\zeta(0)\rangle d\tau \,. \tag{8.40}$$

Now when the self-correlation is slowly varying with x, the diffusion coefficient is not constant and the following result can be demonstrated (see Appendix A.3):

$$\langle\delta x\rangle = \Delta t \partial_x D \,. \tag{8.41}$$

The behavior that we just described is typical of a Brownian motion with a short correlation time. The probability distribution g that describes the particle position is governed by the following equation:

$$\partial_t g = -\partial_x(ug) + \partial_x\left(D\partial_x g\right) \,. \tag{8.42}$$

Indeed the exact evolution of the density g results from the evolution of

$$g(x,t) \equiv \langle\delta(x - x(t))\rangle = \int dx' \langle\delta(x - x' - \Delta x(t,t'))\delta(x' - x(t'))\rangle \tag{8.43}$$

where $\Delta x(t,t')$ is the random jump undergone by the particle located at x' on time t' during the time lapse $\Delta t \equiv t - t'$. In order to obtain the diffusion equation, we must make two essential approximations. First, the evolution from t' to t for a given location x' is independent of the past evolution before t' that led to the arrival at x'. This is the Markovian approximation valid if the correlation time τ_c of the random process is short compared to the time scale of the average evolution, in other words, the evolution of the density g, that will be estimated a posteriori. Secondly, a jump $\Delta x(t,t')$ during a time interval $\Delta t > \tau_c$ remains small and the transition function in the integral can be expanded to second order. Thus, we write

$$g(x,t) \simeq \int dx' \langle\delta(x - x' - \Delta x(t,t'))\rangle g(x',t') \tag{8.44}$$

with

$$\langle \delta(x - x' - \Delta x(t,t')) \rangle = \delta(x - x') + \langle \Delta x(t,t') \rangle \partial_{x'} + \frac{1}{2} \langle (\Delta x(t,t'))^2 \rangle \partial^2_{x'} + \ldots \tag{8.45}$$

Integrating by parts, we obtain the evolution equation for $g(x)$ called the Fokker–Planck equation:

$$\partial_t g = -\partial_x \left(\frac{\langle \Delta x \rangle}{\Delta t} g \right) + \partial^2_x \left(\frac{\langle \Delta x^2 \rangle}{2\Delta t} g \right) . \tag{8.46}$$

The diffusion is inhomogeneous when the correlation function of the random process ζ that perturbs the position such that $\delta \dot{x} = \zeta$ depends slowly of the unperturbed position. The calculation of the diffusion coefficient offers no technical difficulty as long as Δt is sufficiently long compared with the correlation time τ_c that characterized the decrease of the correlation function of the random process ζ, and one obtains:

$$D \equiv \frac{\langle \Delta x^2 \rangle}{2\Delta t} = \int_0^\infty \langle \zeta(\tau)\zeta(0) \rangle \, d\tau . \tag{8.47}$$

However, the calculation of the first order coefficient $\langle \Delta x \rangle / \Delta t$ is more technical (see Appendix A.3). The first contribution u is obvious, the second term in the derivative of D is less obvious:

$$\frac{\langle \Delta x \rangle}{\Delta t} = u + \partial_x D . \tag{8.48}$$

This development can be extended to several dimensions without difficulty, the Fokker–Planck equation now reads:

$$\partial_t g = -\partial_{x_i} \left(\frac{\langle \Delta x_i \rangle}{\Delta t} g \right) + \partial^2_{x_i x_j} \left(\frac{\langle \Delta x_i \Delta x_j \rangle}{2\Delta t} g \right) . \tag{8.49}$$

The diffusion tensor derives from the correlation tensor of a vectorial random process ζ_j:

$$D_{ij} = \int_0^\infty \langle \zeta_i(\tau)\zeta_j(0) \rangle \, d\tau . \tag{8.50}$$

And the first order term is

$$\frac{\langle \Delta x_i \rangle}{\Delta t} = u_i + \partial_{x_j} D_{ij} . \tag{8.51}$$

The Fokker–Planck equation can then be rewritten:

$$\partial_t g = -\partial_{x_i}(u_i g) + \partial_{x_i} \left(D_{ij} \partial_{x_j} g \right) . \tag{8.52}$$

It appears as a continuity equation containing an advection term and a diffusion current.

Momentum Diffusion Now let us consider the energy variable p (this is the radial coordinate in momentum space). The particle energy might experience a regular loss due to radiation, for instance, also a systematic averaged gain due to some acceleration process (as will be seen with the first order Fermi acceleration); these effects are described by the term A, a kind of resulting ordered force pointing in the direction of the momentum. The particle is also supposed to experience random energy variations δp due to deflections on moving magnetic disturbances. So we write that during a time Δt the energy variation is such that

$$\Delta p = A\Delta t + \delta p \,, \tag{8.53}$$

with a diffusion in energy space (that will turns out to be the so-called second order Fermi process):

$$\langle \delta p^2 \rangle = 2\Gamma_\ell \Delta t \,. \tag{8.54}$$

Similar to the previous inhomogeneity effect, the p-dependence of the diffusion coefficient implies that

$$\langle \delta p \rangle = \Delta t \frac{1}{p^2} \partial_p \left(p^2 \Gamma_\ell \right) \,. \tag{8.55}$$

This is, notably, the contribution that describes the energy gain by the second order Fermi process. Indeed, for high-energy particles such that their energy $\epsilon \simeq pc$, the internal energy density of cosmic rays increases under the effect of second order Fermi acceleration according to:

$$d_t e = \int \frac{\langle \Delta \epsilon \rangle}{\Delta t} f 4\pi p^2 dp > 0 \,. \tag{8.56}$$

This growth is linked to the fact that the diffusion coefficient is such that $p^2 \Gamma_\ell$ is an increasing function of p.

The so-called transport equation of cosmic rays is the evolution equation of the isotropic part $\bar{f}(p, x)$ of the complete distribution function, when the anisotropy is weak, which is true when the fluid is in nonrelativistic motions, but not in relativistic motion where strong anisotropic effects develop. The distribution function is normalized such that the number density of cosmic rays $n_* = \int f 4\pi p^2 dp$. The transport equation reads [157]:

$$\partial_t \bar{f} + \partial_x (u \bar{f}) = -\frac{1}{p^2} \partial_p \left(p^2 A \bar{f} \right) + \frac{1}{p^2} \partial_p \left(p^2 \Gamma_\ell \partial_p \bar{f} \right) + \partial_x \left(D \partial_x \bar{f} \right) \,. \tag{8.57}$$

The transverse coefficient of momentum diffusion does not appear since the distribution function is supposed to have been isotropized. The friction term A describes not only ay kind of energy loss, but also the energy gain by the "first order Fermi process" that will be presented further on. The radiative loss experienced by a relativistic electron of Lorentz factor γ stems from the photon emission forwards in a narrow cone of half angle γ^{-1} with respect to the momentum direction, which leads to a friction force in opposite direction of the momentum. The synchrotron and inverse Compton radiative losses contribute to A as follows:

$$A_{\text{rad}} \equiv \frac{\langle \Delta p \rangle}{\Delta t}\Big|_{\text{rad}} = -\frac{4}{3}\sigma_T \left(\frac{m_e}{m}\right)^2 (W_m + W_{ph})\gamma^2 , \tag{8.58}$$

where W_m is the magnetic energy density (synchrotron) and W_{ph} the energy density of low energy photons (inverse Compton effect in the Thomson regime). Usually radiative losses are considered for electrons only because for other particles these losses are very reduced by the square of the mass ratio $(m_e/m)^2$; however, in the case of ultrahigh-energy protons, these losses can become important because the very large Lorentz factor of those particles can compensate the small mass ratio.

Now let us look at the contribution of the first order Fermi process. This acceleration process occurs when a plasma flow carries a magnetic turbulence that insures the scattering of suprathermal particles which tend to be isotropized in the fluid rest frame, and this flow converges (div $\boldsymbol{u} > 0$) somewhere due to some compression effect, as in a shock. In the opposite case of an expansion, this is an energy loss. The contribution of this "first order" effect (in the sense of Fokker–Planck description) is obtained by calculating the average power delivered to the particle by the convergence of the diffusive medium. Indeed, the Fermi process can be described as a noninertial dragging effect due to the deceleration of the diffusive medium. Under that physical condition, the inertial force is $F_j = -p_i \partial_{x_i} u_j$, and its acceleration power is

$$P_{\text{acc}} = -\langle v_j p_i \rangle \partial_{x_i} u_j = -\frac{pv}{3} \nabla \cdot \boldsymbol{u} . \tag{8.59}$$

Only a compressed fluid (div $\boldsymbol{u} < 0$) produces the first order acceleration of particles. Its simplified contribution (8.57) is then

$$A_{\text{acc}} = -\frac{p}{3}\partial_x u . \tag{8.60}$$

The extension of the previous transport equation to several space variables is straightforward.

8.1.5 Distribution of Suprathermal Particles Crossing a Shock

The transport equation of cosmic rays describes how the distribution function is modified across a shock that propagates in a plasma carrying magnetic turbulence. We will undergo that calculation, following [158], and obtain a first important result about Fermi acceleration at shocks.

We will investigate this problem with simplifying assumptions. First, we assume that the ambient plasma carries a prescribed isotropic magnetic turbulence that has no bulk motion. Suprathermal particles are thus transported in that turbulent state and we take into account of the scattering process that induces a spatial diffusion of the particles described by a coefficient D, but (second assumption) the energy diffusion (what will be called second order Fermi process in the next chapter) is negligible. The third assumption is that the radiative losses incurred by the electrons are negligible in the region of interest, and the possible energy losses incurred by ions, as in, for instance, expansion effect, are negligible too. The fourth assumption is that a nonrelativistic shock propagates in the plasma and its thickness is much smaller than the scattering length of the particles on magnetic turbulence. For the problem that we want to solve, the shock is merely described as a sudden change of flow velocity, that transition being characterized by a compression ratio r, $r \equiv u_1/u_2$, where u_1 is the upstream flow velocity measured with respect to the shock front, and u_2 is the downstream flow velocity measured with respect to the shock front. This is the only thing that we need to know about the shock transition. Thus, the description of the flow across the shock front, located at $x = 0$, is simply as follows:

$$u(x) = u_1 + (u_2 - u_1)\theta(x) \tag{8.61}$$

$$\partial_x u = (u_2 - u_1)\delta(x) \, . \tag{8.62}$$

When a stationary state is achieved, it is possible to calculate the distribution function obtained in the downstream flow in term of the incoming distribution function far upstream. This is the solution of the following equation, which is a linear equation as long as the back reaction of the cosmic rays on the flow is negligible:

$$u\partial_x \bar{f} - \frac{1}{3}\partial_x(u)p\partial_p \bar{f} = \partial_x \left(D\partial_x \bar{f}\right) \, . \tag{8.63}$$

The distribution function is necessarily uniform downstream (the other solution is divergent and not physically acceptable). The distribution function upstream exponentially increases up to the shock front over a diffusion length, the matching is achieved by stipulating that the distribution function is continuous at the front. $l_{D1} = D_1/u_1$). In particular, the pressure of suprathermal particles exponentially increases up to the shock and remains uniform downstream. Then one has to stipulate that the distribution function is continuous at the shock transition, which relates upstream and downstream distribution functions. After a simple integration of both sides of the Eq. (8.63) over x from $-\infty$ to $+\infty$, one deduces a differential equation that relates the downstream distribution $\bar{f}_2$ to the upstream distribution $\bar{f}_1$ at $-\infty$:

$$p\partial_p \bar{f}_2 + q\bar{f}_2 = q\bar{f}_1 \, , \tag{8.64}$$

and the integration provides with [158]:

$$\bar{f}_2(p) = qp^{-q} \int_0^p p'^q \bar{f}_1(p') \frac{dp'}{p'} \, . \tag{8.65}$$

We find a power law distribution at high energy with an index

$$q \equiv \frac{3r}{r-1} .$$

Considering the high-energy particles as ultrarelativistic with an energy $\epsilon \simeq pc$, knowing that the energy distribution is proportional to $p^2 \bar{f}(p)$, we finds the following power law energy distribution:

$$F(\epsilon) \propto \epsilon^{-(r+2)/(r-1)} ;$$

which is in ϵ^{-2} for a strong nonrelativistic shock with a compression ratio $r = 4$.

This theory was welcomed with enthusiasm because it predicts a universal power law with an index close to the observational values, both for the cosmic ray spectrum and the synchrotron spectra of nonthermal sources. Also, it can be improved by introducing the cut off due to energy losses and the efficiency of the acceleration process. The theory is good enough to account for the extension of the power law spectra up to the observed high-energy cutoffs. We will see the details in the next chapter where the theory will be presented differently in order to enter more deeply in the physics of the phenomenon. We will look at the extensions of this primer model.

8.1.6 From Transport to Acceleration

In this chapter a large part of the theory of particle transport in disorganized magnetic fields has been presented. It is an important topic for both laboratory physics, space physics, and astrophysics. In space physics and astrophysics this is an essential topic for understanding the behavior of cosmic rays. The results presented in this chapter will be crucial for the development of the Fermi processes that we already introduced briefly, but will be presented more deeply in the next chapter.

8.2 Fermi Acceleration of Cosmic Rays

Nonthermal particle spectra, often observed in the form of a power law energy distribution in ϵ^{-s}, are found in all energetic sources resulting from an explosion or from a continuous ejection of plasma. The suprathermal distributions of relativistic electrons are revealed by synchrotron spectra in several types of radiation sources, such as radio galaxies, active galactic nuclei, X-ray binary systems, pulsar wind nebulae, supernova remnants, gamma ray bursts. These energetic objects are also considered to be sources of high-energy protons and nuclei (so-called cosmic rays). The cosmic ray spectrum, Figure 8.2, recorded by ground detectors (Proton Electron Detector Aix-la-Chapelle), is one of the most impressive results of modern astrophysics. All the events recorded up to energies of 10^{15} eV are dominated

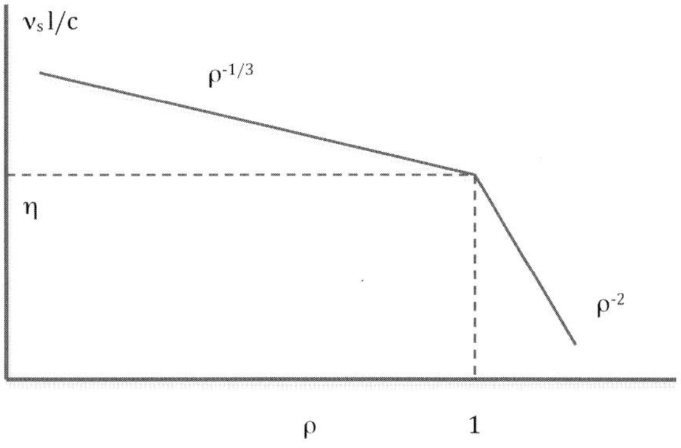

Figure 8.1 Scattering frequency as a function of "rigidity" $\rho \equiv r_L/\ell_c = \epsilon/\epsilon_c$ for a Kolmogorov turbulence of level η (Eq. (8.17)).

by protons accelerated in supernova remnants, and beyond 10^{18} eV these events are thought to be protons and heavy nuclei of extragalactic origin (see Figures 8.2 to 8.4). Other types of neutral high-energy particles are thought to be produced in some of these objects, such as gamma ray photons, very high-energy neutrinos (not generated by nuclear disintegrations but from hadronic collisions).

The generation of very high energy charged particles in astrophysical objects and flows can be thought of as resulting from very intense electric fields. The only situation where we can envisage a large electrostatic field is in the pulsar magnetosphere where the potential drop can reach 10^{12} V in most pulsars and possibly more, up to 10^{14} V in young pulsars. However, as shown in Figure 8.1, the cosmic ray spectrum extends to much higher energies. Moreover, the pulsar population is not numerous enough to yield the measured cosmic ray flux.

All the astrophysical flows carry magnetic fields either propagating at some velocity measured by the Alfvén velocity V_A, possibly modified by compression or nonlinear effects, or frozen in the background plasma. Often, when the magnetic disturbances propagate, there are more or less the same intensities of forward and backward propagating waves, so that at leading approximation in V_A/v, particles of velocity $v \gg V_A$ experience almost frozen disturbances and even static in the flow frame. Nevertheless, it is important to bear in mind that in the more general case, the magnetic disturbances have some motion with respect to the background plasma that supports them. This remark is particularly important when one considers relativistic flows. When we talk about a magnetic disturbance, its magnetic character makes sense in its proper frame; any Lorentz transformation towards another frame, involving a significant Lorentz factor, transforms the magnetic disturbance into a fully electromagnetic wave.

High-energy particles interact very weakly with matter in astrophysical flows, the collisional mean free paths are often larger than the size of the flow. The most efficient interaction with the background medium occurs through the magnetic field it carries. In the frame of a magnetic disturbance, a charged particle experiences a momentum deflection without change of its energy; it is a kind of elastic interaction that one usually calls "scattering". A variation of energy is, nevertheless, measured in a different frame, the frame of a shock front for instance, or the observer frame, and so on. When a magnetic disturbance moves with respect to some frame, the particle scattering off the magnetic disturbance leads to an energy variation that can be interpreted as the effect of the induction electric field present in that frame. The acceleration process designed by Fermi in 1949 [159] is based on the scattering of particles off moving magnetic fields. Several situations are favorable to such acceleration mechanism; they can be classified into three categories:

1. a situation where magnetic disturbances cross each other, for instance, in forward and backward propagation of Alfvén waves;
2. a shear flow carrying frozen-in magnetic perturbation, so that particles gain energy by scattering off magnetic perturbations moving at different speeds in different region of the sheared flow;
3. a plasma compressed by a shock, where upstream magnetic perturbations move faster than downstream perturbation, the particles being scattered back and forth on both sides of the shock front.
4. Fermi acceleration occurs when particles undergo reflections between two mirror points on a magnetic field line and when the points converge toward each other. This is the scenario described in Section 7.3.2. This occurs also in reconnection sites (see Section 7.2) when a sea of magnetic islands has been generated; particles gain energy by bouncing on converging islands or by bouncing inside a contracting island.

8.2.1 The Basic Fermi Process

The Fermi process can be formulated as follows in the spirit of relativity. Let us consider a magnetic perturbation that propagates at a velocity $\boldsymbol{V}_* = \boldsymbol{\beta}_* c$ (one notes $\gamma_* \equiv (1 - \beta_*^2)^{-1/2}$). In its comoving frame the electric field vanishes (this is what "magnetic type" means), and the particles experience a pure magnetic field that only changes the direction of their momentum. The inverse Lorentz transform relates the energy measured in the observatory frame ϵ to the energy ϵ' and momentum $\boldsymbol{p}'$ measured in the proper frame of the magnetic perturbation:

$$\epsilon = \gamma_* \left(\epsilon' + \boldsymbol{V}_* \cdot \boldsymbol{p}'\right) ; \tag{8.66}$$

and similarly the momentum component in the $\boldsymbol{V}_*$-direction (or normal direction) is changed according to

$$p_n = \gamma_* \left(p'_n + \beta_* \epsilon'/c\right) . \tag{8.67}$$

During the interaction ϵ' does not vary and $\boldsymbol{p}'$ changes its direction only. Thus, we deduce a change of energy and momentum measured in the observatory frame:

$$\Delta p_n = \gamma_* \Delta p'_n , \tag{8.68}$$

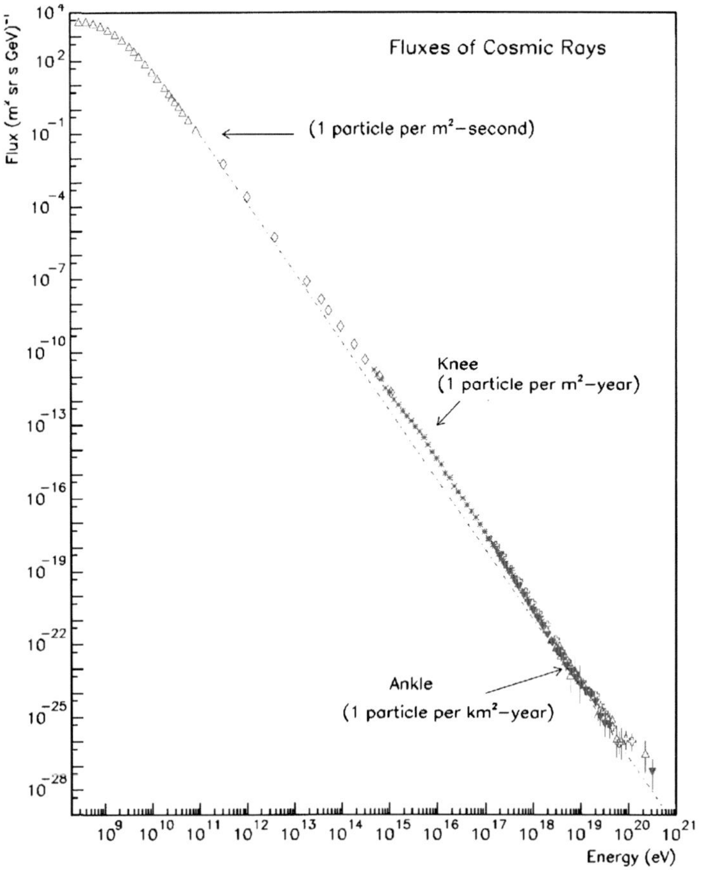

Figure 8.2 The energy spectrum of cosmic rays. Below 10^{17} eV the cosmic rays can be of galactic origin. Beyond that energy their origin is more likely extragalactic (from the PERDaix experiment).

and

$$\Delta\epsilon = \gamma_* \boldsymbol{V}_* \cdot \Delta\boldsymbol{p}' = \boldsymbol{V}_* \cdot \Delta\boldsymbol{p} \; . \tag{8.69}$$

If $\boldsymbol{V}_* \cdot \boldsymbol{p}_1' < 0$ and $\boldsymbol{V}_* \cdot \boldsymbol{p}_2' > 0$ (like for a head-on collision), then we get an energy gain, $\Delta\epsilon > 0$. This is the case for a specular collision where $\boldsymbol{V}_* \cdot \boldsymbol{p}_2' = -\boldsymbol{V}_* \cdot \boldsymbol{p}_1' > 0$. If $\boldsymbol{V}_* \cdot \boldsymbol{p}_1' > 0$ and $\boldsymbol{V}_* \cdot \boldsymbol{p}_2' < 0$ (like for a rear-end collision), then $\Delta\epsilon < 0$. Thus, the statistics over the collisions can lead to an energy increase in some conditions. If we do not specify the elastic collision, there are at least two random variables: the incidence angle of the particle before scattering and the scattering angle. Often the velocity of the magnetic perturbation is determined; however, it can be another random variable as well. Some simple models assume a mirror deflection so that $\boldsymbol{V}_* \cdot \Delta\boldsymbol{p}' = -2\boldsymbol{V}_* \cdot \boldsymbol{p}'$, the pitch angle of the particle before scattering is thus the

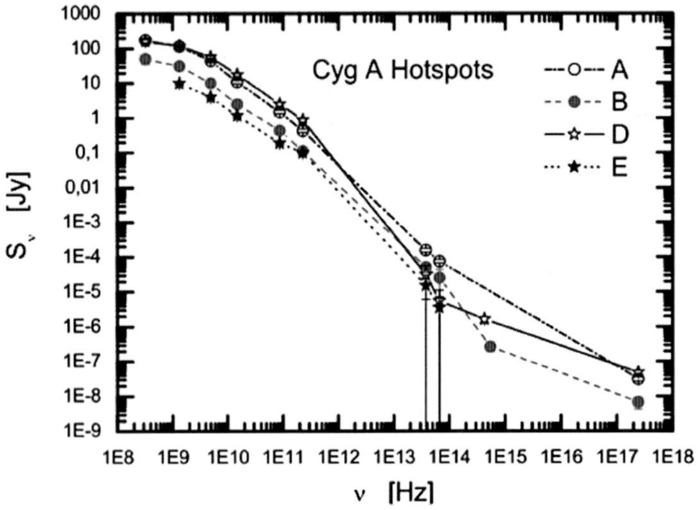

Figure 8.3 Synchrotron spectrum of Cygnus A hot spots. A photon spectrum in $\nu^{-\alpha}$ reveals an energy distribution of relativistic electrons in $\epsilon^{-2\alpha-1}$ (from [160]).

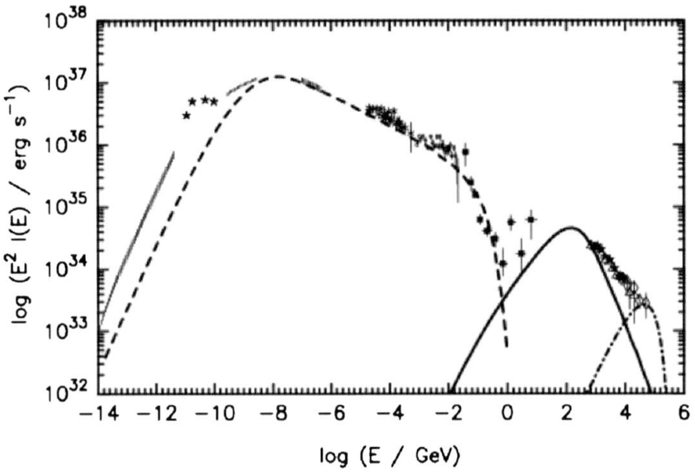

Figure 8.4 Synchrotron and inverse Compton spectrum of the Crab pulsar wind nebula. Synchrotron and inverse Compton emissions are radiated by a nonthermal distribution of relativistic electrons (from [161]).

only random variable. Now the Lorentz transform allows us to write

$$\boldsymbol{\beta}_* \cdot \boldsymbol{p}' = \gamma_* \left(\boldsymbol{\beta}_* \cdot \boldsymbol{p} - \beta_*^2 \epsilon / c\right) , \tag{8.70}$$

so that the energy gain at each scattering is

$$\Delta\epsilon = 2\gamma_*^2 \left(-c\boldsymbol{\beta}_* \cdot \boldsymbol{p} + \beta_*^2 \epsilon\right) = 2\gamma_*^2 \left(\beta_*^2 - \boldsymbol{\beta}_* \cdot \boldsymbol{\beta}\right) \epsilon . \tag{8.71}$$

That expression indicates that there are two kinds of effects, one in β_*^2 that describes a systematic gain, and another one $\boldsymbol{\beta}_* \cdot \boldsymbol{\beta}$ that leads to a gain when the collision is head-on. When the motion of the magnetic disturbances is not relativistic ($\beta_* \ll 1$) this latter effect, at first order in β_* is stronger, and can lead to an efficient energy gain when the geometry of the magnetic flow is converging. However, because this first order effect has a changing sign with the particle pitch angle, a careful statistical theory involving the scattering process must be done. So we will investigate the widely encountered situation of a nonrelativistic regime of Fermi acceleration ($\beta_* \ll 1$).

8.2.1.1 Nonrelativistic Regime: First and Second Order Processes

When we consider suprathermal particles and magnetic disturbances moving at $V_* \ll c$, the Fermi process is in a nonrelativistic regime. Thus, the energy gain can be expanded up to second order in V_*/v. However, because that expansion contains a relativistic correction, it is more convenient to consider the relativistic result presented above and to expand it at second order.

A first order acceleration occurs when during a time Δt the average energy gain $\langle \Delta\epsilon \rangle$ contains a positive contribution in β_*:

$$\langle \Delta\epsilon \rangle = -2\langle \boldsymbol{\beta}_* \cdot \boldsymbol{\beta} \rangle \epsilon > 0 \,.$$

We will see that this occurs essentially at shocks. When this average vanishes at first order and the gain is at second order in β_*, one calls this effect a second order Fermi process:

$$\langle \Delta\epsilon \rangle = -2\left\langle \beta_*^2 - \boldsymbol{\beta}_* \cdot \boldsymbol{\beta} \right\rangle \epsilon = \mathcal{O}\left(\beta_*^2\right) \epsilon \,.$$

However, we have to be careful that the averaging process, for some duration Δt, can introduce velocity dependencies through the scattering frequency as in the next example.

8.2.1.2 A Simple Model of Second Order Fermi Acceleration

We consider a homogeneous and isotropic ensemble of "magnetic targets" with density n_c, having the same cross section for particle scattering, σ, and thus characterized by a mean free path $\bar{\ell}$. The scattering frequency is thus $\nu_s = v_r/\bar{\ell}$, where v_r is the relative velocity of a particle with respect to a magnetic disturbance: $v_r = c|\boldsymbol{\beta} - \boldsymbol{\beta}_*|$. We will calculate the Fokker–Planck coefficients at second order in β_0 supposed to be small and averaging over the particle pitch angle with respect to the motion of the magnetic disturbance.

$$\frac{\langle \Delta\epsilon \rangle}{\Delta t} = \frac{2c}{\bar{\ell}} \left\langle |\boldsymbol{\beta} - \boldsymbol{\beta}_*| \left(\beta_*^2 - \boldsymbol{\beta}_* \cdot \boldsymbol{\beta}\right) \right\rangle \epsilon \tag{8.72}$$

$$\frac{\langle \Delta\epsilon^2 \rangle}{\Delta t} = \frac{2c}{\bar{\ell}} \beta \left\langle (\boldsymbol{\beta}_* \cdot \boldsymbol{\beta})^2 \right\rangle \epsilon^2 + \mathcal{O}\left(\beta_*^3\right) \,. \tag{8.73}$$

One obtains

$$\frac{\langle \Delta\epsilon \rangle}{\Delta t} = \frac{8}{3} \left\langle \beta_*^2 \right\rangle \frac{c}{\bar{\ell}} \beta \epsilon \tag{8.74}$$

$$\frac{\langle \Delta\epsilon^2 \rangle}{2\Delta t} = \frac{2}{3} \langle \beta_*^2 \rangle \frac{c}{\ell} \beta^3 \epsilon^2 \, . \tag{8.75}$$

The Fokker–Planck equation, governing the evolution of the energy distribution $\rho(\epsilon)$ normalized such that $\int \rho(\epsilon) d\epsilon = n_*$ (the number density of cosmic rays), thus reads

$$\partial_t \rho = -\partial_\epsilon \left(\frac{\langle \Delta\epsilon \rangle}{\Delta t} \right) \rho + \partial_\epsilon^2 \left(\frac{\langle \Delta\epsilon^2 \rangle}{2\Delta t} \rho \right) \, . \tag{8.76}$$

It is useful to know another expression of the Fokker–Planck equation operating on the isotropized distribution function in phase space $\bar{f}(p)$. The transition operates through $\rho d\epsilon = \bar{f} 4\pi p^2 dp$, $d\epsilon/dp = v$ and thus $\rho = (4\pi p^2/v)\bar{f}$. After some elementary manipulations we obtain:

$$\partial_t \bar{f} = \frac{1}{p^2} \partial_p \left(p^2 \Gamma_\ell \partial_p \bar{f} \right) \, , \tag{8.77}$$

where the diffusion coefficient Γ_ℓ in momentum space is such that

$$\Gamma_\ell = \frac{\langle \Delta p^2 \rangle}{2\Delta t} = \frac{\beta_*^2}{3} \frac{c}{\ell} \frac{p^2}{\beta} \, . \tag{8.78}$$

As expected in the case of a conservative force, the generalized friction term disappears, for the relation

$$\frac{\langle \Delta p \rangle}{\Delta t} = \frac{1}{\beta p^2} \partial_p \left(\beta p^2 \Gamma_\ell \right) \tag{8.79}$$

allows the friction term to vanish in favor of a transport equation that expresses as a simple diffusion in momentum space. The calculations are done by using the relations $pc = \beta\epsilon$ and $\Delta\epsilon = \beta c \Delta p$). However, when one calculates the global energy gain,

$$\dot{E} = \int \epsilon \partial_t \rho d\epsilon = \int \epsilon \partial_t \bar{f} 4\pi p^2 dp \, , \tag{8.80}$$

with the first integral, when inserting the right hand side of the Fokker–Planck equation, one obtains the contribution of the friction term only:

$$\dot{E} = \int \frac{\langle \Delta\epsilon \rangle}{\Delta t} \rho d\epsilon \, . \tag{8.81}$$

With the second integral, one obtains the contribution of the derivative of the diffusion coefficient only:

$$\dot{E} = \int \frac{1}{p^2} \partial_p \left(v p^2 \Gamma_\ell \right) \bar{f} 4\pi p^2 dp \, . \tag{8.82}$$

The relation previously obtained between the friction and the diffusion coefficient insures the equality of these two last expressions of the global energy gain and it shows that this is the second order contribution of the friction force which is responsible for the energy gain. In the momentum diffusion equation, the global energy gain is expressed by the fact that the diffusion coefficient Γ_ℓ is a growing function of p in p^2.

Because the transport equation is just a diffusion equation in momentum space, the distribution function is spreading forever. The only way to stop that kind of energizing is to take into account some loss effect. Let us consider a loss of the form:

$$P_{\text{loss}} = -\frac{\epsilon}{t_{\text{loss}}} . \tag{8.83}$$

This kind of loss occurs, for instance, when particles are wandering in an expanding wind where they are scattered by a frozen-in magnetic turbulence and thereby lose energy. This is the inverse effect of the Fermi process: $P_{\text{loss}} = (-1/3)\epsilon \operatorname{div} \boldsymbol{u}$ with $\operatorname{div} \boldsymbol{u} > 0$ as, for instance, an inertial outflow of velocity $\boldsymbol{u} = u_0 \boldsymbol{e}_{\text{r}}$ (see Chapter 7). A stationary solution can then be obtained; it satisfies the differential equation:

$$\frac{\epsilon}{t_{\text{loss}}} \bar{f} + \Gamma_\ell \partial_p \bar{f} = 0 . \tag{8.84}$$

For simplicity, we take the ultrarelativistic approximation: $\beta \simeq 1$, $\epsilon \simeq pc$, and we define an acceleration time t_{acc} such that

$$c\Gamma_\ell \equiv \frac{\epsilon^2}{t_{\text{acc}}} \quad \text{with} \quad t_{\text{acc}}^{-1} \simeq \frac{\langle \beta_*^2 \rangle c}{3\bar{\ell}} . \tag{8.85}$$

We can notice that the acceleration time is linked with the scattering time $t_{\text{s}} = \bar{\ell}/c$ with $t_{\text{acc}} = 3t_{\text{s}}/\langle \beta_*^2 \rangle$. The differential equation can then be written:

$$\epsilon \partial_\epsilon \bar{f} + \frac{t_{\text{acc}}}{t_{\text{loss}}} \bar{f} = 0 , \tag{8.86}$$

which leads to the solution $\bar{f} \propto \epsilon^{-t_{\text{acc}}/t_{\text{loss}}}$, or:

$$\rho(\epsilon) \propto \epsilon^{-s} \quad \text{with} \quad s = \frac{t_{\text{acc}}}{t_{\text{loss}}} - 2 . \tag{8.87}$$

It is interesting to learn that we obtain a power law distribution function of the energy; however, the index s is not universal and can vary a lot.

8.2.1.3 Fermi Process in Alfvén Turbulence

Fermi's original idea was that the galaxy was populated with clouds carrying a frozen-in magnetic field and moving with velocities dispersed around the averaged rotation velocity around the bulge. But the estimates for the velocities of clouds and their mean free path led to the conclusion that the Fermi process has a characteristic time scale that is far too long (longer than the age of Universe). This is why the second order Fermi process became obsolete. However, its modern version based on the resonant interaction of cosmic rays with Alfvén waves in a magnetic turbulence has turned out to be much more efficient in some media especially when the Alfvén velocity is large. In this case the parameter β_* is given by the propagation velocity of the incompressible MHD waves:

$$\beta_{\mathrm{A}}^2 \equiv \frac{V_{\mathrm{A}}^2}{c^2} = \frac{B^2}{\mu_0 \rho c^2} . \tag{8.88}$$

As for the mean free path, it is derived from the scattering process of the particle momentum off Alfvén wave packets. As we have seen in the previous chapter, the most effective scattering occurs when particles have a Larmor radius comparable with the wave length of a mode of the wave packet. The factor $c/\bar{\ell}$ is then the scattering frequency ν_{s} of cosmic rays, which depends on their Larmor radius and the characteristics of the spectrum, as derived in the previous chapter:

$$\nu_{\mathrm{s}}(p) = \alpha_0 \frac{c}{\ell_{\mathrm{c}}} \frac{\langle \delta \boldsymbol{B}^2 \rangle}{\langle B^2 \rangle} \left(\frac{r_{\mathrm{L}}}{\ell_{\mathrm{c}}} \right)^{\beta - 2} , \tag{8.89}$$

where β is the turbulence spectrum index ($\beta = 5/3$ in the case of a Kolmogorov spectrum) and the coherence length ℓ_{c} corresponds to a large scale when $1 < \beta \leq 2$ (α_0 is some coefficient of order unity that can be exactly calculated in weak turbulence theory). This formula is valid when the Larmor radii are in the turbulent range, that is, has values corresponding to a wavelength of the spectrum. At larger Larmor radii, as seen in the previous chapter, the scattering frequency decreases proportionally to the inverse of the square of the particle energy: $\nu_{\mathrm{s}} \sim c\ell_{\mathrm{c}}/r_{\mathrm{L}}^2 \propto \epsilon^{-2}$. A general formula has been obtained in agreement with numerical simulations in [153]:

$$\nu_{\mathrm{s}} = \omega_{\mathrm{L}} \frac{\pi}{3} \int\limits_{k r_{\mathrm{L}} > 1} \frac{S(k)}{k r_{\mathrm{L}}} dk , \tag{8.90}$$

for any turbulence spectrum $S(k)$ normalized such that $\int S(k) dk = \langle \delta \boldsymbol{B}^2 \rangle / \langle \boldsymbol{B}^2 \rangle$. We mention that a Bohm conjecture is often used in the astroparticle community, which corresponds to a scaling similar to that obtained for $\beta = 1$, but there is no ground for this assumption.

As we have seen in the chapter devoted to the transport of suprathermal particles and the Fokker–Planck equation, when magnetic disturbances move, as is the case with Alfvén waves, there is a diffusion process in momentum space. Its coefficient Γ_ℓ describes the energy gain of particles under the second order Fermi process and its value is related with the scattering frequency:

$$\Gamma_\ell \sim \nu_{\mathrm{s}} V_{\mathrm{A}}^2 c^2 p^2 . \tag{8.91}$$

Despite the similarity of the Fokker–Planck equation with the previous model, the efficiency of the acceleration process is different because the acceleration time is not increasing so dramatically, since, from the previous formula,

$t_{acc} \sim (\ell_c/c)(c^2/V_A^2)(r_L/\ell_c)^{2-\beta}/\eta$. For instance, consider a Kolmogorov spectrum of a strong magnetic turbulence, then

$$t_{acc} \sim \frac{\ell_c}{c}\frac{c^2}{V_A^2}\left(\frac{\epsilon}{Ze\bar{B}\ell_c}\right)^{1/3} .$$

As the particle energy grows its Larmor radius grows up to the large coherence length of turbulence and reaches a characteristic energy

$$\epsilon_c = Ze\bar{B}\ell_c . \tag{8.92}$$

Beyond that energy, the scattering frequency decreases rapidly in ϵ^{-2} and Γ_ℓ saturates at its maximum value (corresponding to a "Bohm maximum"):

$$\Gamma_{\ell,max} \sim \frac{c}{\ell_c}\frac{V_A^2}{c^2}\frac{\epsilon_c^2}{c^2} . \tag{8.93}$$

This process is not devoid of some efficiency; however, it does not produce a universal power law spectra, as those observed in astrophysics.

8.2.1.4 Fermi Process at a Shear Layer

Consider a shear flow with a velocity $\boldsymbol{u} = f(y/\delta)u_0\boldsymbol{e}_x$, where the function is an odd function of y that tends to +1 (resp. -1) as $y \gg \delta$ (resp. $y \ll -\delta$) (a typical example $f(y/\delta) = \tanh(y/\delta)$). It describes a sheared flow in a layer of width δ, the variation of velocity being $\Delta u = 2u_0$ in the layer. We assume that the flow carries a turbulent magnetic field that makes suprathermal particles to scatter and diffuse everywhere in the layer. Despite the fact that magnetic disturbances do not cross each other, they move forwards for $y > 0$ and backwards for $y < 0$ and then particles scatter off counterstreaming magnetic flows. According to the elementary Fermi process, they gain energy by wandering several times between both half-planes. The physics can be developed by considering two cases, $\bar{\ell} < \delta$ and $\bar{\ell} > \delta$. This process is very similar to a second order Fermi process.

8.2.1.5 First Order Fermi Process at a Nonrelativistic Shock

The physics of a nonrelativistic shock has been presented in Section 6.1. Jump conditions, known as Rankine–Hugoniot relations, relate upstream fluxes, such as matter flux, momentum flux, and energy flux, with downstream fluxes, whatever the kind of dissipation occurring in the transition region; under astrophysical conditions, shocks relevant to high-energy phenomena are collisionless. Turbulence is excited in the transition region, even suprathermal particles participate in generating turbulence. Turbulence tends to isotropize the distribution functions and to heat particle populations; however, a complete thermalization is not necessarily achieved. Rankine–Hugoniot relations are nevertheless fairly well satisfied in collisionless shocks and one characterizes the shock by a Mach number $\mathcal{M}$ and a compression ratio r such that (see Section 6.1.3):

$$r = \frac{\gamma_a + 1}{\gamma_a - 1 + 2/\mathcal{M}^2} . \tag{8.94}$$

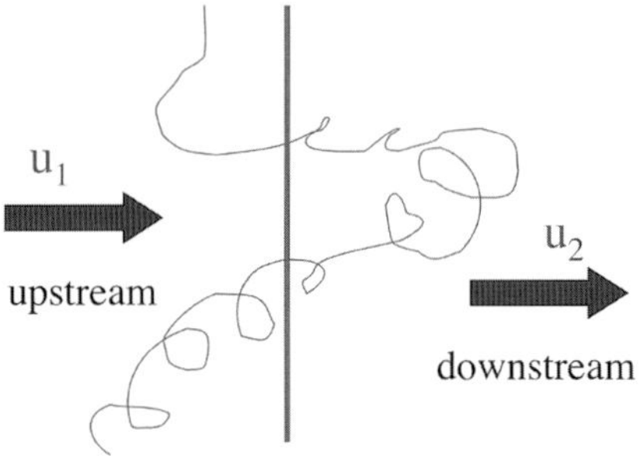

Figure 8.5 Sketch of a shock in a flow carrying magnetic disturbances and suprathermal particles wandering back and forth.

In a perfect mono-atomic plasma, the adiabatic index $\gamma_a = 5/3$ and a strong adiabatic shock has a compression ratio $r = 4$.

Now, an astrophysical flow always carries some amount of turbulent magnetic field. In this chapter, we will assume that the magnetic energy density is much smaller than the kinetic one. For scales large enough, the magnetic disturbances are frozen in the flow. Even if they propagate at Alfvén speed in the flow, we will assume that the Alfvén velocity is much smaller than the flow velocity with respect to the shock front. Under those conditions, magnetic disturbances are carried at the velocity u_1 with respect to the upstream flow, that is, r times faster than their velocity u_2 in the downstream flow. The scattering length of high-energy particles (that replaces their collisional mean free path), is much larger than the thickness of the shock front ($\delta \sim u_1/\omega_{ci}$), so that when crossing the shock they experience magnetic disturbances that move at different speeds upstream and downstream (see Figure 8.5). A first order Fermi process develops with a velocity parameter $\beta_* \equiv (u_1 - u_2)/c$.

8.2.2
Fermi Process at a Nonrelativistic Shock

We state the following assumptions that can be removed one by one in a more realistic theory; however, these assumptions apply in many astrophysical cases. First we assume that the mean magnetic field is weak compared to the turbulent one; since the energy density in the turbulent field is smaller than the incoming kinetic energy density, Alfvén velocity is very small compared to the flow velocities upstream and downstream. The magnetic turbulence is supposed to be frozen in the flow, thus particle scattering is elastic in the frame moving with the flow. The shock thickness is of the order of the ion inertial scale δ_i, whereas the mean free path for high-energy particles is much larger than the ion inertial scale. The suprathermal particles are thus able to wander on both sides of the shock front and to cross the front many times as will be estimated further.

We consider the energy gain at first order in the flow velocities upstream (β_1) and downstream (β_2) for a particle that leaves the upstream flow with a pitch angle cosine μ_1 $(-1 < \mu_1 < -\beta_s)$, crosses the front and comes to the downstream flow and then comes back to the upstream flow with a pitch angle cosine μ_2 $(-\beta_s < \mu_2 < 1)$. Since $\mu_2' \simeq \mu_2$, we get the energy increase:

$$\Delta p = -\frac{\beta_1 - \beta_2}{\beta}(\mu_1 - \mu_2)p \ . \tag{8.95}$$

Now let us estimate the average energy increase, averaging over the intervals defined by $-1 < \mu_1 < -\beta_s$ and $-\beta_s < \mu_2 < 1$;

$$\langle \mu_1 \rangle = \frac{\int_{-1}^{-\beta_s} \mu_1^2 d\mu_1}{\int_{-1}^{-\beta_s} \mu_1 d\mu_1} = -\frac{2}{3} + \mathcal{O}(\beta_s) \tag{8.96}$$

$$\langle \mu_2 \rangle = \frac{\int_{-\beta_s}^{1} \mu_2^2 d\mu_2}{\int_{-\beta_s}^{1} \mu_2 d\mu_2} = \frac{2}{3} + \mathcal{O}(\beta_s) \ . \tag{8.97}$$

Therefore, we obtain the average energy increase per cycle:

$$\langle \Delta p \rangle = \frac{4}{3}\frac{\beta_1 - \beta_2}{\beta} p \ . \tag{8.98}$$

A similar calculation can be done by starting from downstream; this leads to the same result.

Let us estimate the escape probability as a flux ratio:

$$\eta \equiv \frac{\text{escaping flux}}{\text{incoming flux}} = \frac{u_2}{v\frac{1}{2}\int_{-\beta_s}^{1} \mu_2 d\mu_2} = \frac{4u_2}{v} \ . \tag{8.99}$$

An interesting aspect of the process is that the energy gain is linked with the escape probability:

$$\langle \Delta p \rangle = \frac{r-1}{3}\eta p \ . \tag{8.100}$$

Knowing the probability for a particle to undergo n cycles and only n, namely $(1-\eta)^n \eta$ (we assume $v = c$ for simplicity), we can calculate the average number of cycles n_c:

$$n_c = \sum_{n=1}^{\infty} n\eta(1-\eta)^n = \frac{1-\eta}{\eta} \ . \tag{8.101}$$

The larger the particle velocity compared to the flow velocity u, the larger this number (notice that the deviation with respect to that mean can also be large since $\sqrt{\langle (n-n_c)^2 \rangle} = \sqrt{1-\eta/\eta}$). The averaged frequency of Fermi cycle is thus:

$$\nu_{\text{cycle}} = n_c / t_r \ . \tag{8.102}$$

It is now possible to give an estimate of the first order Fermi acceleration at a shock through an effective force derived from a probabilistic viewpoint:

$$\frac{\langle \Delta p \rangle}{\Delta t} = \nu_{\text{cycle}} \Delta p_{\text{cycle}} = \frac{r-1}{3 t_{\text{r}}} p \; ; \tag{8.103}$$

the last expression, quite simply, stems from the insertion of the escape probability Eq. (8.99) and from the energy gain Eq. (8.100). Clearly, the effective accelerating force is proportional to the escape time (or residence time). That remarkable proportionality between the acceleration and the escape is the origin of the formation of universal power law spectrum, as will be seen in the next section.

Let us first estimate the residence time and hence the efficiency of the acceleration that directly depends on it. The residence time of a particle downstream is determined by its diffusion coefficient over magnetic irregularities.

When a particle travels over a distance $u_2 t$ larger than its diffusion distance $\sqrt{2Dt}$, it has less and less probability to come back to the shock front. Thus, the residence time downstream can be estimate as

$$t_{r|d} \simeq \frac{2D}{u_2^2} \, . \tag{8.104}$$

A similar time scale can be estimated upstream, despite the fact that the particle always comes back to the front if no escape is considered upstream. This time is generally quite shorter than the residence time downstream and the total characteristic time of residence in the vicinity of the shock front is thus $t_{\text{r}} = (2D_1/u_1^2) + (2D_2/u_2^2)$ and the acceleration time $t_{\text{acc}} = 3t_{\text{r}}/(r-1)$. The average duration of a Fermi cycle is given by:

$$\tau_{\text{cycle}} \simeq t_{\text{r}}/n_{\text{c}} \simeq \frac{8u_2}{c} \left(\frac{D_1}{u_1^2} + \frac{D_2}{u_2^2} \right) \, . \tag{8.105}$$

A similar result has been obtained in [162]. However, it must be understood that the characteristic acceleration time is not determined by the cycle time but by the residence time, $t_{\text{acc}} \sim t_{\text{r}} \sim \tau_{\text{s}}/\beta_2^2$, which depends on the scattering time. That last estimate, obtained with the diffusion coefficient ($D = v^2 \tau_{\text{s}}/3$), indicates that the Fermi cycles are much less frequent than the scatterings. The cycle frequency is much slower than the scattering frequency since $\nu_{\text{cycle}} \simeq (3u_2/8v)\nu_{\text{s}}$. Whereas the Fermi cycle frequency is of first order in β_2 because $\nu_{\text{cycle}} \sim \beta_2 \nu_{\text{s}}$, the acceleration rate is of second order: $\nu_{\text{acc}} \sim \beta_2^2 \nu_{\text{s}}$. Fermi acceleration of first order at shocks is really more efficient than second order Fermi acceleration only if $\beta_2 \gg \beta_{\text{A}}$.

8.2.2.1 Spectrum

This probabilistic derivation of the spectrum is inspired by Bell [163] and Peacock [164]. The energy spectrum of accelerated particles can be obtained from the probability for a particle of initial energy p_0 to reach an energy larger than p. To achieve such an energy, the considered particle has to undergo at least n Fermi cycles such that $p_n \geq p$, and the probability of such event is $\Pr[> n] = \prod_{k=1}^{k=n}(1-\eta_k)$.

The energy p_n is such that:

$$p_n = p_0 \prod_{k=1}^{k=n} G_k \,, \tag{8.106}$$

where the gain G_k is a random variable depending the pitch angles of the particle as it crosses the front from upstream to downstream (μ_1) and then recrosses the front from downstream to upstream (μ_2):

$$G_k(\mu_1, \mu_2) = 1 + \frac{r-1}{4}\eta_k \Delta\mu_k \,, \tag{8.107}$$

where $\Delta\mu_k = \mu_2 - \mu_1$ at the kth cycle; this is a random variable distributed between 0 and 2. As we saw in the previous section, its average value is

$$\langle G_k(\mu_1, \mu_2)\rangle = 1 + \frac{r-1}{3}\eta_k \,. \tag{8.108}$$

As acceleration proceeds, η_k saturates at a constant value $\eta = 4u_2/c$ and, similarly, $P_{\text{ret},k} \simeq P_{\text{ret}} = 1 - \eta$. To determine the distribution of the variable $\ln p_n/p_0$, we analyze the random variable $X_n \equiv \sum_{k=1}^{k=n} \ln G_k$. Its mean increases with n as $\langle X_n\rangle \simeq n\langle \ln G\rangle$ and its variance

$$\begin{aligned} V(X_n) &\equiv \langle X_n^2\rangle - \langle X_n\rangle^2 \simeq n\langle(\ln G)^2\rangle + n(n-1)\langle \ln G\rangle^2 \\ &- n^2\langle \ln G\rangle^2 \simeq n(\langle(\ln G)^2\rangle - \langle \ln G\rangle^2) \,. \end{aligned} \tag{8.109}$$

Thus, for large n, the distribution of X_n is concentrated around its mean because the ratio of the standard deviation over the mean decreases as $1/\sqrt{n}$. Therefore, the probability that the considered particle reaches an energy $p \geq p_n$ is almost equal to the probability that it makes at least n cycles, for n large enough. Therefore, this probability is given by:

$$\ln \Pr[p > p_n | p_0] \simeq n \ln P_{\text{ret}} \,. \tag{8.110}$$

Now since the random variable $\ln p_n/p_0 \simeq n\langle \ln G\rangle$, we can substitute this variable to the random variable n in the previous equation:

$$\ln \Pr[p > p_n | p_0] \simeq \ln \frac{p_n}{p_0} \frac{\ln P_{\text{ret}}}{\langle \ln G\rangle} \,. \tag{8.111}$$

Thus, the conditional repartition function can be put in the following form:

$$\Pr[> p | p_0] = \left(\frac{p}{p_0}\right)^{-\alpha} \,. \tag{8.112}$$

and the index

$$\alpha \equiv -\frac{\ln P_{\text{ret}}}{\langle \ln G\rangle} \,. \tag{8.113}$$

For the nonrelativistic regime it turns out that this index is a universal constant to an excellent approximation:

$$\alpha = \frac{3}{r-1} \, . \tag{8.114}$$

This is a very important point in the theory of Fermi acceleration at nonrelativistic shock. The energy spectrum measured behind the shock is then

$$\rho_2(p) = r\alpha \int_0^p \left(\frac{p}{p_0}\right)^{-\alpha-1} \rho_1(p_0) \frac{d p_0}{p_0} \, , \tag{8.115}$$

where ρ_1 is the spectrum injected far upstream. For $p \gg p_0$, we obtain a power law spectrum $\propto p^{-s}$ with

$$s = 1 - \frac{\ln P_{\text{ret}}}{\langle \ln G \rangle} \simeq 1 + \frac{3}{r-1} \, , \tag{8.116}$$

which is the same result than that obtained in [158] and presented at the end of Section 8.1.5 by integrating the Fokker–Planck equation. That derivation shows that the result is very robust and derived from elementary arguments of probability only. Nevertheless, the Fokker–Planck equation is useful when one needs to solve more realistic problems involving losses, nonplanar flow and so on. For an adiabatic strong shock, the compression ratio $r = 4$ and then the predicted spectrum index is $s = 2$, which appears to be an universal result.

8.2.3
Astrophysical Application: Cosmic Rays and Supernovae

The previous theoretical result suggests a precise power law spectrum of high-energy particles at a shock as long as various energy losses are not considered. Of course, losses change the shape of the distribution function as we will see later on. High-energy nuclei are advected by the downstream flow and then escape from the environment. This can simply be modeled by taking account of an escape time depending on their energy. For instance, the termination shocks of supernova remnants are considered as the main sources of high-energy cosmic rays up to energies of 10^{15} eV and the rate of supernova explosions in our galaxy would allow us to account for the energy flux of cosmic rays if 10% of the supernova's energy is converted into suprathermal particles. It is commonly thought that about 10% of particles are protons and about 1% electrons. According to the theory, the termination shock generates an energy flux with a power law spectrum close to ϵ^{-2} for adiabatic strong shocks having a compression ratio $r = 4$. These cosmic rays reside during a finite time in the galaxy before escaping in the halo and then in the intergalactic medium. This escape is produced by the magnetic turbulence of the galaxy that transports them, according to the theory developed in the previous chapter. In particular, they are diffused away in a time scale $t_{\text{esc}} = h^2/D$, where h

is the galaxy thickness and it is thought that only cosmic ray of energy smaller than 10^{17} eV can be confined in our galaxy as long as one considers protons. The cosmic ray observatories on earth record high-energy events with a flux that is proportional to the energy flux generated by the supernovae multiplied by the escape time: $J(\epsilon) \propto Q(\epsilon) t_{\text{esc}}(\epsilon)$. Two different interpretations of the data are still under debate: since a power law distribution in $\epsilon^{-2.7}$ is recorded up to 10^{15} eV, and an escape time in $\epsilon^{-0.6}$ has been measured for low energy cosmic rays, one can think that the theory is very successful since it predicts a spectrum close to ϵ^{-2} at the source and a steepening by escape loss such that $s \mapsto s' = s + 0.6$, which is close to the observational result. Another theory considers that the escape time is given by Kolmogorov magnetic turbulence theory that predicts an escape time in $\epsilon^{-1/3}$ and that the index at the source is modified by some nonlinear effect such that $s \simeq 2.3$.

Nonlinear modification of the theory presented in this chapter is relevant, because a spectrum in ϵ^{-2} leads to a diverging energy density and pressure of relativistic particles, as long as no cut off is taken into account:

$$P_* = \frac{1}{3} \int_0^{p_{\max}} p c \rho(p) d p \propto \ln(p_{\max}/p_0) \,. \tag{8.117}$$

The pressure of relativistic particles has a back reaction on the flow and modifies the shock structure and its compression ratio. This effect is a priori significant for supernova remnants since they are supposed to convert 20% of their energy into cosmic rays. Various nonlinear theories has been developed; however, they cannot be yet regarded as sufficiently firm to be presented in a text book. Moreover, the most popular theory predicts an increase of the effective compression ratio and thus an even harder spectrum of concave shape, which makes the nonlinear effect worse.

8.2.4 Astrophysical Application: Synchrotron Sources

Astrophysical shocks generate also a population of relativistic electrons that are observed by their synchrotron radiation on the ambient magnetic field or the turbulent magnetic field generated by instabilities. A relativistic electron of Lorentz factor γ moving in a magnetic field of intensity B radiates photons with energy around $\omega_* \sim \gamma^2 \omega_c$, where ω_c is the electron cyclotron pulsation. When the population of relativistic electron has a power law energy distribution in ϵ^{-s}, then the spectrum of emitted photons is a power law with an index $\alpha = (s-1)/2$. Such power law synchrotron spectra are ubiquitous in astrophysics; the first examples detected by radio astronomy are quasars, radio galaxies, pulsars, supernova remnants. Synchrotron spectra have also been identified at high energies for some powerful sources like X-ray jets, gamma ray bursts, pulsar wind nebulae.

From a theoretical point of view, when a shock generates a power law distribution of electrons with an index s, there is a cut off at some energy that can be explained as follows. The power lost by a relativistic electron due to synchrotron emission is proportional to the square of its energy: $P_{\text{syn}} = (4/3)\sigma_T c(B^2/8\pi)\gamma^2$. As long as the energy lost during a Fermi cycle of duration t_{cycle} is much smaller than the energy gain, the spectrum generated by the Fermi process is not significantly modified. However, as the energy increases, the synchrotron energy loss becomes more important, until some energy at which the energy lost during a Fermi cycle cancels the energy gain. Thus, we obtain a cut off energy ϵ_c such that $\Delta\epsilon = P_{\text{syn}} t_{\text{cycle}}$.

Now what happens to that population of relativistic electrons with its cut off determined at shock when the downstream flow carries it far from the shock region? When a relativistic electron has been advected during a duration Δt, its energy has been declining by an amount $\int_0^{\Delta t} P_{\text{syn}} dt$. Thus, the initial power law distribution is progressively broken with a steeper high-energy part which turns out to be again a power law distribution with an index changed according to $s \mapsto s' = s + 1$. The photon index is then steepened also such that $\alpha \mapsto \alpha' = \alpha + 1/2$. Indeed the distribution steepening stems from the integration of the transport equation in a stationary flow:

$$u\partial_x \rho = \partial_\epsilon \left(P_{\text{syn}}\rho\right) + \partial_x \left(D\partial_x \rho\right) , \tag{8.118}$$

with a boundary condition at $x = 0$ given by $\rho(\epsilon, x = 0) \propto \epsilon^{-s}$. An exact integration can be done and leads to a description of the evolution of the distribution function with x. However, we will provide a simple argument that leads to previous statement. Because of the synchrotron loss and the absence of reenergization of the electrons far from the shock, the distribution function and its derivatives vanishes at large distance. But its integral over x does not vanish. Indeed, let $F(\epsilon, x) \equiv \int_0^x \rho(\epsilon, x')dx'$ and from the transport equation we get (we assume u and D uniform for simplicity and $\partial_x \rho(\epsilon, x = 0) = 0$):

$$u[\rho(\epsilon, x) - \rho_0(\epsilon)] = \partial_\epsilon[P_{\text{syn}} F(\epsilon, x)] + D\partial_x \rho(\epsilon, x) . \tag{8.119}$$

Then for x large enough, namely $x \gg u\tau_{\text{syn}}(\epsilon)$ where the synchrotron time $\tau_{\text{syn}}(\epsilon) \equiv \epsilon / P_{\text{syn}}$,

$$\partial_\epsilon[P_{\text{syn}} F(\epsilon, x)] \simeq u\rho_0(\epsilon) . \tag{8.120}$$

Therefore, the integrated distribution becomes asymptotically

$$F(\epsilon, \infty) = \frac{u}{P_{\text{syn}}} \int_\epsilon^\infty \rho_0(\epsilon')d\epsilon' . \tag{8.121}$$

The observation of a synchrotron source in radio astronomy provides information about the integrated distribution along the line of sight, never about the local distribution, and the latter formula indicates that the spectral index s of the distribution delivered by the shock is modified by synchrotron losses at large distance with an index s' of the integrated distribution along the line of sight such that $s' = s + 1$.

8.2.5
Generation of Magnetic Turbulence

Until now we have considered a prescribed turbulent state. However, for several types of astrophysical sources, there is evidence that the radiation is from a shock where the magnetic field has been amplified compared to its ambient intensity. Astrophysical shocks, as those of supernovae remnants or hot spots of radio jets, are strong collisionless shocks (the mean free path for Coulomb collision is much larger than any scale involved in the shock physics). Those shocks are a particular self-sustaining region of magnetic turbulence generated by cosmic rays themselves. Some of the incoming protons are reflected back to the front and constitute an energetic stream that interacts with the background plasma, generating MHD instabilities and turbulence. The reflection is produced by the magnetic field, either the compressed ambient magnetic field, or the generated turbulence itself. Since the majority of reflected or suprathermal particles that pervades the upstream flow is constituted of protons, there is an intense cosmic ray current generated by the shock in its precursor. That cosmic ray current $\boldsymbol{J}_{\rm cr}$, which flows along the filed lines, needs to be compensated by a current $\boldsymbol{J}_0$ carried by the incoming plasma. This current is responsible for MHD instabilities. In a recent work A. Bell proposed such instability [165] to produce a high level of magnetic turbulence of intensity much larger than the ambient field intensity. The derivation of the instability is simple when one considers only modes of wavelengths smaller than the Larmor radii of the cosmic ray, so that the response of the cosmic ray population can be neglected. The most unstable modes are obtained in the incompressible regime (div $\boldsymbol{u} = 0$ and the mass density remains constant: $\rho = \rho_0$). Like for Alfvén waves, the field lines are perturbed transversally: $\boldsymbol{u} \perp \boldsymbol{B}_0$. The linearized equation of motion is thus:

$$\partial_t \boldsymbol{u} = \delta \boldsymbol{J} \times \frac{\boldsymbol{B}_0}{\rho_0} + \frac{\boldsymbol{J}_0}{\rho_0} \times \delta \boldsymbol{B} \, , \tag{8.122}$$

where we can see a supplementary Lorenz force density due to the compensating current that will turn out to be the cause of the instability. $\delta \boldsymbol{J} = (1/\mu_0)$ rot $\delta \boldsymbol{B}$ and the equation of motion is coupled with the induction equation, whose linearized version is:

$$\partial_t \delta \boldsymbol{B} = B_0 \partial_x \boldsymbol{u} \, . \tag{8.123}$$

We introduce the characteristic wave number $k_{\rm c}$, an algebraic number whose sign depends on the sense of current with respect to the direction of the mean field, defined by

$$k_{\rm c} \equiv \frac{\mu_0 J_0}{B_0} \, . \tag{8.124}$$

We apply a second time-derivation to both sides of the equation of motion and insert the induction equation, so we get the following wave equation:

$$\partial_t^2 \boldsymbol{u} - V_{\rm A}^2 \partial_x^2 \boldsymbol{u} - V_{\rm A}^2 k_0 \boldsymbol{e}_x \times \partial_x \boldsymbol{u} = 0 \, . \tag{8.125}$$

When $k_c = 0$ (no cosmic ray current), we recognize the equation of propagation of Alfvén waves. The modes have to be decomposed into right and left polarized plane waves, and we obtain the dispersion relation for each polarization:

$$\omega^2 = V_A^2 k(k - \varepsilon k_0) , \tag{8.126}$$

where the polarization is $\varepsilon = 1$ for right modes and $\varepsilon = -1$ for left modes. For $k_0 > 0$, modes such that $k < k_c$ and of right polarization are unstable, vice versa for $k_0 < 0$. For $k > k_0$ we found that modified Alfvén waves are stable. It has been shown that this instability can generate an intense turbulent field, larger than the mean field, as suggested by X-ray observations [166].

8.2.6
Why Are Fermi Processes Favored at Shocks?

As explained in the text, the efficiency of Fermi acceleration processes directly depends on the efficiency of particle scattering by magnetic disturbances. In nonrelativistic regime, the first order Fermi process has an acceleration time $\tau_1 \sim \tau_s/\beta_s^2$, where $\beta_s c$ is the velocity of the shock front that propagates in the upstream (ambient) plasma and the second order process $\tau_2 \sim \tau_s/\beta_A^2$ where $\beta_A c = V_A$ is the propagation velocity of Alfvén waves. Only the circumstance $\beta_s > \beta_A$ allows us to assert that the first order Fermi process is more efficient than the second order one. Are there other acceleration processes that are competitive with Fermi processes?

May one imagine large variations of electric potential generated in some cosmic regions? The largest potential differences are expected in pulsar magnetospheres between the magnetic axis and the last open field line: $\Delta V = (\Omega_*/2\pi c)\Delta\Phi$, where Ω_* is the angular velocity of the magnetosphere and $\Delta\Phi$ is the magnetic flux between the axis and the last opened magnetic surface. With the standard parameters,

$$\Delta V \sim 10^{12} \left(\frac{B}{10^{12}\,\mathrm{G}}\right)\left(\frac{P}{1\,\mathrm{s}}\right)^{-1} \mathrm{eV} . \tag{8.127}$$

That potential difference does not allow an acceleration in the vicinity of the polar caps, but one should find again that difference between the foot of the last open line and its end, with the development of an electric field component along the magnetic line. Even with a millisecond pulsar with a strong field close to the critical value of 10^9 T, energies covering the cosmic ray spectrum knee are not easily achieved, except with iron nuclei.

The magnetic reconnection phenomenon is known as a possible site of suprathermal particle generation. In the quasi 2D-configuration where the polarity of the field reverses across the current sheet that separates two region of ideal MHD (with a large magnetic Reynolds number $\mathcal{R}_m = u_0 l_0/\nu_m \gg 1$), a reconnection site is a region of converging plasma under the effect of the Lorenz force. Indeed, a strong current is concentrated in sheet such that $\boldsymbol{J} \times \boldsymbol{B}$ squeezes the sheet to a small width δ characterizing the development of a intense dissipation by Joule effect, such that $\mathcal{R}_m = 1$. On the sheet $J \sim B_0/\delta\mu_0$ and a strong electric field develops in the transverse direction: $E = \eta J \sim \eta B_0/\delta\mu_0$. Along this sheet of length ℓ, often not well determined, a large potential variation is set up $\Delta V \sim u_0 B_0 \ell$ and particles reach energy: $\epsilon_{\max} \sim Zeu_0 B_0 \ell$.

Plasma microturbulence also produces an electric field with components parallel to the mean magnetic field $E_{\parallel}$, but random. They give rise to an energy diffusion: from $\dot{p}_{\parallel} = qE_{\parallel}$, one derives

$$\Gamma_{\parallel} \equiv \frac{\langle \delta p_{\parallel}^2 \rangle}{2\Delta t} = q^2 \int_0^{\infty} \langle E_{\parallel}(t) E_{\parallel}(t-\tau) \rangle d\tau \,. \tag{8.128}$$

Microturbulence also contributes to rapidly isotropizing the distribution functions. For instance, Langmuir turbulence (electrostatic waves of pulsation ω_{pe}) can reach a level of energy density comparable with thermal energy nT and display a very short coherence time ($\sim \omega_{\mathrm{pe}}^{-1}$), when the spectral bandwidth is large. The acceleration time is, therefore, very rapid:

$$\tau_{\mathrm{acc}} \equiv \frac{p^2}{2\Gamma_{\parallel}} \sim \frac{p^2}{\bar{p}^2} \omega_{\mathrm{pe}}^{-1} \,. \tag{8.129}$$

The acceleration process has a large efficiency but at low energy only. This is an interesting process for injecting particles in the cosmic ray population. By the way, the Fermi processes work with particles that already have enough energy in order to interact with magnetic disturbances (otherwise they just follow the field lines and are reflected from time to time, but does not undergo efficient scattering).

Magnetosonic waves can be considered acceleration agents through a kind of second order Fermi process. A priori in ideal MHD, there is no parallel component $E_{\parallel}$, except when one considers the generalized Ohm's law with an electronic pressure gradient. A parallel electric field component is generated for compensating the parallel pressure gradient developed by the magnetosonic wave: $qE_{\parallel} = T_{\mathrm{e}}(\nabla_{\parallel} n_e)/n_e$; this field is absorbed by Landau damping and then its energy is converted into particle energy. The acceleration time can be estimate as follows:

$$\tau_{\mathrm{acc}} \sim \omega_{\mathrm{ci}}^{-1} \frac{P_m}{P} \frac{p^2}{m_{\mathrm{i}} T} \,. \tag{8.130}$$

This process is also a good injector in the cosmic ray population. However, as in the example of the acceleration with Langmuir waves, this time becomes longer and longer with increasing energy, proportionally with the energy square, and Fermi processes, whose time scale grows more slowly with $\tau_{\mathrm{acc}} \propto \tau_{\mathrm{s}} \propto p^{2-\beta}$, become sooner or later dominant.

8.2.7 What about the Relativistic Regime of Fermi Acceleration?

The relativistic regime of Fermi acceleration ($\beta_* \simeq 1$) is an important topic in high-energy astrophysics because the results obtained in the nonrelativistic regime that we have presented in this chapter, indicate that the best efficiency is obtained in this regime. Moreover, the ultrarelativistic shocks produced in gamma ray bursts, for instance, are considered to be the source of intense high-energy radiation, probably the most intense in the Universe. Progress in understanding the physics of relativistic shocks and Fermi acceleration in them have recently been obtained; however, at the date of publication of this text book, important questions are still open. Some important points have already been addressed and will be summarized now.

- After the first cycle that increases the particle energy by a factor Γ_s^2 as measured upstream, or Γ_s as measured downstream, the next cycles lead to an average gain close to 2.
- The probability of return is significantly less than 1 and strongly depends on the pitch angle of the particle momentum at entrance from upstream to downstream. Its average value is a little larger than but close to 0.5.
- The spectrum index is $s = 2.2$–2.3 [167–171] and its estimate using Eq. (8.116) with a probability of return average over angles, as proposed in [164], is quite good.
- Distribution functions are highly anisotropic; gain and probability of return strongly depend on pitch angles, which makes theory more difficult, but not hopeless.

The particle scattering off of magnetic disturbances is an important issue. It has been shown that phase space is locked when magnetic turbulence has a large coherence length, as is usually the case with Kolmogorov turbulence, and then the Fermi process is not operating [172, 173]. The reason is that the suprathermal particles penetrate upstream over a length much shorter than the coherence length. In order to have an operating Fermi process, turbulence must have a short coherence length and a large intensity. This can be generated by a microinstability, such as a Weibel instability, triggered by reflected particles. The weakness of that physical situation is that, because particles have effective Larmor radii larger than the coherence length, they undergo a weak scattering with a scattering length that increases proportional to the square of their energy. For this reason very high energies for protons are difficult to achieve. Mildly relativistic shocks seem more promising to accelerate protons and nuclei to the ultrahigh-energy range from 10^{18} to a few 10^{20} eV.

9
The Kinetic–Fluid Duality

An interesting challenge in the description of collisionless plasmas consists of going as far as possible with the fluid description, analyzing what is missing, and using the microscopic description to correct the fluid model by using appropriate (empirical or not) closures. Finding the appropriate closure may not be too hard in simple toy problems, which we consider in the first section, but may become a much harder task in the case of a real plasma, like the solar wind, which is considered in the second section. The first section on toy models is a compact summary of the courses taught to PhD students which some of the authors have regularly done for about 10 years in the Paris–Meudon Observatory during the period 2003–2013.

9.1
Toy Models

In this section we consider toy models, that is, simplified setups allowing us to define easily the closure problems. The first two cases will be 1D and *ballistic*, that is, consider particles moving along the mean magnetic field without interactions, while the last case, considered more briefly, will deal with a plasma with the mean field perpendicular to the gradients, that is, a 2D case.

Let us consider the ballistic case first. The expression of the velocity distribution function at successive times, for a given initial condition, can be obtained either by using an algorithm written to solve the equations $x = vt + x_0$ and $v = v_0$ for N particles, or also by using a hybrid code in the limit of zero electron temperature. In the case of velocities purely along the Ox axis, this should lead to the same result. We will consider both models in the following. In the case of the hybrid model, we remark that in the limit of the cold electrons, the electric field indeed goes to zero. If the movement is purely one-dimensional (parallel to the mean magnetic field), then we will indeed consider the ballistic limit. Physically, electrons with zero temperature will remain glued onto ions, one per each ion. As a consequence, the electrons will play no role (they only ensure a zero charge everywhere),

Collisionless Plasmas in Astrophysics, First Edition. Gérard Belmont, Roland Grappin, Fabrice Mottez, Filippo Pantellini, and Guy Pelletier.

9.1.1 Small Amplitude Ballistic Fluctuations

In the first case considered, we use only the hybrid model. We start at time $t = 0$ with a plasma at equilibrium, uniform density $\rho = 1$, and temperature T_0. It has zero velocity, but for a perturbation localized in the center of the domain where it is $u = u_0 = 0.1$, small compared to the thermal velocity, more precisely, with a small Mach number $u_0/\sqrt{3T_0/m_i} \ll 1$. The evolution of the fluid moments, x-component of the velocity u_x, density, and temperature is shown in Figure 9.1. The initial velocity peak is seen to split in two secondary peaks, both positive, but with half the initial amplitude u_0, which propagate in opposite directions. The phase velocity is found to be comparable to the sound velocity used above to define the Mach number: $c_s = \sqrt{3T_0/m_i}$. There is a large spreading of the particles in each secondary peak, so that the peak values decrease rapidly (Figure 9.1a). At the same time, density and temperature fluctuations appear and propagate at comparable speed in both directions, with a positive amplitude for fluctuation propagating to the right, and a negative one for fluctuation propagating to the left. Apart from the observed strong damping, these properties are close to the properties of eigenmodes of sound waves found in a collisional fluid, or also the ion acoustic waves found in a collisionless fluid with a high electron temperature. In the present quasi-ballistic case, finding again such properties deserves some explanation.

These acoustic-like properties in our ballistic experiment are now explained with the help of Figure 9.2. Figure 9.2a, we show again the two velocity profiles at times $t = 0$ and $t = 48$. Figure 9.2c, we give a sketch of the distribution function $f(x, v)$ (with two isocontours indicating approximately the values $v = +c_s$ and $v = -c_s$) at the same times. At time $t = 0$, the positive fluid velocity bump is associated with a positive bump (or shift towards positive velocities) of both the upper and lower contours, localized in the center of the domain. Note that the amplitude of the shift, relative to the thermal width of the distribution, has been highly exaggerated for clarity. We also represented as a dashed line the mean particle velocity, that is, the fluid velocity u vs. the spatial coordinate x. Now, the parts of the contour level that are shifted respectively upward and downward (in other words, corresponding initially to the position of the velocity bump) will move in opposite directions. Each contour is indeed transported at the velocity of the nearby particles. Thus (as seen

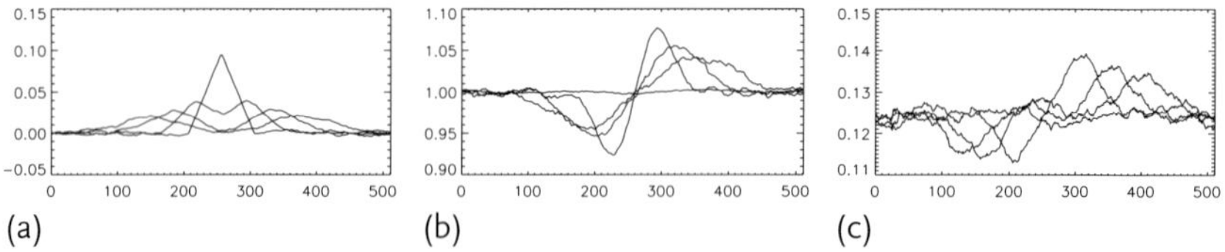

Figure 9.1 Ballistic evolution of a small velocity perturbation (with uniform density and temperature). Profiles of the first moments at times $t = 0, 16, 32, 48$. (a–c) velocity u_x, density ρ, ion temperature T. The initial velocity bump gives rise to fluctuations of the three moments propagating both rightward and leftward.

in Figure 9.2c, right), the upper contour propagates rightward, while the lower contour propagates leftward. As soon as the two bumps are well separated, the lower contour "bump" actually corresponds to an interval which has a deficit in particles with negative velocities, while the position of the upper contour bump corresponds to a location with an increase of particles with positive velocities. The result is the same: the fluid velocity is larger at both places. This explains the first mystery, that is, why a positive velocity bump propagates to the left, as well as to the right.

Simple visual inspection also indicates that the density is lower at the location of the left bump, and larger at the velocity of the right bump. This explains why the density, which is initially uniform, splits into a peak and a trough of equal amplitudes, which also propagate at the sound speed in both directions. Both peaks spread with time, reflecting the thermal velocity spread around the mean speed u_0.

As we see now, this spreading is due to a thermal conduction which is growing with time. To measure this conduction, we measure the heat flux q, which is shown in Figure 9.3 (solid lines). Visual comparison between Figure 9.3 and Figure 9.1c indicates that $q \propto -\partial_x T$. We thus test the classical collisional form for the heat flux:

$$q = -\kappa \partial_x T \tag{9.1}$$

with

$$\kappa = n v_{\text{th}}^2 \tau \tag{9.2}$$

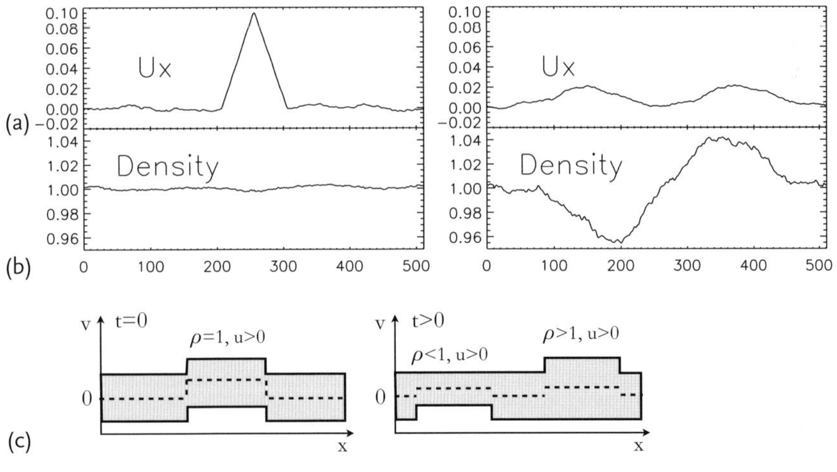

Figure 9.2 Ballistic evolution of a small velocity perturbation with respect to thermal ion velocity. (a,b) Hybrid simulation with negligible electron temperature. Velocity (a) and density profiles (b) at times $t = 0$ (left) and $t = 48$ (right). (c) Schematics of the velocity distribution function $f(x, v)$ (vdf) in (x, v) space (most particles being within the shaded area), with the velocity amplitude exaggerated compared to the vertical width (or thermal velocity) of the vdf. The (approximate) fluid velocity u profile is represented as a dashed line.

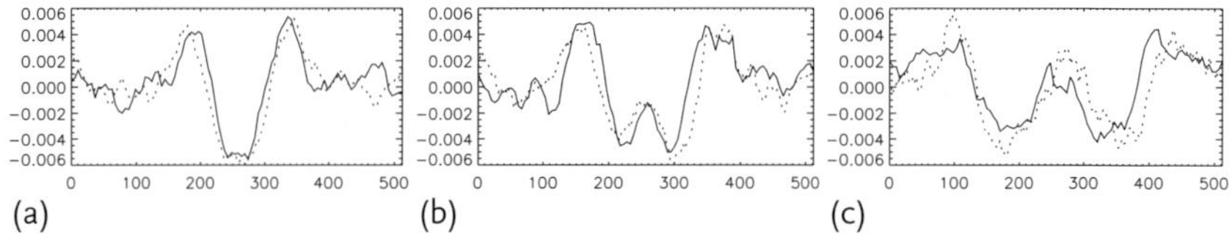

Figure 9.3 Ballistic evolution of a small velocity perturbation. Measured heat flux (solid line) and empirical closure (dotted line, see text, Eqs. (9.1)–(9.3)) at three times: $t = 8$, 24, and 48 (a–c).

where $v_{\text{th}}^2 = T/m_{\text{i}}$. The characteristic time τ is the collision time in a collisional plasma, which is not defined in a ballistic plasma. We actually find that a good fit is obtained when taking

$$\tau = 6.4\,t^{0.4}\,. \tag{9.3}$$

The reconstructed heat flux using this fit for the conductivity and the measured temperature profiles are shown in Figure 9.3 as dotted lines: both curves match reasonably well.

The conductivity κ contributes to the temperature variation (after linearizing to first order in the temperature fluctuation) as follows

$$\partial_t T = \kappa \partial_x^2 T\,. \tag{9.4}$$

In a collisional fluid with $\kappa = \text{const}$, this leads to the temperature characteristic length H_T increasing with time as in a diffusion process: $H_T \propto \sqrt{t}$. More generally, with $\kappa \propto \tau \propto t^{\alpha}$ we obtain $H_T \propto t^{(1+\alpha)/2}$. Here, since $\kappa \propto t^{0.4}$, we should have $H_T \propto t^{0.7}$, which is compatible with the increase of H_T with time visible in Figure 9.3.

When comparing the damping of the velocity, density, and temperature profiles, one sees that all profiles show the same damping rate, as well as the same increase of characteristic size with time. Hence, we have here an example of irreversible dissipation of coherent energy (convective) due to the sole effect of the heat flux. In this particular case, the heat flux leads to diffusion and damping of the gradients of all quantities, not only of the temperature/pressure gradients.

This example is restricted to the ballistic case. The case with finite β_{e} has been investigated by [174]. These authors find that the effective mean free path $\lambda = v_{\text{th}}\tau$ that intervenes in Eq. (9.2) scales as the effective temperature scale length H_T. This is clearly incompatible with the present evolution for which both quantities grow with time at different rates: $\lambda \propto \tau \propto t^{0.4}$ while $H_T \propto t^{0.7}$, as seen above.

Equation (9.1) is an approximate closure for the evolution considered here. But how general is it? Can we apply it to the evolution of any kind of small perturbations? The answer is no. Recall indeed that the system is reversible. This means that we can stop the flow at a any time t and run the flow "backward" by changing the sign of all particle velocities. As a result, the system will reverse its time evolution, adopting successively all previous states, but with opposite signs for the odd moments, in particular, the fluid velocity, and the conductive heat flux.

We know already that the system satisfies approximately the closure Eq. (9.1) during its normal evolution. Since the backward evolution has the same even moments (as T), but odd moments (as q) with opposite signs, the approximate closure satisfied during the backward evolution will be

$$q = \kappa \partial_x T \tag{9.5}$$

which is Eq. (9.1), but with a positive instead of a negative sign in front of the temperature gradient. This is the necessary condition to allow for the temperature, density, velocity profiles to follow the "antidiffusive" behavior which will lead back to the initial conditions at the end of the backward evolution.

Does this mean that closure relations such as Eq. (9.1) are useless? The flow evolution considered here, although reversible, evolves irreversibly, when viewed from a macroscopical level (in other words, considering only the first moments). This irreversible evolution is more or less well described by the approximate empirical closure of Eq. (9.1). So, while it is true that this closure can certainly not be universal, it might be applicable to a large number of small perturbations which share this macroscopic entropy increase.

An interesting comparison is in order. Consider a (purely imaginary) ballistic solar wind which would consist of a rotating source (located around the equator) emitting ballistic particles along radials, with the speed varying with longitude only. At a given longitude, all particles share the same velocity, so the temperature is zero at the emission location. However, at some distance from the source, fast particles would catch up and overtake slower particles emitted by other longitudes, so leading to a finite temperature, thus warmer flow. This entropy increase with distance might also be described by some closure, similar or not to Eq. (9.1), and, as in the example studied above, the closure would not be able to represent the accretion flow corresponding to the reversed "backward" wind. We will see in Section 9.2.7 some less naive examples of closures adapted to the quasi-collisionless evolution of the solar wind.

9.1.2
Large-Amplitude Ballistic Fluctuations

We consider now the case of large-amplitude fluctuations, again in the quasi-ballistic case. Again, we will be able to obtain an approximate closure for the third moment q, with the interesting property that the closure relation will now work as well during a backward evolution. However, this will not mean that the closure is applicable to any initial perturbation of a ballistic fluid. As the previous example, the following example is to be considered merely an exercise introducing the world of closures.

We consider initially a population of ions with uniform density $n = 1$, negligible temperature (and thus also negligible pressure P), with the velocity depending on position x as follows:

$$u(x) = U_0 \sin x \tag{9.6}$$

with $U_0 = 1.5$. The isocontours of the velocity distribution function in phase space (x, v) are shown in Figure 9.4, column 1, as well as the profiles of the first moments: from left to right: density n, velocity u, pressure P, and heat flux q. Three successive times are shown, from top to bottom: $t = 0, 2, 4$.

At time $t = 0$, as the plasma has a vanishing temperature, the distribution function appears as a ribbon in (x, v) space with vanishing vertical (velocity) width, or more precisely with a width very small compared to the global amplitude of the velocity variation along the Ox axis. At each position x, the vdf $f(x; v)$ versus v is obtained by making a vertical cut of the ribbon. The vdf is close to a Dirac distribution $\delta(v - U_0 \sin x)$ at each point x. As seen in Figure 9.4a, the fluid velocity profile $u(x)$ at time $t = 0$ is a almost a copy of the initial "ribbon". This simply expresses the fact that all particles share almost the same velocity $u(x)$ at any given point x.

As time evolves, the fast particles catch up to the slower ones moving in front of them in the same direction. Particles begin to accumulate at the center of the domain, so density grows there, and both the isocontours of the function $f(x, v)$ and the velocity curve $u(x)$ steepen in the central region, with the velocity curve remaining a mere copy of the envelope of the "ribbon" traced by the contours of $f(x, v)$. In hydrodynamics terms, one says that a shock is forming. During a first phase (say for $t < 2$), the deformation remains moderate enough, so that the vdf at

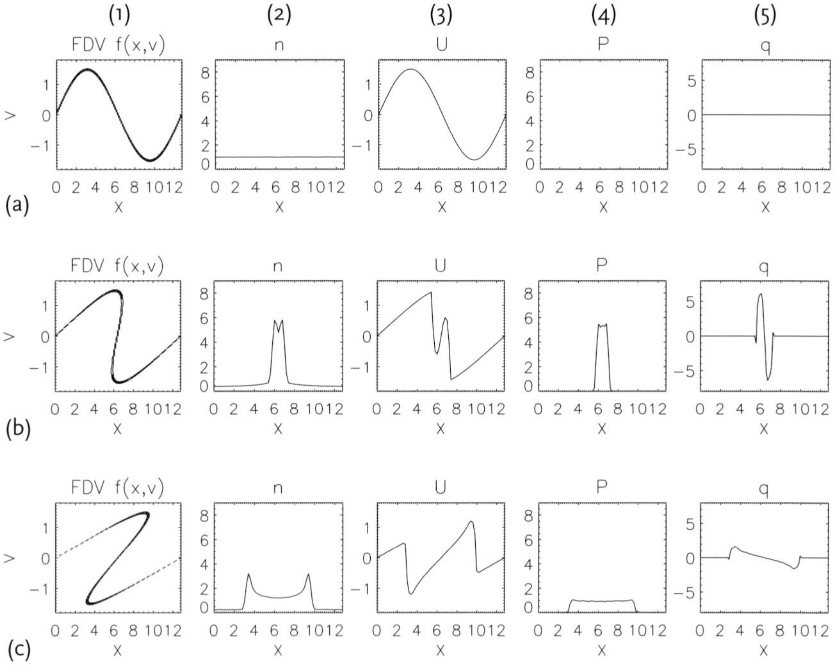

Figure 9.4 Ballistic evolution of a large-velocity perturbation with respect to ion thermal speed, using a hybrid model with cold ions and electrons. (a–c): times $t = 0, 2, 4$. From left to right: (1) isocontours of the velocity distribution function (vdf) $f(x, v_x)$ in (x, v_x) space; (2) density $n(x)$; (3) fluid velocity $u_x(x)$; (4) ion pressure P; (5) heat flux q.

each point shows a single peak, as at $t = 0$. The pressure, temperature, and third moment thus remain negligible everywhere, as the distribution $f(x, v)$ remains Dirac-like at each point x. The fluid equation reduces to

$$D_t u = 0 \,. \tag{9.7}$$

As time approaches $t = 2$, the ribbon becomes closer to a sawtooth, and the situation suddenly changes when a bit before $t = 2$ the derivative of the "ribbon" curve becomes infinite: at that time, a finite number of particles becomes concentrated in a tiny region around the center of the domain, so that the density there grows drastically, while the temperature (which scales as the square of the vertical "width" of the function $f(x_0, v_x)$ where x_0 is the position of the center of the domain) suddenly becomes of order unity. The pressure as well increases suddenly, which pushes the fluid away from the compressed region. Equation (9.7) becomes:

$$n m_\mathrm{i} D_t u = -\partial_x P \,. \tag{9.8}$$

This flow reversal due to the overpressure is quite visible in Figure 9.4b,c. The high pressure zone inflates as the Z-like part of the function $f(x, v)$ increases with time, corresponding to the mixed region where the slow particles coexist with the fast ones that have overtaken them. Note also the two density peaks at the boundary of the expanding region: this particle accumulation corresponds to the points where the velocity curve (or the "ribbon" function $f(x, v)$) has a vertical tangent (see Figure 9.4, columns 1 and 2).

Visual inspection of the heat flux curves (Figure 9.4, column 5) shows that they are zero outside the nonzero pressure region, but that inside the overpressure region, they follow well the profiles of $-u(x)$. A possible (dimensionally correct) closure for the conductive flux is indeed of the form

$$q = A P u \tag{9.9}$$

where A should thus be a negative factor. During the preshock phase of the hybrid calculation, the factor A is zero. During the very beginning of the expanding phase, the heat flux is not well fitted by Eq. (9.9). However, this fit becomes rapidly correct, the value $A = -3/2$ being acceptable during the main phase of expansion (see Figure 9.5).

We now examine how the solution evolves, depending on the model. In Figure 9.6 we show the evolution of the density, velocity, and pressure at times $t = 0, 2, 4$, using, from top to bottom, an "exact" solution (to be defined below), the hybrid code solution already given in Figure 9.4, a Navier–Stokes solution using the closure Eq. (9.9) for the conductive flux, and the standard Navier–Stokes solution with classical conductivity and viscosity.

The "exact" solution is obtained by throwing a certain number of particles in the interval of simulation, with a uniform spatial distribution. The temperature is exactly zero, and each particle is followed at machine precision. However, the resulting solutions (vdf, density, velocity, pressure, conductive flux) depend on the

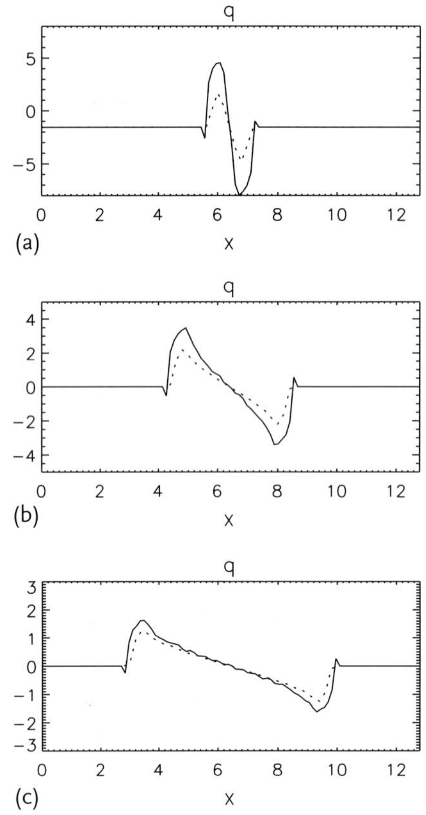

Figure 9.5 Ballistic evolution of a large-velocity perturbation. Measured heat flux (solid line) and empirical closure (dotted line, see Eq. (9.9) with $A = -1.5$) at three times: $t = 2, 3, 4$ (a–c).

number of particles, because the fluid moments (the density in particular) are made partly of Dirac distributions. The solution shown uses 10^3 particles, which is a number large enough to show quasi-Dirac peaks in the density profiles. The hybrid code solution given below in the second row shows artificially smoothed profiles, although the number of particles is much larger ($N = 10^5$). This is mainly due to the fact that building the moments involve binning of the distribution using a small number of bins, here $N = 128$. As a consequence, there is a strong smoothing of the velocity and density peaks and also as well of the associated gradients in the hybrid code results. Note that the ion temperature is very small, with $\beta_\text{i} = 4 \times 10^{-4}$ in the hybrid code calculation.

The Navier–Stokes models which are shown in Figure 9.6c,d share a common modification which is necessary to describe the evolution of the ballistic fluid at times larger than the shock formation. Indeed, if the absence of any modification, a Navier–Stokes code is unable to describe any evolution after the shock formation time $t \simeq 2$. This is because, as mentioned previously, a Dirac-like density profile forms that the fluid code is not able to reproduce. A way out consists of switching

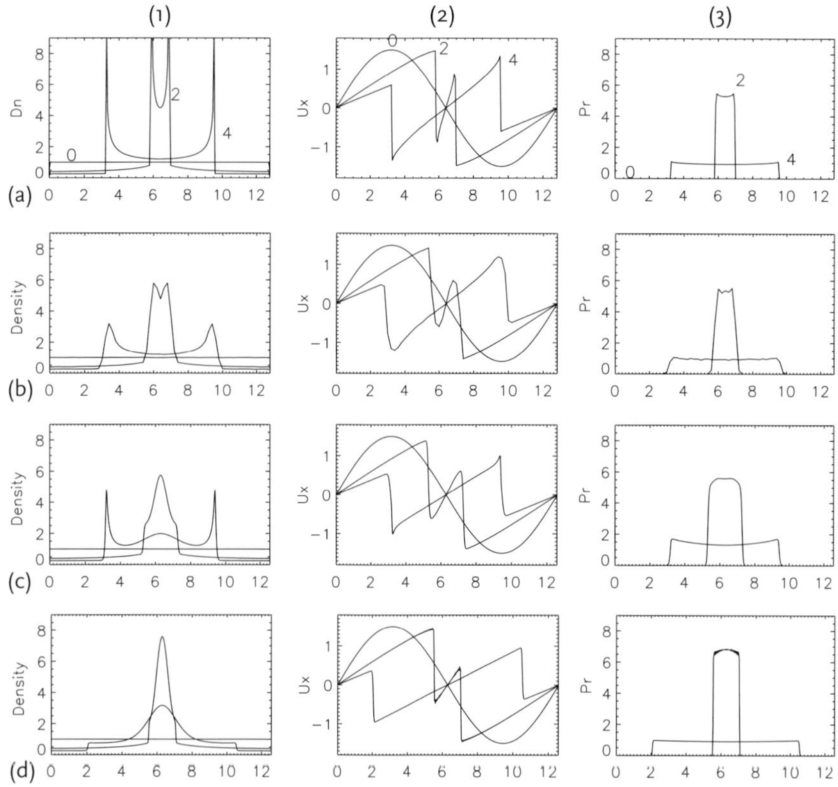

Figure 9.6 Evolution of the fluid quantities (columns 1–3: density, velocity, pressure vs. coordinate x) using four different models of a ballistic plasma. Time (0, 2, 4) is indicated by a number. From top to bottom (a) quasi-exact model with zero temperature; (b) hybrid model with low temperature: $\beta_i = 4 \times 10^{-4}$, $\beta_e = 2 \times 10^{-3}$; (c) Navier–Stokes model with the closure Eq. (9.9) for the conductive flux (see text). (d) quasi-standard Navier–Stokes model.

on a finite temperature T_{max} at the time of the shock formation. In practice, we have chosen to switch the temperature on when the density exceeds a given threshold value ρ_{th}, here $\rho_{th} = 10$. This temperature is computed from the "ribbon" at the shock time, that is, is the temperature associated with a flat vdf with width $2\,U_0$:

$$T_{max}/m_i = U_0^2/3. \tag{9.10}$$

Figure 9.6c thus gives the evolution of the Navier–Stokes model using the above procedure to switch the temperature on, and also using the empirical closure of Eq. (9.9). Note that classical diffusive terms, that is, a conductive term $(\gamma - 1)\kappa \partial_x^2 T$, a viscous term, and also an additional artificial term to damp the density gradients are used, in order to cope with the strong gradients occurring at the shock time. When comparing with the exact and hybrid model results (Figure 9.6a,b), we find that the beginning of the expanding phase (Figure 9.6b, $t = 2$) is not well reproduced: the two peaks are absent from the density structure, and the size of the rebound structure is exaggerated (better seen in the velocity and pressure figures). However, in the subsequent phase ($t = 4$), the density peaks are present (and closer to the exact result than the hybrid code result, thanks to the reduced smoothing due to the higher resolution $N = 512$ in the Navier–Stokes code), and also the size of the high pressure expanding zone is now correct.

Figure 9.6d shows the evolution of the classical Navier–Stokes model. While the early phase is curiously closer to the exact evolution than the modified Navier–Stokes model (apparently the closure Eq. (9.9) is counterproductive there), in the subsequent phase ($t = 4$), both the size of the expanding region and the density profile are incorrect.

We stress that we have used in the Navier–Stokes models the correct adiabatic exponent $\gamma = 3$ for a one-dimensional plasma (1D polarization as well as dependence on the sole x coordinate). If we use instead the value $\gamma = 5/3$ appropriate for a 3D plasma, then the size of the expanding region increases at a very different velocity, which is expected, since the sound speed depends on the value of γ.

9.1.3
Quasi-Fluid Behavior of a Collisionless Plasma: Launching a 2D Plasma Bullet

The motivation for this numerical experiment has to do with the question of the plasma transport through the magnetopause, a magnetic surface which is in principle an impermeable boundary between the Earth and the solar wind, but in practice allows a substantial intermittent flux of particles to pass the boundary. To investigate the energetics of the process of "making a hole in the magnetopause", several authors have done numerical experiments in which a plasma bullet was launched through a model of the terrestrial magnetic envelope, using successively an hybrid code [175] and an MHD code [176].

The chronology of the experiment is thus to first follow the bullet in its trajectory up to the encounter with the magnetic wall, then to follow the bullet as it deforms the magnetic field line of the boundary (so working against the magnetic elastic forces) up to the reconnection of the field lines behind the bullet, which finally allows the bullet to be considered a part of the rest of the plasma on the other side of the magnetopause.

In both the MHD and hybrid experiments, the conclusion was that the kinetic energy of the plasma bullet had to be 50 times larger than necessary to deform the magnetic field up to the point where it could reconnect.

The explanation of this energy loss is actually quite general, and works in various contexts. The kinetic energy which is initially given to the limited plasma volume is immediately largely lost, because it is given to pressure waves (magnetosonic or sound, depending on the context), with a reduced amount remaining within the initial plasma volume (in other words, the bullet). The pressure waves immediately accelerate the plasma around the "bullet", leading to a pair of vortices. Last, the 2D vortices lose their energy rapidly, typically in one turnover time (if in the strong coupling regime), whatever the dissipation mechanism is. Even if the bullet is initially close to the plasma wall, so that the vortices don't have time to lose a substantial part of their energy when they reach the magnetic "wall", the simple initial energy lost in launching the sound waves and vortices around the bullet is enough to explain the dramatic inefficiency of the process. A similar scenario explains the low efficiency of many processes like rowing or mixing, say, hot and cold water in a bathtub; it intervenes in evaluating the limit speed of objects pushed in a fluid either by their own energy source (swimmer), by gravity, buoyancy, and so on.

Figure 9.7 illustrates the scenario in a simple configuration without magnetic "wall", and without magnetic field at all: the plain hydrodynamic equations are integrated, in the isothermal approximation $\gamma = 1$. The isocontours in Figure 9.7b mark the advance of the pressure wavefront: one sees that the wavefront coincides with the boundary of the region filled with vortices: this is a visual proof that the vortices are indeed accelerated by the advancing pressure wavefront. No big change is produced if we change the γ or if we add a magnetic field *perpendicular* to the plane: both changes lead to a plain increase of the phase speed of the pressure waves. The reason why adding a mean field perpendicular to the plane changes only that lies in the fact that the elastic magnetic forces can play no role in that case.

No big change is to be seen either when solving the same problem by using the hybrid code with a mean field perpendicular to the plane of the simulation, as is seen in Figure 9.8. Indeed, the gyration of particles around the mean forbids any important deviations from the initial Maxwellian distribution of velocities. It would be interesting to undergo the same comparison in the full 3D case. Indeed, in this case, whatever the orientation of the mean field with respect to the initial momentum of the plasma bullet, the magnetic field effect is no longer restricted to a plain contribution to the global pressure, and in this case the absence or presence of collisions plays an important role in the dynamics, as we know.

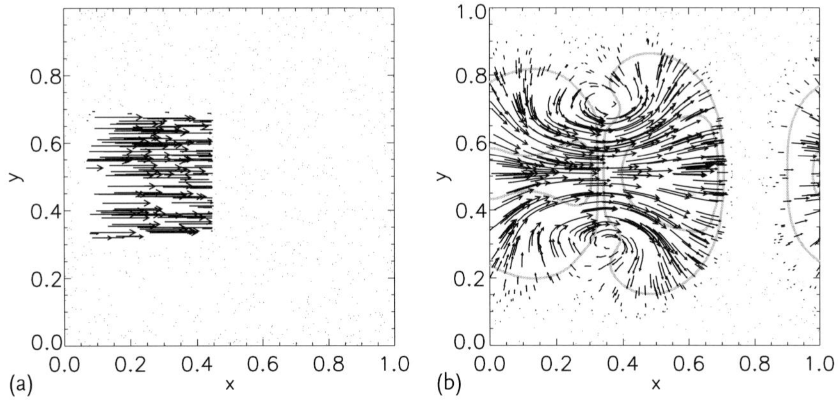

Figure 9.7 Launching a square bullet of plasma with Mach number 0.3 (isothermal flow, that is, $\gamma = 1$). Compressible 2D hydrodynamics simulation. (a) Starting configuration (velocity flow). (b) Configuration after a short time, velocity flows with isocontours of density. Boundary conditions are periodic.

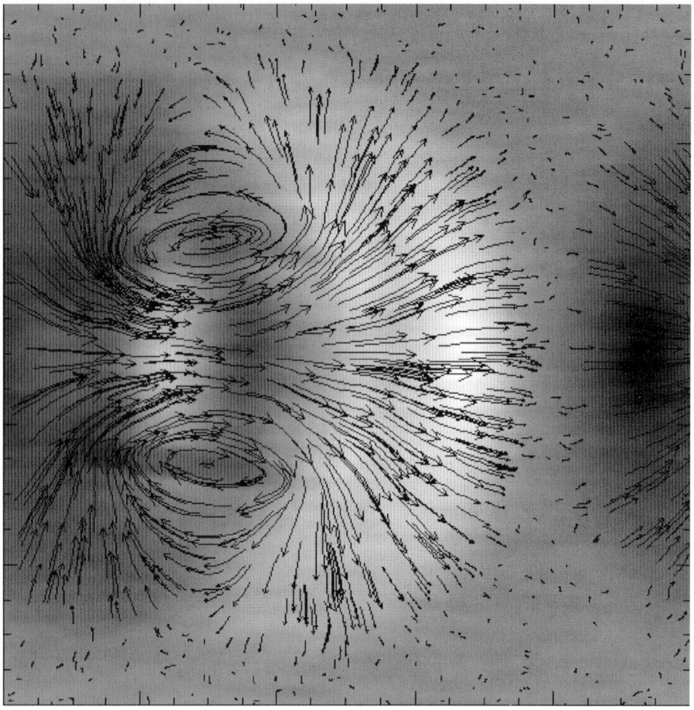

Figure 9.8 Launching a square bullet of plasma with Mach number 0.3 (hybrid code), with mean field perpendicular to the plane of the simulation. The same parameters as in the fluid simulation are shown in Figure 9.7. Velocity flow with density. Boundary conditions are periodic.

9.2 Solar and Stellar Wind Expansion

9.2.1 A Simple Noncollisional Wind

It is tempting to generate a stationary wind by injecting noninteracting particles at the level of the stellar surface or somewhat above the surface, at a height where the density of the fluid has reached such a low level that the collisional mean free path and the characteristic scales of variations of the density or the temperature are of the same order. Let us denote r_0 as the radial distance from the center of the star where the mean free path equals the scale of variation of the density, which is often the fastest varying macroscopic quantity. Lest us make the drastic approximation that the fluid is completely collisionless for $r > r_0$ and strongly collisional for $r < r_0$. This is the so-called exospheric assumption and r_0 is called the exobase. Let us further assume that the Universe is empty, that is, there are no particles flowing towards the star at infinity, or, stated otherwise, all particles in the system originate from r_0 where we take the freedom to specify the velocity distribution function. Given that we assume that collisions dominate for $r < r_0$, we may chose a Maxwellian velocity distribution function at r_0 except that the Maxwellian must be truncated somewhere on the side of negative velocities to account for the just mentioned no inflow condition at infinity. If gravity is the only force acting on the particles and assuming purely radial velocities (no angular momentum), the truncation must occur at $v = -v_{esc}$ as only particles with velocities smaller than the escape velocity $v_{esc} = (2G\mathcal{M}/m r_0)^{1/2}$ will ultimately come back to r_0. This will fill the negative part of the distribution symmetric of the positive part (see figure), up to the escape velocity $-v_{esc}$ (see Figure 9.9). So finally the distribution function $f(r_0, v)$ imposed at the surface is

$$f_0(v) = n_0(\pi v_0)^{-1/2} \exp\left(-v^2/v_0^2\right) v > -v_{esc} \tag{9.11}$$

$$f_0(v) = 0 v < -v_{esc}. \tag{9.12}$$

The mass flux (mass per unit time and unit surface) at the surface $r = r_0$ is

$$nv = \int_{-v_{esc}}^{\infty} v\, f_0(v) dv. \tag{9.13}$$

The mass flux $n v r^2$ is invariant along the r axis. At higher altitudes, the distribution function should lose a growing part of its central population with too low velocity to reach this height. This corresponds to a plain shift of the wings towards the center of the distribution. The resulting variation with height of the first two moments are given in Figure 9.10.

The temperature and heat flux profiles for the model are shown in Figure 9.11.

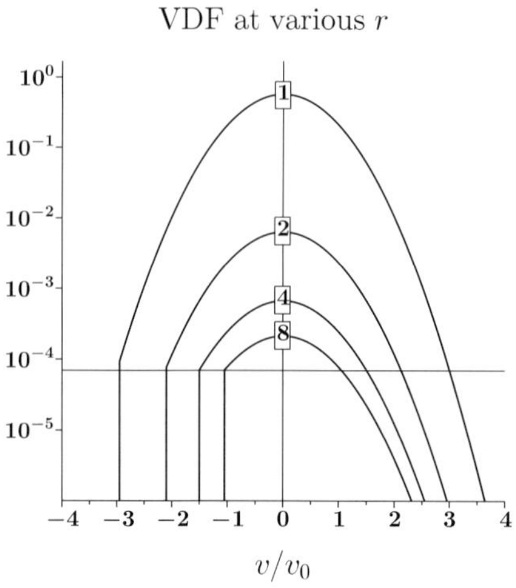

Figure 9.9 Velocity distribution function at various distances from the star center. The fraction of particles with negative velocities decreases with distance and vanishes at $r = \infty$. Velocities are normalized by the thermal velocity $v_0 = (2T_0/m)^{1/2}$, where T_0 is the temperature of the untruncated Maxwellian distribution.

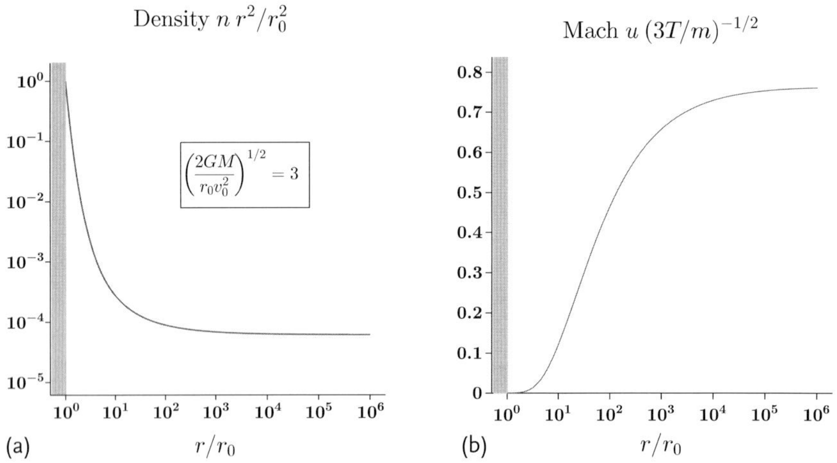

Figure 9.10 Density and velocity profiles for the one species collisionless stellar wind model with all particles having only radial velocities. Note that the velocity remains subsonic at all distances.

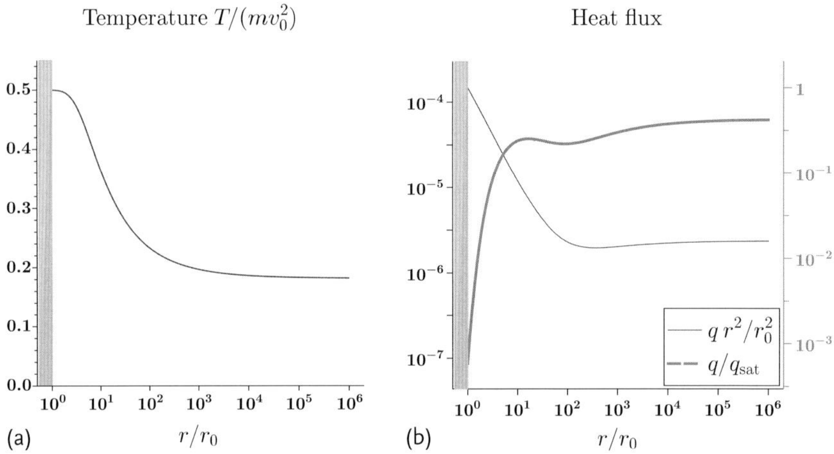

Figure 9.11 (a) Temperature and heat flux profiles for the one-species collisionless stellar wind model with all particles having only radial velocities. Note the monotonically decreasing temperature profile. (b) The dashed profile with the dimensionless flux q/q_{sat}, where $q_{\mathrm{sat}} \equiv 0.5 n T v_0$, corresponds to the right-hand axis and q to the left-hand axis.

9.2.2
More Sophisticated Noncollisional Wind Models

The model presented in Section 9.2.1 is unable to produce a supersonic wind and is clearly not, at least in its simplest form, a pertinent paradigm for the solar wind. The first, obvious, shortcoming of the model is that only one species of particles is considered if the solar wind was made of neutral gas particles, whereas the real wind is a fully ionized plasma made of ions (mainly protons and alpha particles) and electrons. It may appear intriguing that the one-species approximation allows for a transonic wind in the fluid approximation (see Section 7.1.6) whereas a transonic wind is impossible in the exospheric approximation. The paradox is particularly striking if one considers that the solar wind is a rather collisionless medium which one expects to more closely follow the behavior of a collisionless model rather than the behavior of a fluid model, in principle, based on a high collisionality assumption. In the 1960s, J. Chamberlain proposed a model similar to the one of the preceding section for a population of ions (for simplicity we may assume protons only) and a population of electrons. If electrons and protons were allowed to move independently of each other and if one assumes equal temperatures in the corona, the asymptotic bulk velocity of the electrons would be larger than the asymptotic bulk velocity of the protons by a factor $(m_p/m_e)^{1/2} = 42.85$. Measured in terms of the proton sound speed $(3T_0/m_p)^{1/2}$ the electron fluid is, therefore, strongly supersonic opening the possibility, at least in principle, of a supersonic wind model in the exospheric approximation. Of course, electrons and protons are charged point particles, which interact via the long distance Coulomb field. Under such circumstances, it is no longer possible to consider gravitation as the only force in addition to the pressure gradient force $-\partial P/\partial r$. An electric field must exist to prevent, that most of the electrons, on average 42.85 times faster then the protons, to escape with no difficulty from the solar gravitational field while only an extremely small fraction of the protons are sufficiently fast to do so. Let us compute the strength of the electric field required for the plasma in a central gravitational field to remain neutral.

9.2.3 Charge Neutralizing Field for a Plasma in a Gravitational Field

The easiest way to compute the electric field required to ensure quasi-neutrality in a plasma plunged in an external gravitational field $E(r)$ is to first write two separate stationary momentum equations for both electrons and protons. In the spherical approximation, retaining only pressure, gravitational and electric forces, the equations can be written as

$$m_e u_e \frac{\partial u_e}{\partial r} = -\frac{1}{n_e}\frac{\partial P_e}{\partial r} + g(r)m_e - eE(r) \tag{9.14}$$

$$m_p u_p \frac{\partial u_p}{\partial r} = -\frac{1}{n_p}\frac{\partial P_p}{\partial r} + g(r)m_p + eE(r) \tag{9.15}$$

where $g(r) = -G\mathcal{M}_*/r^2$ is the gravitational acceleration. As we shall see later, the quasi-neutrality condition $n_e = n_p$ holds to a very good approximation in the solar wind. Thus, assuming $n \equiv n_e = n_p$, equal proton and electron temperatures $T \equiv T_e = T_p$ (a rather crude approximation as we will see), a perfect gas equation of state $P_j = nT$ for $j = \text{e, p}$, and equal bulk velocities $u \equiv u_e = u_p$ so that there is no net current in the system, we can eliminate the pressure term by subtracting Eq. (9.14) from Eq. (9.15):

$$(m_\mathrm{p} - m_\mathrm{e})u\frac{\partial u}{\partial r} = (m_\mathrm{p} - m_\mathrm{e})g(r) + 2eE(r) \,. \tag{9.16}$$

If gravity at r_0 is strong enough to confine most of the plasma near r_0, that is, if the gravitational energy exceeds the thermal energy $G\mathcal{M}(m_\mathrm{p} - m_\mathrm{e})/(r_0 T_0) \gg 1$ we expect the Mach number of the flow to be $\ll 1$ at the exobase r_0. The right-hand side of Eq. (9.16) can then be neglected with respect to the gravitational term which must be balanced by the electrostatic field term alone. The resulting neutralizing electric field, also called the Pannekoek–Rosseland field, reads

$$E_\mathrm{PR} = -\frac{(m_\mathrm{p} - m_\mathrm{e})}{2e}g(r) \simeq \frac{m_\mathrm{p}}{2e}\frac{G\mathcal{M}_*}{r^2} \,. \tag{9.17}$$

This field pushes the electrons inwards towards the star's center and the protons outwards, away from the star so that the total (electric + gravitational) force on a proton is exactly the same as the total force on an electron. In 1960 Chamberlain used this expression for the electric field in order to compute a collisionless model for the solar wind. Chamberlain came to the conclusion that the solar wind must be subsonic even at large distances, thus contradicting Parker's proposition of a transonic solution, later confirmed by *in situ* observations. The failure of Chamberlain's exospheric model to produce a transonic solution was soon recognized to be because the Pannekoek–Rosseland field is inadequate at large distances where the small Mach number assumption underlying the derivation of Eq. (9.17) fails. Let us relax the zero velocity assumption $u = 0$ underlying the derivation of Eq. (9.17), but not the assumptions of equal density and temperature for both proton and electrons. Using Eqs. (9.14) and (9.15) to eliminate $u\partial u/\partial r$ and neglecting terms proportional to the small parameter $m_\mathrm{e}/m_\mathrm{p} \ll 1$ leads to an expression for the electric field more general than Eq. (9.17)

$$E(r) \simeq -\frac{P'}{en} \tag{9.18}$$

where the prime sign $'$ stands for $\partial/\partial r$. This equation provides an implicit expression for the electric field profile which must approach the Pannekoek–Rosseland field E_PR in the hydrostatic limit. However, for nonzero fluid velocities the field Eq. (9.18) decreases less rapidly than r^{-2}. For example, in the isothermal case $T = T_0$, the pressure gradient transforms to $P' = T_0 n'$. Defining the Mach number as $\mathcal{M} = u/(2T_0)^{1/2}$ the equation for the proton motion Eq. (9.15) can be written as

$$\mathcal{M}\mathcal{M}' = -\frac{n'}{n} - \frac{G\mathcal{M}_*}{c_1^2 r^2} \tag{9.19}$$

where $c_1^2 = 2T_0/m_\mathrm{p}$. Equation (9.19) illustrates the competing effect of electric field and gravity on the acceleration of the protons, the first term on the right-hand side

$-n'/n = eE/T > 0$ being proportional to the outwards pointing electric field E whereas the second term is the inwards pointing gravitational acceleration. On the other hand, the continuity equation $(\mathcal{M}nr^2)' = 0$ implies

$$\frac{\mathcal{M}'}{\mathcal{M}} + \frac{n'}{n} + \frac{2}{r} = 0 \tag{9.20}$$

and the equation of motion for the proton reduces to the familiar equation

$$\left(\mathcal{M} - \frac{1}{\mathcal{M}}\right)\mathcal{M}' = \frac{2}{r} - \frac{2r_s}{r^2} \tag{9.21}$$

where $r_s \equiv G\mathcal{M}_*/2c_1^2$ is the sonic radius. It is instructive to recast the electric field from Eq. (9.18) for the isothermal wind using the continuity equation (9.20)

$$E = \frac{T}{e}\left(\frac{2}{r} + \frac{\mathcal{M}'}{\mathcal{M}}\right). \tag{9.22}$$

which illustrates that in spherical geometry the electric field results both from expansion (first term on the right-hand side of Eq. (9.22)) and acceleration (second term on the right-hand side of Eq. (9.22)). The static fluid approximation underlying the derivation of the Pannekoek–Rosseland field Eq. (9.17) suggests that the latter is a valid approximation in the low Mach number regime $\mathcal{M} \ll 1$ where the left-hand side of Eq. (9.21) may be approximated as $-\mathcal{M}'/\mathcal{M}$, in which case Eq. (9.21), again using the continuity equation (9.20), transforms to $n'/n \simeq -2r_s/r^2$ and the electric field, as expected, to the Pannekoek–Rosseland field

$$E \simeq 2\frac{r_s}{r^2}\frac{T}{e} = E_{PR} . \tag{9.23}$$

We note in passing that from Eq. (9.21) the plasma obeys the same equations as the equations for a neutral gas with mass $m = m_p/2$, that is, $m = (m_p + m_e)/2 \simeq m_p/2$. We do already know that the transonic solution to Eq. (9.21) is an attractor. The asymptotic Mach number can be deduced from the above equation to grow with distance as $\mathcal{M}\mathcal{M}' \approx 2/r$, that is $\mathcal{M}^2 \propto \ln r$. Thus, at large distances $r \to \infty$ the continuity equation (9.20) reduces to $n'/n = -2/r$ corresponding to a density variation $n(r) \propto r^{-2}$. We can then integrate Eq. (9.18) in the isothermal regime and for $r \to \infty$:

$$E(r) = \frac{2}{r}\frac{T_0}{e} . \tag{9.24}$$

Even though Eq. (9.24) has been obtained at the cost of some approximations not representative of the real solar wind (isothermal approximation, equal proton and electron temperatures, isotropic pressure, and so on), it shows that as a consequence of the plasma moving at close to sonic or supersonic velocities, the electric field decreases less steeply than suggested by the static approximation (9.17). This is the main reason for the wind obtained by Chamberlain in his pioneering exospheric model to remain subsonic. Here we see that the electric field, implicitly underlying Parker's fluid model of Section 7.1.6, goes as r^{-2} in the subsonic region where the Mach number is small $\mathcal{M} \ll 1$ but as r^{-1} beyond the sonic point, where $\mathcal{M} \gg 1$.

9.2.4
Qualitative Radial Profile of the Total Proton Potential

The fact that the electric field decreases slower than the gravitational field at large distances has an important consequence on the total potential (gravitational + electrostatic) of a proton

$$\Psi_{\text{p}}(r) = -m_{\text{p}}\frac{G\mathcal{M}_*}{r} + e\phi(r) . \tag{9.25}$$

In the subsonic region where velocities are small compared to the sound speed, the electric field is given by the Pannekoek–Rosseland field Eq. (9.17) and the electrostatic potential ϕ, obtained by integrating $E_{\text{PR}} = -\partial\phi/\partial r$ gives a total potential

$$\Psi_{\text{p}}(r) = -\frac{1}{2}(m_{\text{p}} + m_{\text{e}})\frac{G\mathcal{M}_*}{r} . \tag{9.26}$$

This spatial dependence of the proton potential corresponding to a positive gradient $\Psi_{\text{p}}' > 0$. Equation (9.26) is expected to provide a good approximation of the real electric field as long as the basic assumptions of small velocities and approximately constant temperature hold. Probably a fair approximation at the base of the solar corona. The electric potential at large distances, where the flow is close to sonic or supersonic, the temperature cannot be assumed to be constant and the expression (9.24), despite being a better guess than the hydrostatic approximation, is certainly an overestimate of the real electric field in this region. An obvious problem with Eq. (9.24) is that it predicts a diverging potential $\Psi_{\text{p}}(r \to \infty) \propto -2T_0 \ln(r)$ corresponding to an endless, unrealistic, acceleration of the protons. The reason for the paradox obviously stems from the isothermal assumption which implicitly implies that energy is injected everywhere in the system to counterbalance the natural cooling of the wind as it expands. The amount of injected energy is so high that it not only allows for the temperature to keep spatially constant, but also to accelerate the wind to infinite velocity. We shall generalize the expression of the electric field at large distances for generic variations of the temperature with distance later in this section. However, the expression for the isothermal case Eq. (9.24) does already anticipate the main point: the proton potential is not a monotonically growing function of distance as the electron potential. Indeed, as already stated, Eq. (9.24) implies that $\Psi_{\text{p}}' < 0$ for $r \to \infty$. Since the gradient of Ψ_{p} must be positive near the base of the corona r_0 we conclude that the proton potential must have a maximum r_{Ψ} somewhere around the sonic point as illustrated in Figure 9.12.

The existence of a maximum for the proton potential has been inferred using a two-fluid (one for the electrons another for the protons). If we adopt the exospheric point of view, qualitatively illustrated in Figure 9.12, we conclude that if a maximum in the proton potential is postulated it must be located at a point where the wind velocity is of the order of the local sound speed $c_1 = (2T/m_{\text{p}})^{1/2}$. We note that even though one may expect from the above qualitative arguments that the one-fluid sonic radius r_{s} from Eq. (7.51) and the position r_{Ψ} of the maximum of the proton potential to be equal up to a order unity multiplicative constant the relative

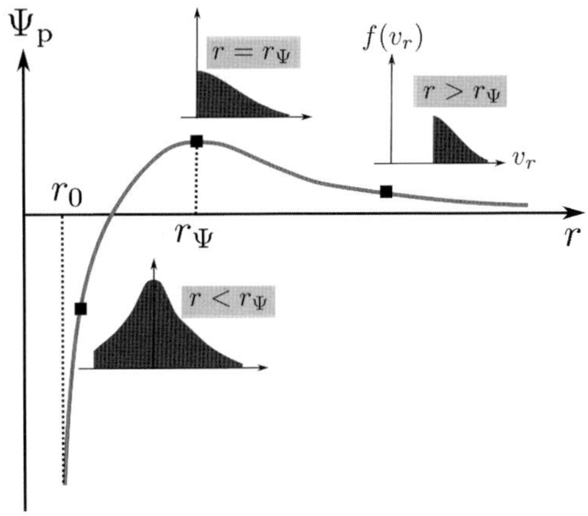

Figure 9.12 Characteristic profile of the proton total potential for a transonic wind. Also plotted are the proton's radial velocity distribution functions at three different locations: below, at, and above the position of the proton potential maximum r_Ψ. Protons are injected at $r = r_0$ with only radial and Gaussian distributed velocities. Note that at r_Ψ the velocity distribution is a truncated Maxwellian with no particles for $v_r < 0$ a fluid velocity $u = (2T_0/\pi m_p)^{1/2}$, a temperature $T = T_0(1 - 2/\pi) = 0.36T_0$ and an order unity Mach number $\mathcal{M} = 0.94$.

position may not be easily established in case of a weakly collisional plasma as in the case of the solar wind. The obvious reason is that the lack of collisions implies a decoupling of the electron and ion temperatures. Now, on one side, the position of the one-fluid sonic point depends on the variation of the polytropic sound speed $c_\Gamma = (\Gamma T/m)^{1/2}$ where $m = (m_p + m_e)/2$ and where T is the temperature of both electrons and protons which are supposed to be equal in all parts of the system. On the other side, the maximum of the proton potential r_Ψ is characterized by the requirements for a local positive maximum, that is $\Psi_p > 0$, $\Psi_p' = 0$ and $\Psi_p'' = 0$. The potential Ψ_p being a function of the electron pressure gradient as expressed by Eq. (9.18) the position of the maximum is sensitive in the first instance to the electron temperature profile and only indirectly to the proton fluid. The existence of a local maximum for the proton potential Ψ_p was first suggested by Jockers (1970). A detailed discussion on the maximum of the proton potential Ψ_p and its relation to the sonic point was published by Scudder (1996).

9.2.4.1 **Asymptotic Electric Field for a Polytropic Temperature Profile**

In the previous section it was established that in the isothermal two-fluid approximation the electric field must decrease asymptotically as r^{-1} which then implies that the electrostatic potential diverges as $\ln(r)$, leading to a never ending growth of the wind speed as $r \to \infty$. This unpleasant feature is a consequence of the computationally convenient, but drastic assumption of an isothermal wind which, according to Eq. (7.46) requires an infinite energy flux to flow through the plasma. Indeed, in the real solar wind the temperature is observed to decrease with distance from 10^6 K in the corona to some 10^5 K near earth orbit. We may, therefore, relax the isothermal assumption by assuming that the pressure P and the density n are related by a polytropic relation $P \propto n^{\Gamma}$. The electric field is still implicitly determined by the pressure gradient of the electrons according to Eq. (9.18). The equations describing the expansion are then given by the polytropic wind equation (7.47) and the corresponding continuity equation (7.50) by setting the mass of the "wind particles" to the mean of the proton and electron mass $m = m_p/2$.

Using the polytropic relation $P/P_0 \propto (n/n_0)^{\Gamma}$ where P_0 and n_0 are the pressure and density at some arbitrary reference radius r_0, the electric field is easily computed from Eq. (9.18)

$$e E = -\frac{\Gamma P_0}{n_0} \frac{n'}{n} = \Gamma T_0 \frac{n'}{n} . \tag{9.27}$$

Assuming a radially decreasing density $n = n_0 (r/r_0)^{-\beta}$, with $\beta > 0$ the electric field can than be written as

$$e E = \beta \frac{\Gamma P_0}{n_0} \left(\frac{r}{r_0} \right)^{\beta(1-\Gamma)-1} . \tag{9.28}$$

For the potential $\phi = -\int_a^{\infty} E dr$ not to diverge for $r \to \infty$ and the fluid velocity not to grow indefinitely one has to ask for $\beta(1-\Gamma) < 1$, or, equivalently $\Gamma > 1$. We do indeed already know that the case $\Gamma = 1$ corresponds to a slow but endless logarithmic growth of the fluid velocity for large r.

9.2.5
Charge Neutralizing Electric Field and Dreicer Field

The electric field strength may be best appreciated when compared to the Dreicer field E_D, a common yardstick in weakly collisional plasmas

$$e E_D \equiv \frac{T}{\lambda} \tag{9.29}$$

where λ is the mean free path for a thermal electron. As density is generally the fastest changing macroscopic fluid quantity in the solar wind we may use the density based Knudsen number

$$K_n = \lambda \left| \frac{n'}{n} \right| \tag{9.30}$$

in order to rewrite the electric field E given by Eq. (9.27) in terms of the Dreicer field

$$\frac{E}{E_D} = \Gamma K_n . \tag{9.31}$$

This equation immediately shows that for standard values of Γ in the allowed range $[1, 3/2]$ the electric field approaches the Dreicer field when the Knudsen number approaches unity. Estimates for the Sun, indicate that the Knudsen number is of order unity in the corona and above the corona, well before the wind has actually reached the sonic radius.

9.2.6
Electric Field Intensity at the Sonic Radius r_s

At the sonic radius r_s the electric field may be computed implicitly from Eq. (9.27) and the continuity equation

$$E(\Gamma, r_s) = \frac{2\Gamma}{\Gamma + 1} T(r_s) \left[\frac{2}{r_s} + \mathcal{M}'(r_s) \right] . \tag{9.32}$$

The derivative of the Mach number at the sonic point can be obtained through linearization of the wind equation for arbitrary polytropic index Γ (cf. Eq. (7.48)) near the sonic point. Thus, setting $r = r_s + \delta r$, $\mathcal{M} = \mathcal{M}(r_s) + \mathcal{M}'(r_s)\delta r$ and $c_\gamma = c_\gamma(r_s) + c'_\gamma(r_s)\delta r$ in Eq. (7.48) we obtain an implicit expression for the derivative of the Mach number at the sonic point

$$\left[\mathcal{M}'(r_s)\right]^2 = -\frac{n'}{n} \frac{1}{r_s} \left(\frac{\Gamma - 1}{2} \right) + \frac{1}{r_s^2} . \tag{9.33}$$

In the isothermal case the first term on the right-hand side vanishes and we recover the spatial derivative of the Mach number at the sonic point for the transonic wind solution $\mathcal{M}'(r_s) = r_s^{-1}$ and the associated electric field

$$E(r_s) = 3\frac{T}{e r_s} , \quad \text{isothermal} \quad \Gamma = 1 . \tag{9.34}$$

In the general case we have to use the continuity equation (7.50) to eliminate density from Eq. (9.33) and compute $\mathcal{M}'(r_s)$ explicitly as a function of Γ. We do not need to go into such details here as we are only interested in order of magnitude estimates. From the continuity equation (7.49) and Eq. (9.33) we do indeed deduce that

$$2\mathcal{M}' r_s = \alpha + [\alpha^2 + 4(1 + 2\alpha)]^{1/2} , \quad \text{with} \quad \alpha \equiv \frac{\Gamma - 1}{\Gamma + 1} \tag{9.35}$$

where the right-hand side ranges from 2 for $\Gamma = 1$ to a somewhat stepper gradient 2.575 for $\Gamma = 3/2$ leading to a coefficient 3.945 instead of a coefficient 3 for the isothermal expression for the electric field Eq. (9.34).

9.2.7
Effective Closure for the Solar Wind

The solar wind is a nearly collisionless plasma with a collisional mean free path on the order of the scale of variation of density or temperature. Under such conditions it would be a mere coincidence if the collisional heat conduction Eq. (5.73) was the appropriate description for the heat conduction in the solar wind. Indeed, measurements of the electron heat conduction in the solar wind indicate that the collisional expression does often provide an inappropriate description of the real conductive flux. In Figure 9.13 measures of the electron heat flux at earth orbit is compared

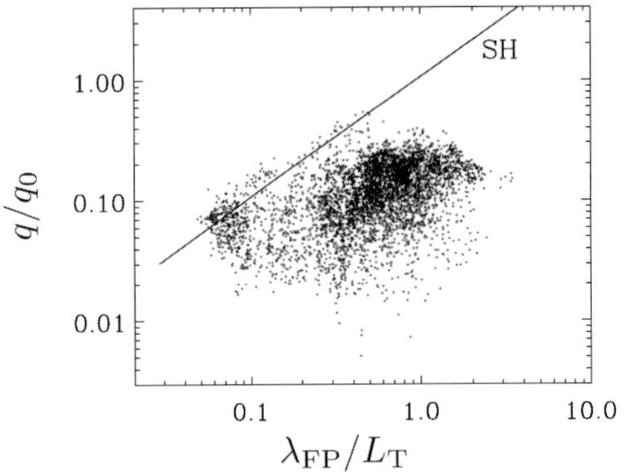

Figure 9.13 Electron heat flux in solar wind at 1 AU from the Sun as a function of the thermal Knudsen number λ_{FP}/L_T. The heat flux is normalized to the saturation flux $q_0 \equiv 3/2 n T v_0$. The SH denoted line is the collisional prediction (5.73). Measurements are from the Wind satellite (adapted from [177]).

with the prediction of the collisional expression (5.73) which can be conveniently written in terms of the thermal Knudsen number $K_T = \lambda/L_T$ and the saturation flux $q_0 \equiv 3/2 n T v_0$ as

$$\frac{q_{SH}}{q_0} = 1.07\, K_T\,. \tag{9.36}$$

This expression gives the absolute value of the conductive flux which always flows down the temperature gradient. Each experimental measure of the flux is plotted against an estimate of the thermal Knudsen number based on a Fokker–Planck estimate λ_{FP} of the electron mean free path. The figure shows that only for Knudsen numbers $K_T \lesssim 0.1$ the measured flux tends to be close to the collisional prediction SH. For larger values of K_T the collisional flux does overestimate the real flux. The departure of the measured flux from the collisional curve for K_T approaching unity is not surprising as the collisional result heavily rests on the small Knudsen number approximation $K_T \ll 1$. The reason the real flux is an underestimate (as opposed to an overestimate) of the collisional flux is sort of obvious since for the K_T approaching unity the expected SH flux approaches the saturation flux q_0 implying an extreme (not observed) departure of the electron velocity distribution function from a Maxwellian velocity distribution.

Figure 9.13 seems to indicate that the electron conduction in the solar wind may be satisfactorily described by the collisional expression provided that the thermal Knudsen number is of the order $K_T \approx 0.1$ or smaller. In general this is the case for the so called slow wind, which is the dominant wind in the ecliptic plane, but not for the fast wind which is generally dominant at high heliospheric latitudes above 30° from the equator. Some theoretical works (for example, [178]) suggest that the collisional expression for the heat conduction may sometimes fail for K_T as small as 10^{-3}. The invoked reason is that the heat flux is a relatively high-order moment of the distribution function which is very sensitive to the high velocity portions of the velocity distribution function. Now, since the mean free path of a particle increases as the fourth power of its velocity, a particle at three times the thermal velocity covers a distance 81 times larger than a thermal particle between successive collisions. Thus, in case of an order 0.1 Knudsen number only particles up to roughly two times the thermal velocity undergo a collision within a distance less than L_T.

Having established that the collisional approximation for the conductive heat flux is suitable at best for $K_T \lesssim 0.1$ and by noting that in the solar wind, and even in the solar corona, the plasma is generally characterized by $K_T \gtrsim 0.1$ we seek an alternative formulation for the conductive flux in order to define a better closure of the fluid equations than the collisional closure based on the collisional conductive flux Eq. (5.73).

9.2.7.1 **Closure Based on the Steep Temperature Gradient Approximation**

A steep temperature gradient is characterized by an order unity thermal Knudsen number. Accordingly, the temperature gradient in the solar wind must be viewed as steep, that is, the variation of the temperature δT over a distance equal to the mean free path λ is on the order of the temperature T itself. From a microscopic point of view one may imagine that a particle from a region at temperature T travels without collisions over a distance of order λ, it then makes a collision with a particle coming from a substantially colder region at temperature $T_0 \ll T$. On average the first high-energy particle transfers an energy δT to the low energy particle. Assuming an efficient transfer of energy βT from the most energetic particle to the other (with no momentum exchange as we are reasoning in the plasma rest frame) we can evaluate the flux as $q_{FS} \approx n\beta T v_0$. Including all order unity factors into β we then obtain that the conductive flux is proportional to the saturation flux q_0

$$\frac{q_{FS}}{q_0} = \beta \,. \tag{9.37}$$

Figure 9.13 suggest that for the solar wind at 1 AU β ranges from 0.01 to 0.5 with a clear clustering around $\beta \approx 0.1$ and a rather weak dependence on the Knudsen number.

The collisionless expression for the conductive flux Eq. (5.77) for a plasma bounded by two thermostats at temperature T_0 and $T_1 = \zeta T_0$ (cf. Figure 5.13) may be the clue to understand the relative insensitivity of the conductive flux in the solar wind with respect to the temperature gradient, that is, the relative insensitivity of β with respect to the Knudsen number. We thus normalize Eq. (5.77) with respect to the saturation flux between the two thermostats. Noting that the temperature between the thermostats is a constant $T = (T_0 T_1)^{1/2}$ we obtain

$$\frac{q_{NC}}{q_0} = 0.376\left(\zeta^{1/4} - \zeta^{-1/4}\right). \tag{9.38}$$

Thus, if we consider that the box size corresponds to the mean free path we may use q_{NC} in Eq. (9.38) to evaluate q_{FS}. Assuming a factor two variation (in other words, $\zeta = 2$) of the temperature over a mean free path we obtain a flux $0.13 q_0$

and even in the case of an exceptionally large $\zeta = 5$ the flux is merely $0.31 q_0$. This slow growth is of course due to the weak dependence $q_{NC} \propto \zeta^{1/4}$.

9.2.7.2 **Closure Based on the Exospheric Approximation**

In 1974 J. Hollweg [179] proposed a description of the conductive electron heat flux in the solar wind using an exospheric model for the velocity distribution function similar to the distributions shown in Figure 9.9 and again in Figure 9.14.

The crucial point in the exospheric description of the solar wind is that there are more electrons (or ions) leaving the Sun than there are electrons falling towards the Sun. Assuming that the Sun is the only source of particles in the universe, there are only two populations of electrons: (i) the trapped electrons, unable to escape from the gravitational and electrostatic potential of the Sun and (ii) the escaping electrons which have a high enough kinetic energy to escape to infinity and which will never fall back to onto the Sun. At any given height above the surface of the Sun the population of trapped electrons is necessarily symmetric in that stationary conditions the flux of outward flowing electrons at any given velocity must be equal to the flux of inward flowing electrons at the same velocity. Taking into account the effect of a radially decreasing (not necessarily radial) magnetic field and the effect of the mirror force does not alter the qualitative picture of Figure 9.14.

Making the reasonable assumption that the Sun does not charge itself, the mean velocity of the electron distribution must be equal to the solar wind velocity u_{SW} which of course is positive because the flux of escaping particles is not compensated by a flux of particles with opposite velocity. The solar wind velocity being generally much smaller than the thermal velocity of the electrons, the truncation velocity must be located at a relatively high velocity with respect to the thermal velocity (cf. Figure 9.14).

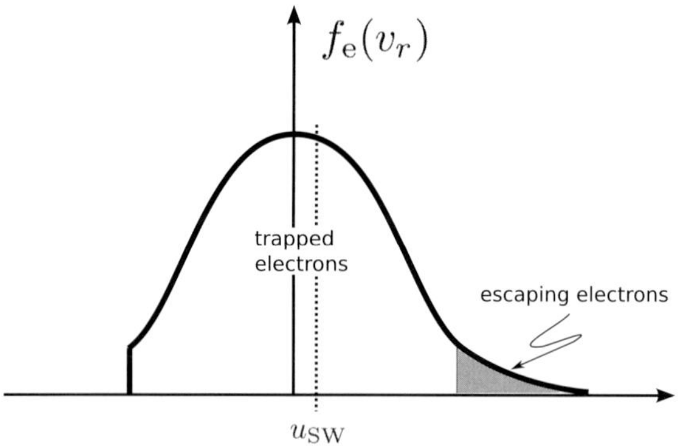

Figure 9.14 Characteristic distribution of the radial component of the electron velocity in a collisionless (exospheric) model of the solar wind. In stationary conditions, the distribution of trapped particles must be symmetric around $v_r = 0$ in the Sun's frame of reference. The average velocity $u_{SW} > 0$ of the distribution is positive due to uncompensated contribution of the escaping particles. The heat flux for such a function is given by Eq. (9.39).

Further assuming that the distribution in Figure 9.14 is a truncated Maxwellian and computing the third moment of the distribution function one obtains the following remarkable expression for the conductive flux in the solar wind

$$q_{\mathrm{H}} = \alpha \frac{3}{2} n T u_{\mathrm{SW}} \tag{9.39}$$

where α is an order unity constant which depends on the electronic Mach number of the wind velocity u_{SW}/v_0 (see [180, Figure 3]). One may argue that the closure Eq. (9.39) rests on a particular choice of the distribution of both trapped and escaping electrons. It is indeed true that the computation of the coefficient α depends on the detailed shape of the distribution. However, as shown in Section 1.3 an expression for the heat flux similar to Eq. (9.39) with an order unity α can be used to describe the heat flux for a strongly non-Maxwellian distribution (with a negative sign though). It is still not clear whether the closure Eq. (9.39) provides a better description of the conductive electron flux in the solar wind than the saturation flux Eq. (9.37).

Appendix

A.1
Notation

A.1.1
Vectors and Tensors

The vectors and the tensors are noted with bold characters without distinction concerning their orders. $\boldsymbol{A}$ can be a vector, but $\boldsymbol{M}$ can be a tensor as well (or its associated matrix), whatever the number of subscripts.

A.1.2
Derivatives

Scalar operators Total and partial derivatives can be notated on a single line, for instance $D_t(f)$ and $\partial_t(f)$. The function to be derived is put into parentheses, except when there is no possible ambiguity on it. It can then (and only then) be implied. For instance:

$$D_t(fg) = f D_t(g) + g D_t(f) \tag{A1}$$

can be written without ambiguity:

$$D_t(fg) = f D_t g + g D_t f \ . \tag{A2}$$

The notation d_x is used for the derivative with respect to x of a single variable function, ∂_x for the partial derivative of a function of several variables, and $D_t = \partial_t + \boldsymbol{u} \cdot \boldsymbol{\nabla}$ for the "convective derivative", $\boldsymbol{u}$ being the flow velocity.

Vector differential operators They are noted thanks to the symbol $\boldsymbol{\nabla}$. For the gradient, divergence and curl operators:

$$\text{grad}(f) = \boldsymbol{\nabla}(f) \quad \text{or} \quad \boldsymbol{\nabla} f \tag{A3}$$

$$\text{div}(\boldsymbol{A}) = \boldsymbol{\nabla} \cdot (\boldsymbol{A}) \quad \text{or} \quad \text{div}(\boldsymbol{A}) = \boldsymbol{\nabla} \cdot \boldsymbol{A} \tag{A4}$$

$$\text{rot}(\boldsymbol{A}) = \boldsymbol{\nabla} \times (\boldsymbol{A}) \quad \text{or} \quad \text{rot}(\boldsymbol{A}) = \boldsymbol{\nabla} \times \boldsymbol{A} \tag{A5}$$

Collisionless Plasmas in Astrophysics, First Edition. Gérard Belmont, Roland Grappin, Fabrice Mottez, Filippo Pantellini, and Guy Pelletier.

Nablas: parenthesis use The nabla operators have both a vector and a differential character. To distinguish both characters, one can explicitly indicate what is derived by parentheses and use brackets to indicate the groups with respect to the scalar and vector products. For instance, for the curl of a cross product, the safest way of writing is:

$$\begin{aligned}\nabla \times \left[(\boldsymbol{A} \times \boldsymbol{B})\right] &= \nabla \cdot \left[(\boldsymbol{BA} - \boldsymbol{AB})\right] \\ &= \nabla \cdot (\boldsymbol{B})\boldsymbol{A} + \boldsymbol{B} \cdot \nabla(\boldsymbol{A}) - \nabla \cdot (\boldsymbol{A})\boldsymbol{B} - \boldsymbol{A} \cdot \nabla(\boldsymbol{B}) \,. \end{aligned} \tag{A6}$$

Here, $\boldsymbol{A}$ and $\boldsymbol{B}$ are vectors, and $\boldsymbol{AB}$ (without $\cdot$ or $\times$) is their dyadic product, a second order tensor. In a term such as $\boldsymbol{B} \cdot \nabla(\boldsymbol{A})$, the dot is the sign for the scalar product (vectorial aspect), and the parentheses delineate the function on which the nabla operator is applied (differential aspect).

Of course, some parentheses and brackets can be implied. For a familiar user, the following notation is sufficient:

$$\nabla \times (\boldsymbol{A} \times \boldsymbol{B}) = \nabla \cdot (\boldsymbol{B})\boldsymbol{A} + \boldsymbol{B} \cdot \nabla \boldsymbol{A} - \nabla \cdot (\boldsymbol{A})\boldsymbol{B} - \boldsymbol{A} \cdot \nabla \boldsymbol{B} \,. \tag{A7}$$

Note that all the parentheses cannot be suppressed, however, since it is quite necessary to distinguish, for instance, $\nabla \cdot (\boldsymbol{A})\boldsymbol{B}$ from $\nabla \cdot (\boldsymbol{AB})$. Note also that this notation excludes the usual notation convention $(\boldsymbol{B} \cdot \nabla)\boldsymbol{A}$, where the parentheses around the scalar product are useless. If parentheses are to be used, they should be around $\boldsymbol{A}$.

The Vlasov equation is given below with the use of this notation convention:

$$\partial_t(f) + \boldsymbol{v} \cdot \nabla_r(f) + \boldsymbol{F}/m \cdot \nabla_v(f) = 0 \,. \tag{A8}$$

In this equation, one has to distinguish the gradient in real space $\nabla_r(f)$ and the gradient in velocity space $\nabla_v(f)$. However, the parentheses are not mandatory in this case since there is no ambiguity.

A.1.3 List of Notation

$\boldsymbol{B}$	Magnetic field
c	Speed of light
c_S	Sound speed
$\boldsymbol{E}$	Electric field
P or $\boldsymbol{P}$	Pressure (scalar or tensor)
$\boldsymbol{p}$	Momentum
Q or $\boldsymbol{Q}$	Heat flux (scalar or tensor)
T	Temperature (in energy units: to be divided by the Boltzmann constant k_B to obtain kelvin)
$\boldsymbol{u}$	Plasma velocity
$\boldsymbol{v}$	Particle velocity
$V_A = B/\sqrt{(\mu_0 \rho)}$	Alfvén speed
$V_{the} = \sqrt{T_e/m_e}$	Electron thermal speed (without factor 2)

$V_{\text{thi}} = \sqrt{T_\text{i}/m_\text{i}}$	Ion thermal speed (without factor 2)
λ_D	Debye length
ρ	Mass density
ϱ	Charge density
ω	Pulsation (in s^{-1}; $\omega = 2\pi f$, the frequency f being measured in Hz)
ω_{pe}	Electron plasma frequency
ω_{pi}	Ion plasma frequency
ω_p	Total plasma frequency
e	Absolute value of the electron charge
$\omega_{\text{ce}} = eB/m_\text{e}$	Electron cyclotron frequency
$\omega_{\text{ci}} = eB/m_\text{i}$	Ion cyclotron frequency

When there is an ambiguity, an explicit formula is provided.

Symmetrized tensors The notation $\{X\}^\circ$ symbolizes the sum of all the tensors obtained from the tensor $\boldsymbol{X}$ by the cyclic permutations over the indices. If $\boldsymbol{X}$ is a tensor of order p, it is:

$$X^\circ_{i_1,i_2,\ldots,i_p} = X_{i_1,i_2,\ldots,i_p} + X_{i_2,i_3,\ldots,i_1} + \ldots + X_{i_p,i_1,\ldots,i_{p-1}} \,. \tag{A9}$$

This notation is particularly useful when calculating the moment equations (Chapter 2).

When the original tensor $\boldsymbol{X}$ is a symmetric one, the result is of course symmetric as well. All terms obtained by permutation are then equal and one has simply $\{\boldsymbol{X}\}^\circ = p\boldsymbol{X}$, p being the order of $\boldsymbol{X}$.

Some of the tensors involved in the moment equations are actually of the form $\boldsymbol{X} = \boldsymbol{S}\boldsymbol{u}$, which is the tensor product of a symmetric tensor by a vector. It can be checked that the result $\{\boldsymbol{X}\}^\circ$ (which is of order $(p+1)$), is indeed symmetric: the original tensor, as all the tensors obtained by permutation, is asymmetric, but this asymmetry only comes from the position of one single index, the index of the vector. As all these indices are scanned in the sum, the result is indeed symmetric.

In the particular case of a second order tensor, the tensor X° is simply the sum of the original tensor with its transposed since: $X^\circ_{ij} = X_{ij} + X_{ji}$.

A.2
Asymptotic Expansions and Adiabatic Invariants

A.2.1
Multiscale Expansion

We propose to solve the equation

$$d_t\zeta = f(\zeta, t, \epsilon) \tag{A10}$$

where the ζ variable can be a scalar, or a vector function, f is a nonlinear function that depends on a small parameter ϵ, and we suppose that the solution of Eq. (A10) when $\epsilon = 0$ is a known function that oscillates at a frequency ω_0 that is considered to be the reference of fast time scale. The aim is, if possible, to show that the solution is a combination of fast varying and slow varying functions that can be described by decoupled equations. As a consequence, we accept approximate solutions that are valid only when ϵ is small.

Various ways of solving this problem are introduced in [181], in the context of the nonlinear wave theory. Among the various methods presented in this book, the multiscale expansion method is of quite general interest, and is applicable to the equations of motion that are our concern in the present book.

The base of the multiscale expansion is the transformation of the dependence in t into a dependence on a series of time variables that correspond to variations $t_0, t_1, \ldots$ of increasing slowness.

$$t_0 = t\,, \quad t_1 = t\epsilon\,, \quad t_2 = t\epsilon^2\,, \quad t_3 = t\epsilon^3\,, \ldots \tag{A11}$$

The first variable t_0 is called the "fast time", and the others $t_1, t_2, \ldots$ are the "slow times". The time derivative of any variable a is transformed into the following expansion

$$d_t a \to d_t(t_0)\partial_{t_0} a + d_t(t_1)\partial_{t_1} a + d_t(t_2)\partial_{t_2} a + \ldots \to \partial_{t_0} a + \epsilon\partial_{t_1} a + \epsilon^2\partial_{t_2} a + \ldots \tag{A12}$$

The variable ζ is cut in two part, the second part depending only on the slow times,

$$\zeta(t) = z(t_0, t_1, t_2, \ldots) + Z(t_1, t_2, \ldots)\,. \tag{A13}$$

The fast varying part z of ζ also depends on the slow times; therefore, the parameters that characterize the fast oscillations (including the frequency ω_0) evolve slowly. The degrees of arbitrariness introduced with the expansion in fast and slow times are compensated by a constraint on whatever depends on t_0. For any function $a(t_0, t_1, t_2 \ldots)$ we impose

$$\langle a(t_0, t_1, t_2 \ldots)\rangle = \frac{1}{T} \int_{0, t_1, t_2 \ldots = \mathrm{cst}}^{T} a\left(t_0', t_1, t_2 \ldots\right) dt_0' = 0\,, \tag{A14}$$

where $T = 2\pi/\omega_0$. From Eq. (A13)

$$\langle Z(t_1, t_2, \ldots)\rangle = Z(t_1, t_2, \ldots)\,. \tag{A15}$$

This variable, as well as the known function f are expanded in the vicinity of Z according to the powers of ϵ and Eq. (A10) becomes

$$\left(\sum_i \epsilon^i \partial_{t_i}\right) \times \left[\sum_j \epsilon^j (Z_j + z_j)\right] = \sum_k \epsilon^k f_k(Z, t_0, t_1, \ldots)\,. \tag{A16}$$

Here are the first terms of this development

$$\begin{aligned} \partial_{t_0} z_0 &= f_0(Z, t_0, t_1, \ldots) , \\ \partial_{t_1}[Z_0 + z_0] + \partial_{t_0} z_1 &= f_1(Z, t_0, t_1, \ldots) . \end{aligned} \tag{A17}$$

The average of these equations, in the sense of Eq. (A14), provides the slow motion equations

$$\begin{aligned} 0 &= \langle f_0 \rangle (Z, t_1, \ldots) , \\ \partial_{t_1} Z_0(t_1, \ldots) &= \langle f_1 \rangle (Z, t_1, \ldots) , \end{aligned} \tag{A18}$$

and the subtraction of these two sets of equations provides the equations of the fast motion

$$\begin{aligned} \partial_{t_0} z_0(t_0, t_1, \ldots) &= f_0(Z, t_0, t_1, \ldots) - \langle f_0(Z, t_0, t_1, \ldots) \rangle , \\ \partial_{t_0} z_1(t_0, t_1, \ldots) &= f_1(Z, t_0, t_1, \ldots) - \langle f_1(Z, t_0, t_1, \ldots) \rangle - \partial_{t_1} z_0(t_0, t_1, \ldots) . \end{aligned} \tag{A19}$$

This method can be applied for the resolution of the equation of motion of charged particles. Then

$$\zeta(t) = (\boldsymbol{r}(t), \boldsymbol{v}(t)) \quad \text{and} \quad d_t \boldsymbol{r}(t) = \boldsymbol{v}(t) \quad \text{and} \quad d_t \boldsymbol{v}(t) = \boldsymbol{F}(\boldsymbol{r}, \boldsymbol{v}, t) . \tag{A20}$$

If the velocity is finite, an expansion of positive powers of ϵ can do well. But the position can go to infinity for a finite velocity; therefore, a term $\boldsymbol{U}_0 t_0$ must be included in the expansion of the particle position. Then

$$\begin{aligned} \boldsymbol{r}(t) &= t_0 \boldsymbol{U}_0(t_1, \ldots) + \boldsymbol{R}_0(t_1, \ldots) + \boldsymbol{r}_0(t_0, \ldots) + \epsilon(\boldsymbol{R}_1(t_1, \ldots) + \boldsymbol{r}_1(t_0, \ldots)) + \ldots \\ \boldsymbol{v}(t) &= \boldsymbol{U}_0(t_1, \ldots) + \boldsymbol{u}_0(t_0, \ldots) + \epsilon(\boldsymbol{U}_1(t_1, \ldots) + \boldsymbol{u}_1(t_0, \ldots)) + \ldots \end{aligned} \tag{A21}$$

where $\boldsymbol{u}_0, \boldsymbol{u}_1, \ldots$ is the notation for the fast part of the velocity with a null average value. The first equations of the slowly varying position are

$$\partial_{t_1} \boldsymbol{R}_0 = \boldsymbol{U}_1, \quad \text{and} \quad \partial_{t_1} \boldsymbol{R}_1 + \partial_{t_2} \boldsymbol{R}_0 = \boldsymbol{U}_2 , \tag{A22}$$

and for the fast varying position

$$\partial_{t_0} \boldsymbol{r}_0 = \boldsymbol{u}_0 . \tag{A23}$$

The first velocity equations are established in a similar way:

$$\partial_{t_0} \boldsymbol{u}_0 = \boldsymbol{F}_0(\boldsymbol{R}, \boldsymbol{U}, t_0, \ldots) , \tag{A24}$$

$$\partial_{t_1}(\boldsymbol{U}_0 + \boldsymbol{u}_0) + \partial_{t_0} \boldsymbol{u}_1 = \boldsymbol{F}_1(\boldsymbol{R}, \boldsymbol{v}, t_0, \ldots) ; \tag{A25}$$

they depend on the development $\boldsymbol{F}_0, \boldsymbol{F}_1, \boldsymbol{F}_2, \ldots$ of the force $\boldsymbol{F}$.

A.2.2 The Adiabatic Invariants

We now consider a Hamiltonian system, where the Hamiltonian has slow temporal variations. We suppose that on the vicinity of any point in a domain of the phase space, the Hamiltonian can be expanded relatively to a small parameter ϵ that characterizes its slow variations. The evolution of the system is periodic when the slow variations are neglected. We apply the multiscale time expansion, setting a series of time variables $t_0 \dots t_n$, as defined by Eq. (A11). The time derivative

$$d_t = \partial_{t_0} + \epsilon \partial_{t_1} + \epsilon^2 \partial_{t_2} + \dots = \sum_{i=0}^{n} \epsilon^i \partial_{t_i} \tag{A26}$$

is applied to the generalized impulsion and the generalized coordinates. The impulsion and the coordinates are indexed following the multiscale expansion

$$p = \sum_{i=0}^{n} \epsilon^i p_i \quad \text{and} \quad q = \sum_{i=0}^{n} \epsilon^i q_i \,. \tag{A27}$$

The Hamilton equations are

$$\frac{\partial H}{\partial p_{i,a}} = d_t q_{i,a} \tag{A28}$$

$$\frac{\partial H}{\partial q_{i,a}} = -d_t p_{i,a} \tag{A29}$$

$$\partial_t H = -d_t L \tag{A30}$$

where $p_{i,a}$ is the component of index a of the vector p_i. Now, we omit the direct consideration of the index a. The hypothesis can be now reformulated in the context of the multiscale time expansion. First hypothesis: the Hamiltonian does not depend on t_0,

$$H = H(p_0, p_1, \dots, p_n, q_0, q_1, \dots, q_n, t_1, \dots, t_n) \,. \tag{A31}$$

Second hypothesis: when the long times $t_1, t_2, \dots$ are fixed, the variation of the Hamiltonian, the generalized impulsion $p_i = p_i(t_0, t_1, \dots, \epsilon)$ and the generalized velocity dq_i/dt are periodic with a period that we note T. This is expressed in the following way:

$$\int_{t_0=0, t_1, \dots = \text{cst}}^{t_0=T} dA = \int_{t_0=0, t_1, \dots = \text{cst}}^{t_0=T} \partial_{t_0} A dt_0 = 0 \,. \tag{A32}$$

for $A = H$, $A = p_i$ and $A = d_t q_i$. We have not supposed that $\epsilon = 0$. The generalized coordinates q_i do not have to be periodic.

We now compute the portion of area I_i described in the phase spaces of the slow variables p_i, q_i by the system over one period, when the long times are fixed. We shall show that under the two above hypotheses, these numbers are conserved. They are called the adiabatic invariants. By definition

$$I_i = \int_{t_0=0,t_1,\ldots=\text{cst}}^{t_0=T} \boldsymbol{p}_i \cdot d\boldsymbol{q}_i = \int_{t_0=0,t_1,\ldots=\text{cst}}^{t_0=T} \boldsymbol{p}_i \partial_{t_0} \boldsymbol{q}_i d t_0 \,. \tag{A33}$$

For one degree of freedom,

$$I_i = \int_{t_0=0,t_1,\ldots=\text{cst}}^{t_0=T} p_i d q_i = \int_{t_0=0,t_1,\ldots=\text{cst}}^{t_0=T} p_i \partial_{t_0} q_i d t_0 \,. \tag{A34}$$

According to Eq. (A14), we can write $\langle A \rangle$ for $\int_{t_0=0,t_1,\ldots=\text{cst}}^{t_0=T} A d t_0$, when it is not needed to make an explicit reference to the integration procedure. With the multiscale time expansion

$$\begin{aligned} d_t I_i &= \sum_{l=0}^{n} \epsilon^l \langle \partial_{t_l} (p_i \partial_{t_0} q_i) \rangle \\ &= \sum_{l=0}^{n} \epsilon^l \int_{t_0=0,t_1,\ldots=\text{cst}}^{t_0=T} \left(\partial_{t_l} p_i \partial_{t_0} q_i + p_i \partial^2_{t_0 t_l} q_i \right) d t_0 \,. \end{aligned}$$

It is integrated by parts,

$$d_t I_i = \sum_{l=0}^{n} \epsilon^l \int_{t_0=0,t_1,\ldots=\text{cst}}^{t_0=T} \left(\partial_{t_l} p_i \partial_{t_0} q_i - \partial_{t_0} p_i \partial_{t_l} q_i \right) d t_0 + R \tag{A35}$$

where

$$R = \sum_{l=0}^{n} \epsilon^l \left[\partial_{t_l} q_i p_i \right]_{t_0}^{t_0+T} \,. \tag{A36}$$

The expansion is reordered

$$d_t I_i = \left\langle \left(\sum_{l=0}^{n} \epsilon^l \partial_{t_l} p_i \right) \partial_{t_0} q_i - \left(\sum_{l=0}^{n} \epsilon^l \partial_{t_l} q_i \right) \partial_{t_0} p_i \right\rangle + R \,. \tag{A37}$$

The multiscale time expansion terms are removed and the Hamilton equations are applied

$$d_t I_i = \langle \partial_t p_i \partial_{t_0} q_i - \partial_t q_i \partial_{t_0} p_i \rangle + R = -\left\langle \frac{\partial H}{\partial q_i} \partial_{t_0} q_i + \frac{\partial H}{\partial p_i} \partial_{t_0} p_i \right\rangle + R \,. \tag{A38}$$

The last term in brackets is a part of the time derivative of the Hamiltonian,

$$d_t I_i = \langle d_{t_0} H - \partial_{t_0} H \rangle + R \,. \tag{A39}$$

From the first hypothesis expressed in Eq. (A31) $\partial_{t_0} H = 0$. Moreover, from Eq. (A32) for $A = H$, the variation for fixed long times of H over one period is null. Therefore, the integral on the right-hand side is null. By the hypothesis in Eq. (A32) for $A = p_i$ and $A = d_t q_i$, $R = 0$. Therefore,

$$d_t I_i = 0 \,. \tag{A40}$$

This result is exact at any finite order n, as long the hypotheses are valid. If we consider that the second hypothesis is not exactly valid, we can develop R, and find

$$R = [d_t q_i p_i]_{t_0, t_1, t_2 \ldots}^{t_0 + T, t_1, t_2 \ldots} \,. \tag{A41}$$

Considering Eq. (A28),

$$d_t I_i = [-L + d_t q_i p_i]_{t_0, t_1, t_2 \ldots}^{t_0 + T, t_1, t_2 \ldots} \,. \tag{A42}$$

The equation between braces is the variation of the energy. In other words, we can conclude that I_i does not vary over long times provided that the energy does not vary over short time scales. This is why I_i is called an *adiabatic* invariant.

It is important to notice that when the quasi-adiabatic conditions are fulfilled, I_i is conserved at any order in ϵ. Consequently, it is invariant over arbitrary long durations.

A.2.3
Derivation of the Guiding Center Equations

The velocity equation (Lorentz force) is

$$d_t \boldsymbol{v} = \boldsymbol{e}(\boldsymbol{r}, t_1, \ldots) + \boldsymbol{v} \times \boldsymbol{b}(\boldsymbol{r}, t_1, \ldots) \,, \tag{A43}$$

where we note $\boldsymbol{e} = q\boldsymbol{E}/m$ and $\boldsymbol{b} = q\boldsymbol{B}/m$. We accept a contribution of order 0 of the magnetic field and the perpendicular electric field. As mentioned above, the parallel electric field must be of order 1

$$\boldsymbol{e}(\boldsymbol{r}, t_1, \ldots) = \boldsymbol{e}_\perp(\boldsymbol{r}, t_1, \ldots) + \epsilon \boldsymbol{e}_{\parallel 1}(\boldsymbol{r}, t_1, \ldots) \,, \tag{A44}$$

where $e_\parallel = \epsilon e_{\parallel 1}$.

The guiding center theory requires that the electromagnetic field has smooth spatial variations. To the electric field (or the magnetic field) $\boldsymbol{e}(\boldsymbol{r})$ we can associate a field $\boldsymbol{e}(\boldsymbol{l}_0, \boldsymbol{l}_1, \boldsymbol{l}_2, \ldots)$ where the spatial variables $\boldsymbol{l}_i$ are defined in a way analogous to the times t_i,

$$\boldsymbol{l}_i = \epsilon^i \boldsymbol{r} \,. \tag{A45}$$

Then, the nabla operator $\nabla = \partial_r$ is transformed into a sum of specific operators $\nabla_i = \partial_{l_i}$ associated to the l_i variables,

$$\nabla e = \sum_i \partial_r l_i \partial_{l_i} e = \sum_i \epsilon^i \partial_{l_i} e \,. \tag{A46}$$

The consideration of a smooth electromagnetic field is then equivalent to the hypothesis that it does not depend on l_0. The development of e and b in the vicinity of R becomes

$$e(r, t_1, \ldots) = e_\perp(R, t_1, \ldots) + \epsilon(r_0 \cdot \nabla) e_\perp + \epsilon e_{\parallel 1} + \ldots \tag{A47}$$

$$b(r, t_1, \ldots) = b(R, t_1, \ldots) + (r_0 \cdot \nabla) b + \ldots \tag{A48}$$

where the gradients are computed at R. The development in terms of l_i implies, at first order,

$$e(r, t_1, \ldots) = e_\perp(R, t_1, \ldots) + \epsilon \left[(r_0 \cdot \nabla_1)\, e_\perp + e_{\parallel 1} \right] + \ldots \tag{A49}$$

$$b(r, t_1, \ldots) = b(R, t_1, \ldots) + \epsilon\, (r_0 \cdot \nabla_1)\, b + \ldots \tag{A50}$$

where the gradients are computed for R.

The velocity equation at orders ϵ^0 and ϵ^1 is

$$\begin{aligned} \frac{1}{\omega_c} d_t v = {} & [e_\perp + (U_0 + u_0) \times b] \\ & + \epsilon \left\{ e_{\parallel 1} + (r_0 \cdot \nabla_1)\, e_\perp + (U_1 + u_1) \times b \right. \\ & \left. + (U_0 + u_0) \times [(r_0 \cdot \nabla_1)\, b] \right\} . \end{aligned} \tag{A51}$$

To the lowest order, from Eq. (A24),

$$\partial_{t_0} u_0(t_0, \ldots) = e_\perp + (U_0 + u_0) \times b \,. \tag{A52}$$

The average part of this equation is

$$0 = e_\perp + U_0 \times b \tag{A53}$$

equivalent to

$$U_0 = \frac{E \times B}{B^2} + U_{\parallel 0} b \,, \tag{A54}$$

with $U_{\parallel 0}$ still undetermined at this stage. The fast part of the equation is

$$\partial_{t_0} u_0(t_0, t_1, \ldots) = u_0 \times b(R, t) \,. \tag{A55}$$

Let us notice at this stage that a slow parallel electric field would lead to the divergent solution $u_0(t_0, \ldots) = e_{\parallel 0} t_0$. This is a posteriori why we do not keep the possibility of a zero order parallel electric field in Eq. (A44). Equation (A55) describes a linear oscillator, whose solution is

$$\begin{aligned} u_{0x} &= +u_{\perp 0} \cos(\omega_c t_0 + \psi) = +u_{\perp 0} \cos(\omega_c t + \psi) \\ u_{0y} &= -u_{\perp 0} \sin(\omega_c t_0 + \psi) = -u_{\perp 0} \sin(\omega_c t + \psi) \\ u_{0z} &= 0 \,. \end{aligned} \tag{A56}$$

The slow part of Eq. (A25) is

$$\partial_{t_1} \boldsymbol{U}_0(t_1,\ldots) = \boldsymbol{e}_{\|1} + \boldsymbol{U}_1 \times \boldsymbol{b} + \langle \boldsymbol{u}_0 \times (\boldsymbol{r}_0 \cdot \boldsymbol{\nabla}_1)\, \boldsymbol{b} \rangle \tag{A57}$$

The fast solution shown above is now introduced into Eq. (A57) and the average value is computed,

$$\partial_{t_1} \boldsymbol{U}_0(t_1,\ldots) = \boldsymbol{e}_{\|1} + \boldsymbol{U}_1 \times \boldsymbol{b} - \frac{u_{\perp 0}^2}{2\omega_c} \boldsymbol{\nabla} b_z \,.$$

The last term is

$$\frac{u_{\perp 0}^2}{2\omega_c} \boldsymbol{\nabla} b_z = \frac{\mu}{m} \boldsymbol{\nabla} B_z = \frac{\mu}{m} \boldsymbol{\nabla} B \,, \tag{A58}$$

where we have introduced the magnetic moment μ of the electron,

$$\mu = \frac{m u_{\perp 0}^2}{2 B_0} \,. \tag{A59}$$

We have also used the equation

$$d B = d\left(B_x^2 + B_y^2 + B_z^2\right)^{1/2} = B^{-1}(B_x d B_x + B_y d B_y + B_z d B_z) = d B_z \,, \tag{A60}$$

that is true because locally, $B = B_z$. Then, one can get rid of the terms that are specific to the asymptotic expansion, going back to the ordinary variables, and considering that the electric field is $\boldsymbol{E}_{\|} = \epsilon \boldsymbol{E}_{\|1}$ and the first order slow component of the velocity is $\epsilon \boldsymbol{U}_1$, and $\boldsymbol{\nabla}_1 = \epsilon^{-1}\nabla$ and $\partial_{t_1} = \epsilon^{-1}\partial_t$,

$$d_t \boldsymbol{U}_0 = \frac{q}{m}[\boldsymbol{E}_{\|} + \epsilon \boldsymbol{U}_1 \times \boldsymbol{B}] - \frac{\mu}{m} \boldsymbol{\nabla} B \,. \tag{A61}$$

Inserting $\boldsymbol{U}_0$ from Eq. (A54) into Eq. (A61), one can deduce $\boldsymbol{U}_1$,

$$\begin{aligned} \boldsymbol{U}_\perp &= \boldsymbol{U}_0 + \epsilon \boldsymbol{U}_1 \\ &= \frac{\boldsymbol{E} \times \boldsymbol{B}}{B^2} + \epsilon \frac{m \boldsymbol{b}}{q B} \times \left[\partial_{t_1}\left(\frac{\boldsymbol{E} \times \boldsymbol{B}}{B^2}\right) + U_{\|0} \partial_{t_1} \boldsymbol{b} + \frac{\mu}{m} \boldsymbol{\nabla}_1 \boldsymbol{B}\right] . \end{aligned} \tag{A62}$$

In terms of ordinary time and space variables, one finds Eq. (1.118). The time derivative of Eq. (A54) provides the following relation,

$$(\partial_{t_1} \boldsymbol{U}_0)_{\|} = \partial_{t_1} U_{\|0} + \partial_{t_1}\left(\frac{\boldsymbol{E} \times \boldsymbol{B}}{B^2}\right) \cdot \boldsymbol{b} + U_{\|0} \partial_{t_1} \boldsymbol{b} \cdot \boldsymbol{b} \tag{A63}$$

$$= \partial_{t_1} U_{\|0} - \frac{\boldsymbol{E} \times \boldsymbol{B}}{B^2} \cdot \partial_{t_1} \boldsymbol{b} \,. \tag{A64}$$

This combined with Eq. (A61) gives the equation of evolution of the parallel velocity Eq. (1.122).

A.3 Fokker–Planck Equation, First Order Term

Under relatively wide assumptions, one can establish the following property:

$$\langle \zeta(t_1)\partial_x \zeta(t_2)\rangle = \langle \partial_x \zeta(t_1)\zeta(t_2)\rangle = \frac{1}{2}\partial_x \langle \zeta(t_1)\zeta(t_2)\rangle \ . \tag{A65}$$

Indeed, let the random function $\zeta(t) = F(x(t), t)$ where

$$F(x,t) = \int \frac{dk}{2\pi} A_k \exp[ikx - \omega_k t + \phi_k] \tag{A66}$$

ϕ_k being an initial phase randomly distributed between 0 and 2π. The self-correlation function of $\zeta(t)$ is of the form:

$$\begin{aligned}\langle \zeta(t_1)\zeta(t_2)\rangle = &\int \frac{dk}{2\pi}\int \frac{dk'}{2\pi} A_k A_{k'} e^{i(k+k')x}\\ &\times \big\langle \exp\big[ik\delta x(t_1) + ik'\delta x(t_2) - i(\omega_k t_1 + \omega_{k'} t_2)\\ &\quad + i(\phi_k + \phi_{k'})\big]\big\rangle \ .\end{aligned} \tag{A67}$$

The second exponential is supposed to contain only fast variations almost homogeneous over the coherence length, implying $k' \simeq -k$. Since the self-correlation function is approximatively of the form:

$$\langle \zeta(t_1)\zeta(t_2)\rangle = \int \frac{dk}{2\pi} S(k;x)\langle \exp[ik(\delta x(t_1) - \delta x(t_2)) - i\omega_k(t_1 - t_2)]\rangle \ , \tag{A68}$$

where $S(k;x) \equiv \int (dk'/2\pi) A_k A_{k'} e^{i(k+k')x}$ is a spectrum slowly varying with x. The relations (Eq. (A65)) are easily derived from this result.

Let us come back to the estimate of $\langle \Delta x\rangle$. Even if u varies with x, its contribution to the variation of Δx is taken at leading order $u\Delta t$, the other contribution being on order larger than Δt. However, $\langle \delta x\rangle$ has an inhomogeneous contribution at order Δt; indeed

$$\delta x = \int_{t_0}^{t_0+\Delta t} F(x_0 + u(x_0)\tau + \delta x(\tau), \tau) d\tau \tag{A69}$$

can be expanded as

$$\delta x = \int_{t_0}^{t_0+\Delta t} F(x_0, \tau) d\tau + \int_{t_0}^{t_0+\Delta t} (u(x_0)\tau + \delta x(\tau))\partial_x F(x_0, \tau) d\tau \ . \tag{A70}$$

Thus,

$$\langle \delta x\rangle = \int_{t_0}^{t_0+\Delta t} d\tau \int_{t_0}^{\tau} d\tau' \langle \zeta(\tau')\partial_x \zeta(\tau)\rangle \ . \tag{A71}$$

Taking into account (Eq. (A65)) and the parity of the self-correlation function, one obtains

$$\langle \delta x \rangle = \frac{1}{2}\partial_x \int_{t_0}^{t_0+\Delta t} d\tau \int_{t_0}^{t_0+\Delta t} d\tau' \langle \zeta(\tau)\zeta(\tau') \rangle \,, \tag{A72}$$

which leads to the expected result:

$$\langle \delta x \rangle = \Delta t \partial_x D \,. \tag{A73}$$

Inserting this result into the Fokker–Planck equation, the equation takes the expected form (Eq. (8.42)).

A similar analysis can be generalized for the diffusion in momentum space. Thus, when a momentum component experiences a jump during Δt that exceeds the correlation time of the random force,

$$\Delta p_i = A_i \Delta t + \delta p_i \,, \tag{A74}$$

the contribution A_i accounts for the drag force or the systematic acceleration, non-random, and δp_i is the jump in the random force whose average does not vanishes when its diffusion tensor Γ_{ij} depends on the slow variable $\boldsymbol{p}$. Therefore,

$$\langle \delta p_i \rangle = \Delta t \partial_{p_j} \Gamma_{ij} \,. \tag{A75}$$

When the diffusion process is isotropic, $\langle \delta p \rangle = \Delta t \partial_p (p^2 \Gamma)/p^2$ and the diffusive part of the Fokker–Planck equation can take a self-adjoint form as indicated in the main text, this is how it describes the second order acceleration process of cosmic rays.

References

1 Tsyganenko, N.A. and Stern, D.P. (1996) Modeling the global magnetic field of the large-scale Birkeland current systems. *Journal of Geophysical Research (Space Physics)*, **101**, 27187–27198, doi:10.1029/96JA02735.

2 Northrop, T. (1963) *The Adiabatic Motion of Charged Particles*, Interscience Publishers, New York.

3 Mottez, F., Adam, J.C., and Heron, A. (1998) A new guiding centre PIC scheme for electromagnetic highly magnetized plasma simulation. *Computer Physics Communications*, **113**, 109–130.

4 Mangeney, A., Salem, C., Lacombe, C., Bougeret, J.L., Perche, C., Manning, R., Kellogg, P.J., Goetz, K., Monson, S.J., and Bosqued, J.M. (1999) WIND observations of coherent electrostatic waves in the solar wind. *Annales Geophysicae*, **17**, 307–320, doi:10.1007/s00585-999-0307-y.

5 Brittin, W.E. (1962) New Series in Physics (Book Reviews: Studies in Statistical Mechanics. Vol. 1). *Science*, **138**, 965, doi:10.1126/science.138.3544.965.

6 Klimontovich, Y.L. (1964) Statisticheskaya Teoriya Neravnovesnykh Protsessov v Plazme. Izd. MGU, Moscow.

7 Chew, G.F., Goldberger, M.L., and Low, F.E. (1956) The Boltzmann equation and the one-fluid hydromagnetic equations in the absence of particle collisions. *Royal Society of London Proceedings Series A*, **236**, 112–118, doi:10.1098/rspa.1956.0116.

8 Tanaka, M. (1988) Macroscale implicit electromagnetic particle simulation of magnetized plasmas. *Journal of Computational Physics*, **79**, 209–226, doi:10.1016/0021-9991(88)90012-5.

9 Mason, R.J. (1981) Implicit moment particle simulation of plasmas. *Journal of Computational Physics*, **41**, 233–244, doi:10.1016/0021-9991(81)90094-2.

10 Hess, S. and Mottez, F. (2009) How to improve the diagnosis of kinetic energy in δf PIC codes. *Journal of Computational Physics*, **228**, 6670–6681, doi:10.1016/j.jcp.2009.05.035.

11 Marsch, E., Schwenn, R., Rosenbauer, H., Muehlhaeuser, K.H., Pilipp, W., and Neubauer, F.M. (1982) Solar wind protons – Three-dimensional velocity distributions and derived plasma parameters measured between 0.3 and 1 AU. *Journal of Geophysics Research*, **87**, 52–72, doi:10.1029/JA087iA01p00052.

12 Avrett, E.H. and Loeser, R. (2008) Models of the solar chromosphere and transition region from SUMER and HRTS Observations: Formation of the extreme-ultraviolet spectrum of hydrogen, carbon, and oxygen. *Astrophysical Journal, Supplement*, **175**, 229–276, doi:10.1086/523671.

13 Landau, L. (1946) On the vibrations of the electronic plasma. *Journal of Physics USSR*, **10**, 25.

14 Chandrasekhar, S. (1951) The fluctuations of density in isotropic turbulence. *Proceedings of the Royal Society of London A*, **210** (1), 18–25.

15 Grappin, R., Cavillier, E., and Velli, M. (1997) Acoustic waves in isothermal winds in the vicinity of the sonic point. *Astronomy and Astrophysics*, **322**, 659–670.

Collisionless Plasmas in Astrophysics, First Edition. Gérard Belmont, Roland Grappin, Fabrice Mottez, Filippo Pantellini, and Guy Pelletier.

16 Alazraki, G. and Couturier, P. (1971) Solar wind accejeration caused by the gradient of Alfvén wave pressure. *Astronomy and Astrophysics*, **13**, 380.

17 Grappin, R., Léorat, J., and Habbal, S. (2002) Large-amplitude Alfvén waves in open and closed coronal structures: A numerical study. *Journal of Geophysical Research (Space Physics)*, **107**, 1380.

18 Esser, R., Fineschi, S., Dobrzycka, D., Habbal, S.R., Edgar, R.J., Raymond, J.C., Kohl, J.L., and Guhathakurta, M. (1999) Plasma properties in coronal holes derived from measurements of minor ion spectral lines and polarized white light intensity. *The Astrophysical Journal*, **510** (1), L63–L67.

19 Stix, T.H. (1992) *Waves in Plasmas*, American Institute of Physics, New York.

20 Roennmark, K. (1982) Waves in homogeneous, anisotropic multicomponent plasmas (WHAMP). Tech. Rep., Kiruna Geophysical Institute.

21 LaBelle, J. and Anderson, R.R. (2011) Ground-level detection of auroral kilometric radiation. *Geophysical Research Letters*, **38**, L04104, doi:10.1029/2010GL046411.

22 Snyder, P.B., Hammett, G.W., and Dorland, W. (1997) Landau fluid models of collisionless magnetohydrodynamics. *Physics of Plasmas*, **4**, 3974–3985, doi:10.1063/1.872517.

23 Passot, T. and Sulem, P.L. (2004) A Landau fluid model for dispersive magnetohydrodynamics. *Physics of Plasmas*, **11**, 5173–5189, doi:10.1063/1.1780533.

24 van Kampen, N.G. and Felderhof, B.U. (1967) *Theoretical Methods in Plasma Physics*, North Holland Publication Co., Amsterdam.

25 Belmont, G., Mottez, F., Chust, T., and Hess, S. (2008) Existence of non-Landau solutions for Langmuir waves. *Physics of Plasmas*, **15** (5), 052310, doi:10.1063/1.2921791.

26 Chust, T., Belmont, G., Mottez, F., and Hess, S. (2009) Landau and non-Landau linear damping: Physics of the dissipation. *Physics of Plasmas*, **16** (9), 092104.

27 Whitham, G.B. (1974) *Linear and Nonlinear Waves*, Wiley-Interscience, New York.

28 Landau, L.D. and Lifshitz, E.M. (1960) *Electrodynamics of Continuous Media*, Pergamon Press, Oxford, New York.

29 Jeffrey, A. and Taniuti, T. (1964) *Nonlinear Wave Propagation*, Academic Press, New York, London.

30 Harris, E.G. (1962) On a plasma sheath separating regions of oppositely directed magnetic field. *Il Nuovo Cimento*, **23**, 115–121, doi:10.1007/BF02733547.

31 Belmont, G., Aunai, N., and Smets, R. (2012) Kinetic equilibrium for an asymmetric tangential layer. *Physics of Plasmas*, **19** (2), 022108, doi:10.1063/1.3685707.

32 Horbury, T.S., Cargill, P.J., Lucek, E.A., Balogh, A., Dunlop, M.W., Oddy, T.M., Carr, C., Brown, P., Szabo, A., and Fornaçon, K.H. (2001) Cluster magnetic field observations of the bowshock: Orientation, motion and structure. *Annales Geophysicae*, **19**, 1399–1409, doi:10.5194/angeo-19-1399-2001.

33 Lucek, E.A., Horbury, T.S., Dunlop, M.W., Cargill, P.J., Schwartz, S.J., Balogh, A., Brown, P., Carr, C., Fornacon, K.H., and Georgescu, E. (2002) Cluster magnetic field observations at a quasi-parallel bow shock. *Annales Geophysicae*, **20**, 1699–1710, doi:10.5194/angeo-20-1699-2002.

34 Alexandrova, O., Lacombe, C., Mangeney, A., Grappin, R., and Maksimovic, M. (2012) Solar wind turbulent spectrum at plasma kinetic scales. *The Astrophysical Journal*, **760** (2), 121.

35 Sahraoui, F., Goldstein, M.L., Belmont, G., Canu, P., and Rezeau, L. (2010) Three dimensional anisotropic k spectra of turbulence at subproton scales in the solar wind. *Physical Review Letters*, **105** (13), 131101, doi:10.1103/PhysRevLett.105.131101.

36 Bruno, R. and Carbone, V. (2005) The solar wind as a turbulence laboratory. *Living Reviews in Solar Physics*, **2**, 4.

37 Frisch, U., Sulem, P., and Nelkin, M. (1978) A simple dynamical model of intermittent fully developed turbulence. *Journal of Fluid Mechanics*, **87**, 719–736.

38 Desnianskii, V.N. and Novikov, E.A. (1974) Simulation of cascade processes

in turbulent flows. *Prikladnaia Matematika i Mekhanika*, **38**, 507.

39 Obukhov, A. (1971) Some general characteristic equations of the dynamics of the atmosphere. *Journal of Atmospheric and Oceanic Europhysics*, **7**, 41.

40 Plunian, F., Stepanov, R., and Frick, P. (2013) Shell models of magnetohydrodynamic turbulence. *Physics Reports*, **523** (1), 1–60.

41 Pereira, T.M.D., De Pontieu, B., and Carlsson, M. (2013) The effects of spatio-temporal resolution on deduced spicule properties. *Monthly Notices of the Royal Astronomical Society*, **764** (1), 69.

42 Belcher, J.W. and Davis, L. (1971) Large-amplitude Alfvén waves in the interplanetary medium, 2. *Journal of Geophysical Research*, **76** (16), 3534–3563.

43 Bigot, B., Galtier, S., and Politano, H. (2008) Development of anisotropy in incompressible magnetohydrodynamic turbulence. *Physical Review E*, **78** (6), 66301.

44 Bigot, B., Galtier, S., and Politano, H. (2008) Energy decay laws in strongly anisotropic magnetohydrodynamic turbulence. *Physical Review Letters*, **100** (7), 74502.

45 Müller, W.C. (2009) Magnetohydrodynamic Turbulence. *Lectures Notes in Physics*, **756**, 223–254.

46 Lee, E., Brachet, M.E., Pouquet, A., Mininni, P.D., and Rosenberg, D. (2010) Lack of universality in decaying magnetohydrodynamic turbulence. *Physical Review E*, **81** (1), 16318.

47 Iroshnikov, P.S. (1963) Turbulence of a conducting fluid in a strong magnetic field. *Astronomicheskii Zhurnal*, **40**, 742.

48 Kraichnan, R.H. (1965) Inertial-range spectrum of hydromagnetic turbulence. *Physics of Fluids*, **8**, 1385.

49 Pouquet, A., Sulem, P., and Meneguzzi, M. (1988) Influence of velocity-magnetic field correlations on decaying magnetohydrodynamic turbulence with neutral X points. *Physics of Fluids*, **31**, 2635–2643, (ISSN 0031-9171).

50 Biskamp, D. and Welter, H. (1989) Dynamics of decaying two-dimensional magnetohydrodynamic turbulence. *Physics of Fluids B*, **1**, 1964, (ISSN 0899-8221).

51 Müller, W.C., Biskamp, D., and Grappin, R. (2003) Statistical anisotropy of magnetohydrodynamic turbulence. *Physical Review E*, **67** (6), 66302.

52 Müller, W.C. and Grappin, R. (2005) Spectral energy dynamics in magnetohydrodynamic turbulence. *Physical Review Letters*, **95** (11).

53 Grappin, R., Léorat, J., and Pouquet, A. (1983) Dependence of MHD turbulence spectra on the velocity field-magnetic field correlation. *Astronomy and Astrophysics*, **126**, 51–58, (ISSN 0004-6361).

54 Dobrowolny, M., Mangeney, A., and Veltri, P. (1980) Fully developed anisotropic hydromagnetic turbulence in interplanetary space. *Physical Review Letters*, **45**, 144–147.

55 Grappin, R., Frisch, U., Pouquet, A., and Léorat, J. (1982) Alfvénic fluctuations as asymptotic states of MHD turbulence. *Astronomy and Astrophysics*, **105**, 6–14.

56 Biskamp, D. (1994) Cascade models for magnetohydrodynamic turbulence. *Physical Review E*, **50** (4), 2702–2711.

57 Gloaguen, C., Léorat, J., Pouquet, A., and Grappin, R. (1985) A scalar model for MHD turbulence. *Physica D*, **17** (2), 154–182.

58 Biskamp, D. (2003) *Magnetohydrodynamic Turbulence*, Cambridge University Press, New York.

59 Buchlin, E. (2004) *Signatures et modélisations du chauffage coronal turbulent à petite échelle*, Ph.D. thesis, Université Paris-Sud and Univ. Firenze.

60 Nigro, G., Malara, F., Carbone, V., and Veltri, P. (2004) Nanoflares and MHD turbulence in coronal loops: A hybrid shell model. *Physical Review Letters*, **92** (1), 194501.

61 Buchlin, E. and Velli, M. (2007) Shell models of RMHD turbulence and the heating of solar coronal loops. *The Astrophysical Journal*, **662**, 701.

62 Verdini, A. and Grappin, R. (2012) Transition from weak to strong cascade in MHD turbulence. *Physical Review Letters*, **109** (2), 025004.

63 Goldreich, P. and Sridhar, S. (1995) Toward a theory of interstellar turbulence.

2: Strong alfvenic turbulence. *Astrophysical Journal*, **438**, 763–775.
64 Galtier, S., Nazarenko, S.V., Newell, A.C., and Pouquet, A. (2000) A weak turbulence theory for incompressible magnetohydrodynamics. *Journal of Plasma Physics*, **63** (0), 447–488.
65 Dmitruk, P., Gomez, D.O., and Matthaeus, W.H. (2003) Energy spectrum of turbulent fluctuations in boundary driven reduced magnetohydrodynamics. *Physics of Plasmas*, **10** (9), 3584–3591.
66 Rappazzo, A.F., Velli, M., Einaudi, G., and Dahlburg, R.B. (2007) Coronal heating, weak MHD turbulence, and scaling laws. *The Astrophysical Journal*, **657** (1), L47–L51.
67 Perez, J.C. and Boldyrev, S. (2008) On weak and strong magnetohydrodynamic turbulence. *The Astrophysical Journal*, **672** (1), L61–L64.
68 Boldyrev, S., Mason, J., and Cattaneo, F. (2009) Dynamic alignment and exact scaling laws in magnetohydrodynamic turbulence. *The Astrophysical Journal Letters*, **699** (1), L39–L42.
69 Mason, J., Cattaneo, F., and Boldyrev, S. (2006) Dynamic alignment in driven magnetohydrodynamic turbulence. *Physical Review Letters*, **97** (2), 255 002.
70 Boldyrev, S. (2005) On the spectrum of magnetohydrodynamic turbulence. *The Astrophysical Journal*, **626** (1), L37–L40.
71 Oughton, S., Matthaeus, W.H., and Ghosh, S. (1998) Scaling of spectral anisotropy with magnetic field strength in decaying magnetohydrodynamic turbulence. *Physics of Plasmas*, **5** (1), 4235–4242.
72 Grappin, R. and Müller, W.C. (2010) Scaling and anisotropy in magnetohydrodynamic turbulence in a strong mean magnetic field. *Physical Review E*, **82** (2), 26406.
73 Cho, J. and Vishniac, E.T. (2000) The anisotropy of magnetohydrodynamic Alfvénic turbulence. *The Astrophysical Journal*, **539** (1), 273–282.
74 Grappin, R., Müller, W.C., and Gürcan, Ö. (2012) Quasi-isotropic cascade in MHD turbulence with mean field. archiv: 1209.4450 [physics.plasm-ph].
75 Salem, C. (2000) *Ondes, turbulence et phénomènes dissipatifs dans le vent solaire á partir des observations de la sonde WIND*, Ph.d. thesis, Université Paris 7, Paris, France.
76 Salem, C., Mangeney, A., Bale, S.D., and Veltri, P. (2009) Solar wind magnetohydrodynamics turbulence: Anomalous scaling and role of intermittency. *The Astrophysical Journal*, **702**, 537.
77 Grappin, R., Velli, M., and Mangeney, A. (1991) 'Alfvénic' versus 'standard' turbulence in the solar wind. *Annales Geophysicae*, **9**, 416–426, (ISSN 0939-4176).
78 Dasso, S., Milano, L.J., Matthaeus, W.H., and Smith, C.W. (2005) Anisotropy in fast and slow solar wind fluctuations. *The Astrophysical Journal*, **635** (2), L181–L184.
79 Horbury, T.S., Wicks, R.T., and Chen, C.H.K. (2011) Anisotropy in space plasma turbulence: Solar wind observations. *Space Science Reviews*, **1**, 293.
80 Muschietti, L., Roth, I., Carlson, C.W., and Ergun, R.E. (2000) Transverse instability of magnetized electron holes. *Physical Review Letters*, **85**, 94–97, doi:10.1103/PhysRevLett.85.94.
81 Temerin, M., Cerny, K., Lotko, W., and Mozer, F.S. (1982) Observations of double layers and solitary waves in the auroral plasma. *Physical Review Letters*, **48**, 1175–1179, doi:10.1103/PhysRevLett.48.1175.
82 Muschietti, L., Ergun, R.E., Roth, I., and Carlson, C.W. (1999) Phase-space electron holes along magnetic field lines. *Geophysical Research Letters*, **26**, 1093–1096, doi:10.1029/1999GL900207.
83 Matsumoto, H., Kojima, H., Miyatake, T., Omura, Y., Okada, M., Nagano, I., and Tsutsui, M. (1994) Electrotastic Solitary Waves (ESW) in the magnetotail: BEN wave forms observed by GEOTAIL. *Geophysical Research Letters*, **21**, 2915–2918, doi:10.1029/94GL01284.
84 Cattell, C., Dombeck, J., Wygant, J., Drake, J.F., Swisdak, M., Goldstein, M.L., Keith, W., Fazakerley, A., André, M., Lucek, E., and Balogh, A. (2005) Cluster observations of electron holes in association with mag-

netotail reconnection and comparison to simulations. *Journal of Geophysical Research (Space Physics)*, **110**, A01211, doi:10.1029/2004JA010519.

85 Henri, P., Meyer-Vernet, N., Briand, C., and Donato, S. (2011) Observations of Langmuir ponderomotive effects using the Solar TErrestrial RElations Observatory spacecraft as a density probe. *Physics of Plasmas*, **18** (8), 082308, doi:10.1063/1.3622667.

86 Briand, C., Soucek, J., Henri, P., and Mangeney, A. (2010) Waves at the electron plasma frequency associated with solar wind magnetic holes: STEREO/Cluster observations. *Journal of Geophysical Research (Space Physics)*, **115** (A14), A12113, doi:10.1029/2010JA015849.

87 Bernstein, I.B., Greene, J.M., and Kruskal, M.D. (1957) Exact nonlinear plasma oscillations. *Physical Review*, **108**, 546–550, doi:10.1103/PhysRev.108.546.

88 Turikov, V.A. (1984) Electron phase space holes as localized BGK solutions. *Physica Scripta*, **30**, 73–77, doi:10.1088/0031-8949/30/1/015.

89 Berk, H.L., Nielsen, C.E., and Roberts, K.V. (1970) Phase space hydrodynamics of equivalent nonlinear systems: Experimental and computational observations. *Physics of Fluids*, **13**, 980–995, doi:10.1063/1.1693039.

90 Mottez, F. (2001) Instabilities and formation of coherent structures. *Astrophysics and Space Science*, **277**, 59–70.

91 Sagdeev, R.Z. and Galeev, A.A. (1969) *Nonlinear Plasma Theory*, Benjamin, New York.

92 Hasegawa, A. (1975) *Plasma Instabilities and Nonlinear Effects*, Springer, New York.

93 Melrose, D.B. (1980) *Plasma Astrohysics. Nonthermal Processes in Diffuse Magnetized Plasmas*, Vol. 1: The emission, absorption and transfer of waves in plasmas; Vol. 2: Astrophysical applications, Gordon and Breach, New York.

94 Thorne, R.M., Ni, B., Tao, X., Horne, R.B., and Meredith, N.P. (2010) Scattering by chorus waves as the dominant cause of diffuse auroral precipitation. *Nature*, **467**, 943–946, doi:10.1038/nature09467.

95 Aschwanden, M.J. (1990) Relaxation of the loss-cone by quasi-linear diffusion of the electron-cyclotron maser instability in the solar corona. *Astronomy and Astrophysics Supplement Series*, **85**, 1141–1177.

96 Li, Y., Yoon, P.H., Wu, C.S., Weatherwax, A.T., Chao, J.K., and Wu, B.H. (1997) Ion pitch-angle scattering by Alfvén waves. *Physics of Plasmas*, **4**, 4103–4117, doi:10.1063/1.872530.

97 Beutier, T. and Boscher, D. (1995) A three-dimensional analysis of the electron radiation belt by the Salammbô code. *Journal of Geophysical Research*, **100**, 14853–14862, doi:10.1029/94JA03066.

98 Velli, M. (1994) From supersonic winds to accretion: Comments on the stability of stellar winds and related flows. *Astrophysical Journal*, **432**, L55–L58, doi:10.1086/187510.

99 Del Zanna, L., Velli, M., and Londrillo, P. (1998) Dynamical response of a stellar atmosphere to pressure perturbations: numerical simulations. *Astronomy and Astrophysics*, **330**, L13–L16.

100 Parker, E.N. (1958) Dynamics of the interplanetary gas and magnetic fields. *The Astrophysical Journal*, **128**, 664.

101 Matteini, L., Landi, S., Hellinger, P., Pantellini, F., Maksimovic, M., Velli, M., Goldstein, B.E., and Marsch, E. (2007) Evolution of the solar wind proton temperature anisotropy from 0.3 to 2.5 AU. *Geophysical Research Letters*, **34**, L20105, doi:10.1029/2007GL030920.

102 Chamberlain, J.W. (1960) Interplanetary Gas.II. expansion of a model solar corona. *The Astrophysical Journal*, **131**, 47.

103 Lemaire, J. and Scherer, M. (1971) Kinetic models of the solar wind. *Journal of Geophysical Research*, **76**, 7479–7490.

104 Grappin, R. and Léorat, J. (2001) Turbulent mixing in a non-magnetic corona: Physical and numerical factors. *Astronomy & Astrophysics*, **365**, 228–240, doi:10.1051/0004-6361:20000033.

105 Rappazzo, A.F., Velli, M., Einaudi, G., and Dahlburg, R.B. (2005) Diamagnetic and expansion effects on the observable properties of the slow solar wind in a coronal streamer. *Monthly Notices of the Royal Astronomical Society*, **633** (1), 474–488.

106 Sweet, P.A. (1958) The Neutral Point Theory of Solar Flares, in *Electromagnetic Phenomena in Cosmical Physics.* IAU Symposium, Vol. 6, paper 14 (ed. B. Lehnert), Cambridge University Press, pp. 123–134.

107 Parker, E.N. (1957) Sweet's mechanism for merging magnetic fields in conducting fluids. *Journal of Geophysics Research*, **62**, 509–520, doi:10.1029/JZ062i004p00509.

108 Loureiro, N.F., Samtaney, R., Schekochihin, A.A., and Uzdensky, D.A. (2012) Magnetic reconnection and stochastic plasmoid chains in high-Lundquist-number plasmas. *Physics of Plasmas*, **19** (4), 042303, doi:10.1063/1.3703318.

109 Bhattacharjee, A., Huang, Y.M., Yang, H., and Rogers, B. (2009) Fast reconnection in high-Lundquist-number plasmas due to the plasmoid Instability. *Physics of Plasmas*, **16** (11), 112102, doi:10.1063/1.3264103.

110 Petschek, H.E. (1964) Magnetic field annihilation. *NASA Special Publication*, **50**, 425.

111 Shay, M.A., Drake, J.F., Rogers, B.N., and Denton, R.E. (2001) Alfvénic collisionless magnetic reconnection and the hall term. *Journal of Geophysical Research: Space Physics*, **106** (A3), 3759–3772, doi:10.1029/1999JA001007.

112 Aunai, N., Belmont, G., and Smets, R. (2011) Energy budgets in collisionless magnetic reconnection: Ion heating and bulk acceleration. *Physics of Plasmas*, **18** (12), 122901, doi:10.1063/1.3664320.

113 Furth, H.P., Killeen, J., and Rosenbluth, M.N. (1963) Finite-resistivity instabilities of a sheet pinch. *Physics of Fluids*, **6**, 459–484, doi:10.1063/1.1706761.

114 Masson, S., Aulanier, G., Pariat, E., and Klein, K.L. (2012) Interchange slip-running reconnection and sweeping SEP beams. *Solar Physics*, **276**, 199–217, doi:10.1007/s11207-011-9886-3.

115 Knight, S. (1973) Parallel electric fields. *Planetary and Space Science*, **21**, 741–750.

116 Ergun, R.E., Andersson, L., Main, D., Su, Y.J., Newman, D.L., Goldman, M.V., Carlson, C.W., McFadden, J.P., and Mozer, F.S. (2002) Parallel electric fields in the upward current region of the aurora: Numerical solutions. *Physics of Plasmas*, **9**, 3695–3704.

117 Andersson, D. (1981) Double layer formation in a magnetised laboratory plasma. *Journal of Physics D Applied Physics*, **14**, 1403–1418.

118 Block, L.P. (1972) Acceleration of auroral particles by electric double layers, in *Earth's Magnetospheric Processes*, Vol. 32, (ed. B.M. McCormac), D. Reidel Publishing Company, Dordrecht, Netherlands, pp. 259–267.

119 Block, L.P. (1978) A double layer review. *Astrophysics and Space Science*, **55**, 59–83.

120 Raadu, M.A. (1989) The physics of double layers and their role in astrophysics. *Physics Reports*, **178**, 25–97.

121 Swift, D.W. (1975) On the formation of auroral arcs and acceleration of auroral electrons. *Journal of Geophysical Research (Space Physics)*, **80**, 2096–2108.

122 Wagner, J.S., Kan, J.R., Akasofu, S.I., Tajima, T., Leboeuf, J.N., and Dawson, J.M. (1980) V-potential double layers and the formation of auroral arcs. *Physical Review Letters*, **45**, 803–806, doi:10.1103/PhysRevLett.45.803.

123 Singh, N. and Khazanov, I. (2005) Planar double layers in magnetized plasmas: Fine structures and their consequences. *Journal of Geophysical Research (Space Physics)*, **110**, 4209–4222, doi:10.1029/2004JA010620.

124 Goertz, C.K. (1979) Double layers and electrostatic shocks in space. *Reviews of Geophysics and Space Physics*, **17**, 418–426.

125 Temerin, M. and Carlson, C.W. (1998) Current-voltage relationship in the downward auroral current region. *Geophysical Research Letters*, **25**, 2365–2368.

126 Ergun, R.E., Carlson, C.W., McFadden, J.P., Mozer, F.S., and Strangeway, R.J. (2000) Parallel electric fields in discrete arcs. *Geophysical Research Letters*, **27**, 4053–4056.

127 Heacock, R.R. and Hessler, V.P. (1967) Polarization characteristics of Pc 1 micropulsations at College. *Planetary and Space Science*, **15**, 1361–1364, doi:10.1016/0032-0633(67)90110-9.

128 Sakurai, T. and McPherron, R.L. (1983) Satellite observations of Pi 2 activity at synchronous orbit. *Journal of Geophysical Research (Space Physics)*, **88**, 7015–7027, doi:10.1029/JA088iA09p07015.

129 Baumjohann, W. and Glassmeier, K.H. (1984) The transient response mechanism and Pi2 pulsations at substorm onset – Review and outlook. *Planetary and Space Science*, **32**, 1361–1370, doi:10.1016/0032-0633(84)90079-5.

130 Lessard, M.R., Lund, E.J., Kim, H.M., Engebretson, M.J., and Hayashi, K. (2011) Pi1B pulsations as a possible driver of Alfvénic aurora at substorm onset. *Journal of Geophysical Research (Space Physics)*, **116** (A15), A06203, doi:10.1029/2010JA015776.

131 Liang, J., Liu, W.W., and Donovan, E.F. (2009) Ion temperature drop and quasi-electrostatic electric field at the current sheet boundary minutes prior to the local current disruption. *Journal of Geophysical Research (Space Physics)*, **114** (A13), A10215, doi:10.1029/2009JA014357.

132 Radoski, H.R. (1967) A note on oscillating field lines. *Journal of Geophysical Research*, **72**, 418–419.

133 Cummings, W.D., O'Sullivan, R.J., and Coleman, Jr., P.J. (1969) Standing Alfvén waves in the magnetosphere. *Journal of Geophysical Research*, **74**, 778–793.

134 Lee, D.H. and Lysak, R.L. (1989) Magnetospheric ULF wave coupling in the dipole model – The impulsive excitation. *Journal of Geophysical Research*, **941**, 17097–17103, doi:10.1029/JA094iA12p17097.

135 Southwood, D.J. and Kivelson, M.G. (1990) The magnetohydrodynamic response of the magnetospheric cavity to changes in solar wind pressure. *Journal of Geophysical Research (Space Physics)*, **95**, 2301–2309, doi:10.1029/JA095iA03p02301.

136 Streltsov, A. and Lotko, W. (1995) Dispersive field line resonances on auroral field lines. *Journal of Geophysical Research (Space Physics)*, **100**, 19457–19472, doi:10.1029/95JA01553.

137 Young, M.A., Lessard, M., Engebretson, M., Woodroffe, J.R., and Oksavik, K. (2012) Spectral enhancements associated with Pi1B events observed at high latitude. *Journal of Geophysical Research (Space Physics)*, **117** (A16), A09314, doi:10.1029/2012JA017940.

138 Lysak, R.L. and Song, Y. (2003) Nonlocal kinetic theory of Alfvén waves on dipolar field lines. *Journal of Geophysical Research (Space Physics)*, **108** (8), 9–1.

139 Andersson, L., Ergun, R.E., Newman, D.L., McFadden, J.P., Carlson, C.W., and Su, Y.J. (2002) Characteristics of parallel electric fields in the downward current region of the aurora. *Physics of Plasmas*, **9**, 3600–3609.

140 Streltsov, A. and Lotko, W. (1996) The fine structure of dispersive, nonradiative field line resonance layers. *Journal of Geophysical Research (Space Physics)*, **101**, 5343–5358, doi:10.1029/95JA03762.

141 Goertz, C.K. (1984) Kinetic Alfvén waves on auroral field lines. *Planetary and Space Science*, **32**, 1387–1392.

142 Hasegawa, A. and Mima, K. (1978) Anomalous transport produced by kinetic Alfvén wave turbulence. *Journal of Geophysical Research (Space Physics)*, **83** (12), 1117–1123.

143 Lysak, R.L. and Lotko, W. (1996) On the kinetic dispersion relation for shear Alfvén waves. *Journal of Geophysical Research (Space Physics)*, **101**, 5085–5094, doi:10.1029/95JA03712.

144 Gérard, J.C., Dols, V., Paresce, F., and Prangé, R. (1993) Morphology and time variation of the Jovian far UV aurora: Hubble Space Telescope observations. *Journal of Geophysics Research*, **98**, 18.

145 Bunce, E.J., Arridge, C.S., Clarke, J.T., Coates, A.J., Cowley, S.W.H., Dougherty, M.K., GéRard, J.C., Grodent, D., Hansen, K.C., Nichols, J.D., Southwood, D.J., and Talboys, D.L. (2008) Origin of saturn's aurora: Simultaneous observations by Cassini and the Hubble Space Telescope. *Journal of Geophysical Research (Space Physics)*, **113**, A09209, doi:10.1029/2008JA013257.

146 Su, Y.J., Ergun, R.E., Jones, S.T., Strangeway, R.J., Chaston, C.C., Parker, S.E., and Horwitz, J.L. (2007) Generation of short-burst radiation through Alfvénic acceleration of auroral electrons. *Journal of Geophysical Research (Space Physics)*, **112**, A06209, doi:10.1029/2006JA012131.

147 Mottez, F., Hess, S., and Zarka, P. (2010) Explanation of dominant oblique radio emission at Jupiter and comparison to the terrestrial case. *Planetary and Space Science*, **58**, 1414–1422, doi:10.1016/j.pss.2010.05.012.

148 Pelletier, G. (1999) Cosmic ray acceleration and nonlinear relativistic wavefronts. *Astronomy and Astrophysics*, **350**, 705–718.

149 Zaslavskiĭ, G.M. and Chirikov, B.V. (1972) Reviews of topical problems: Stochastic instability of nonlinear oscillations. *Soviet Physics Uspekhi*, **14**, 549–568, doi:10.1070/PU1972v014n05ABEH004669.

150 Chirikov, B.V. (1979) A universal instability of many-dimensional oscillator systems. *Physics Reports*, **52**, 263–379, doi:10.1016/0370-1573(79)90023-1.

151 Jones, F.C., Jokipii, J.R., and Baring, M.G. (1998) Charged-particle motion in electromagnetic fields having at least one ignorable spatial coordinate. *Astrophysical Journal*, **509**, 238–243, doi:10.1086/306480.

152 Jokipii, J.R. (1966) Cosmic-ray propagation. I. Charged particles in a random magnetic field. *Astrophysical Journal*, **146**, 480, doi:10.1086/148912.

153 Jokipii, J.R. (1987) Rate of energy gain and maximum energy in diffusive shock acceleration. *Astrophysical Journal*, **313**, 842–846, doi:10.1086/165022.

154 Casse, F., Lemoine, M., and Pelletier, G. (2002) Transport of cosmic rays in chaotic magnetic fields. *Physical Review D*, **65** (2), 023002, doi:10.1103/PhysRevD.65.023002.

155 Plotnikov, I., Pelletier, G., and Lemoine, M. (2011) Particle transport in intense small–scale magnetic turbulence with a mean field. *Astronomy and Astrophysics*, **532**, A68.

156 Schlickeiser, R. (2002) *Cosmic Ray Astrophysics*, Springer, Berlin.

157 Skilling, J. (1975) Cosmic ray streaming. I – Effect of Alfvén waves on particles. *Monthly Notices of the RAS*, **172**, 557–566.

158 Blandford, R.D. and Ostriker, J.P. (1978) Particle acceleration by astrophysical shocks. *Astrophysical Journal Letters*, **221**, L29–L32, doi:10.1086/182658.

159 Fermi, E. (1949) On the origin of the cosmic radiation. *Physical Review Letters*, **75**, 1169–1174, doi:10.1103/PhysRev.75.1169.

160 Stawarz, Ł, Cheung, C.C., Harris, D.E., and Ostrowski, M. (2007) The electron energy distribution in the hotspots of Cygnus A: Filling the gap with the Spitzer Space Telescope. *The Astrophysical Journal*, **662** (1), 213–223.

161 Bednarek, W. and Bartosik, M. (2003) Gamma-rays from the pulsar wind nebulae. *Astronomy and Astrophysics*, **405**, 689–702.

162 Drury, L.O. (1983) An introduction to the theory of diffusive shock acceleration of energetic particles in tenuous plasmas. *Reports on Progress in Physics*, **46**, 973–1027, doi:10.1088/0034-4885/46/8/002.

163 Bell, A.R. (1978) The acceleration of cosmic rays in shock fronts. I. *Monthly Notices of the RAS*, **182**, 147–156.

164 Peacock, J.A. (1981) Fermi acceleration by relativistic shock waves. *Monthly Notices of the RAS*, **196**, 135–152.

165 Bell, A.R. (2004) Turbulent amplification of magnetic field and diffusive shock acceleration of cosmic rays. *Monthly Notices of the RAS*, **353**, 550–558, doi:10.1111/j.1365-2966.2004.08097.x.

166 Cassam-Chenaï, G., Decourchelle, A., Ballet, J., Hwang, U., Hughes, J.P., and Petre, R. (2004) Xmm-newton observation of Kepler's supernova remnant. *Astronomy and Astrophysics*, **414** (2), 545–558, doi:10.1051/0004-6361:20031551.

167 Gallant, Y.A. and Achterberg, A. (1999) Ultra-high-energy cosmic ray acceleration by relativistic blast waves. *Monthly Notices of the RAS*, **305**, L6–L10, doi:10.1046/j.1365-8711.1999.02566.x.

168 Achterberg, A., Gallant, Y.A., Kirk, J.G., and Guthmann, A.W. (2001) Particle acceleration by ultrarelativistic shocks: Theory and simulations. *Month-*

ly Notices of the RAS, **328**, 393–408, doi:10.1046/j.1365-8711.2001.04851.x.

169 Bednarz, J. and Ostrowski, M. (1998) Energy spectra of cosmic rays accelerated at ultrarelativistic shock waves. *Physical Review Letters*, **80**, 3911–3914, doi:10.1103/PhysRevLett.80.3911.

170 Vietri, M. (2003) On particle acceleration around shocks. I. *Astrophysical Journal*, **591**, 954–961, doi:10.1086/375534.

171 Lemoine, M. and Pelletier, G. (2003) Particle transport in tangled magnetic fields and Fermi acceleration at relativistic shocks. *Astrophysical Journal, Letters*, **589**, L73–L76, doi:10.1086/376353.

172 Lemoine, M., Pelletier, G., and Revenu, B. (2006) On the efficiency of Fermi acceleration at relativistic shocks. *Astrophysical Journal, Letters*, **645**, L129–L132, doi:10.1086/506322.

173 Pelletier, G., Lemoine, M., and Marcowith, A. (2009) On Fermi acceleration and magnetohydrodynamic instabilities at ultra-relativistic magnetized shock waves. *Monthly Notices of the RAS*, **393**, 587–597, doi:10.1111/j.1365-2966.2008.14219.x.

174 Luciani, J.F., Mora, P., and Virmont, J. (1983) Nonlocal heat transport due to steep temperature gradients. *Physical Review Letters*, **51**, 1664, (ISSN 0031-9007).

175 Ma, Z.W., Hawkins, J.G., and Lee, L.C. (1991) A simulation study of impulsive penetration of solar wind irregularities into the magnetosphere at the dayside magnetopause. *Journal of Geophysical Research*, **96**, 15751, (ISSN 0148-0227).

176 Savoini, P., Scholer, M., and Fujimoto, M. (1994) Two-dimensional hybrid simulations of impulsive plasma penetration through a tangential discontinuity. *Journal of Geophysical Research*, **99**, 19377, (ISSN 0148-0227).

177 Salem, C., Hubert, D., Lacombe, C., Bale, S.D., Mangeney, A., Larson, D.E., and Lin, R.P. (2003) Electron properties and coulomb collisions in the solar wind at 1 AU: *Wind* observations. *The Astrophysical Journal*, **585** (2), 1147–1157, doi:10.1086/346185.

178 Shoub, E.C. (1983) Invalidity of local thermodynamic equilibrium for electrons in the solar transition region. I – Fokker–Planck results. *The Astrophysical Journal*, **266**, 339–369.

179 Hollweg, J.V. (1974) On electron heat conduction in the solar wind. *Journal of Geophysical Research*, **79**, 3845, doi:10.1029/JA079i025p03845.

180 Hollweg, J.V. (1976) Collisionless electron heat conduction in the solar wind. *Journal of Geophysical Research*, **81**, 1649–1658, doi:10.1029/JA081i010p01649.

181 Jeffrey, A. and Kawahara, T. (1982) *Asymptotic Methods in Nonlinear Wave Theory*, Applicable Mathematics Series, Boston: Pitman.

Index

Collisionless Plasmas in Astrophysics, First Edition. Gérard Belmont, Roland Grappin, Fabrice Mottez, Filippo Pantellini, and Guy Pelletier.

D